NATURALIZED Parrots *of* *the* World

NATURALIZED Parrots of the World

Distribution, Ecology, and Impacts of the World's Most Colorful Colonizers

Edited by Stephen Pruett-Jones

Princeton University Press

Princeton and Oxford

Copyright © 2021 by Princeton University Press

Princeton University Press is committed to the protection of copyright and the intellectual property our authors entrust to us. Copyright promotes the progress and integrity of knowledge. Thank you for supporting free speech and the global exchange of ideas by purchasing an authorized edition of this book. If you wish to reproduce or distribute any part of it in any form, please obtain permission.

Requests for permission to reproduce material from this work
should be sent to permissions@press.princeton.edu

Published by Princeton University Press
41 William Street, Princeton, New Jersey 08540
6 Oxford Street, Woodstock, Oxfordshire OX20 1TR

press.princeton.edu

All Rights Reserved

Library of Congress Cataloging-in-Publication Data
Names: Pruett-Jones, Stephen, editor.
Title: Naturalized parrots of the world : distribution, ecology, and impacts of the world's most colorful colonizers / edited by Stephen Pruett-Jones.
Description: Princeton : Princeton University Press, [2021] | Includes bibliographical references and index.
Identifiers: LCCN 2020045356 (print) | LCCN 2020045357 (ebook) | ISBN 9780691204413 (hardback) | ISBN 9780691220710 (ebook)
Subjects: LCSH: Parrots. | Parrots—Colonization—Case studies. | Introduced birds—Case studies.
Classification: LCC SF473.P3 N38 2021 (print) | LCC SF473.P3 (ebook) | DDC 636.6/865--dc23
LC record available at https://lccn.loc.gov/2020045356
LC ebook record available at https://lccn.loc.gov/2020045357

British Library Cataloging-in-Publication Data is available

Editorial: Robert Kirk and Abigail Johnson
Production Editorial: Ellen Foos
Jacket Design: Wanda España
Production: Steven Sears
Publicity: Matthew Taylor and Caitlyn Robson
Copyeditor: Amy K. Hughes
Typset and design: D & N Publishing, Wiltshire, UK

Front cover image: Rose-ringed Parakeet (*Psittacula krameri*) in Yala National Park, Sri Lanka. Photo by Dave Stamboulis / Alamy Stock Photo. Back cover image: Monk Parakeets (*Myiopsitta monachus*) enjoying water from a birdbath, Port Salerno, Florida USA, May 2019. Photo by Jon-Mark Davey

This book has been composed in Avenir Book and Objectiv

Printed on acid-free paper. ∞

Printed in Italy

10 9 8 7 6 5 4 3 2 1

CONTENTS

Preface — 7
Contributors — 8

Part I. Background and Ecology

1. The World Parrot Trade — 13
   Laura Cardador, Pedro Abellán, José D. Anadón, Martina Carrete, and José L. Tella

2. The Distribution of Naturalized Parrot Populations — 22
   Kay Royle and Wayne B. Donner

3. Parrots and People: Human Dimensions of Naturalized Parrots — 41
   Sarah L. Crowley

4. Genetics of Invasive Parrot Populations — 54
   Michael A. Russello, Grace Smith-Vidaurre, and Timothy F. Wright

5. Naturalized Parrots: Conservation and Research Opportunities — 71
   Simon Kiacz and Donald J. Brightsmith

6. The Ecological Impacts of Introduced Parrots — 87
   Emiliano Mori and Mattia Menchetti

7. Decision-Making Models and Management of the Monk Parakeet — 102
   Juan Carlos Senar, Michael Conroy, and Tomás Montalvo

8. Management of Human-Parrot Conflicts: The South American Experience — 123
   Enrique H. Bucher

9. Are Naturalized Parrots Priority Invasive Species? — 133
   Donald J. Brightsmith and Simon Kiacz

Part II. Case Studies

10. Global Invasion Success of the Rose-ringed Parakeet — 159
    Hazel A. Jackson

11. Monk Parakeets as a Globally Naturalized Species — 173
    Carlos E. Calzada Preston, Stephen Pruett-Jones, and Jessica R. Eberhard

12. Introduced and Naturalized Parrots in the Contiguous United States — 193
    Jennifer J. Uehling, Jason Tallant, and Stephen Pruett-Jones

13. Status of Naturalized Parrots in the Hawaiian Islands — 211
    Eric A. VanderWerf and Nicholas P. Kalodimos

14. Introduced and Naturalized Parrots in Europe — 227
    Michael P. Braun

15. The Fate of Multistage Parrot Invasions in Spain and Portugal — 240
    Martina Carrete, Pedro Abellán, Laura Cardador, José D. Anadón, and José L. Tella

CONTENTS

16. Naturalized Parrots in the United Kingdom   249
    Christopher J. Butler

17. Introduced and Naturalized Parrots of South Africa: Colonization and the Wildlife Trade   260
    Craig T. Symes, Ielyzaveta M. Ivanova, Caroline G. Howes, and Rowan O. Martin

18. Australia's Urban Cavity Nesters and Introduced Parrots: Patterns, Processes, and Impacts   277
    Andrew M. Rogers and Salit Kark

19. The Future for Naturalized Parrots   293
    Stephen Pruett-Jones

Index   299

PREFACE

As a group, parrots are among the most endangered birds in the world. A number of species have already been driven to extinction, and many species are very close to that point. Originally, the major threat to wild parrots was humans shooting or trapping the birds for the millinery trade and/or as a result of real or perceived threats to agriculture. Within the last hundred years, however, and although parrots are still killed as a result of threats to agriculture, the greater concern has become and remains habitat loss or modification and legal or illegal trapping for the pet trade. The pet trade has contributed to population declines of many species, but it has also ironically led to the inadvertent introduction of parrots in novel geographical regions and habitats. The transport and introduction of non-native species of plants and animals is a widespread phenomenon involving thousands of species. In the case of parrots, approximately three-quarters of all species have been exported to new countries through the pet trade.

In some instances, parrots held as pets escape captivity or are purposefully released into the wild. If the habitat or climate regime is inhospitable, escaped or released parrots survive for a short period or season but then perish. However, if the habitat is suitable, parrots are quite capable of surviving in novel regions and habitats, as are many species. If the novel population increases in numbers, and reproduction is sufficient to maintain the population size, a self-sustaining population can be established. Parrots have been remarkably successful at adapting themselves to new and novel environments in introduced ranges. This is almost certainly the result of their behavioral flexibility, cognitive abilities that rival those of some primates, and high degree of sociality. Given the worldwide distribution of naturalized parrots, the increasing and expanding population sizes of some species, and escalating conflict with humans in some areas, it seemed appropriate to attempt a summary of the major issues surrounding naturalized parrots. That is the focus of this book: *Naturalized Parrots of the World:*
Distribution, Ecology, and Impacts of the World's Most Colorful Colonizers.

Although different terminology can be found in different publications, in this book the authors refer to *naturalized parrots* as species that have been introduced to a novel area and have established self-sustaining breeding populations. The terms *established* and *naturalized* are thus used interchangeably here. Although not all authors discuss the impacts of introduced parrots, those that do will use the term *invasive* where the data suggest that a species is having a negative impact on native species, ecosystems, agricultural activities, or other economic activities (e.g., electrical transmission).

This book is divided into two parts, Part I, Background and Ecology; and Part II, Case Studies. The chapters in Part I deal with issues related to all introduced and naturalized parrots, and those in Part II deal with specific species or the community of parrots in individual geographic regions. The two species that have their own case studies are the Rose-ringed Parakeet (*Psittacula krameri*) and the Monk Parakeet (*Myiopsitta monachus*). The most widely distributed and abundant of the naturalized parrots in the world, these two species highlight the relevant social and ecological issues surrounding introduced parrots.

This book builds upon the work, both professional and volunteer, of hundreds of people worldwide whose efforts focus on the conservation of wild parrots and censuses of introduced species. Additionally, without the observations available in online databases from citizen science efforts, the chapters in this book would not have been possible. I wish to thank everyone who contributes to such databases. Additionally, I wish to thank each of the authors contributing to this book; my family and friends for support; and my graduate adviser, Dr. Clayton M. White, for giving me my first opportunity to see parrots in the wild.

SP-J
Chicago, Illinois, 2020

CONTRIBUTORS

Pedro Abellán
Department of Zoology
University of Seville
Seville, Spain

José D. Anadón
Pyrenean Institute of Ecology
Zaragoza, Spain

Michael P. Braun
Institute of Pharmacy and Molecular
　Biotechnology
University of Heidelberg
Heidelberg, Germany

Donald J. Brightsmith
Department of Veterinary Pathobiology
Texas A&M University
College Station, Texas, US

Enrique H. Bucher
Center of Applied Zoology
National University of Córdoba
Córdoba, Argentina

Christopher J. Butler
Department of Biology
University of Central Oklahoma
Edmond, Oklahoma, US

Carlos E. Calzada Preston
Department of Ecology and Evolution
University of Chicago
Chicago, Illinois, US

Laura Cardador
Center for Ecological Research and Forestry
　Applications (CREAF)
Bellaterra, Spain

Martina Carrete
Department of Physical, Chemical, and Natural
　Systems
University Pablo de Olavide
Seville, Spain

Michael Conroy
Warnell School of Forestry and Natural Resources
University of Georgia
Athens, Georgia, US

Sarah L. Crowley
Centre for Geography and Environmental Science
University of Exeter
Exeter, England

Wayne B. Donner
School of Science and the Environment
Manchester Metropolitan University
Manchester, England

Jessica R. Eberhard
Department of Biological Sciences and Museum
　of Natural Science
Louisiana State University
Baton Rouge, Louisiana, US

Caroline G. Howes
School of Animal, Plant and Environmental
　Sciences
University of the Witwatersrand
Johannesburg, South Africa

Ielyzaveta M. Ivanova
School of Biological Sciences
Monash University
Melbourne, Victoria, Australia

Hazel A. Jackson
Durrell Institute of Conservation and Ecology
　(DICE)
University of Kent
Kent, England

Nicholas P. Kalodimos
Department of Natural Resources and
　Environmental Management
University of Hawaii
Honolulu, Hawaii, US

CONTRIBUTORS

Salit Kark
The Biodiversity Research Group
School of Biological Sciences
University of Queensland
Brisbane, Queensland, Australia

Simon Kiacz
Ecology and Evolutionary Biology Program
Texas A&M University
College Station, Texas, US

Rowan O. Martin
Fitzpatrick Institute of African Ornithology
Department of Biological Sciences
University of Cape Town
Cape Town, South Africa
Also:
Africa Conservation Programme
World Parrot Trust
Cornwall, England

Mattia Menchetti
Department of Biology
University of Florence
Florence, Italy
Also:
Institute of Evolutionary Biology (CSIC-UPF)
Barcelona, Spain

Tomás Montalvo
Barcelona Public Health Agency
Barcelona, Spain

Emiliano Mori
Institute of Research on Terrestrial Ecosystems (IRET)
National Research Council (CNR)
Sesto Fiorentino (Florence), Italy

Stephen Pruett-Jones
Department of Ecology and Evolution
University of Chicago
Chicago, Illinois, US

Andrew M. Rogers
The Biodiversity Research Group
School of Biological Sciences
University of Queensland
Brisbane, Queensland, Australia

Kay Royle
School of Science and the Environment
Manchester Metropolitan University
Manchester, England

Michael A. Russello
Department of Biology
University of British Columbia
Kelowna, British Columbia, Canada

Juan Carlos Senar
Museum of Natural Sciences of Barcelona
Barcelona, Spain

Grace Smith-Vidaurre
Department of Biology
New Mexico State University
Las Cruces, New Mexico, US

Craig T. Symes
School of Animal, Plant and Environmental Sciences
University of the Witwatersrand
Johannesburg, South Africa

Jason Tallant
University of Michigan Biological Station
University of Michigan
Ann Arbor, Michigan, US

José L. Tella
Doñana Biological Station
Seville, Spain

Jennifer J. Uehling
Department of Ecology and Evolutionary Biology
Cornell University
Ithaca, New York, US

Eric A. VanderWerf
Pacific Rim Conservation
Honolulu, Hawaii, US

Timothy F. Wright
Department of Biology
New Mexico State University
Las Cruces, New Mexico, US

PART I
BACKGROUND AND ECOLOGY

1

THE WORLD PARROT TRADE

Laura Cardador, Pedro Abellán, José D. Anadón, Martina Carrete, and José L. Tella

INTRODUCTION

International trade is recognized as an important and rapidly growing source of introduction of non-native species worldwide (Hulme 2009). Particularly, the trade in wildlife has been directly related to the introduction of non-native, sometimes invasive bird species (Carrete and Tella 2008; Cardador et al. 2019), among which those belonging to the order Psittaciformes play a dominant role (Beissinger 2001; Blackburn and Duncan 2001). The Psittaciformes include parrots, parakeets, lovebirds, cockatoos, macaws, etc., which we collectively refer to as simply "parrots." Almost two-thirds of all existing parrot species have been commonly transported outside their native ranges as pets, and several more are traded locally (Cassey et al. 2004). This trade has strongly contributed to the decline of many parrot species in their native ranges (Collar and Juniper 1992; Tella and Hiraldo 2014). Nearly one-third of all parrot species are threatened under criteria set forth by the International Union for Conservation of Nature (IUCN 2016), but parrots are also among the most widespread introduced birds in the world (Cassey et al. 2004; Strubbe and Matthysen 2009a; Cardador et al. 2016). Almost one-quarter of all transported species find their way into exotic environments, and at least 10% of parrot species have established naturalized populations (Cassey et al. 2004; Abellán et al. 2017), sometimes with undesirable effects on native fauna and human socioeconomic activities (Strubbe and Matthysen 2009b; Hernández-Brito et al. 2014; Menchetti and Mori 2014; Peck et al. 2014).

As with other bird species, the human interest in and transport of parrot species date back centuries (Blackburn et al. 2009; Tella 2011). There is consistent evidence, for example, that pre-Columbian cultures in Central America already transported and kept parrot species in captivity, as the birds' brightly colored feathers were prized for their use in rituals and decoration. The interest in and transport of alien parrots was also present in ancient Old World civilizations. As far back as the year 326 BCE, Alexander the Great's expedition transported parakeets to Europe from India, and it is well known that Romans too kept parrots in captivity (Blackburn et al. 2009; Tella 2011).

The interest in parrot species has been maintained throughout history. However, for most of the time, parrot keeping has been mostly a luxury, confined to the wealthiest people,

Mosaic, c. 500 CE, of Rose-ringed Parakeets (*Psittacula krameri*). The collars indicate that these birds were kept as pets, likely having been traded from Africa. Israel Museum, Jerusalem. Photo by Salit Kark.

THE WORLD PARROT TRADE

thus involving low numbers of individuals and species (Cassey et al. 2015). It was during the 20th century that the activity became especially popular, stimulated by the accelerated economic growth and the development of efficient transport infrastructures. Today, bird keeping is one of the world's most popular hobbies (Carrete and Tella 2008), and the demand for birds as pets involves millions of individual parrots annually (Cardador et al. 2017; Reino et al. 2017).

In this chapter we describe how increases in international parrot trade demand have contributed to increased invasion risks worldwide in recent decades, and we describe some of the factors that might contribute to the success of the traded species. We then analyze temporal trends in international parrot trade numbers, the species' geographic origins, and their destinations in recent decades. To conclude, we discuss potential conservation implications of parrot trade and propose alternative strategies to combat them.

PARROT TRADE AND INVASIONS

The increased demand for parrots has been accompanied by a dramatic increase in the number of new parrot introductions in non-native areas during the 20th century (Fig. 1.1). These introductions have been related mostly to accidental or unplanned releases by pet owners (Blackburn et al. 2009; Abellán et al. 2016). Nearly 80% of all known parrot introductions occurred after 1950 and are directly linked to annual parrot trade numbers (Cardador et al. 2019) (Fig. 1.2). As biological invasions are a dynamic process, introduced species constitute the pool of species from which establishment, and thereafter spread, of self-sustaining populations in new non-native areas can take place (Blackburn et al. 2011). In turn, the likelihood of establishment and spread of species traded at larger numbers is expected to be higher, because a greater number of individuals of such species are likely to escape.

Several studies have suggested that propagule pressure (understood as a composite measure of the number of individuals released into a region in which they are not native) is one of the most important factors explaining invasion success (Colautti et al. 2006; Blackburn et al. 2009, 2015). Increased propagule pressure favors invasions by overcoming environmental and demographic stochasticity, avoiding genetic bottlenecks (decreased genetic diversity), and increasing standing genetic variation on which selection can act (Facon et al. 2006; Simberloff 2009). Greater propagule pressure increases, first, the likelihood that an alien population will establish a self-sustaining population at a location and, second, the likelihood that an alien population subsequently will spread toward other places to become invasive (Blackburn et al. 2015; Abellán et al. 2017).

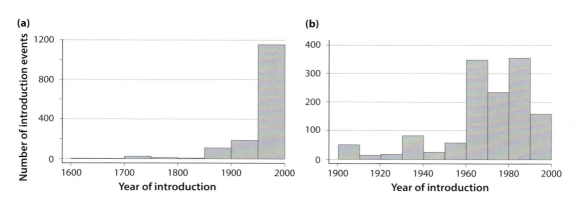

Figure 1.1. Frequency distribution of year of introduction for (a) 1,491 parrot introduction events since 1600 for which an estimate of introduction date was available, and (b) 1,352 parrot introduction events that occurred after 1900. Data source: Dyer et al. (2017).

PARROT TRADE AND INVASIONS

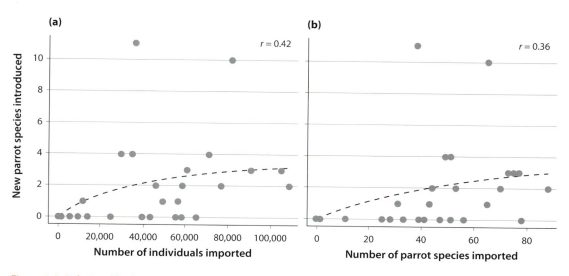

Figure 1.2. Relationships between new parrot species introduced in the wild in Spain and Portugal and total number of wild-caught parrots (a) and species (b) imported. Each dot represents a single year in the period 1976–2005. A Michaelis–Menten curve (dashed lines) was fitted to account for the potential saturation of new species introduced at large import values. The goodness of fit between observed data and the fitted curve is indicated by the Pearson's correlation coefficient (r). Data sources: CITES (2015) and Cardador et al. (2019).

Two Yellow-crested Cockatoos (*Cacatua sulphurea*) for sale at Yuen Po Street bird market in Hong Kong. The orange-crested individual on the left is of the subspecies *C. s. citrinocristata*. Yellow-crested Cockatoos are critically endangered due to habitat loss and the pet trade but are regularly seen in Hong Kong's bird markets. January 2018. Photo by Astrid Andersson.

Additionally, other factors beyond trade numbers, such as breeding origin of traded species, can also affect the probability of invasion of traded species. There is indeed evidence that although captive-bred species, such as Budgerigars (*Melopsittacus undulatus*), often greatly outnumber species caught in the wild and traded on the pet market, the wild-caught birds are more successful invaders (Carrete and Tella 2008). The greater success of wild-caught cage birds compared to captive-bred ones could be related to the loss in captive-bred birds of the ability to cope with new environments. This hypothesis is supported by experiments with parrots showing that exotic wild-caught individuals have longer physiological responses to acute stress than captive-bred ones (Cabezas et al. 2013). This may facilitate wild-caught individuals in both escaping captivity and facing challenges in new environments. Additionally, wild-caught birds, including parrots, maintain their antipredatory behavior and escape responses in captivity (Carrete and Tella 2015). In contrast, captive-bred birds lose these traits and thus have a lower probability of escaping captivity as well as a lower probability of success if they do escape (Carrete and Tella 2015).

THE WORLD PARROT TRADE

Environmental similarity between the regions of origin and destination of traded species is also a key determinant of the effect of trade on invasion risks. Recent evidence shows that most invasive species within different taxa occupy areas with similar environmental characteristics to those in their native ranges (Petitpierre et al. 2012; Strubbe and Matthysen 2014). A significant effect of environmental suitability (i.e., similarity with native ranges) on invasion success of parrot species has also been found (Cardador et al. 2016, 2017). While some species can establish alien populations in truly novel areas (Abellán et al. 2017), their capacity to spread and become invasive is expected to be strongly influenced by environmental matching (Duncan et al. 2001; Abellán et al. 2017). In this sense, not only global trade numbers but also trade routes can have important effects on global invasion risks.

THE INTERNATIONAL TRADE IN LIVE PARROTS

Almost all parrot species are listed in the appendices of the Convention on International Trade in Endangered Species of Wild Fauna and Flora (CITES 2015), and thus, their international trade requires permits detailing the number, origin, and destination of the individuals involved. According to CITES data, the total number of live parrots legally traded among countries from 1975 (when CITES entered in force) to 2015 was 19,607,897 individuals (annual mean of 490,197) belonging to 336 species. This figure corresponds to around 25% of the total registered legal bird trade. Only seven species—Grey Parrot (*Psittacus erithacus*), Senegal Parrot (*Poicephalus senegalus*), Monk Parakeet (*Myiopsitta monachus*), Rose-ringed Parakeet (*Psittacula krameri*), Rosy-faced Lovebird (*Agapornis roseicollis*), Fischer's Lovebird (*A. fischeri*), and Yellow-collared Lovebird (*A. personatus*)—represented half of all of the parrots traded during this period. Importantly, at least 40% of the total parrot trade involved wild-caught individuals, whose invasive potential is significantly higher than that of captive-bred ones (Carrete and Tella 2008; Abellán et al. 2017). Considering individuals with a known wild-caught origin, only four species represented over 60% of total trade in the period 1975–2015: Grey Parrot, Senegal Parrot, Monk Parakeet, and Rose-ringed Parakeet. It is worth mentioning that Monk Parakeets and Rose-ringed Parakeets are considered two of the most widespread naturalized species of the world (Strubbe and Matthysen 2009a; Cardador et al. 2016; Jackson, chap. 10 this vol.; Uehling et al., chap. 12 this vol.).

Grey Parrots (*Psittacus erithacus*), originally bound for the illegal pet trade, confiscated from a local bird trapper. Maniema Province, Democratic Republic of Congo, May 2016. Photo by Cintia Garai.

ORIGIN AND DESTINATION OF PARROTS IN THE WILDLIFE TRADE

The sources of parrots in the wildlife trade in recent decades (1975–2015) are, not surprisingly, located within the main native areas of parrot species, namely South and Central America, sub-Saharan Africa, the Indian subcontinent, Indochina, and Indonesia (Fig. 1.3). The geographic focus of the parrot trade is also widespread, although it has undergone notable changes throughout time, as economic growth in developing countries has increased (Weber and Li 2008) and the main importing countries have imposed bans on wild bird imports (Cardador et al. 2017; Reino et al. 2017). In the recent past, the United States (US) and European countries dominated international parrot trade imports (Fig. 1.4). However, importations in these areas have been drastically reduced since the application of two bans. First, the US Wild Bird Conservation Act, enacted in 1992, prohibited the importation of wild birds unless they were collected in accordance with predefined management plans for sustainable use of the species. Second, the Wild Bird Declaration prohibited wild-caught bird importations in the European Union (EU). It was adopted in 2005, first as a temporal measure to prevent the spread of avian flu and other diseases

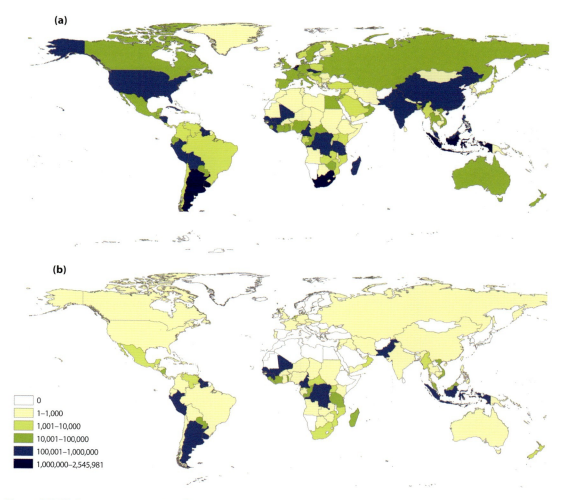

Figure 1.3. Main exporter countries of parrot species from 1975 to 2015. Total number of individuals traded (a) and total number of individuals traded with a known wild origin (b) are shown. Data source: CITES (2015).

and since 2007 as an indefinite measure also focused on conservation and animal welfare.

The reduction of the US and EU wild bird markets after the bans reduced invasion risks (Cardador et al. 2019) but also led to important redirections of commercial species. Notable increases in importations occurred in Mexico, South America, countries of the former Soviet Union, and across Southeast Asia after the application of the US ban, whereas Mexico and the Middle East became the main destinations of commercial parrot species after the EU ban (Fig. 1.4). As a result of such redirections, the cumulative world surface exposed to alien parrots actually increased after the successive bans. Additionally, as new destinations include areas environmentally similar to those of the species' origins, they are susceptible to invasion, which has led to a general increase in global invasion risk (Cardador et al. 2017). In fact, since the redirection of commerce, some new invasions have been described, such as the introduction of Monk Parakeets in Mexico (MacGregor-Fors 2011), and additional invasion events are likely in the near future (Cassey et al. 2015; Cardador et al. 2017).

From a socioeconomic point of view, changes in trade routes observed after the application of the US and EU bans have resulted in a change in the relationship between importations and countries' socioeconomic development. Thus, considering parrot importation data from the past few decades, it can be seen that while a clear positive relationship between parrot importations and gross domestic product existed at a global scale before 1992, this relationship became weaker after the application of the US ban and almost null after the application of the EU ban in 2005 (Fig. 1.5). Such changes are of considerable concern, since the redirection of trade from developed countries, where knowledge and resources to combat invasive species are available

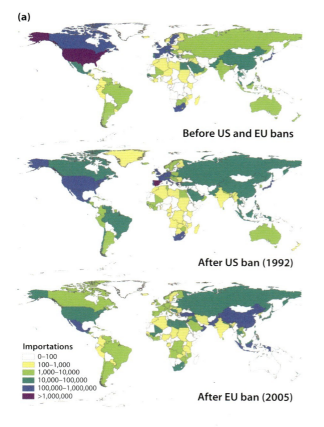
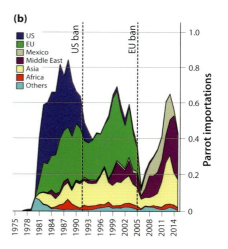

Figure 1.4. Trade patterns for parrot species in different geographic areas in three main periods between 1975 and 2015 (CITES 2015). Spatial (a) and temporal (b) patterns are shown. Imports are based on live individuals from the direct international trade to reduce double counting of reexported specimens. Note that in (b), importations represent cumulative values across regions and are expressed in million individuals. In (b), vertical dashed lines represent wild-bird trade bans of the United States in 1992 and the European Union in 2005.

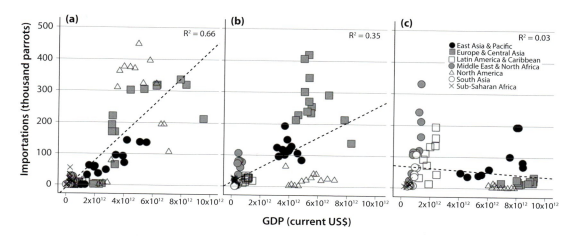

Figure 1.5. Relationships between the number of parrot species imported and gross domestic product (GDP) in the periods (a) 1975–1992, (b) 1993–2005, and (c) 2006–2015. Each dot represents a single year in a given region. Linear regressions (dashed lines) were fitted and the coefficient of determination (R^2) is provided. Parrot imports and GDP per region represent cumulative values across countries included in those same regions. Data sources: CITES (2015) and World Bank (2019).

and social awareness is high, to developing countries, which are less well equipped to deal with invasions, may strongly increase invasion risks and impacts in these areas (Nuñez and Pauchard 2010; Early et al. 2016).

Interestingly, private demand for pet animals in countries subject to wildlife trade bans is now mainly satisfied by domestic trade of captive-bred species (Cassey et al. 2015; Cardador et al. 2019). As mentioned, captive-bred species are less successful than wild-caught species at establishing themselves in novel environments due to behavioral and physiological adaptations to captivity (Carrete and Tella 2008, 2015; Cabezas et al. 2013). This contributes to the invasion risk imbalance between developed and non-developed countries.

FINAL REMARKS

It is increasingly acknowledged that human activities interact with ecological processes in multiple ways. Increasing international trade, in particular, is accelerating the dispersal of alien species across the globe and inherently contributing to the homogenization of species assemblages at a global scale (McKinney and Lockwood 1999; Capinha et al. 2015). While it is widely recognized that wildlife trade is currently the most important and increasing source of alien vertebrates, such as parrot species, trade regulations are usually not instituted because of invasion risks. Additionally, although biological invasions are a global issue, responsibility for protection against invaders lies mostly with national governments. This has led to important differences in legislation among countries, even for those signatories of the Convention on Biological Diversity (CBD), which includes prevention, eradication, and control of invasive species as a commitment (McGeoch et al. 2010). Additionally, when applied, this legislation takes the form mostly of defensive measures (mainly bans and quarantines) to protect particular importer countries or regions against the potentially harmful effects of imported species. While these regulations offer an option to reduce the invasion likelihood in countries or regions of implementation, they do not tackle the problem of invasive species as a global issue, as risky species can still be exported to other countries. To avoid unintended consequences of national or regional regulations, such as unexpected geographic redirections or taxonomic changes, in the international pet trade (Cardador et al. 2017; Reino et al. 2017), more global, intercontinental strategies addressing biological invasions as a global issue are required. Applying the precautionary principle, blanket bans, such as the EU ban, should be seriously considered at a global scale. However, blanket bans are widely debated, as they can be difficult to apply, can be counterproductive—by promoting illegal trade or the development of new markets to support demand—and may produce negative impacts on the livelihoods of local human communities from the exporting countries (Cooney and Jepson 2005; Roe 2006; Rivalan et al. 2007). Alternatively, a trade regulation framework similar to that developed by CITES—the primary international instrument available to monitor and control wildlife trade of threatened species—should be created with the aim of creating binding international standards to regulate, monitor, and control the trade of potentially harmful species in both importer and exporter countries.

REFERENCES

Abellán, P., Carrete, M., Anadón, J. D., Cardador, L., and Tella, J. L. 2016. Non-random patterns and temporal trends (1912–2012) in the transport, introduction and establishment of exotic birds in Spain and Portugal. *Diversity and Distributions* 22:263–273.

Abellán, P., Tella, J. L., Carrete, M., Cardador, L., and Anadón, J. D. 2017. Climate matching drives spread rate but not establishment success in recent unintentional bird introductions. *Proceedings of the National Academy of Sciences (USA)* 114:9385–9390.

Beissinger, S. 2001. Trade of live wild birds: Potentials, principles and practices of sustainable use. In *Conservation of Exploited Species*, ed. Reynolds, J. D., Mace, G. M., Redford, K. H., and Robinson, J. G., 182–202. Cambridge: Cambridge Univ. Press.

Blackburn, T. M., and Duncan, R. P. 2001. Establishment patterns of exotic birds are constrained by non-random patterns in introduction. *Journal of Biogeography* 28:927–939.

Blackburn, T. M., Lockwood, J. L., and Cassey, P. 2009. *Avian Invasions: The Ecology and Evolution of Exotic Birds*. Oxford: Oxford Univ. Press.

Blackburn, T. M., Lockwood, J. L., and Cassey, P. 2015. The influence of numbers on invasion success. *Molecular Ecology* 24:1942–1953.

Blackburn, T. M., Pyšek, P., Bacher, S., Carlton, J. T., Duncan, R. P., Jarošík, V., Wilson, J.R.U., and Richardson, D. M. 2011. A proposed unified framework for biological invasions. *Trends in Ecology & Evolution* 26:333–339.

Cabezas, S., Carrete, M., Tella, J. L., Marchant, T. A., and Bortolotti, G. R. 2013. Differences in acute stress responses between wild-caught and captive-bred birds: A physiological mechanism contributing to current avian invasions? *Biological Invasions* 15:521–527.

Capinha, C., Essl, F., Seebens, H., Moser, D., and Pereira, H. M. 2015. The dispersal of alien species redefines biogeography in the Anthropocene. *Science* 348:1248–1251.

Cardador, L., Carrete, M., Gallardo, B., and Tella, J. L. 2016. Combining trade data and niche modelling improves predictions of the origin and distribution of non-native European populations of a globally invasive species. *Journal of Biogeography* 43:967–978.

Cardador, L., Lattuada, M., Strubbe, D., Tella, J. L., Reino, L., Figueira, R., and Carrete, M. 2017. Regional bans on wild-bird trade modify invasion risks at a global scale. *Conservation Letters* 10:717–725.

Cardador, L., Tella, J., Anadón, J., Abellán, P., and Carrete, M. 2019. The European trade ban on wild birds reduced invasion risks. *Conservation Letters* 12:e12631.

Carrete, M., and Tella, J. 2008. Wild-bird trade and exotic invasions: A new link of conservation concern? *Frontiers in Ecology and the Environment* 6:207–211.

Carrete, M., and Tella, J. L. 2015. Rapid loss of antipredatory behaviour in captive-bred birds is linked to current avian invasions. *Science Reports* 5:18274.

Cassey, P., Blackburn, T. M., Russell, G. J., Jones, K. E., and Lockwood, J. L. 2004. Influences on the transport and establishment of exotic bird species: An analysis of the parrots (Psittaciformes) of the world. *Global Change Biology* 10:417–426.

Cassey, P., Vall-llosera, M., Dyer, E., and Blackburn, T. M. 2015. The biogeography of avian invasions: History, accident and market trade. In *Biological Invasions in Changing Ecosystems: Vectors, Ecological Impacts, Management and Predictions*, ed. J. Canning-Clode, 37–54. Warsaw: De Gruyter Open.

CITES. 2015. CITES Trade Database. Accessed Dec. 2019. http://trade.cites.org/.

Colautti, R. I., Grigorovich, I. A., and MacIsaac, H. J. 2006. Propagule pressure: A null model for biological invasions. *Biological Invasions* 8:1023–1037.

Collar, N. J., and Juniper, A. T. 1992. Dimensions and causes of the parrot conservation crisis. In *New World Parrots in Crisis: Solutions from Conservation Biology*, ed. S. R. Beissinger and N.F.R. Snyder, 1–24. Washington, DC: Smithsonian Institution Press.

Cooney, R., and Jepson, P. 2005. The international wild bird trade: What's wrong with blanket bans? *Oryx* 40:18.

Duncan, R. P., Bomford, M., Forsyth, D. M., and Conibear, L. 2001. High predictability in introduction outcomes and the geographical range size of introduced

REFERENCES

Australian birds: A role for climate. *Journal of Animal Ecology* 70:621–632.

Dyer, E. E., Redding, D. W., and Blackburn, T. M. 2017. The Global Avian Invasions Atlas: A database of alien bird distributions worldwide. *Scientific Data* 4:170041.

Early, R., Bradley, B. A., Dukes, J. S., Lawler, J. J., Olden, J. D., Blumenthal, D. M., Gonzalez, P., Grosholz, E. D., Ibáñez, I., Miller, L. P., Sorte, C.J.B., and Tatem, A. J. 2016. Global threats from invasive alien species in the twenty-first century and national response capacities. *Nature Communications* 7:12485.

Facon, B., Genton, B. J., Shykoff, J., Jarne, P., Estoup, A., and David, P. 2006. A general eco-evolutionary framework for understanding bioinvasions. *Trends in Ecology & Evolution* 21:130–135.

Hernández-Brito, D., Carrete, M., Popa-Lisseanu, A. G., Ibáñez, C., and Tella, J. L. 2014. Crowding in the city: Losing and winning competitors of an invasive bird. *PLoS One* 9:e100593.

Hulme, P. E. 2009. Trade, transport and trouble: Managing invasive species pathways in an era of globalization. *Journal of Animal Ecology* 46:10–18.

IUCN. 2016. The IUCN Red List of Threatened Species (v.2016-1). https://www.iucnredlist.org/resources/summary-statistics.

MacGregor-Fors, I., Calderón-Parra, R., Meléndez-Herrada, A., López-López, S., and Schondube, J. E. 2011. Pretty, but dangerous! Records of non-native monk parakeets (*Myiopsitta monachus*) in Mexico. *Revista Mexicana de Biodiversidad* 82:1053–1056.

McGeoch, M. A., Butchart, S.H.M., Spear, D., Marais, E., Kleynhans, E. J., Symes, A., Chanson, J., and Hoffmann, M. 2010. Global indicators of biological invasion: Species numbers, biodiversity impact and policy responses. *Diversity and Distributions* 16:95–108.

McKinney, M., and Lockwood, J. 1999. Biotic homogenization: A few winners replacing many losers in the next mass extinction. *Trends in Ecology & Evolution* 14:450–453.

Menchetti, M., and Mori, E. 2014. Worldwide impact of alien parrots (Aves Psittaciformes) on native biodiversity and environment: A review. *Ethology Ecology & Evolution* 26:172–194.

Nuñez, M. A., and Pauchard, A. 2010. Biological invasions in developing and developed countries: Does one model fit all? *Biological Invasions* 12:707–714.

Peck, H. L., Pringle, H. E., Marshall, H. H., Owens, I.P.F., and Lord, A. M. 2014. Experimental evidence of impacts of an invasive parakeet on foraging behavior of native birds. *Behavioral Ecology* 25:582–590.

Petitpierre, B., Kueffer, C., Broennimann, O., Randin, C., Daehler, C., and Guisan, A. 2012. Climatic niche shifts are rare among terrestrial plant invaders. *Science* 335: 1344–1347.

Reino, L., Figueira, R., Beja, P., and Araújo, M. B. 2017. Networks of global bird invasion altered by regional trade ban. *Science Advances* 11:e1700783.

Rivalan, P., Delmas, V., Angulo, E., Bull, L. S., Hall, R. J., Courchamp, F., Rosser, A. M., and Leader-Williams, N. 2007. Can bans stimulate wildlife trade? *Nature* 447:529–530.

Roe, D. 2006. Blanket bans—conservation or imperialism? A response to Cooney and Jepson. *Oryx* 40:27.

Simberloff, D. 2009. The role of propagule pressure in biological invasions. *Annual Review Ecology and Systematics* 40:81–102.

Strubbe, D., and Matthysen, E. 2009a. Establishment success of invasive ring-necked and monk parakeets in Europe. *Journal of Biogeography* 36:2264–2278.

Strubbe, D., and Matthysen, E. 2009b. Experimental evidence for nest-site competition between invasive ring-necked parakeets (*Psittacula krameri*) and native nuthatches (*Sitta europaea*). *Biological Conservation* 142:1588–1594.

Strubbe, D., and Matthysen, E. 2014. Patterns of niche conservatism among non-native birds in Europe are dependent on introduction history and selection of variables. *Biological Invasions* 16:759–764.

Tella, J. L. 2011. The unknown extent of ancient bird introductions. *Ardeola* 58:399–404.

Tella, J. L., and Hiraldo, F. 2014. Illegal and legal parrot trade shows a long-term, cross-cultural preference for the most attractive species increasing their risk of extinction. *PLoS One* 9:e107546.

Weber, E., and Li, B. 2008. Plant invasions in China: What is to be expected in the wake of economic development. *Bioscience* 58:437.

World Bank. 2019. World Development Indicators (WDI) Database. https://databank.worldbank.org.

2

THE DISTRIBUTION OF NATURALIZED PARROT POPULATIONS

Kay Royle and Wayne B. Donner

INTRODUCTION

Humans have been intentionally introducing birds and other animals outside their native ranges for approximately 5,000 years for a variety of reasons, including for sport, for food, for plumage, as a biological control, for conservation, and even out of curiosity (Lever 2005; Cassey et al. 2015). Sometimes this process has been an intentional introduction and release of species, and at other times it has been accidental. Regardless of the process, however, the outcome has been the same: new species are introduced into new environments and ecosystems. If the introduced species becomes established—that is, has a self-sustaining population in an area outside its native range—it can be classified as being a naturalized species in that area (Blackburn and Duncan 2001a; Dyer et al. 2017). Previously, the term *naturalized* has been used in as a synonym for *invasive* or *alien*, but it is used throughout this review, and this volume generally, to describe a species that is established in a new environment (Colautti and MacIsaac 2004).

Among vertebrates, birds have been transported around the world more than any other group, and the order Psittaciformes (parrots) has been especially successful (Ancillotto et al. 2016). Parrots have experienced higher numbers of introductions across the globe than would be expected by chance alone (Blackburn and Duncan 2001a), driven, for the most part, by the pet trade (Cassey et al. 2015). One of the challenges in tracking the global transport of birds is that changes in taxonomic treatment result in legislative lags, as conservation organizations catch up with taxonomic amendments (Garnett and Christidis 2017). A recent example of this is the African Grey Parrot (*Psittacus erithacus*), classified as a single species up until 2011. According to the Convention on International Trade in Endangered Species of Wild Fauna and Flora (CITES), this was, and remains, one of the most widely traded of all the parrot species. Since 2011, however, the species' taxonomy has been amended, and it is now split into the Grey Parrot (*Psittacus erithacus*) and the Timneh Parrot (*P. timneh*), both initially classified as vulnerable and now as endangered (BirdLife International 2019). Such splitting and lumping of species is undoubtedly confounding scientific population studies and the simplification of environmental legislation (Garnett and Christidis 2017).

The current number of extant parrot species (excluding subspecies) is 381, according to the International Ornithologists' Union checklist (Gill and Donsker 2019), up from the previous classification of 356 (Marsden and Royle 2015). This increase reflects a new focus on genetic differences among taxa as well as separation of ecologically distinct taxa. As a result of taxonomic changes in the classification of species, the number and distribution of introduced and naturalized species will also change. Thus, the number of recognized naturalized parrots is constantly changing, varying from study to study. In 2005, Lever suggested that of the 350 parrot species then recognized, 34 had established non-native populations. Two years later, Runde et al. (2007) stated that there were 39 naturalized parrot species. In 2014, Menchetti and Mori stated that 60 parrot species had breeding populations outside their natural range. One of the most recent studies, by Avery and Shiels (2018), states that 54 species have been introduced, and 38 of these have become established. Furthermore, the chapters in this volume suggest that the number of naturalized parrots is much larger than any of these earlier numbers suggest. What is the true number? The uncertainty of the answer highlights the problem with attempting to answer the most basic of questions: How many parrot species have populations outside their natural range?

NATURALIZED PARROTS

In this chapter we use records from one of the most recently established databases on invasive bird species: the Global Avian Invasions Atlas (GAVIA) (Dyer et al. 2017). The GAVIA database contains over 27,000 records of bird species, denoting each species' invasive status as: established, breeding, unsuccessful, died out, extirpated, or unknown. We extracted studies dating between 1993 and 2012, which resulted in 2,941 records for invasive parrots, revealing a total of 129 species, recorded across 106 countries or territories. The majority of records belong to populations classified as either unknown or established (Table 2.1); however, their status may have changed across studies in the database. GAVIA follows the International Union for Conservation of Nature's (IUCN) taxonomy; we made the taxonomy consistent with that of the IOC World Bird List (Gill and Donsker 2019) by reclassifying the genus of seven species.

Of the current extant parrot species, 47 species in 21 genera have at least one naturalized population. According to the IUCN (2019), six of these species are endangered, one of them critically, the Yellow-crested Cockatoo (*Cacatua sulphurea*). The six endangered species are now present in eight countries or territories. The genus *Amazona* contains the most endangered species (four) and also the most established species (eight). The Rose-ringed Parakeet (*Psittacula*

TABLE 2.1

Categories of invasion status of parrots based on the GAVIA database, excluding introductions. The fact that there are more "unknown" records than any other highlights the lack of historical knowledge for invasive parrots. See Appendix 2.1 for more details.

CATEGORY	% OF ALL RECORDS
Established	1,110 (37.58%)
Breeding	419 (14.18%)
Unsuccessful	161 (5.45%)
Died Out	58 (1.96%)
Extirpated	8 (0.27%)
Unknown	1,198 (40.56%)

Monk Parakeets (*Myiopsitta monachus*) being fed in a park in Barcelona, Spain. In the city, up to 40% of the food Monk Parakeets consume may be provided by humans. The birds are marked with numbered collars, as part of Juan Carlos Senar's research of the parrots' life history and movements. January 2019. Photo by Alba Ortega-Segalerva.

krameri), Monk Parakeet (*Myiopsitta monachus*), and Budgerigar (*Melopsittacus undulatus*) have the most records of all the species in the GAVIA database. In addition to the 47 naturalized species, 16 other species have had at least one established population recorded outside their native range but within their native country and so are being classed as reintroductions in this review (e.g., Kakapo, *Strigops habroptila*).

POPULATION STATUS OF NATURALIZED PARROTS

Alien species are now the second-largest threat to native species after biological resource use (Bellard et al. 2016) and are linked to negative economic impacts and the transportation of diseases (Shirley and Kark 2009; Menchetti and Mori 2014; Evans et al. 2016). Some parrot species are highly invasive and compete with native species for food, while others seem to utilize alternative food sources (Mori et al. 2017). A large number of parrot species are considered pests where they occur naturally (Lever 2005; Russello et al. 2008), but it is difficult to estimate the true extent of damage in introduced ranges because it is rarely quantified (Menchetti and Mori 2014). Some species cause ecological damage (Cassey et al. 2004a), while others come into conflict with humans by causing significant damage to crops, orchards, and even ornamental gardens (Menchetti and Mori 2014). These issues highlight why empirical data on the threat and spread of naturalized species is an important conservation tool, for both the species in question and also the habitats and native species found in their new ranges.

Of the 47 established parrot species listed in GAVIA, nine have populations classified as increasing, a further seven are stable, and the rest have decreasing populations (Table 2.2) (Birdlife International 2019).

The United States (US), including Hawaii, is particularly well studied, with 304 established records, almost a third (28.6%) of the global records in GAVIA for Psittaciformes. Based on the GAVIA database, 19 species have been recorded in the US, four of which account for the majority of records, each with over 40 established populations recorded: Red-crowned Amazon (*Amazona viridigenalis*), Budgerigar, Monk Parakeet, and Rose-ringed Parakeet. A more complete analysis by Uehling et al. (2019), however, has shown that the numbers of naturalized parrots in the US is actually larger than evidenced from the GAVIA records. Uehling et al. (2019) document that for the contiguous US, there are now 25 different species breeding in 23 states, with Florida, California, and Texas supporting the largest numbers of naturalized parrots.

In the years between 1986 and 1991, a total of 17 invasive parrot species were counted in Florida alone, with nine of these found breeding in more than one location (Pranty and Epps 2002; Butler 2005). By 2003, this number had risen to 22 breeding species and an additional three breeding but not yet established (Pruett-Jones et al. 2005; Uehling et al. 2019). One of the most frequently released species, either accidentally or intentionally, is the Budgerigar (Uehling et al. 2019). It was first reported as breeding in Florida in 1963 and had established a population by 1977. It reached a peak population of between 6,000 and 8,000 individuals in 1978, followed by a sharp decline, from which it apparently never fully recovered (Butler 2005). Dramatic declines of Budgerigars have been reported in both the US and the United Kingdom (UK), with a sudden reduction of bird feeders as the suspected cause for the UK population crash (Butler 2005). Sudden declines in other parrot populations in the US have been reported—e.g., Red-masked Parakeet (*Psittacara erythrogenys*) and White-winged Parakeet (*Brotogeris versicolurus*)—but without obvious causes (Aagaard and Lockwood 2016), highlighting the difficulty in determining long-term population dynamics.

Rose-ringed Parakeets and Monk Parakeets both established populations in the US in the 1970s (Strubbe and Matthysen 2009) and have been the subjects of several studies (e.g., South and Pruett-Jones 2000; Pruett-Jones et al. 2012; Avery and Shiels 2018). Population estimates of Rose-ringed Parakeets differ among studies (Butler 2005), but naturalized populations generally appear to be increasing and expanding (Avery and Shiels 2018). While many species of naturalized parrots are rapidly increasing, some

TABLE 2.2

Species of naturalized parrots whose populations are stable, increasing, or decreasing in size. Almost one-sixth of all naturalized parrots have increasing populations, according to Birdlife International, including the three most common species: Rose-ringed Parakeet, Monk Parakeet, and Budgerigar. Species are listed in alphabetical order by common name.

STABLE

Grey-headed Lovebird (*Agapornis canus*)	Senegal Parrot (*Poicephalus senegalus*)
Mitred Parakeet (*Psittacara mitratus*)	White-winged Parakeet (*Brotogeris versicolurus*)
Niam-niam Parrot (*Poicephalus crassus*)	Yellow-collared Lovebird (*Agapornis personatus*)
Orange-fronted Parakeet (*Eupsittula canicularis*)	

INCREASING

Brown-throated Parakeet (*Eupsittula pertinax*)	Little Corella (*Cacatua sanguinea*)
Budgerigar (*Melopsittacus undulatus*)	Monk Parakeet (*Myiopsitta monachus*)
Dusky-headed Parakeet (*Aratinga weddellii*)	Nanday Parakeet (*Aratinga nenday*)
Eastern Rosella (*Platycercus eximius*)	Rose-ringed Parakeet (*Psittacula krameri*)
Galah (*Eolophus roseicapilla*)	

DECREASING

Alexandrine Parakeet (*Psittacula eupatria*)	Orange-winged Amazon (*Amazona amazonica*)
Blue-crowned Parakeet (*Thectocercus acuticaudatus*)	Red Lory (*Eos bornea*)
Blue-naped Parrot (*Tanygnathus lucionensis*)	Red-breasted Parakeet (*Psittacula alexandri*)
Chestnut-fronted Macaw (*Ara severus*)	Red-crowned Amazon (*Amazona viridigenalis*)
Coconut Lorikeet (*Trichoglossus haematodus*)	Red-crowned Parakeet (*Cyanoramphus novaezelandiae*)
Crimson Rosella (*Platycercus elegans*)	
Crimson Shining Parrot (*Prosopeia splendens*)	Red-lored Amazon (*Amazona autumnalis*)
Eclectus Parrot (*Eclectus roratus*)	Red-masked Parakeet (*Psittacara erythrogenys*)
Fischer's Lovebird (*Agapornis fischeri*)	Rosy-faced Lovebird (*Agapornis roseicollis*)
Green Parakeet (*Psittacara holochlorus*)	Sulphur-crested Cockatoo (*Cacatua galerita*)
Green-rumped Parrotlet (*Forpus passerinus*)	Tanimbar Corella (*Cacatua goffiniana*)
Hispaniolan Amazon (*Amazona ventralis*)	Turquoise-fronted Amazon (*Amazona aestiva*)
Kuhl's Lorikeet (*Vini kuhlii*)	White Cockatoo (*Cacatua alba*)
Lilac-crowned Amazon (*Amazona finschi*)	Yellow-chevroned Parakeet (*Brotogeris chiriri*)
Lilian's Lovebird (*Agapornis lilianae*)	Yellow-crested Cockatoo (*Cacatua sulphurea*)
Maroon Shining Parrot (*Prosopeia tabuensis*)	Yellow-crowned Amazon (*Amazona ochrocephala*)
Olive-throated Parakeet (*Eupsittula nana*)	Yellow-headed Amazon (*Amazona oratrix*)

of their ranges seem to be restricted, possibly by climate or food sources but also in part because of their natural limited dispersal habits (Butler 2005). Rose-ringed and Monk Parakeets suffer heavy losses in colder weather, so it is likely that their numbers will increase with climate change, as this will reduce the extent of winter cold, which seems to be a limiting factor (Strubbe and Matthysen 2009).

Population changes may be attributable to differences in survey effort, observation methods, or seasonal fluctuations (Simberloff 1995; Garrett 1997; Bibby et al. 2000). Over the years, citizen science increasingly has been used across various disciplines as a source of large data sets (e.g., eBird.org or the British Trust for Ornithology's Breeding Bird Survey) (Tulloch et al. 2013; Sullivan et al. 2014). Citizen science allows for the collection of data that would otherwise be impossible due to a lack of scientific resources (Tulloch et al. 2013). Parrots are often brightly colored, conspicuous, and vocal, and therefore are easily seen by members of the public (Cassey et al. 2004a). However, data of this nature should be used with caution because projects differ in their yield of usable data (Tulloch et al. 2013). Misidentification may occur when morphologically similar species are found in close proximity, and studies that do not span long time frames may also produce limited results for long-lived birds like parrots. Determining whether there is an established, self-sustaining population, or whether the adults present are nonbreeding individuals that will die out, can be complex (Long 1981; Runde et al. 2007). Many invasive species exhibit time lags between their release and exponential population growth, so their presence may go undetected for many years (Sakai et al. 2001; Crooks 2005; Aagaard and Lockwood 2014). Although this delay is a common feature of alien invasions, the reason for it still remains unclear (Sakai et al. 2001; Crooks 2005). A more sensible approach, therefore, is to investigate individual species and areas over time. This, however, is unfeasible given the amount of time and effort needed for micro-scale studies. However, while data are deficient for the majority of species (Mori et al. 2017), some have more data available than others. A greater quantity of data is available for those birds that are easier to observe; however, its quality is not always suitable for reliable statistical analysis (Lodge 1993).

Definitive conclusions are hard to draw from existing data on introduction, success, or failure rates (Mack et al. 2000). The GAVIA database reveals that ~41% of all species' records have

Scarlet-fronted Parakeets (*Psittacara wagleri*) inspecting drainpipes for potential nest sites. Miami Springs, Florida, US, May 2011. Photo by Roelant Jonker.

Native but urbanized White-winged Parakeets (*Brotogeris versicolurus*) coming into roost at the Praça Santuário de Nazaré in Belém, Brazil. January 2009. Photo by Roelant Jonker.

the status "unknown," highlighting the lack of knowledge in studies (Table 2.1). Unsuccessful invasions may often go unnoticed, are harder to detect, and may be underrepresented (Lodge 1993); therefore, data on these species are simply not available (Simberloff 1995; Sol and Lefebvre 2000).

SUCCESS OF NATURALIZED SPECIES

One of the key research subjects in avian invasions is determining the factors that drive the success of one species over another in becoming naturalized. It has been suggested that some species, e.g., Rose-ringed Parakeet, seem to be successful wherever they are introduced, while others, e.g., Tepui Parrotlet (*Nannopsittaca panychlora*), Plain Parakeet (*Brotogeris tirica*), and Grey-cheeked Parakeet (*Brotogeris pyrrhoptera*), seem unable to succeed at all (Mori et al. 2013). The reasons for this may be biological, or they may be environmental, as even closely related species seem to differ in their ability to become established once introduced. The closely related naturalized species Yellow-headed (*Amazona oratrix*) and White-fronted (*A. albifrons*) Amazons provide one example. Both are native to Central America and Mexico and have been introduced into other areas of Mexico, Puerto Rico, and the US. The more widespread Yellow-headed Amazon is classified as endangered, due to a decreasing global population, whereas the White-fronted Amazon is classified as a species of "least concern" and is increasing in numbers (Birdlife International 2019).

Discovering whether any common biological traits are shared among naturalized or invasive species could be advantageous in future conservation efforts (Griffith et al. 1989; Strubbe and Matthysen 2009). However, there is a lack of historical data on biological factors affecting success rates of establishment (Sol and Lefebvre 2000). Failed attempts often go unreported, so most data come from successfully established species, which can lead to interpretative bias if not accounted for (Lodge 1993).

Several studies have investigated the biological traits of introduced bird species, including body weight (Green 1997; Cassey et al. 2004a), fecundity (Cassey et al. 2004a), migratory habits (Kolar and Lodge 2001; Cassey et al. 2004a), and forebrain size and adaptability (Sol and Lefebvre 2000). Many regional studies suggest that biology can affect introduction success (e.g., Cassey 2002), but this may be because specific traits are influenced by the nonrandom distribution of birds (Lockwood 1999; Blackburn and Duncan 2001a).

Studies have shown that certain traits appear to affect introduction and establishment differently. Body weight has been suggested as a successful trait for invasive birds, but its effect on introduction and establishment may differ among families. There are large morphological variations within naturalized psittaciforms, which range in length from 15 cm to 95 cm (Juniper and Parr 2010). The largest of these is the Red-and-green Macaw (*Ara chloropterus*), which can weigh up to 1.7 kg (Collar et al. 2020), and among the smallest is the Grey-headed Lovebird (*Agapornis canus*), which weighs only 25–35.5 g (Collar and Kirwan 2020). Green (1997) found that greater body weight had a positive effect on the introduction success of New Zealand exotics, but these results varied, based on whether statistics were carried out within a family or not. Cassey et al. (2004a) found similar results with parrots but noted that morphology may be indirectly related to the chance of establishment: longer-lived or larger, more colorful birds may be more likely to be traded. Birds that have a longer life span are also more likely to be deliberately released, because of the amount of care needed in captivity (Cassey et al. 2004a).

The chance of a species being introduced increases with higher fecundity (e.g., greater clutch size and fledging success, etc.) (Cassey 2002; Cassey et al. 2004a). However, a slower rate of population growth (e.g., longer generation times, lower fecundity, and longer fledging periods, etc.) appears to be a factor in becoming established for introduced species (Cassey et al. 2004b; Blackburn et al. 2009a). These findings are contradicted somewhat by Blackburn and Duncan (2001b), who found that family-level traits (e.g., body mass, clutch size, and incubation period) had no significant effect on the chances of naturalization success.

Regional studies show that sedentary species are more likely to establish in new areas following release than migratory birds (Kolar and Lodge 2001; Cassey 2002), especially with the same introductory effort (Veltman et al. 1996). However, once a migratory species is established in a new range, it is more likely to spread widely, due to natural dispersal habits (Kolar and Lodge 2001). It is thought that dispersal behaviors reduce the chances of a breeding individual finding a suitable mate during the initial phases of colonization (Veltman et al. 1996).

A broader diet in the native range of an invasive parrot species may also contribute to success among naturalized populations (Cassey et al. 2004a). Indeed, a greater diet breadth is correlated with successful establishment (Lockwood 1999). Species that are habitat or diet specialists are far less likely to colonize than those that are generalists (Brooks 2001). Therefore, species that readily exploit novel food sources are more likely to persist in new environments (Sol and Lefebvre 2000; South and Pruett-Jones 2000; Nicolakakis et al. 2003). Behavioral flexibility can be an advantage in colonizing new areas (Sol and Lefebvre 2000; Nicolakakis et al. 2003). Parrots are sociable birds, and there is some evidence that Rose-ringed and Monk Parakeets have formed mixed flocks, which may be mutually beneficial, at the expense of native species (Ancillotto et al. 2016). With expanding urbanization, species that are able to exploit these areas and new niches are more likely to succeed.

FACTORS AFFECTING SPECIES TRANSPORT

One of the biggest factors influencing the transport of non-native species is the pet trade (Carrete and Tella 2008; Smith et al. 2009), especially since globalization closed the gaps in geography and facilitated international trade (Hulme 2009; Mori et al. 2017). An estimated two-thirds of parrot species have been transported to areas outside their native range (Beissinger 2001; Cassey et al. 2004a; Ancillotto et al. 2016), and these numbers seem to be increasing (Lockwood et al. 2019). Escape from suppliers, breeders, or owners (Carrete and Tella 2008; Hulme 2009; Lockwood et al. 2019) is likely to be one of the reasons parrots have so many populations in urban areas and why their numbers and diversity peak around ports of entry (Runde et al. 2007).

As well as being an indirect contributor of non-native parrots, the pet trade may also be confounding research into the distribution and abundance of established species. Perceived increases in numbers could simply be because of the increase of parrot imports (particularly from the Neotropics to the US) and the subsequent accidental releases (Garrett 1997). It may also be possible to draw incorrect conclusions about specific traits because the phylogenetic associations that are leading to taxonomic patterns in data may be caused by human influences on their distribution (Lockwood 1999; Blackburn and Duncan 2001a).

There seems to be a general trend for naturalized parrots to have strong associations with humans (Avery and Shiels 2018), and the proportion of naturalized species is higher in urban and suburban areas (Runde et al. 2007). Many parrots are tolerant of a wide variety of ecological factors and show a high level of synanthropy (Mori et al. 2013), becoming established in urban areas with greater amounts of exotic flora, which may be a suitable foraging source (Garrett 1997). This may be because of a specific reliance on humans or because urban areas are where the majority of releases and escapes occur (Butler 2005; Menchetti and Mori 2014). Particularly successful species, such as the Rose-ringed Parakeet and the Monk Parakeet, have close associations with humans and inhabit areas with greater human populations (Strubbe et al. 2015; Domènech et al. 2003; Crowley, chap. 3 this vol.). Both parakeets are synanthropic and are often found in close association with humans in their native ranges (Minor et al. 2012; Strubbe et al. 2015). Monk Parakeets have a diverse diet and have adapted to use bird feeders over winter (South and Pruett-Jones 2000; Minor et al. 2012).

MacArthur and Wilson (1967) determined that founder population size was important in the success of island colonizations. This led to a now a widely used theory in the study of both invasive and naturalized bird species. The theory is that one of the primary factors that determine success of naturalized species is related to the numbers of birds released (known as introduction effort) (Green 1997; Lockwood 1999; Sol and Lefebvre 2000; Kolar and Lodge 2001; Cassey et al. 2005; Runde et al. 2007). Introduction effort relates to propagule size and number; propagule size is the total number of individuals in a single release event, and the propagule number is the amount of separate releases (Cassey et al. 2004b). Greater introduction success rates are correlated with larger numbers of a species being released into a site and with the species' repeated release (Blackburn and Duncan, 2001a). If propagule size is excluded from statistical modeling, then other factors appear to be more important—e.g., habitat generalism, migratory behavior, range size, and the proximity of source and destination latitudes (Cassey et al. 2004b).

Results of studies into influencing factors of colonization are likely to be confounded by the nonrandom nature of the species being introduced and the location of that introduction (Lockwood 1999; Blackburn and Duncan 2001a; Cassey et al. 2004b). This may lead to the conclusion that introduction appears to be correlated with specific traits, whereas in fact, the traits relate more to which species are selected for the pet trade (Blackburn and Duncan 2001a). The influence of humans has a far greater effect on establishment than any taxonomic trait (Lockwood 1999).

ECOLOGICAL FACTORS IN NATURALIZATION

As we previously posed the question of biological traits that affect naturalization success, so we must ask whether there are any ecological factors in the areas of introduction that could indicate a greater or lesser success rate. Factors including latitude (Green 1997; Blackburn and Duncan 2001b), community assemblage (Lodge 1993), biotic resistance (e.g., Elton 1958; Jeschke 2014), and competition (e.g., Grarock et al. 2013) may play roles in the ability of introduced parrots to become established.

Parrots have native and non-native ranges on six of the seven continents and are absent only from Antarctica; around one-third of parrot species are native to Australia or South America (Runde et al. 2007). The 47 species that are naturalized have at least one population recorded in 82 countries or territories (Fig. 2.1) and in every biogeographic realm (Olson et al. 2001) except the Antarctic. Most of the records are from across the Palearctic, and just three species account for 74.3% of these: Rose-ringed Parakeet, Monk Parakeet, and Alexandrine Parakeet (*Psittacula eupatria*). The most species-rich area is the Nearctic region, with 20 species recorded (Fig. 2.2).

Four species are particularly widespread: Rose-ringed Parakeet, Monk Parakeet, Alexandrine Parakeet, and Budgerigar (46, 26, 13, and 11 countries, respectively); 59 countries record at least one of these species. (Fig. 2.3).

The distribution of alien birds, however, is nonrandom and has been influenced by humans, and the majority of introductions are to islands and temperate areas (Blackburn and Duncan 2001a). Temperate regions and species-poor islands are traditionally viewed as easier to invade than other areas. Islands in particular seem to have a large number of invasions (Simberloff 1995), despite comprising only 3% of the planet's ice-free landmass (Blackburn and Duncan 2001a). It has been suggested that introduced species succeed in these areas because they are likely to encounter less "biotic resistance" from the native biota. This observation gave rise to the "biotic resistance theory," which hypothesized that areas of reduced species richness were easier to colonize (Elton 1958; Jeschke 2014). Blackburn and Duncan (2001b), however, found evidence to contradict this traditional view in two highly species-rich areas, the Afrotropics and Central and South America. In these areas, introduced species have been very successful at becoming naturalized despite native species-rich parrot faunas (Blackburn and Duncan 2001b).

The effect of latitude upon the introduction and establishment of species has been widely studied. Green (1997) suggested that latitude of native range affects a species' chance of establishment success. Other studies show that the latitudes of origin and sites of introduction do not individually affect the establishment success, but their proximity to each other does (e.g., Blackburn and Duncan 2001b). This is believed

THE DISTRIBUTION OF NATURALIZED PARROT POPULATIONS

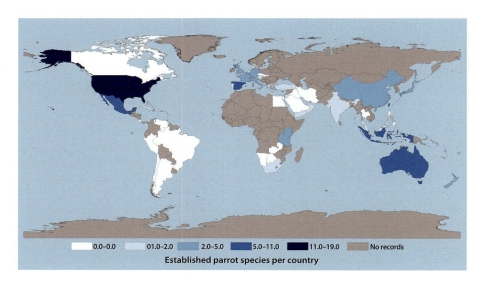

Figure 2.1. The distribution of naturalized species, based on the GAVIA database (Dyer et al. 2017). The map shows all countries that have had at least one species recorded as naturalized. All areas designated "No records" have not had any naturalized species recorded. The map was produced using QGIS 3.4 (www.qgis.org) and the World Countries base map downloaded from ESRI (www.arcgis.com). Image © 2019 Esri, DeLorme Publishing Company, Inc.

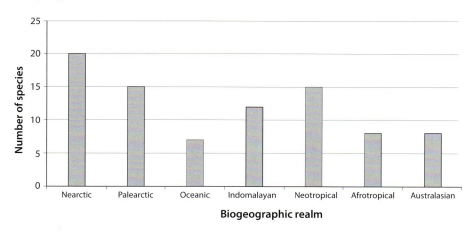

Figure 2.2. Numbers of naturalized parrot species by biogeographic realm. The Nearctic region clearly has the greatest number of invasive parrots (20); the Palearctic and Neotropical realms are next, with 15 species each. The United States, the country with the highest number of naturalized parrot species, is in the Nearctic realm.

to be due to abiotic factors, which are more likely to be similar if the origin and introduction sites are closer to each other (Blackburn and Duncan 2001b). Green (1997) admitted that the strength of latitudinal effects might be reliant on other factors and latitude may have been acting as a proxy for climate. If a species is introduced into an area outside its range but with a similar climate, then it stands a greater chance of becoming established (Duncan et al. 2001, 2003). Parrots do not generally cope well when temperatures drop (Butler 2005), but Monk Parakeets seem to be more tolerant of colder weather, showing increasing numbers in colder northern parts of the US as well as the warmer South (Butler 2005; Pruett-Jones et al. 2012). One additional factor could be that they roost not in the open but inside their nests. However, the role

THE FUTURE OF THREATENED PARROT SPECIES

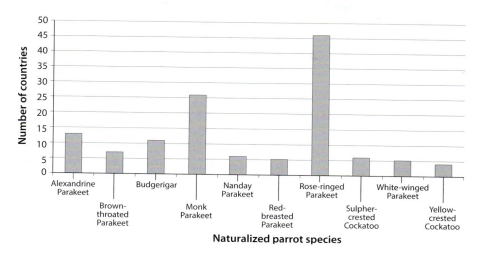

Figure 2.3. The top 10 naturalized parrots (in alphabetical order); these species have colonized the most countries.

of climate alone is too weak to be the main factor in predicting establishment success for all species (Duncan et al. 2001).

The success of an introduction, whether intentional or not, is likely to be related to the community assemblage at the release site. Factors such as species richness, predation, or interspecific competition may determine naturalization (Lodge 1993; Blackburn et al. 2009b). The "enemy release hypothesis" (ERH), whereby an alien species thrives because it has escaped the predators, parasites, and pathogens of its native range (Williamson 1996; Keane and Crawley 2002), is offered as one explanation for a species' success. However, studies supporting the ERH tend to overlook the impact novel enemies have on the introduced species, the effects of which, at the community level, appear to be inconclusive (Colautti et al. 2004). While there are reports of alien parrots being depredated by native species, they are very rare (Menchetti and Mori 2014), and such predation does not seem to have an impact on establishment.

Once an alien species is present within a community, its establishment may be hindered by competition with native birds for similar resources. One such resource is nesting sites, which could be a limiting factor for establishment, especially for cavity-nesting birds (Grarock et al. 2013; Martens and Woog 2017). With the exception of a few species, psittaciforms are cavity nesters (Juniper and Parr 2010), and it is thought that native cavity-nesting species suffer the most impact from the presence of naturalized parrots (Menchetti and Mori 2014). Nest cavities may be especially scarce in urban and suburban areas, leading to increased competition (Menchetti and Mori 2014). Several parrots are known to enlarge existing nests, but this is common only when the wood is soft enough or no alternative nest sites are available. Some alien parrots also start breeding earlier than native species (Menchetti and Mori 2014), giving them an advantage on claiming nesting resources. It seems that it is neither a specific climatic nor environmental factor that makes an area suitable for colonization; rather it is a composition of resources similar to those found in the colonizing species' native range, or that the colonizing species is adaptable or flexible.

THE FUTURE OF THREATENED PARROT SPECIES

Current conservation plans often include efforts to reintroduce species back into their native ranges. The problems and solutions facing purposeful reintroductions are similar to those for invasions (Blackburn and Cassey 2004; Lever 2005). Therefore, we should be able to learn from the successes and failures of both to better inform future legislation for endangered species (Blackburn and Cassey 2004).

Although parrots, overall, are at a higher extinction risk than other families of birds (Bennett and Owens 1997), many have increasing populations. The most successful appear to be synanthropes, and this supports the findings that alien bird species increase toward urban centers (Marzluff 2012). The greatest body of scientific literature on alien parrot species concerns the Rose-ringed (reviewed by Jackson, chap. 10 this vol.) and Monk Parakeets (reviewed by Calzada Preston et al., chap. 11 this vol.). Both of these species are closely associated with humans, so as urbanization continues to expand, so will populations of both (Strubbe and Matthysen 2009). As humans continue to encroach into natural areas, those species that exhibit behavioral flexibility may be more likely to prosper in those habitats. However, flexibility is unlikely to reduce extinction risk unless it is related to opportunistic feeding behaviors (Nicolakakis et al. 2003).

One of the key points appears to be that, no matter what other factors are involved, if an insufficient number of individuals is released, then failure is more likely. One of only two native US psittaciforms, the Thick-billed Parrot (*Rhynchopsitta pachyrhyncha*) was eradicated from most of its native range by hunting and habitat loss. Although a reintroduction program took place in the US between 1986 and 1992 (Snyder et al. 1994), it did not result in a self-sustaining population, and the only population that remains is in Mexico. Snyder et al. (1994) admit that an insufficient number of individuals was released, which may have put initial populations at an increased risk of predation owing to small flock size.

The greatest factor in success is introduction effort; therefore, this has implications for the reintroduction of species back to their previous range (Green 1997). All introduced species must grow in number from a small initial colonizing population (Cassey et al. 2004a). Even after the initial causes for a species' extirpation have been resolved, introduction effort may still be the most important factor for success (Veltman et al. 1996) and must be accounted for in future action plans.

CAN NATURALIZED POPULATIONS BE RESERVOIRS FOR ENDANGERED SPECIES?

The threats from invasive species upon the environment and biodiversity are well documented, and the numbers of invasive species

One member of Hong Kong's population of 150-plus wild, naturalized Yellow-crested Cockatoos (*Cacatua sulphurea*). This individual is occupying a nest cavity in Hong Kong Park, in the heart of the city's financial district. September 2016. Photo by Astrid Andersson.

are increasing (Mack et al. 2000); it is possible that interspecific competition may play a role in limiting further establishment (Moulton and Pimm 1983). Unfortunately, most of the time we are unable to predict those species that will thrive and those that will perish (Cassey et al. 2004a).

The eradication of a species from its newly colonized range is a common practice, and there have been successful programs conducted (Simberloff et al. 2013). If steps are not taken to limit the transport of parrots, then it is likely that more of them will become invasive. The recommended management of invasive species should focus on prevention, rather than mitigation and restoration after the event (Pyšek and Richardson 2010).

In growing instances, the invading species itself is endangered (Marchetti and Engstrom 2016; Gibson and Yong 2017). Since 1993, six species of parrots that have been recorded as being established outside their native ranges are also currently classified as endangered (IUCN 2019). Within Europe, eight species of parrot have been highlighted as likely to become more readily established and spread farther, with the Alexandrine Parakeet suggested as one of the worst threats (Roy et al. 2015). Although the Alexandrine Parakeet is still fairly common in some of its native range, there has been a general decline in its population across its whole range, resulting in it being listed as "near threatened" by the IUCN in 2015 (Ancillotto et al. 2016). The causes for the decline are thought to be due to the demands of the pet trade and habitat loss (Ancillotto et al. 2016). There may come a point when some endangered species have greater populations outside their native ranges.

It has been suggested that invasive populations could act as an insurance pool for endangered species (Marchetti and Engstrom 2016; Gibson and Yong 2017) and also increase genetic diversity within the native populations via reintroductions (Bradshaw et al. 2006). Greater genetic variation may be found in an invasive population if the individuals originate from across the species' native range (Blackburn et al. 2009b). However, there is also the risk of reduced variation due to low propagule size, especially in smaller founder populations (Blackburn et al. 2009b). Genetic diversity may also be reduced through the selective breeding seen in aviculture, particularly for traits considered attractive by breeders (Forstmeier et al. 2007).

One of the major challenges of understanding invasive biology is the lack of knowledge (Jeschke et al. 2012). Unfortunately, detailed population studies are no more likely to have been done on threatened parrots than on nonthreatened parrots, and many genera are not well studied (Marsden and Royle 2015). At least one alien parrot species has been recorded in GAVIA as being present, but with its population status recorded as "unknown," in over half the countries of the world. Although there is a paucity of data, and the data are constantly evolving, there are valuable lessons that can be learned for future conservation, not only of parrots but also other threatened species. Reintroducing a species back into its natural range is more likely to succeed after eradication caused by hunting or harvest for the illegal pet trade rather than by habitat loss or climate change. This is due to the fact that suitable habitat already exists, but the initial threats must be removed in order for a species to survive. Where species are not limited by specific habitat, climate, or dietary requirements, then it seems that colonization is more likely if introduction effort is greater.

THE DISTRIBUTION OF NATURALIZED PARROT POPULATIONS

APPENDIX 2.1

Distribution of records, by parrot species and category of establishment, in the GAVIA database. Each column contains the percentage value of each category for each species; e.g., Grey-headed Lovebird accounts for 28% of all "died out" records.

SPECIES	ESTABLISHED	BREEDING	UNSUCCESSFUL	DIED OUT	EXTIRPATED	UNKNOWN
Alexandrine Parakeet	1.82%	4.53%	0.62%	0.00%	0.00%	3.68%
Antipodes Parakeet	0.00%	0.00%	0.00%	3.51%	25.00%	0.00%
Australian Ringneck	0.09%	0.00%	0.00%	0.00%	0.00%	0.75%
Black Lory	0.00%	0.00%	0.00%	0.00%	0.00%	0.08%
Black-headed Parrot	0.00%	0.00%	0.00%	0.00%	0.00%	0.17%
Blossom-headed Parakeet	0.00%	0.00%	0.62%	0.00%	0.00%	0.75%
Blue Lorikeet	0.00%	0.00%	0.00%	5.26%	0.00%	0.00%
Blue-and-yellow Macaw	0.18%	2.63%	0.00%	0.00%	0.00%	0.75%
Blue-crowned Hanging Parrot	0.00%	0.00%	0.62%	0.00%	0.00%	0.00%
Blue-crowned Parakeet	0.91%	6.21%	0.00%	0.00%	12.50%	1.84%
Blue-eyed Cockatoo	0.00%	0.00%	0.00%	0.00%	0.00%	0.08%
Blue-headed Parrot	0.00%	0.00%	0.00%	0.00%	0.00%	0.08%
Blue-naped Parrot	0.91%	0.00%	0.00%	0.00%	0.00%	0.17%
Blue-streaked Lory	0.36%	0.00%	0.00%	0.00%	0.00%	0.17%
Blue-winged Parrotlet	0.00%	0.00%	0.00%	0.00%	0.00%	0.08%
Bourke's Parrot	0.00%	0.00%	0.00%	0.00%	0.00%	0.08%
Brown-throated Parakeet	1.27%	0.24%	0.62%	5.26%	0.00%	2.17%
Budgerigar	5.19%	4.53%	27.95%	12.28%	0.00%	5.01%
Burrowing Parrot	0.00%	0.48%	1.24%	0.00%	0.00%	0.92%
Chattering Lory	0.00%	0.00%	0.00%	0.00%	0.00%	0.58%
Chestnut-fronted Macaw	0.09%	1.91%	0.00%	0.00%	0.00%	1.09%
Cockatiel	0.18%	0.48%	1.24%	0.00%	0.00%	2.26%
Coconut Lorikeet	0.82%	0.24%	0.62%	0.00%	0.00%	0.75%
Crimson Rosella	1.00%	0.00%	0.62%	0.00%	0.00%	0.75%
Crimson Shining Parrot	0.27%	0.00%	0.00%	0.00%	0.00%	0.00%
Cuban Amazon	0.00%	0.00%	0.00%	0.00%	0.00%	0.08%
Dusky Lory	0.00%	0.00%	0.00%	0.00%	0.00%	0.17%
Dusky-headed Parakeet	0.09%	0.48%	0.62%	0.00%	0.00%	0.75%
Eastern Bluebonnet	0.00%	0.00%	0.00%	0.00%	0.00%	0.08%
Eastern Rosella	1.27%	0.00%	0.00%	0.00%	0.00%	0.42%
Eclectus Parrot	0.45%	0.00%	1.86%	0.00%	0.00%	1.09%
Festive Amazon	0.00%	0.00%	0.00%	0.00%	0.00%	0.33%
Finsch's Parakeet	0.00%	0.24%	0.00%	0.00%	0.00%	0.25%
Fischer's Lovebird	0.73%	0.24%	0.00%	0.00%	0.00%	0.84%

APPENDIX 2.1

SPECIES	ESTAB-LISHED	BREEDING	UNSUC-CESSFUL	DIED OUT	EXTIRPATED	UNKNOWN
Galah	0.36%	0.00%	0.62%	0.00%	0.00%	0.67%
Gang-gang Cockatoo	0.18%	0.00%	0.00%	0.00%	0.00%	0.33%
Golden-collared Macaw	0.00%	0.00%	0.00%	0.00%	0.00%	0.25%
Great-billed Parrot	0.00%	0.00%	0.62%	0.00%	0.00%	0.33%
Greater Vasa Parrot	0.00%	0.00%	0.00%	0.00%	0.00%	0.08%
Green Parakeet	0.45%	0.24%	0.00%	0.00%	0.00%	1.34%
Green Rosella	0.00%	0.00%	0.00%	0.00%	0.00%	0.08%
Green-cheeked Parakeet	0.00%	0.24%	0.00%	0.00%	0.00%	0.50%
Green-rumped Parrotlet	0.73%	0.00%	6.21%	1.75%	0.00%	0.67%
Grey Parrot	0.00%	0.00%	0.00%	0.00%	0.00%	1.34%
Grey-cheeked Parakeet	0.00%	0.00%	0.00%	1.75%	0.00%	0.08%
Grey-headed Lovebird	0.91%	0.00%	5.59%	28.07%	0.00%	1.17%
Hispaniolan Amazon	0.55%	0.48%	0.62%	0.00%	0.00%	0.50%
Hispaniolan Parakeet	0.00%	0.00%	0.00%	0.00%	0.00%	0.50%
Hyacinth Macaw	0.00%	0.00%	0.00%	0.00%	0.00%	0.33%
Jandaya Parakeet	0.00%	0.00%	0.62%	0.00%	0.00%	0.17%
Kakapo	0.27%	0.00%	6.83%	5.26%	0.00%	0.75%
Kuhl's Lorikeet	0.27%	0.00%	0.00%	0.00%	0.00%	0.17%
Layard's Parakeet	0.00%	0.00%	0.00%	0.00%	0.00%	0.08%
Lesser Vasa Parrot	0.00%	0.00%	0.62%	0.00%	0.00%	0.25%
Lilac-crowned Amazon	0.45%	1.43%	0.62%	0.00%	0.00%	0.25%
Lilac-tailed Parrotlet	0.00%	0.00%	0.00%	0.00%	0.00%	0.08%
Lilian's Lovebird	0.18%	0.00%	0.00%	0.00%	0.00%	0.25%
Little Corella	0.45%	0.00%	0.00%	0.00%	0.00%	0.00%
Long-billed Corella	0.36%	0.00%	0.00%	0.00%	0.00%	0.67%
Major Mitchell's Cockatoo	0.09%	0.00%	0.62%	0.00%	0.00%	0.08%
Maroon Shining Parrot	0.36%	0.00%	0.00%	3.51%	0.00%	0.33%
Maroon-bellied Parakeet	0.00%	0.95%	0.00%	0.00%	0.00%	0.25%
Maroon-fronted Parrot	0.00%	0.00%	0.00%	0.00%	0.00%	0.25%
Meyer's Parrot	0.00%	0.00%	0.00%	1.75%	0.00%	0.17%
Military Macaw	0.00%	0.24%	0.00%	0.00%	0.00%	0.50%
Mitred Parakeet	1.91%	2.86%	0.00%	0.00%	0.00%	1.42%
Monk Parakeet	13.28%	17.90%	14.29%	12.28%	50.00%	9.27%
Musk Lorikeet	0.09%	0.00%	0.00%	0.00%	0.00%	0.00%
Nanday Parakeet	4.19%	3.82%	1.24%	0.00%	0.00%	1.75%
Niam-niam Parrot	0.09%	0.00%	0.00%	0.00%	0.00%	0.17%
Northern Rosella	0.00%	0.00%	0.00%	0.00%	0.00%	0.08%
Olive-throated Parakeet	0.00%	0.00%	0.00%	0.00%	0.00%	0.08%

THE DISTRIBUTION OF NATURALIZED PARROT POPULATIONS

SPECIES	ESTAB-LISHED	BREEDING	UNSUC-CESSFUL	DIED OUT	EXTIRPATED	UNKNOWN
Orange-chinned Parakeet	0.00%	0.00%	0.00%	0.00%	0.00%	0.92%
Orange-fronted Parakeet	0.36%	0.24%	0.00%	0.00%	0.00%	1.09%
Orange-winged Amazon	0.27%	3.34%	0.00%	0.00%	0.00%	1.25%
Ornate Lorikeet	0.00%	0.00%	0.00%	0.00%	0.00%	0.25%
Pale-headed Rosella	0.00%	0.00%	0.62%	1.75%	0.00%	0.50%
Palm Cockatoo	0.00%	0.00%	0.62%	0.00%	0.00%	0.33%
Peach-fronted Parakeet	0.00%	0.00%	0.00%	0.00%	0.00%	0.42%
Plain Parakeet	0.00%	0.00%	0.00%	1.75%	0.00%	0.08%
Plum-headed Parakeet	0.00%	1.19%	0.00%	0.00%	0.00%	1.34%
Red Lory	0.18%	0.72%	0.62%	0.00%	0.00%	0.67%
Red-and-green Macaw	0.00%	0.00%	0.00%	0.00%	0.00%	0.17%
Red-breasted Parakeet	1.18%	1.19%	1.86%	0.00%	0.00%	1.17%
Red-crowned Amazon	4.82%	3.10%	0.00%	0.00%	0.00%	0.84%
Red-crowned Parakeet	0.55%	0.00%	0.62%	0.00%	0.00%	0.00%
Red-fronted Macaw	0.00%	0.00%	0.00%	0.00%	0.00%	0.08%
Red-headed Lovebird	0.00%	0.00%	0.00%	0.00%	0.00%	0.17%
Red-lored Amazon	0.55%	0.24%	0.62%	0.00%	0.00%	0.84%
Red-masked Parakeet	0.82%	3.82%	0.62%	0.00%	0.00%	0.92%
Red-rumped Parrot	0.00%	0.00%	0.00%	0.00%	0.00%	0.25%
Red-shouldered Macaw	0.00%	0.00%	0.00%	0.00%	0.00%	0.33%
Red-spectacled Amazon	0.00%	0.00%	0.00%	0.00%	0.00%	0.33%
Red-winged Parrot	0.00%	0.00%	0.00%	0.00%	0.00%	0.25%
Rose-ringed Parakeet	36.58%	16.23%	8.70%	7.02%	12.50%	16.71%
Rosy-faced Lovebird	0.45%	0.95%	0.62%	0.00%	0.00%	2.01%
Rüppell's Parrot	0.00%	0.48%	0.00%	0.00%	0.00%	0.08%
Salmon-crested Cockatoo	0.00%	0.24%	1.24%	0.00%	0.00%	0.92%
Scaly-breasted Lorikeet	0.36%	0.00%	0.00%	0.00%	0.00%	0.25%
Scaly-headed Parrot	0.00%	0.00%	0.00%	0.00%	0.00%	0.08%
Scarlet Macaw	0.00%	0.00%	0.00%	0.00%	0.00%	0.67%
Scarlet-fronted Parakeet	0.00%	0.48%	0.00%	0.00%	0.00%	0.33%
Senegal Parrot	0.09%	0.95%	0.62%	0.00%	0.00%	1.17%
Southern Mealy Amazon	0.00%	0.00%	0.00%	0.00%	0.00%	1.00%
St. Vincent Amazon	0.00%	0.00%	0.00%	0.00%	0.00%	0.08%
Sulphur-crested Cockatoo	2.46%	1.19%	0.62%	3.51%	0.00%	2.76%
Superb Parrot	0.00%	0.00%	0.00%	0.00%	0.00%	0.17%
Swift Parrot	0.00%	0.00%	0.00%	0.00%	0.00%	0.08%
Tanimbar Corella	1.00%	0.48%	0.62%	0.00%	0.00%	0.92%
Tepui Parrotlet	0.00%	0.00%	0.00%	1.75%	0.00%	0.00%

SPECIES	ESTAB-LISHED	BREEDING	UNSUC-CESSFUL	DIED OUT	EXTIRPATED	UNKNOWN
Thick-billed Parrot	0.09%	0.00%	0.00%	0.00%	0.00%	0.08%
Tui Parakeet	0.00%	0.00%	0.00%	0.00%	0.00%	0.25%
Turquoise-fronted Amazon	0.09%	2.39%	0.62%	1.75%	0.00%	1.25%
Ultramarine Lorikeet	0.09%	0.00%	0.00%	0.00%	0.00%	0.00%
Vernal Hanging Parrot	0.00%	0.00%	1.86%	0.00%	0.00%	0.00%
White Cockatoo	0.18%	0.24%	0.00%	1.75%	0.00%	0.58%
White-crowned Parrot	0.00%	0.00%	0.00%	0.00%	0.00%	0.25%
White-eyed Parakeet	0.00%	0.24%	0.00%	0.00%	0.00%	0.50%
White-fronted Amazon	0.36%	0.48%	0.00%	0.00%	0.00%	0.33%
White-winged Parakeet	3.09%	5.97%	0.62%	0.00%	0.00%	0.58%
Yellow-chevroned Parakeet	1.36%	1.67%	0.00%	0.00%	0.00%	0.58%
Yellow-collared Lovebird	0.82%	0.72%	0.00%	0.00%	0.00%	0.92%
Yellow-crested Cockatoo	1.55%	0.00%	0.62%	0.00%	0.00%	0.92%
Yellow-crowned Amazon	0.27%	1.67%	1.24%	0.00%	0.00%	1.00%
Yellow-eared Parrot	0.00%	0.00%	0.00%	0.00%	0.00%	0.08%
Yellow-headed Amazon	1.55%	0.48%	0.00%	0.00%	0.00%	1.00%
Yellow-naped Amazon	0.00%	0.95%	0.62%	0.00%	0.00%	0.25%
Yellow-shouldered Amazon	0.00%	0.00%	1.24%	0.00%	0.00%	0.42%
Yucatan Amazon	0.00%	0.00%	0.00%	0.00%	0.00%	0.08%

REFERENCES

Aagaard, K., and Lockwood, J. 2014. Exotic birds show lags in population growth. *Diversity and Distributions* 20:547–554.

Aagaard, K., and Lockwood, J. L. 2016. Severe and rapid population declines in exotic birds. *Biological Invasions* 18:1667–1678.

Ancillotto, L., Strubbe, D., Menchetti, M., and Mori, E. 2016. An overlooked invader? Ecological niche, invasion success and range dynamics of the Alexandrine parakeet in the invaded range. *Biological Invasions* 18:583–595.

Avery, M. L., and Shiels, A. B. 2018. Monk and rose-ringed parakeets. In *Ecology and Management of Terrestrial Vertebrate Invasive Species in the United States*, ed. W. C. Pitt, J. C. Beasley, and G. W. Witmer, 333–357. Boca Raton, FL: CRC Press.

Beissinger, S. 2001. Trade of live wild birds: Potentials, principles, and practices of sustainable use. In *Conservation of Exploited Species*, ed. J. D. Reynolds, G. M. Mace, K. H. Redford, J. G. Robinson, 182–202. Cambridge: Cambridge Univ. Press.

Bellard, C., Cassey, P., and Blackburn, T. M. 2016. Alien species as a driver of recent extinctions. *Biology Letters* 12:1–4.

Bennett, P. M., and Owens, I.P.F. 1997. Variation in extinction risk among birds: Chance or evolutionary predisposition? *Proceedings of the Royal Society of London B: Biological Sciences* 264:401–408.

Bibby, C. J., Burgess, N. D., Hill, D. A., and Mustoe, S. H. 2000. *Bird Census Techniques*. London: Academic Press.

BirdLife International. 2019. Date Zone. Accessed 19 Apr. 2019. http://datazone.birdlife.org.

Blackburn, T. M., and Cassey, P. 2004. Are introduced and re-introduced species comparable? A case study of birds. *Animal Conservation* 7:427–433.

Blackburn, T. M., Cassey, P., and Lockwood, J. L. 2009a. The role of species traits in the establishment success of exotic birds. *Global Change Biology* 15:2852–2860.

Blackburn, T. M., and Duncan, R. P. 2001a. Establishment patterns of exotic birds are constrained by non-random patterns in introduction. *Journal of Biogeography* 28:927–939.

Blackburn, T. M., and Duncan, R. P. 2001b. Determinants of establishment success in introduced birds. *Nature* 414:195–197.

Blackburn, T. M., Lockwood, J., and Cassey, P. 2009b. *Avian Invasions: The Ecology and Evolution of Exotic Birds*. Oxford: Oxford Univ. Press.

Bradshaw, C.J.A., Isagi, Y., Kaneko, S., Bowman, D.M.J.S., and Brook, B. W. 2006. Conservation value of non-native banteng in northern Australia. *Conservation Biology* 20:1306–1311.

Brooks, T. 2001. Are unsuccessful avian invaders rarer in their native range than successful invaders? In *Biotic Homogenization*, ed. J. L. Lockwood and M. L. McKinney, 125–155. Boston, MA: Springer.

Butler, C. J. 2005. Feral parrots in the continental United States and United Kingdom: Past, present, and future. *Journal of Avian Medicine and Surgery* 19:142–149.

Carrete, M., and Tella, J. 2008. Wild-bird trade and exotic invasions: A new link of conservation concern? *Frontiers in Ecology and the Environment* 6:207–211.

Cassey, P. 2002. Life history and ecology influences establishment success of introduced land birds. *Biological Journal of the Linnean Society* 76:465–480.

Cassey, P., Blackburn, T. M., Duncan, R. P., and Gaston, K. J. 2005. Causes of exotic bird establishment across exotic islands. *Proceedings of the Royal Society of London B: Biological Sciences* 272:2059–2063.

Cassey, P., Blackburn, T. M., Russell, G. J., Jones, K. E., and Lockwood, J. L. 2004a. Influences on the transport and establishment of exotic bird species: An analysis of the parrots (Psittaciformes) of the world. *Global Change Biology* 10:417–426.

Cassey, P., Blackburn, T. M., Sol, D., Duncan, R. P., and Lockwood, J. L. 2004b. Global patterns of introduction effort and establishment success in birds. *Proceedings of the Royal Society of London B: Biological Sciences* 271:S405–S408.

Cassey, P., Vall-llosera, M., Dyer, E., and Blackburn, T. M. 2015. The biogeography of avian invasions: History, accident and market trade. In *Biological Invasions in Changing Ecosystems. Vectors, Ecological Impacts, Management and Predictions*, ed. J. Canning-Clode, 37–54. Warsaw and Berlin: De Gruyter Open.

CITES. 2019. CITES Trade Database. http://trade.cites.org/.

Colautti, R. I., and MacIsaac, H. J. 2004. A neutral terminology to define "invasive" species. *Diversity and Distributions* 10:135–141.

Colautti, R. I., Ricciardi, A., Grigorovich, I. A., and MacIsaac, H. J. 2004. Is invasion success explained by the enemy release hypothesis? *Ecology Letters* 7:721–733.

Collar, N., Boesman, P.F.D., and Sharpe, C. J. 2020. Red-and-green Macaw (*Ara chloropterus*) (v.1.0). In *Birds of the World*, ed. J. del Hoyo, A. Elliott, J. Sargatal, D. A. Christie, and E. de Juana. Ithaca, NY: Cornell Lab of Ornithology. https://doi.org/10.2173/bow.ragmac1.01.

Collar, N., and Kirwan, G. M. 2020. Gray-headed Lovebird (*Agapornis canus*) (v.1.0). In *Birds of the World*, ed. J. del Hoyo, A. Elliott, J. Sargatal, D. A. Christie, and E. de Juana. Ithaca, NY: Cornell Lab of Ornithology. https://doi.org/10.2173/bow.gyhlov1.01.

Crooks, J. A. 2005. Lag times and exotic species: The ecology and management of biological invasions in slow motion. *Ecoscience* 12:316–329.

Domènech, J., Carrillo, J., and Senar, J. S. 2003. Population size of the monk parakeet *Myiopsitta monachus* in Catalonia. *Revista Catalana d'Ornitologia* 20:1–9.

Duncan, R. P., Blackburn, T. M., and Sol, D. 2003. The ecology of bird introductions. *Annual Review of Ecology, Evolution and Systematics* 34:71–98.

Duncan, R. P., Bomford, M., Forsyth, D. M., and Conibear, L. 2001. High predictability in introduction outcomes and the geographical range size of introduced Australian birds: A role for climate. *Journal of Animal Ecology*, 70:621–632.

Dyer, E. E., Redding, D. W., and Blackburn, T. M. 2017. The Global Avian Invasions Atlas: A database of alien bird distributions worldwide. *Scientific Data* 4:170041.

Elton, C. S. 1958. *The Ecology of Invasions by Animals and Plants*. Chicago, IL: Univ. of Chicago Press.

Evans, T., Kumschick, S., and Blackburn, T. M. 2016. Application of the Environmental Impact Classification for Alien Taxa (EICAT) to a global assessment of alien bird impacts. *Diversity and Distributions*, 22:919–931.

Forstmeier, W., Segelbacher, G., Mueller, J. C., and Kempenaers, B. 2007. Genetic variation and differentiation in captive and wild zebra finches (*Taeniopygia guttata*). *Molecular Ecology* 16:4039–4050.

Garnett, S. T., and Christidis, L. 2017. Taxonomy anarchy hampers conservation. *Nature* 546:25–27.

Garrett, K. L. 1997. Population status and distribution of naturalized parrots in southern California. *Western Birds* 28:181–195.

Gibson, L., and Yong, D. L. 2017. Saving two birds with one stone: Solving the quandary of introduced, threatened species. *Frontiers in Ecology and the Environment* 15:35–41.

Gill, F., and Donsker, D., eds. 2019. IOC World Bird List (v.9.1). Accessed 15 Jan. 2019. https://www.worldbirdnames.org.

Grarock, K., Lindenmayer, D. B., Wood, J. T., and Tidemann, C. R. 2013. Does human-induced habitat modification influence the impact of introduced species? A case study on cavity-nesting by the introduced common myna (*Acridotheres tristis*) and two Australian native parrots. *Environmental Management* 52:958–970.

Green, R. E. 1997. The influence of numbers released on the outcome of attempts to introduce exotic bird species to New Zealand. *Journal of Animal Ecology* 66:25–35.

Griffith, B., Scott, J. M., Carpenter, J. W., and Reed, C. 1989. Translocation as a species tool: Status and strategy. *Science* 245:477–480.

Hulme, P. E. 2009. Trade, transport and trouble: Managing invasive species pathways in an era of globalization. *Journal of Applied Ecology* 46:10–18.

IUCN. 2019. The IUCN Red List of Threatened Species. Accessed 7 June 2019. https://www.iucnredlist.org.

Jeschke, J. M. 2014. General hypotheses in invasion ecology. *Diversity and Distributions* 20:1229–1234.

Jeschke, J. M., Gómez Aparicio, L., Haider, S., Heger, T., Lortie, C. J., Pyšek, P., and Strayer, D. L. 2012. Support for major hypotheses in invasion biology is uneven and declining. *NeoBiota* 14:1–20.

Juniper, T., and Parr, M. 2010. *Parrots: A Guide to Parrots of the World*. New Haven, CT, and London: Yale Univ. Press.

REFERENCES

Keane, R. M., and Crawley, M. J. 2002. Exotic plant invasions and the enemy release hypothesis. *Trends in Ecology & Evolution* 17:164–170.

Kolar, C. S., and Lodge, D. M. 2001. Progress in invasion biology: Predicting invaders. *Trends in Ecology & Evolution* 16:199–204.

Lever, C. 2005. *Naturalised Birds of the World*. London: A & C Black.

Lockwood, J. L. 1999. Using taxonomy to predict success among introduced avifauna: Relative importance of transport and establishment. *Conservation Biology* 13:560–567.

Lockwood, J. L., Welbourne, D. J., Romagosa, C. M., Cassey, P., Mandrak, N. E., Strecker, A., Leung, B., Stringham, O. C., Udell, B., Episcopio-Sturgeon, D. J., et al. 2019. When pets become pests: The role of the exotic pet trade in producing invasive vertebrate animals. *Frontiers in Ecology and the Environment* 17:323–330.

Lodge, D. M. 1993. Biological invasions: Lessons for ecology. *Trends in Ecology & Evolution* 8:133–137.

Long, J. L. 1981. *Introduced Birds of the World: The Worldwide History, Distribution and Influence of Birds Introduced to New Environments*. London: David & Charles.

Mabb, K. T. 1997. Roosting behavior of naturalized parrots in the San Gabriel valley, California. *Western Birds* 28:202–208.

MacArthur, R. H., and Wilson, E. O. 1967. *The Theory of Island Biogeography*. Princeton, NJ: Princeton Univ. Press.

Mack, R. N., Simberloff, D., Lonsdale, W. M., Evans, H., Clout, M., and Bazzaz, F. A. 2000. Biotic invasions: Causes, epidemiology, global consequences, and control. *Ecological Applications* 10:689–710.

Marchetti, M. P., and Engstrom, T. 2016. The conservation paradox of endangered and invasive species. *Conservation Biology* 30:434–437.

Marsden, S. J., and Royle, K. 2015. Abundance and abundance change in the world's parrots. *Ibis* 157:219–229.

Martens, J. M., and Woog, F. 2017. Nest cavity characteristics, reproductive output and population trend of naturalized Amazon parrots in Germany. *Journal of Ornithology* 158:823–832.

Marzluff, J. M. 2012. Worldwide urbanization and its effects on birds. In *Avian Ecology and Conservation in an Urbanizing World*, ed. J. M. Marzluff, R. Bowman, and R. Donnelly, 19–47. New York: Springer Science & Business Media.

Menchetti, M., and Mori, E. 2014. Worldwide impact of alien parrots (Aves Psittaciformes) on native biodiversity and environment: A review. *Ethology Ecology & Evolution* 26:172–194.

Minor, E. S., Appelt, C. W., Grabiner, S., Ward, L., Moreno, A., and Pruett-Jones, S. 2012. Distribution of exotic monk parakeets across an urban landscape. *Urban Ecosystems* 15:979–991.

Mori, E., Di Febbraro, M., Foresta, M., Melis, P., Romanazzi, E., Notari, A., and Boggiano, F. 2013. Assessment of the current distribution of free-living parrots and parakeets (Aves: Psittaciformes) in Italy: A synthesis of published data and new records. *Italian Journal of Zoology* 80:158–167.

Mori, E., Grandi, G., Menchetti, M., Tella, J. L., Jackson, H. A., Reino, L., van Kleunen, A., Figueira, R., and Ancillotto, L. 2017. Worldwide distribution of non-native Amazon parrots and temporal trends of their global trade. *Animal Biodiversity and Conservation* 40:49–62.

Moulton, M. P., and Pimm, S. L. 1983. The introduced Hawaiian avifauna: Biogeographic evidence for competition. *American Naturalist* 121:669–690.

Nicolakakis, N., Sol, D., and Lefebvre, L. 2003. Behavioural flexibility predicts species richness in birds, but not extinction risk. *Animal Behaviour* 65:445–452.

Olson, D. M., Dinerstein E., Wikramanayake, E. D., Burgess, N. D., Powell, G.V.N., Underwood, E. C., D'Amico, J. A., Itoua, I., Strand, H. E., Morrison, J. C., et al. 2001. Terrestrial ecoregions of the world: A new map of life on earth. *BioScience* 51:933–938.

Pranty, B., and Epps, S. 2002. Distribution, population status, and documentation of exotic parrots in Broward County, Florida. *Florida Field Naturalist* 30:111–131.

Pruett-Jones, S., Appelt, C. W., Sarfaty, A., Van Vossen, B., Leibold, M. A., and Minor, E. S. 2012. Urban parakeets in northern Illinois: A 40-year perspective. *Urban Ecosystems* 15:709–719.

Pruett-Jones, S., Newman, J. R., Newman, C. M., and Lindsay, J. R. 2005. Population growth of monk parakeets in Florida. *Florida Field Naturalist* 33:1–14.

Pyšek, P., and Richardson, D. M. 2010. Invasive species, environmental change and management, and health. *Annual Review of Ecology and Systematics* 35:25–55.

Pyšek, P., Richardson, D. M., Pergl, J., Jarosík, V., Sixtová, Z., and Weber, E. 2008. Geographical and taxonomic biases in invasion ecology. *Trends in Ecology & Evolution*, 23:237–244.

Roy, H. E., Adriaens, T., Aldridge, D. C., Bacher, S., Bishop, J.D.D., Blackburn, T. M., Branquart, E., Brodie, J., Carboneras, C., Cook, E. J., et al. 2015. *Invasive Alien Species: Prioritising Prevention Efforts through Horizon Scanning*. ENV.B.2/ETU/2014/0016. Brussels: European Commission.

Runde, D. E., Pitt, W. C., and Foster, J. 2007. Population ecology and some potential impacts of emerging populations of exotic parrots. In *Managing Vertebrate Invasive Species: Proceedings of an International Symposium*, ed. G. W. Witmer, W. C. Pitt, and K. A. Fagerstone, 338–360. Fort Collins, CO: USDA/APHIS Wildlife Services, National Wildlife Research Center.

Russello, M. A., Avery, M. L., and Wright, T. F. 2008. Genetic evidence links invasive monk parakeet populations in the United States to the international pet trade. *BMC Evolutionary Biology* 8:217.

Sakai, A. K., Allendorf, F. W., Holt, J. S., Lodge, D. M., Molofsky, J., With, K. A., Baughman, S., Cabin, R. J., Cohen, J. E., Ellstrand, N. C., et al. 2001. The population biology of invasive species. *Annual Review of Ecology and Systematics* 32:305–332.

Shirley, S. M., and Kark, S., 2009. The role of species traits and taxonomic patterns in alien bird impacts. *Global Ecology and Biogeography* 18:450–459.

Simberloff, D. 1995. Why do introduced species appear to devastate islands more than mainland areas? *Pacific Science* 49:87–97.

Simberloff, D., Martin, J.-L., Genovesi, P., Maris, V., Wardle, D. A., Aronson, J., Courchamp, F., Galil, B., Garcia-Berthou, E., Pascal, M. et al. 2013. Impacts of biological invasions: What's what and the way forward. *Trends in Ecology & Evolution* 28:58–66.

Smith, K. F., Behrens, M., Schloegel, L. M., Marano, N., Burgiel, S., and Daszak, P. 2009. Reducing the risks of the wildlife trade. *Science* 324:594–595.

Snyder, N.F.R., Koenig, S. E., Koschmann, J., Snyder, H. A., and Johnson, T. B. 1994. Thick-billed parrot releases in Arizona. *Condor* 96:845–862.

Sol, D., and Lefebvre, L. 2000. Behavioural flexibility predicts invasion success in birds introduced in New Zealand. *Oikos* 90:599–605.

South, J. M., and Pruett-Jones, S. 2000. Patterns of flock size, diet, and vigilance of naturalized monk parakeets in Hyde Park, Chicago. *Condor* 102:848–854.

Strubbe, D., Jackson, H., Groombridge, J., and Matthysen, E. 2015. Invasion success of a global avian invader is explained by within-taxon niche structure and association with humans in the native range. *Diversity and Distributions* 21:675–685.

Strubbe, D., and Matthysen, E. 2009. Establishment success of invasive ring-necked and monk parakeets in Europe. *Journal of Biogeography* 36:2264–2278.

Sullivan, B. L., Aycrigg, J. L., Barry, J. H., Bonney, R. E., Bruns, N., Cooper, C. B., Damoulas, T., Dhondt, A. A., Dietterrich, T., Farnsworth, A., et al. 2014. The eBird enterprise: An integrated approach to development and application of citizen science. *Biological Conservation* 169:31–40.

Tulloch, A.I.T., Possingham, H. P., Joseph, L. N., Szabo, J., and Martin, T. G. 2013. Realising the full potential of citizen science monitoring programs. *Biological Conservation* 165:128–138.

Uehling, J. J., Tallant, J., and Pruett-Jones, S. 2019. Status of naturalized parrots in the United States. *Journal of Ornithology* 160:907–921.

Veltman, C. J., Nee, S., and Crawley, M. J. 1996. Correlates of introduction success in exotic New Zealand birds. *American Naturalist* 147:542–557.

Williamson, M. 1996. *Biological Invasions*. London: Chapman and Hall.

3

PARROTS AND PEOPLE: HUMAN DIMENSIONS OF NATURALIZED PARROTS

Sarah L. Crowley

INTRODUCTION

The global movement of parrots, and the variable success of their establishment in new regions, provide an excellent illustration of how animal and human social worlds are inextricably entangled. Human activity is the primary reason that naturalized parrot populations exist. Moreover, how people respond to the arrival and establishment of parrots plays a significant role in determining the future of introduced populations. People and parrots are caught up together in complex social-ecological systems that include agricultural practices, population dynamics, market trends, urban microclimates, international trade regulations, interspecies competition, and human aesthetic preferences. The ecological elements of these systems are increasingly well understood, thanks to research that, for example, maps the climatic suitability of new regions for introduced parrots (e.g., Di Febbraro and Mori 2015; Ancilloto et al. 2016) or examines their potential impacts on resident wildlife (see White et al. 2019). Research addressing the social dimensions of parrot naturalization, however, is comparatively scarce.

Fortunately, there is increasing recognition that "biological invasions" are as much a human social phenomenon as they are a biological one (McNeely 2013). In recent years, there has been a shift toward greater attention to the social factors that prevent, enable, and affect the success of species introductions (e.g., Vaz et al. 2017; Shackleton et al. 2019a). Introduced parrots and parakeets are charismatic and often highly visible additions to urban and suburban ecosystems. Though many human residents of these areas seem to appreciate the presence of parrots, others are concerned about potential impacts and prefer to see them removed. Consequently, understanding human behaviors and responses to parrots (both within and beyond their native ranges) is important not only for interpreting and predicting patterns of naturalization but also for determining whether and in what form introduced populations might be subject to management. In this chapter, I review the emerging literature on human responses to introduced parrots, highlight the importance of understanding human perceptions in management planning, and propose prospective avenues for future research into the human dimensions of parrot naturalization.

SOCIOECONOMIC INFLUENCES ON PARROT INTRODUCTIONS AND ESTABLISHMENT

Parrots have primarily been translocated beyond their native ranges to supply a worldwide demand for exotic pets and aviculture. To understand how parrots have arrived and become established in new regions, it is therefore vital to examine the economics and impacts of both domestic and international markets for exotic animals (see Cardador et al., chap. 1 this vol.). It is also important to understand what drives this kind of demand: Why do people purchase and keep parrots?

Parrots are considered sociable, intelligent, and interesting pets, and are often compared with companion animals such as domestic cats and dogs. However, unlike dogs and (to some extent) cats, parrots are not domesticated; most birds in the pet trade are separated from wild ancestors by only one or two generations (Engebretson 2006). They are consequently classified as exotic pets. Many of the motivations for keeping exotic species such as parrots are held in common among most pet owners, including a desire for proximity, interaction, and companionship with animals, or childhood experience of having particular animals in the home (Anderson 2003, 2014). There may also be other, more particular reasons people choose to purchase and keep exotic species. Suggestions include the animals' possible role as status symbols, their novelty, and their cultural representation and normalization as "suitable" pets (including their ready availability in pet stores). Nevertheless, the strength and relevance of these motivations for owning parrots have not been empirically studied. Cultural factors certainly play a role, however: Italian households, for example, are 14 times more likely than British households to own a pet bird (UNODC 2016), and the frequency and style in which different species are portrayed—for example, in visual media—can affect market popularity (Silk et al. 2018).

Irrespective of specific motivations, the popularity of parrots as pets has increased: international market demand for psittacines and other exotic birds has grown substantially since the mid-20th century. The import of wild-caught birds has been banned in the United States (US) since the implementation of the Wild Bird Conservation Act of 1992, and in Europe since 2005, reducing the market for imported animals in these regions. Conversely, a 2008 ban on the sale of native parrot species in Mexico created an increase in the popularity of exotic parrots, as they replaced native species in the market (MacGregor-Fors et al. 2011). Many nations, including those that have banned the majority of live imports, have also or alternatively established domestic markets for breeding parrots in captivity, though in the US and Australia, restrictions on ownership, breeding, and trade vary between states (Tillman et al. 2000; Moscatello 2003).

Factors other than market trends also influence the species and numbers of parrots traded. Smaller species with wide native distributions are traded more frequently and sold at lower prices (Menchetti and Mori 2014). The abundance and pest status of parrot species in their native range also positively correlate with the extent of their harvest and export as pets (Cassey et al. 2004; Herrera and Henessey 2008; Pires 2012). Lever (2005) reported that around 70% of introduced parrot species are considered pests in their native ranges.

A good illustration is the Monk Parakeet (*Myiopsitta monachus*), which is known to eat and damage a number of crops, in particular corn and sunflowers (Canavelli et al. 2013). Monk Parakeets have a reputation as pests throughout their native range in South America (Spreyer and Bucher 1998; Simberloff 2003; Canavelli et al. 2012), and control has been carried out at least since Darwin's visit to the region during the *Beagle* voyage (Darwin 1845). Though the extent and regional significance of damage caused by Monk Parakeets has never been systematically studied (Spreyer and Bucher 1998), and recent research suggests that impacts may often be overstated (Canavelli et al. 2013), effects of crop predation can be severe at a local level. Accordingly, Monk Parakeets were heavily exported in the latter half of the 20th century.

Although threatened species are sometimes kept as pets, the most endangered species are less frequently found in parrot markets, likely due to their low abundance and the comparative ease of catching abundant, non-protected species (Pires and Clarke 2012; Olah et al. 2016). The most common parrot species in the exotic bird trade are generally also the cheapest, creating saturated markets (Pires 2012); in the 1970s, Monk Parakeets were being sold in the US for $8 each (Bull 1973). Similarly, until recently, the species has remained common and inexpensive in Spanish pet stores (Muñoz and Real 2006). An abundance of cheap, readily available parrots has two potential effects on the likelihood of wild populations establishing. First, it increases propagule pressure—the number of introduced animals with the potential to end up in the wild—and second, it increases the chances of impulse or ill-considered purchases of pet parrots and therefore the likelihood of their relinquishment or release.

Anderson (2003, 2014), in her ethnographic research, notes that potential parrot owners may not fully understand the captive needs, behavior, or longevity of these birds, or the expense of keeping them. When bored, many parrot species vocalize for attention and/or demonstrate abnormal behaviors. Some—such as Monk Parakeets and Rose-ringed Parakeets (*Psittacula krameri*)—are very vocal as part of normal social interactions. Consequently, although most introduced parrot populations come with one or more mythic "origin stories," the majority of introduced populations are more likely the result of intentional releases or accidental escapes from aviaries or domestic properties (Heald et al. 2019). Vall-llosera and Cassey (2017) studied how escapes from private captivity act as a source of introduced birds in Australia and proposed several factors that could increase the likelihood of escape. Larger-bodied, long-lived species may inspire stronger personal attachment, and be more valuable monetarily, and consequently may be kept more carefully. Conversely, cheaper species and native Australian parrots were thought more likely candidates for escape, due to their greater abundance in captivity and lower cost, potentially resulting in more "leaky doors" (but note that the latter trend may be reversed in countries such as Mexico, where trading native parrots is now banned). Finally, more "docile" species and those able to form closer bonds with humans were thought more likely to be carefully kept.

Once a parrot is "out the window," the likelihood of its establishment success depends on a complex set of factors. In Spain, wild-caught species have been found to be more likely to establish self-sustaining populations than captive-bred parrots (Carrete and Tella 2008; Abellán et al. 2016), and certain species have ecological traits that make them better able to adapt to new latitudes, climates, and habitats (Menchetti and Mori 2014). However, multiple sociocultural and economic factors can also influence the likelihood of successful establishment. These include both the degree of a new population's reliance on anthropogenic habitats and human activity (particularly supplementary feeding) and the extent of both active population management and protectionism.

Introduced parrot populations are most often found in urban and peri-urban areas. Dense human populations are likely to increase propagule pressure (i.e., it is assumed that more human residents correlate with a higher frequency of pet ownership and release) (Strubbe and Matthysen 2009; Minor et al. 2012; Rodríguez-Pastor et al. 2012). Davis et al. (2013) modeled Monk Parakeet distribution in the US to identify the relative importance of anthropic and biophysical variables and found that populations could be split into northern and southern regions. In the southern range, distribution was best predicted by climatic and landscape variables (January dew-point temperature and forest cover), as it is in the native range, and Monk Parakeets were found both in and away from large cities. In contrast, northern populations were apparently limited to urban and suburban areas. The success of populations in the northern US was best explained not by climatic factors but by the human-activity hypothesis, which suggests that anthropic factors can promote the establishment success of introduced species (Davis et al. 2013). Accordingly, synanthropic species (which live alongside humans in their native ranges) are more likely to establish introduced populations. In addition, and importantly in northerly latitudes, introduced populations may be sustained in winter months by human provisioning. The Chicago Monk Parakeet population, for example, has been found to rely almost entirely on bird feeders during the winter (Hyman and Pruett-Jones 1995; South and Pruett-Jones 2000). In urban Barcelona, Rodríguez-Pastor et al. (2012) found a positive correlation between parakeet abundance and density of residents over 65 years of age. They suggest that this is because older and retired people are more likely to feed birds (at least in Western cultures), a hypothesis supported by Cox and Gaston's (2016) study of urban bird-feeding practices in the United Kingdom (UK).

Minor et al. (2012) modeled the distribution of the Chicago Monk Parakeet population in relation to landscape features (e.g., tree canopy

PARROTS AND PEOPLE: HUMAN DIMENSIONS OF NATURALIZED PARROTS

Blue-crowned Parakeets (*Thectocercus acuticaudatus*) at a feeder. Melbourne, Florida, US, 2018. Photo by Diana L. Charland.

cover and distance to nearest water body) and indicators of human activity (e.g., human population density and residential zoning as a proxy for bird feeders). At larger scales, parakeet presence was correlated with human population density. Within the urban landscape, however, parakeet nests were *less* likely to be found in areas with >83% residential cover, suggesting that very high human population density can negatively affect parakeet success or at least discourage nesting. Problematic nests are regularly removed by businesses, residents, and utility companies, and Minor et al. (2012) suggest that increased frequency of these activities in densely populated areas may serve to limit parakeet establishment.

This indicates that the success of introduced populations also depends on the extent and form of human management (e.g., removal of birds or access to their breeding sites), particularly in urban and semi-urban areas. Populations targeted early for eradication or intensive control are clearly less likely to successfully establish. Some populations are purposefully left unmanaged or are simply monitored. In Oregon, for example, a species risk assessment predicted that the state's small resident Monk Parakeet population was likely to be transient and therefore did not pose a significant threat (Stafford 2003). In this instance, the population in question failed to establish, but lack of management can just as often improve the viability of newly founded populations (Pruett-Jones et al. 2012). How new parrot populations are treated by resident humans will depend on a range of factors, such as legal or statutory regulations and requirements (including both those that protect free-living birds and those that target introduced species for removal), the presence and activity of suitable authorities and/or professionals to plan and carry out management, and—perhaps most importantly—the responses of resident humans to having parrots "move in."

HUMAN RESPONSES TO INTRODUCED PARROTS

Decisions (or lack thereof) about management, protection, or acceptance of naturalized parrots are a product of different ways in which human residents, authorities, and the wider public respond to the presence of a new species. (For an overview of different influences on people's perceptions of introduced species, see Shackleton et al. 2019b.) Despite the significance of these responses for determining the future of naturalized populations, they have received comparatively little academic attention, though several quantitative surveys and qualitative case studies of Rose-ringed and Monk Parakeets are now shedding light on the diversity of ways people perceive introduced parrots. In this section, I review first "negative" responses to parrots (those indicating dislike of or concern about introduced populations and, often, desire for their removal), followed by "positive" responses (those indicating people might like or approve of introduced parrot populations and want them to remain).

Negative Responses

The major drivers of scientific (and indeed, social) research into naturalized parrot populations are: (1) ecological curiosity about the traits and distributions of successfully established populations, and (2) concern among conservation and pest management scientists about the impacts introduced populations might have on native species and economic interests (particularly agricultural interests). Brightsmith and Kiacz (chap. 9 this vol.) and Mori and Menchetti (chap. 6 this vol.) review the evidence of ecological interactions and impacts produced by introduced parrots. These impacts will vary depending on location, species, population sizes, and other contextual factors. However, it is worth noting that much of the concern about introduced parakeets arises not only from observed impacts on biodiversity or agriculture but also in relation to potential or predicted impacts.

Based on extensive evidence that introduced species can have unforeseen and sometimes dramatic effects on extant ecologies, the "precautionary principle" in conservation science is the proposition that, when faced with uncertainty, acting now to avoid future impacts is the most effective and least risky option (Cooney 2004). Adhering to this principle means not waiting to see whether negative impacts arise following introductions but ideally preventing introductions in the first place and—if that fails—taking swift and decisive action to remove newly introduced populations before they can establish. Although invasion scientists are aware that not all introduced species go on to have negative impacts, application of the precautionary principle's "guilty until proven innocent" assumption is widespread (Ruesink et al.1995; Davidson et al. 2013). Consequently, the dominant view from invasion science is that introduced parrot populations are ecologically undesirable due to either their known or potential future impacts on extant ecosystems.

Outside of scientific and conservation circles, this response to parrots is less common, not least because many people will have no interest in or awareness of academic debates surrounding the significance of species origin in predicting detrimental impacts. Indeed, social research has shown that the ways biological scientists and the broader public engage with the concept of an invasive species are quite different: to nonscientists, evidence of impact is much more salient than species origin in determining whether a population is considered invasive (Selge and Fischer 2011). Similarly, Lindemann-Matthies (2016) found that providing information about a species' status as invasive did not affect respondents' perceptions of it; this decontextualized classification is less pertinent than apparent ecological and ecological impacts (van der Wal et al. 2015). However, concerns about the potential current and future impacts of parrots do exist outside of scientific and conservation communities. In the Berthier et al. (2017) survey of Parisian residents' perceptions of Rose-ringed Parakeets, the most numerous concerns raised among those respondents who didn't like parakeets related to either ecological impacts ("J'adore, c'est exotique, mais est-ce bien pour l'écosystème?"; "I love [parakeets], they're exotic, but are they good for the ecosystem?"). Other, more nebulous concerns that parrots are simply out of place in European, urban, domestic Paris were also expressed. Berthier et al. (2017) found that respondents who were more familiar with Rose-ringed Parakeets were more likely to express reservations about them, though this could be due to either concerns about impact or to negative experiences of noise disturbance.

Another negative response to parrot presence that may be restricted to scientific and possibly public health communities is concern that parrots are potential vectors of zoonotic diseases. Specifically, concerns have been raised that parrot populations established in the vicinity of human residences may increase risk of disease transmission (Menchetti and Mori 2014). Parrots can carry avian chlamydiosis, which can spread between birds as well as between birds and humans via the inhalation of droppings, dried saliva, feathers, or mucous from infected animals. In humans, the disease is called psittacosis (aka "parrot fever"), which is normally a mild illness that can nevertheless sometimes result in severe pneumonia. Psittacosis had a high mortality rate when first discovered but can now be effectively treated with antibiotics (Stewardson and Grayson 2010). There are a number of strains with different levels of severity. The disease name (psittacosis) refers to parrots because the disease was initially identified in the parrot trade, passing between parrots, owners, and traders. The disease is not, however, exclusive to parrots; it is also found in pigeons, gulls, and other birds. Introduced parrots don't necessarily have a greater chance of carrying this disease than other species, but they serve as an additional vector and may be more likely to transmit the bacteria to humans, both because they are still kept widely as pets and because of their frequent use of garden bird feeders (Pisanu et al. 2018).

Parrots (along with many other bird species) can also carry avian influenza and exotic Newcastle disease (Fletcher and Askew 2007; Boseret et al. 2013; Jones et al. 2014), which are usually spread by direct contact. Both can, however, be transmitted via bird droppings, which has implications for populations of species such as Rose-ringed and Monk Parakeets; they often flock in large numbers in city parks and gardens, for example, and therefore deposit large quantities of droppings in accessible public spaces. However, there are no records of people contracting these diseases via wild parrot populations in Europe (they are more commonly transmitted by captive birds), and there is little evidence to suggest that the majority of people encountering parakeets respond to them as harbingers of disease in the same way that they might respond to a rat, for example. There are also potential indirect disease risks posed by introduced parrots to pet birds or farmed poultry, but again, there are currently no records of this having occurred.

Of more concern to agriculturalists is the potential for introduced parrots to eat or damage crops, an issue particularly noted in relation to Rose-ringed and Monk Parakeets in parts of their introduced range (particularly Spain and Israel) (Brightsmith and Kiacz, chap. 9 this vol.). When it comes to economic damage, the fact that these species are introduced is less relevant, given that they are also considered crop pests in their native ranges, but they present a novel, additive, and potentially difficult challenge in terms of implementing crop protection or population management measures.

In North America, Monk Parakeets have additionally caused damage to electrical infrastructure by building their large communal nests on pylons and transformers (Brightsmith and Kiacz, chap. 9 this vol.), which can in some cases cause electrical fires. Removal of these and other problematically placed nests (e.g., on floodlights of sports fields) can be costly (Avery et al. 2006; Newman et al. 2008) and time-consuming and causes negative perceptions of parakeets among those people affected. This issue has not arisen to the same extent in European cities with Monk Parakeet populations, however, because of differences in electrical infrastructure there.

An important consideration when it comes to management planning is that negative economic impacts created by naturalized parrot populations are unevenly distributed. While impacts may be severe at a local level and a significant problem for some individuals and industries, they don't affect all parties equally and therefore may have differing degrees of influence on people's perceptions of parakeets. This unevenness carries over to other forms of "nuisance" behaviors performed by parrots ("nuisance" is in quotes here because not all people perceive the sounds parrots make as unpleasant noise). Berthier et al.'s (2017) survey found that Parisians who had private green spaces or balconies were more likely to have reservations about or dislike parakeets. They suggest that because gardens are perceived as extensions of private homes, people are less tolerant of disturbance or damage within them. People with parakeets regularly visiting their gardens were also forced to encounter them (and their associated noises, fruit eating, and defecation) more regularly and persistently than those who encounter parrots only while walking through urban parks, for example. It is worth noting that these differences have implications when it comes to weighing up the costs and benefits of parakeet presence for different socioeconomic groups. For wealthier people with private outdoor space to protect, parrots might be considered a nuisance, whereas for those with no private outdoor space—and potentially little access to it—they may be perceived differently.

The most often reported nuisance behavior in relation to introduced parakeets is the generation of "noise pollution" (Menchetti and Mori 2014) in urban and suburban areas because of their rowdy social communications and continuous vocalizations in flight. Noise complaints have also been raised in relation to introduced macaw populations in Miami (Thomson 2009), and the pitch, volume, and persistence of parrot calls may be a key factor in influencing how people respond to their presence. In the Berthier et al. (2017) survey, 35% of respondents who disliked Rose-ringed Parakeets attributed this to the noise the parrots made. Whether parrots are louder, more persistent, or more disruptive than other urban bird species (e.g., gulls), or indeed other noisy wildlife (e.g., frogs, cicadas), is uncertain. It may be that it is the novelty of the sounds made by introduced parrots that draws particular attention. Responses to parrot vocalizations also depend on the tolerance and preferences of the individual listening: nuisance is to some extent in the ear of the beholder, and our research into perceptions of Monk Parakeets identified real variation in the extent to which people found listening to their chatter infuriating or enjoyable (Crowley et al. 2019).

Returning to the data from Paris, Berthier et al. (2017) found that residents of areas with the longest-standing, most populous Rose-ringed Parakeet populations had the most negative assessments of them, and that this was often related directly to their abundance. Larger populations were perceived to be "too many," proliferating, and spreading. Consequently, as predicted by the precautionary principle, it is often only when an introduced population has established and spread that negative impacts are perceived. This demonstrates that people's perceptions of parrots are subject to change in response to shifting circumstances and experiences. Berthier et al. (2017) also note that the spread and increasing abundance of introduced populations may reduce the extent to which they are considered exotic and novel. If a rare and interesting animal is encountered with increasing frequency, it can become common and banal.

Positive Responses

Potential negative impacts of introduced parrots are often considered in great detail as part of risk assessments designed to determine whether an introduced species may pose a threat, and how great, to existing ecosystems and societies. These assessments are rarely designed to meaningfully explore any positive impacts parrots might have, and even reviews focused on impact assessments often tend to skim over the details of positive responses to parrots, such as, "Common people enjoy the sight of bright parrots flying in courtyards or urban parks" (Menchetti and Mori 2014). Yet Braun and Wegener (2008), examining public opinions of Rose-ringed Parakeets in Heidelberg, Germany, found that 79% of respondents reported enjoying having exotic birds in the city. Similarly, survey respondents in Essonne, France, reported that they enjoyed seeing and feeding Rose-ringed Parakeets and found them attractive (Wolff and Touratier 2010), and 49% of the Berthier et al. (2017) respondents who had encountered Rose-ringed Parakeets expressed unequivocally positive views about them.

Appreciation of parrots is often put down to their being so-called charismatic animals. However, what constitutes a charismatic species is often poorly defined, or simply equated with "attractive" or "popular." Charisma is generally referred to as either an inherent characteristic of a species (e.g., rapid movements, bright colors) or an attribute determined by people (e.g., "cuteness"). Lorimer (2007), however, argues that charisma is neither of these things but something that is produced through different kinds of human–nonhuman encounters. He describes a three-part typology of charisma: the ecological, the aesthetic, and the corporeal. Ecological charisma refers to species characteristics that intersect with human senses and ecology in ways that make them particularly detectable or recognizable. Parrots are highly visible to human eyes and highly audible to human ears, making them distinctive and conspicuous. Aesthetic charisma refers to how species attributes evoke affective responses in humans. For example, parrots are often described as attractive or beautiful, even by those who may be otherwise ambivalent about them (Berthier et al. 2017). Color plays a role again here; body shape and plumage color have been identified as key influences on people's assessments of a bird's attractiveness; green and blue (common parrot colors) are particularly preferred (Lišková and Frynta 2013; Lišková et al. 2015). This fascination with parrot color persists even within the parrot breeding and husbandry world, with unusual color mutations considered more desirable and demanding higher prices than common green (Anderson 2003). This effect may speak more to the exoticism of color mutations, however, than to particular color preferences.

Exoticism is also key to the charisma of naturalized parrots. Lorimer identifies corporeal charisma as "affections and emotions engendered by different organisms in their practical interactions with humans" (Lorimer 2007), and we identified "dissonance" as a type of corporeal charisma relevant to introduced parrots (Crowley et al. 2019). Dissonance describes a feeling of interested surprise on encountering something unexpected, a response also identified by respondents to the Berthier et al. (2017) survey. While this feeling that something is out of place can provoke a negative response, it can also inspire curiosity and wonder, as described by this journalist in relation to Chicago's Monk Parakeets:

> It surprises and delights many observers to find that parakeets aren't entirely confined to warm climates. One cold winter day I went for a walk in Chicago's Hyde Park. ... Flurries were dusting the deep snow already on the ground. ... To then see a half dozen emerald-green birds with lazuli primaries flying around the park was like witnessing apparitions escaped from some travel agency's promotional posters. (Friederici 2005)

Berthier et al. (2017) expand on this idea of a positive exoticism and suggest that parrots provide a "taste of the elsewhere" or—as Friederici also noted—an "invitation to travel." In European and North American settings, at least, it appears that parrots evoke a sense of distant (warm) tropics and are partly valued due to their stark contrast with more ordinary, common city birds such as pigeons. Broader studies on human interactions with urban birds suggest that people may prefer species they have opportunities to encounter, and that interacting with birds in urban and suburban areas both evokes the feeling of being "connected to nature" (Cox and Gaston 2016; Harris et al. 2016) and enhances well-being (see Cox et al. 2017). The evidence in Berthier et al. (2017) and the often less effusive responses people have to pigeons (perhaps the most commonly encountered urban bird) also suggests that other factors—such as whether the bird is considered a pest—attenuate these assumed benefits. Parakeets are also unusual in that they often directly interact with the people they encounter, for example, by visiting bird feeders at regular times. Parrots and people can therefore develop longer-term "friendships," in which both species start to recognize and respond more personally to one another. A well-known example of this kind of relationship was documented in *The Wild Parrots of Telegraph Hill*, a film that followed the relationship between San Francisco resident Mark Bittner and a flock of naturalized Red-masked Parakeets (*Psittacara erythrogenys*), known in the film as Cherry-headed Conures.

It has also been suggested that engaging with naturalized parrots may inspire interest and concern for parrot conservation (Kiacz and Brightsmith, chap. 5 this vol.). Conversely, however, it may motivate residents to protect naturalized populations, a response that often is not appreciated by conservationists and wildlife managers. In some areas (e.g., Hyde Park, Chicago; Brooklyn, New York), localized parrot populations have developed into points of interest. Seymour (2013), studying human supporters of Monk Parakeet colonies, argues that in these situations, parrots become conceptualized as "urban assets," providing ecotourism and recreational viewing opportunities.

Our research on human responses to parakeet populations and management found that over time, these personal and cultural links between parrots and people can develop into meaningful, affective attachments, in which residents associate naturalized parrot populations with their personal, community, and/or cultural identities and narratives (Crowley et al. 2019). Across the US and also in the UK, localized Monk Parakeet colonies have come to represent local people (e.g., immigrant communities, Brooklynites, Eastenders) and places, marking them as unique. We defined these wild

SOCIAL CONFLICT SURROUNDING INTRODUCED PARROT MANAGEMENT

Red-masked Parakeets (*Psittacara erythrogenys*) getting handouts in a park in San Francisco, California, US. Red-masked Parakeets are part of the group known locally as The Wild Parrots of Telegraph Hill. Photo by Dawn Endico, Creative Commons.

attachments as "a complex interaction between charisma (ecological and affective relations between humans and parrots), interpersonal relationships (situated, often repeated interactions between particular humans and parrots) and how parrot presence reflects and feeds into the continuous formation of individual, community and cultural identities" (Crowley et al. 2019). We found wild attachments to recur where small, localized populations of parrots had established and remained over long periods (in human terms—in ecological time these establishment periods are very short, which creates challenges for management; see Beever et al. 2019). In northerly latitudes, parrots face challenging climates, potentially limiting their growth. Comparatively, in areas such as Florida and Spain, where populations have rapidly expanded, these links between parrots, people, and places appear not to have emerged. This is consistent with the findings in Paris that identify a potential loss of interest and value when exotic species become commonplace. However, the degree of attachment to their parrot populations that individuals and communities develop has significant implications when it comes to management.

Indifference

While some people feel very strongly about introduced parrot populations, it is worth noting that many others respond neither positively nor negatively to naturalized parrots, either because they never encounter them or because when they do, they are uninterested or ambivalent. In the Berthier et al. (2017) study, 17% of respondents who had encountered parakeets expressed indifference about them. Of these, 43% were generally uninterested in nature and wildlife, and 57% expressed more surprise and confusion at the parrots' presence than a particularly positive or negative attitude. Again, however, these people's views may be subject to change in response to their experiences (or lack thereof) with parrots over time.

Cultural Variation

All the examples from this chapter have been drawn from research and case studies carried out in North America and Europe. Views of naturalized parrots may differ, however, under different sociocultural and ecological conditions, including, for example, where there are existing parrot species, such as in Australia, South America, and central Asia. Cultural attitudes to releasing animals into the wild can also vary; for example, New Zealand maintains stringent biosecurity measures and a strong stance on managing introduced species, while in Taiwan, birds may be mass-released as part of religious practices (Su et al. 2016).

SOCIAL CONFLICT SURROUNDING INTRODUCED PARROT MANAGEMENT

When people with divergent interests, priorities, and ideas disagree about how to respond to introduced species, social conflicts can arise (Estévez et al. 2015; Crowley et al. 2017a, 2017b). Often, these conflicts are triggered by a specific management action or proposal, such as the removal of birds (or their nests, in the case of Monk Parakeets). There are several reasons public and private bodies might decide to control parrots. Some management activities respond to problems that have already arisen; for example, an introduced population may be causing economic damage or negatively affecting a native species. Others are undertaken proactively, in line with the precautionary principle, or as a means of complying with regional or international conservation agreements. For instance, a 2015 European Union directive includes a list of introduced species "of Union concern". There are currently no naturalized parrots on this list, but should they be added in the future, member states would be required to take action to prevent their establishment or control their spread.

Eradication and population control efforts, however—and particularly those involving lethal control—have regularly been met with opposition. Seymour (2013) and Crowley et al. (2019) have both investigated why and how interested parties have resisted management initiatives in the US and the UK. Seymour (2013) studied the language and rhetoric employed in disputes surrounding Monk Parakeet management in the New York City metropolitan area. She found that opponents of management considered parrots worthy of protection by portraying them as symbolically, instrumentally, and intrinsically valuable and/or unfairly persecuted. Parrots were depicted as "urban assets" (positive, ornamental additions to the city landscape); "admirable animals" (characterizing and symbolizing traits such as industriousness and resilience); and "community members," capable of integrating into urban ecologies through their seemingly companionable interactions (mixed flocking, nest sharing) with other urban birds such as sparrows and pigeons. They were also favorably compared with human communities, described as a stereotypical "Brooklynite" or "New Yorker." Seymour showed how parakeets were portrayed both positively and negatively as a marginalized (immigrant) population, cast by their detractors as "illegal avians" and by their supporters as persecuted refugees.

Crowley et al. (2019) examined a case study of conflict surrounding an attempted Monk Parakeet eradication in the UK, identifying common themes between this and previous cases of conflict in the US. We found that both protectionism and conflict were to some extent predictable; when naturalized populations become integrated into the identities of people and communities, there is likely to be opposition to their removal. The chance of conflict is enhanced where management is planned and implemented by external bodies (such as governing authorities or private companies) without the inclusion or agreement of resident communities, but it could also be driven by concerns about animal rights and welfare, or simply opposition in principle to authoritarian, top-down approaches to wildlife management.

Opposition may be particularly vociferous when lethal control is proposed as the primary means of managing bird populations, and there is evidence of widespread ethical concern about eradicating introduced parrots in both North America and Europe. Verbrugge et al. (2013) conducted a questionnaire survey of people in the Netherlands, and 95% of respondents preferred to either "accept" or "control" Rose-ringed Parakeets, as opposed to efforts to "eradicate" the birds. This was much higher than for a range of other non-native species across different taxonomic groups, but is consistent with studies that have found that control or eradication is least supported for birds, compared to other taxonomic groups (Bremner and Park 2007; Verbrugge et al. 2013; Vane and Runhaar 2016).

Crowley et al. (2017a, 2017b, 2019) have argued that it is preferable to anticipate and avoid the development of challenging social conflicts, rather than attempt to resolve them once they have arisen. We have proposed that management measures for introduced populations, including parrots, should be designed with due regard for the specific social and ecological context of the problem and include early, meaningful consultation with interested and affected parties to determine the most appropriate management option. The importance of including people with diverse interests and values in conservation and environmental management decision making is increasingly recognized and will be key to responding to naturalized parrot populations in an effective, equitable, and sustainable manner.

FUTURE RESEARCH

The potential ecological, economic, and social problems caused by parrot introductions and their management would be best avoided by preventing the establishment of new populations. Human activity is the root cause of most species introductions, and as such, future research needs to be reoriented toward considering social drivers of parrot naturalization at least as much as, if not more than, ecological factors. In addition to existing and proposed measures to control trade, we require a better understanding and regulation of exotic pet *keeping* and both the factors that cause parrots to be inadvisably purchased and those that lead to their release or escape. In New Zealand, for example, measures to prevent new escapes and self-maintaining populations include education initiatives and breeding controls in aviaries (Menchetti and Mori 2014), and in Florida there are now established amnesty days on which people can relinquish unwanted pets rather than releasing them into the wild.

Parrots are charismatic, highly visible, and often live in close proximity to humans, making them an excellent candidate for the use of citizen science to monitor populations and build understanding of the naturalization process. Data for species other than Rose-ringed Parakeets and Monk Parakeets are currently deficient and could be improved by making use of reliable data from birdwatchers and ornithologists (Menchetti and Mori 2014). Further research might also focus more specifically on the factors that affect public responses to naturalized parrots, such as the role of population size and spread, and probe the question: Does familiarity really breed contempt?

There is additionally a need for increased research and public engagement that deliberates on the desirability and form of future management initiatives: Should naturalized parrot populations be removed or controlled, and if so, when, how, and to what extent? Given the ethical and public acceptability challenges of lethal control, what are the alternatives for population or impact management? How can newly identified populations be responded to in an effective yet proportionate manner? And are there any occasions when naturalized parrots might be accepted or, if endangered in their native range, even supported?

In conclusion, while we have now generated a substantial amount of knowledge and understanding of naturalized parrot populations (much of which is summarized in this volume), when it comes to the complex relationships between parrots and people, there is still a great deal to learn.

REFERENCES

Abellán, P., Carrete, M., Anadón, J. D., Cardador, L., and Tella, J. L. 2016. Non-random patterns and temporal trends (1912–2012) in the transport, introduction and establishment of exotic birds in Spain and Portugal. *Diversity and Distributions* 22:263–273.

Ancillotto, L., Strubbe, D., Menchetti, M., and Mori, E. 2016. An overlooked invader? Ecological niche, invasion success and range dynamics of the Alexandrine parakeet in the invaded range. *Biological Invasions* 18:583–595.

Anderson, P. K. 2003. A bird in the house: An anthropological perspective on companion parrots. *Society and Animals* 11:393–418.

Anderson, P. K. 2014. Social dimensions of the human–avian bond: Parrots and their persons. *Anthrozoös* 27:371–387.

Avery, M. L., Lindsay, J. R., Newman, J. R., Pruett-Jones, S., and Tillman, E. A. 2006. Reducing monk parakeet impacts to electric utility facilities in south Florida. In *Advances in Vertebrate Pest Management*, ed. C. J. Feare, and D. P. Cowan, 4:125–136. Furth, Germany: Filander Verlag.

Beever, E. A., Simberloff, D., Crowley, S. L., Al-Chokhachy, R., Jackson, H. A., and Petersen, S. L. 2019. Social–ecological mismatches create conservation challenges in introduced species management. *Frontiers in Ecology and the Environment* 17:117–125.

Berthier, A., Clergeau, P., and Raymond, R. 2017. From beautiful exotic to beautiful invasive: Perceptions and appreciations of the rose-ringed parakeet *Psittacula krameri* in the metropolis of Paris. *Annales de Géographie* 4:408–434.

Boseret, G., Losson, B., Mainil, J. G., Thiry, E., and Saegerman, C. 2013. Zoonoses in pet birds: Review and perspectives. *Veterinary Research* 44:36.

Braun, M. P., and Wegener, S. 2008. It's not as bad as all that! The public perception of the rose-ringed parakeet in Heidelberg. *Natur und Landschaft* 9:452–455.

Bremner, A., and Park, K. 2007. Public attitudes to the management of invasive non-native species in Scotland. *Biological Conservation* 139:306–314.

Bull, J. 1973. Exotic birds in the New York City area. *Wilson Bulletin* 85:501–505.

Canavelli, S. B., Aramburú, R. M., and Zaccagnini, M. E. 2012. Aspectos a considerar para disminuir los conflictos originados por los daños de la cotorra *Myiopsitta monachus* en cultivos agrícolas. *Hornero* 27:89–101.

Canavelli, S. B., Swisher, M. E., and Branch, L. C. 2013. Factors related to farmers' preferences to decrease monk parakeet damage to crops. *Human Dimensions of Wildlife* 18:124–137.

Carrete, M., and Tella, J. 2008. Wild-bird trade and exotic invasions: A new link of conservation concern? *Frontiers in Ecology and the Environment* 6:207–211.

Cassey P., Blackburn, T. M., Russell, G. J., Jones, K. E., and Lockwood, J. L. 2004. Influences on the transport and establishment of exotic bird species: An analysis of the parrots (Psittaciformes) of the world. *Global Change Biology* 10:417–426.

Cooney, R. 2004. *The Precautionary Principle in Biodiversity Conservation and Natural Resource Management*. Gland, Switzerland, and Cambridge, UK: IUCN.

Cox, D. T., and Gaston, K. J. 2016. Urban bird feeding: Connecting people with nature. *PLoS One* 11:e0158717.

Cox, D. T., Shanahan, D. F., Hudson, H. L., Plummer, K. E., Siriwardena, G. M., Fuller, R. A., Anderson, K., Hancock, S., and Gaston, K. J. 2017. Doses of neighborhood nature: The benefits for mental health of living with nature. *BioScience* 67:147–155.

Crowley, S. L., Hinchliffe, S., and McDonald, R. A. 2017a. Conflict in invasive species management. *Frontiers in Ecology and the Environment* 15:133–141.

Crowley, S. L., Hinchliffe, S., and McDonald, R. A. 2017b. Invasive species management will benefit from social impact assessment. *Journal of Applied Ecology* 54:351–357.

Crowley, S. L., Hinchliffe, S., and McDonald, R. A. 2019. The parakeet protectors: Understanding opposition to introduced species management. *Journal of Environmental Management* 229:120–132.

Darwin, C. R. 1845. *Journal of Researches into the Natural History and Geology of the Countries Visited during the Voyage of H.M.S.* Beagle *round the World, under the Command of Capt. Fitzroy*. London: John Murray.

Davidson, A. D., Campbell, M. L., and Hewitt, C. L. 2013. The role of uncertainty and subjective influences on consequence assessment by aquatic biosecurity experts. *Journal of Environmental Management* 127:103–113.

Davis, A. Y., Malas, N., and Minor, E. S. 2013. Substitutable habitats? The biophysical and anthropogenic drivers of an exotic bird's distribution. *Biological Invasions* 16:415–427.

Di Febbraro, M., and Mori, E. 2015. Potential distribution of alien parakeets in Tuscany (central Italy): A bioclimatic model approach. *Ethology Ecology & Evolution* 27:116–128.

Engebretson, M. 2006. The welfare and suitability of parrots as companion animals: A review. *Animal Welfare* 15:263–276.

Estévez, R. A., Anderson, C. B., Pizarro, J. C., and Burgman, M. A. 2015. Clarifying values, risk perceptions, and attitudes to resolve or avoid social conflicts in invasive species management. *Conservation Biology* 29:19–30.

Fletcher, M., and Askew, N. 2007. *Review of the Status, Ecology and Likely Future Spread of Parakeets in England*. York, UK: Central Science Laboratory.

Friederici, P. 2005. Loud, new neighbors. *Audubon Magazine*, Jan.

Harris, E., De Crom, E. P., and Wilson, A. 2016. Pigeons and people: Mortal enemies or lifelong companions? A case study on staff perceptions of the pigeons on the University of South Africa, Muckleneuk campus. *Journal of Public Affairs* 16:331–340.

Heald, O.J.N., Fraticelli, C., Cox, S. E., Stevens, M.C.A., Faulkner, S. C., Blackburn, T. M., and Le Comber, S. C. 2019. Understanding the origins of the ring-necked parakeet in the UK. *Journal of Zoology* 312:1–11.

Herrera, M., and Henessey, B. 2008. Monitoring results of the illegal parrot trade in the Los Pozos market, Santa Cruz de la Sierra, Bolivia. In *Proceedings of the Fourth International Partners in Flight Conference: Tundra to Tropics*, 232–234. McAllen, TX: Partners in Flight.

Hyman, J., and Pruett-Jones, S. 1995. Natural history of the monk parakeet in Hyde Park, Chicago. *Wilson Bulletin* 107:510–517.

Jones, J. C., Sonnberg, S., Koçer, Z. A., Shanmuganatham, K., Seiler, P., Shu, Y., Zhu, H., Guan, Y., Peiris, M., Webby, R. J., and Webster, R. G. 2014. Possible role of songbirds and parakeets in transmission of influenza A H7N9 virus to humans. *Emerging Infectious Diseases* 20:380–385.

Lever, C. 2005. *Naturalised Birds of the World*. London: A & C Black.

Lindemann-Matthies, P. 2016. Beasts or beauties? Laypersons' perception of invasive alien plant species in Switzerland and attitudes towards their management. *NeoBiota* 29:15.

Lišková, S., and Frynta, D. 2013. What determines bird beauty in human eyes? *Anthrozoös* 26:27–41.

Lišková, S., Landová, E., and Frynta, D. 2015. Human preferences for colorful birds: Vivid colors or pattern? *Evolutionary Psychology* 13:1339–1359.

Lorimer, J. 2007. Nonhuman charisma. *Environment and Planning D: Society and Space* 25:911–932.

MacGregor-Fors, I., Calderón-Parra, R., Meléndez-Herrada, A., López-López, S., and Schondube, J. E. 2011. Pretty, but dangerous! Records of non-native monk parakeets (*Myiopsitta monachus*) in Mexico. *Revista Mexicana de Biodiversidad* 82:1053–1056.

McNeely, J. A. 2013. Xenophobia or conservation: Some human dimensions of invasive alien species. In *Invasive and Introduced Plants and Animals*, ed. I. D. Rotherham, and R. A. Lambert, 19–38. Abingon, UK: Routledge.

Menchetti, M., and Mori, E. 2014. Worldwide impact of alien parrots Aves Psittaciformes on native biodiversity and environment: A review. *Ethology Ecology & Evolution* 26:172–194.

Minor, E. S., Appelt, C. W., Grabiner, S., Ward, L., Moreno, A., and Pruett-Jones, S. 2012. Distribution of exotic monk parakeets across an urban landscape. *Urban Ecosystems* 15:979–991.

Moscatello, B. 2003. Preliminary review of the status of monk parakeets in New Jersey. *New Jersey Birds* 14:3–5.

REFERENCES

Muñoz, A. R., and Real, R. 2006. Assessing the potential range expansion of the exotic monk parakeet in Spain. *Diversity and Distributions* 12:656–665.

Newman, J. R., Newman, C. M., Lindsay, J. R., Merchant, B., Avery, M. L., and Pruett-Jones, S. 2008. Monk parakeets: An expanding problem on power lines and other electrical utility structures. In *8th International Symposium on Environmental Concerns in Rights-of-Way Management*, ed. J. W. Goodrich-Mahoney, L. P. Abrahamson, J. L. Ballard, and S. M. Tikalsky, 343–354. New York: Elsevier Science.

Olah, G., Butchart, S. H., Symes, A., Guzmán, I. M., Cunningham, R., Brightsmith, D. J., and Heinsohn, R. 2016. Ecological and socio-economic factors affecting extinction risk in parrots. *Biodiversity and Conservation* 25:205–223.

Pires, S. F. 2012. The illegal parrot trade: A literature review. *Global Crime* 13:176–190.

Pires, S., and Clarke, R. V. 2012. Are parrots craved? An analysis of parrot poaching in Mexico. *Journal of Research in Crime and Delinquency* 49:122–146.

Pisanu, B., Laroucau, K., Aaziz, R., Vorimore, F., Le Gros, A., Chapuis, J. L., and Clergeau, P. 2018. Chlamydia avium detection from a ring-necked parakeet (*Psittacula krameri*) in France. *Journal of Exotic Pet Medicine* 27:68–74.

Pruett-Jones, S., Appelt, C. W., Sarfaty, A., Vossen, B., Leibold, M. A, and Minor, E. S. 2012. Urban parakeets in northern Illinois: A 40-year perspective. *Urban Ecosystems* 15:709–719.

Rodríguez-Pastor, R., Senar, J. C., Ortega, A., Faus, J., Uribe, F., and Montalvo, T. 2012. Distribution patterns of invasive monk parakeets in an urban habitat. *Animal Biodiversity and Conservation* 35:107–117.

Ruesink, J. L., Parker, I. M., Groom, M. J., and Kareiva, P. M. 1995. Reducing the risks of nonindigenous species introductions: Guilty until proven innocent. *BioScience* 45:465–477.

Selge, S., and Fischer, A. 2011. How people familiarize themselves with complex ecological concepts—anchoring of social representations of invasive non-native species. *Journal of Community and Applied Social Psychology* 21:297–311.

Seymour, M. 2013. "Support your local invasive species": Animal protection rhetoric and nonnative species. *Society and Animals* 21:54–73.

Shackleton, R. T., Larson, B. M., Novoa, A., Richardson, D. M., and Kull, C. A. 2019a. The human and social dimensions of invasion science and management. *Journal of Environmental Management* 229:1–9.

Shackleton, R. T., Richardson, D. M., Shackleton, C. M., Bennett, B., Crowley, S. L., Dehnen-Schmutz, K., Estévez, R. A., Fischer, A., Kueffer, C., Kull, C. A., et al. 2019b. Explaining people's perceptions of invasive alien species: A conceptual framework. *Journal of Environmental Management* 229:10–26.

Silk, M. J., Crowley, S. L., Woodhead, A. J., and Nuno, A. 2018. Considering connections between Hollywood and biodiversity conservation. *Conservation Biology* 32:597–606.

Simberloff, D. 2003. How much information on population biology is needed to manage introduced species? *Conservation Biology* 17:83–92.

South, J. M., and Pruett-Jones, S. 2000. Patterns of flock size, diet, and vigilance of naturalized monk parakeets in Hyde Park, Chicago. *Condor* 102:848.

Spreyer, M. F., and Bucher, E. H. 1998. Monk parakeet *Myiopsitta monachus*. In *The Birds of North America*, ed. A. Poole. Ithaca, NY: Cornell Lab of Ornithology.

Stafford, T. 2003. *Pest Risk Assessment for the Monk Parakeet in Oregon*. Salem, OR: Oregon Dept. of Agriculture.

Stewardson, A. J., and Grayson, M. L. 2010. Psittacosis. *Infectious Disease Clinics* 24: 7–25.

Strubbe, D., and Matthysen, E. 2009. Establishment success of invasive rose-ringed and monk parakeets in Europe. *Journal of Biogeography* 36:2264–2278.

Su, S., Cassey, P., and Blackburn, T. M. 2016. The wildlife pet trade as a driver of introduction and establishment in alien birds in Taiwan. *Biological Invasions* 18:215–229.

Thomson, T. 2009. Macaws on campus "awesome" but noisy. *Miami Hurricane*, Mar. 4.

Tillman, E. A., Van Doorn, A., and Avery, M. L. 2000. Bird damage to tropical fruit in south Florida. In *The Ninth Wildlife Damage Management Conference Proceedings*, ed. M. C. Brittingham, J. Kays, and R. McPeake, 47–59. State College, PA.

UNODC. 2016. *World Wildlife Crime Report—Trafficking in Protected Species*. United Nations Office on Drugs and Crime. https://www.unodc.org/documents/data-and-analysis/wildlife/World_Wildlife_Crime_Report_2016_final.pdf.

Vall-llosera, M., and Cassey, P. 2017. "Do you come from a land down under?" Characteristics of the international trade in Australian endemic parrots. *Biological Conservation* 207:38–46.

Van der Wal, R., Fischer, A., Selge, S., and Larson, B.M.H. 2015. Neither the public nor experts judge species primarily on their origins. *Environmental Conservation* 42:349–355.

Vane, M., and Runhaar, H. A. 2016. Public support for invasive alien species eradication programs: Insights from the Netherlands. *Restoration Ecology* 24:743–748.

Vaz, A. S., Kueffer, C., Kull, C. A., Richardson, D. M., Schindler, S., Muñoz-Pajares, A. J., Vicente, J. R., Martins, J., Hui, C., Kühn, I., Honrado, J. P. 2017. The progress of interdisciplinarity in invasion science. *Ambio* 46:428–442.

Verbrugge, L. N., Van den Born, R. J., and Lenders, H. R. 2013. Exploring public perception of non-native species from a visions of nature perspective. *Environmental Management* 52:1562–1573.

White, R. L., Strubbe, D., Dallimer, M., Davies, Z. G., Davis, A. J., Edelaar, P., Groombridge, J., Jackson, H. A., Menchetti, M., Mori, E., and Nikolov, B. P. 2019. Assessing the ecological and societal impacts of alien parrots in Europe using a transparent and inclusive evidence-mapping scheme. *NeoBiota* 48:45–69.

Wolff, T., and Touratier, G. 2010. *Recensement et étude des espèces dites "invasives" et "envahissantes" en Essonne*. Savigny-sur-Orge, France: NaturEssonne, Association d'Etude et de Protection de la Nature en Essonne.

4

GENETICS OF INVASIVE PARROT POPULATIONS

Michael A. Russello, Grace Smith-Vidaurre, and **Timothy F. Wright**

INTRODUCTION

Since publication of *The Genetics of Colonizing Species* (Baker and Stebbins 1965), genetics has played a key role in the study of invasive species, from the standpoint of using such systems both for investigating ecological and evolutionary principles related to colonization of novel environments and for informing management strategies. Documenting the population genetics of invasive lineages has been instrumental for testing evolutionary theory related to adaptation and population bottlenecks, reconstructing the complex history of biological invasions to provide insights on the invasion process, and more recently, uncovering the genetic basis of features associated with invasion success (Barrett 2015). The sheer impact of invasion genetics over this period was recently highlighted by a symposium marking the 50th anniversary of the seminal work that covered the current state of the field, which was subsequently communicated in a special issue of *Molecular Ecology* (2015) as well as an expanded edited volume (Barrett 2015). Aided by our ever-improving ability to mine the genomes of living organisms, such work has been carried out over a wide taxonomic breadth, from zooplankton, terrestrial plants, and fungi to fish, amphibians, reptiles, birds, and mammals. Despite the relative frequency with which psittacines invade new habitats (Joseph 2014), only a limited number of invasion genetics studies have been conducted in parrots. These have focused largely on understanding the factors contributing to invasion success: namely, investigating the source(s) of introduction, identifying the factors involved in the establishment of self-sustaining populations, and reconstructing dispersal pathways and modes of spread (Russello et al. 2008; Gonçalves da Silva et al. 2010; Edelaar et al. 2015; Jackson et al. 2015a, 2015b). While these studies have provided new insights into these topics in a select number of species, the relative dearth of studies on the genetics of invasive parrot species represents an open challenge for investigators seeking to understand how and why parrots are so successful at colonizing new habitats.

In this chapter, we will first summarize the broader literature regarding how genetic information can be used for studying the origin(s), establishment, and spread of invasive species, with an emphasis on avian invaders. We will then review recent studies demonstrating the utility of genetics for enhancing our understanding of parrot population history, structure, and behavior; these provide an important baseline for understanding genetic patterns in invasive parrots. This section is followed by a case study of Monk Parakeets (*Myiopsitta monachus*), arguably the most intensively studied parrot species with regard to the genetics of invasion success. We conclude with a discussion of future directions afforded by genomic technologies that could provide deeper insights into genetic and epigenetic mechanisms of invasion success as well as enhance our tool kit for monitoring and managing invasive species.

GENETICS OF INVASION HISTORY

Genetic information has proven instrumental in studying all phases of the invasion process, including origin(s), establishment, and spread of invasive species (Sakai et al. 2001). All these features are critical for improving our understanding of the ecological and evolutionary mechanisms underlying invasion success. As such, molecular approaches have allowed us to identify: (1) native source population(s) and number of introductions from each source (Lockwood et al. 2005), (2) number of individuals introduced from each event, and (3) dynamics of expansion following establishment (Estoup and Guillemaud 2010; Cristescu 2015). Such information is critical for understanding how invasions occur, which in turn can inform strategies for prevention, eradication, and/or containment (Mack et al. 2000).

Origin: Source and Number of Introductions

The number and origin(s) of source populations are key considerations for reconstructing invasion history. For example, multiple introductions have been frequently correlated with the success of invasive species' establishment and spread (Barrett and Husband 1990; Sakai et al. 2001). Indeed, two of North America's most successful invasive bird species, Common Starlings (*Sturnus vulgaris*) and House Sparrows (*Passer domesticus*), became established only after repeated introductions (Ehrlich 1989). For parrots, genetic reconstructions of the number and origin(s) of source populations have been conducted for Monk Parakeets (reviewed in this chapter) and Rose-ringed Parakeets (*Psittacula krameri*). Native to Asia and sub-Saharan Africa, the Rose-ringed Parakeet has established self-sustaining breeding populations in over 35 countries (Lever 2005) and is considered one of Europe's 100 worst alien species (DAISIE 2009). Using mitochondrial DNA haplotypic and nuclear microsatellite genotypic data from across the native and invasive ranges, Jackson et al. (2015b) identified native populations in Asia, specifically from Pakistan and northern India, as the predominant source for invasive populations in Europe and on islands in the Indian Ocean (Mauritius and Seychelles) (Fig. 4.1a–d). Moreover, the most common haplotypes in the invasive populations in Europe were those found in the more northerly native populations, where Rose-ringed Parakeets naturally tolerate colder parts of their range (Fig. 4.1a, b), leading the authors to hypothesize that cold tolerance may be a factor in the birds' establishment success (Jackson et al. 2015b). Interestingly, a number of haplotypes (n = 30) were identified in the invasive range that were not previously detected in the native range. This disparity was most likely due to unequal sampling (96 native vs. 696 invasive samples). Yet, in tandem with results based on nuclear genotypic data, the authors also suggested this pattern could be attributed to admixture between invasive Rose-ringed Parakeets from widely distributed sources in Asia and Africa (Fig. 4.1d) (Jackson et al.

Rose-ringed Parakeets (*Psittacula krameri*) are the most common and widely distributed naturalized parrot. Here, a pair engages in beak locking, a common practice in a mated pair. Jerusalem Mountains, Israel, November 2013. Photo by Salit Kark.

2015b). Admixture between divergent genotypes originating from different source populations have been found in other systems (reviewed in Dlugosch and Parker 2008), offering the potential for positive genetic interactions among previously isolated alleles and for adaptive evolution that can increase fitness (Keller and Taylor 2010; Kolbe et al. 2004), although such benefits may be constrained by negative genetic interactions (Barker et al. 2018; Bertelsmeier and Keller 2018).

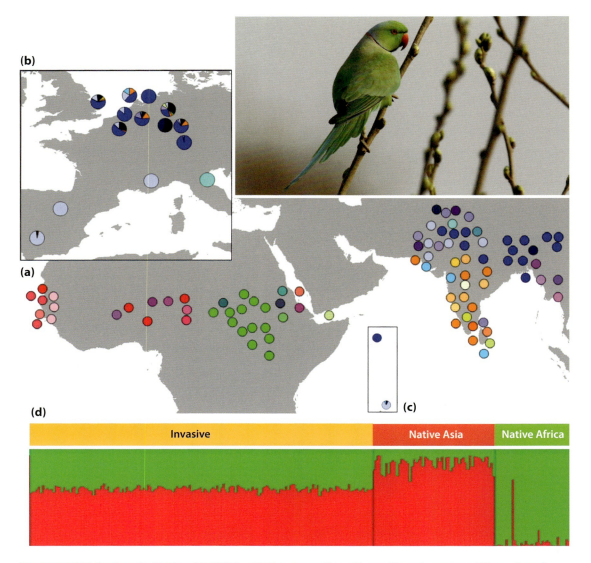

Figure 4.1. Distribution of mitochondrial DNA variation across the native and invasive ranges of Rose-ringed Parakeets (*Psittacula krameri*). (a). Colored circles represent the approximate locations sampled in the Asian and African native ranges. Different colors relate to different haplotypes. (b). Pie charts represent the frequency of native mitochondrial DNA haplotypes detected within each invasive population across Europe. Black proportions indicate mitochondrial DNA haplotypes detected in invasive populations that were not detected in the native range. (c). Pie charts represent the frequency of native mitochondrial DNA haplotypes detected within each invasive population in Mauritius and Seychelles. (d). Bayesian clustering plot based on microsatellite genotypic data collected in the native range in Africa and Asia, and the invasive range in Europe. Modified from Figs. 4 and 5 of Jackson et al. (2015b). Rose-ringed Parakeet image used under CC BY-SA 2.0 (Imran Shah/Flickr).

Establishment: Propagule Pressure

In addition to identifying the geographic location(s) and number of source populations, the number of individuals per introduction event is a key consideration in the ultimate establishment of self-sustaining breeding populations outside a species' native range. Together with the number of independent introductions, the number of individuals introduced is collectively known as propagule pressure (Carlton 1996). This has been a particularly important factor in the establishment success of invasive birds (Blackburn and Duncan 2001). The simultaneous introduction of many individuals has been shown to buffer against genetic bottlenecks (Simberloff 2009), with some introduced populations having levels of genetic variation similar to or higher than native populations (Kolbe et al. 2004). As a result, propagule pressure may help guard against the effects of demographic stochasticity (Lockwood et al. 2005). Propagule pressure exerted by the legal and illegal pet trade has been implicated in parrots' successful establishment outside their native ranges (Cassey et al. 2004). For example, most Rose-ringed Parakeet populations in Europe likely resulted from intentional or unintentional releases of caged birds (Strubbe and Matthysen 2009). Subsequent genetic investigation revealed that the relative proportion of Asian-origin vs. African-origin birds was consistent with European importation records, further implicating the pet trade as a source of propagule pressure (Jackson et al. 2015b), similar to what has been found for Monk Parakeets.

Spread: Dynamics of Geographical Expansion

Once non-native individuals have been introduced and established, genetic studies can reconstruct the dynamics of geographical expansion. Although multiple introductions of invasive species have been demonstrated to be a common phenomenon, including for many of the examples noted above, an alternative scenario for expansion has been shown by way of widespread secondary invasions from a particularly successful invasive population that acts as a source for establishing populations in new areas (dubbed "invasive bridgehead effect" by Estoup and Guillemaud 2010). A host of genetic studies have demonstrated the invasive bridgehead effect in a range of species, including the Harlequin Ladybeetle (*Harmonia axyridis*; Lombaert et al. 2010), Western Corn Rootworm (*Diabrotica virgifera virgifera*; Miller et al. 2005), Colorado Potato Beetle (*Leptinotarsa decemlineata*; Grapputo et al. 2005), American Bullfrog (*Rana catesbeiana*; Ficetola et al. 2008), and European Green Crab (*Carcinus maenas*; Darling et al. 2008). Although to our knowledge this phenomenon has not been demonstrated genetically in invasive parrots, House Sparrows do appear to have expanded due to the invasive bridgehead effect. Using nuclear microsatellite genotypic data across the native and invasive ranges, Schrey et al. (2011) found that introduced North American populations of House Sparrows likely originated from only a few sources, and that ongoing gene flow is occurring in the invasive range. It is important to note that this House Sparrow study was not explicitly designed to test the invasive bridgehead effect, but future work could employ genetic data and approximate Bayesian computation methods to evaluate different invasion scenarios. Given the potential for the self-accelerating process whereby invasion begets invasion, some authors recommend heightened vigilance against invasive bridgehead populations (Lombaert et al. 2010; Bertelsmeier and Keller 2018). We argue that, at the very least, genetic studies examining this and other invasion scenarios are warranted in other systems, including for volant species with high dispersal capacity, such as parrots.

GENETIC VARIATION AND STRUCTURE IN NATURAL PARROT POPULATIONS

Given the dearth of studies of parrots within their naturalized ranges, understanding patterns of genetic variation in the native range may represent the best alternative for inferring how invasion shapes the genetics of naturalized parrots. As these native populations are typically the source of individuals that eventually establish naturalized populations through the invasion process, genetic patterns of native populations provide an informative baseline for the degree of genetic variation and structure that might be expected in the naturalized range. For example, if source populations typically have low levels of genetic variability, then low levels of variability would be expected in the derived naturalized populations, which might limit their potential for evolutionary responses to changing selection regimes in their new range. There might also be interesting interactions between the genetic patterns in the native range population and the manner in which individuals are captured for the pet trade that would impact genetic patterns in naturalized populations. For example, if natural populations exhibit a high degree of genetic structure, pet trade sourcing from only one of these subpopulations may cause naturalized populations to be more genetically homogeneous. On the other hand, if sourcing of individuals for the pet trade was more geographically widespread, then admixture between sources might result in any given naturalized population having more variability than a single native population (e.g., Rose-ringed Parakeets; Jackson et al. 2015b).

In this section, we review studies of the population genetics of wild parrot populations to address the following questions: (1) How structured is genetic variation in wild parrot populations? (2) At what scale is this structure observed? (3) What does this tell us about gene flow among parrot populations? (4) How genetically diverse are typical parrot populations? It is important to note at the outset that many (but by no means all) studies of population genetics in parrots have been motivated by conservation concerns for the species studied. This focus could potentially introduce a bias, because species of conservation concern tend to have small and declining populations and are less genetically diverse than nonthreatened populations due to genetic drift. Conclusions about genetic diversity and degree of structure might also be influenced by the geographic scale and intensity of sampling of a particular study. Consequently, we organize our review below by geographic scale, focusing first on range-wide studies of genetic structure, then on studies of variation among populations on a smaller scale, and finally with studies of genetic diversity within populations.

Phylogeography and Cryptic Diversity

Studies of phylogeography typically examine patterns of genetic variation across large parts of a species range or among members of a closely related species complex. They may have as their goal a better understanding of the evolutionary history of a species or of relationships among component subspecies. Another common goal of studies at this scale is the detection of cryptic species, that is, subspecies or populations within a species that are sufficiently differentiated from other such populations that they warrant full species status. Such studies often have conservation implications, since levels of endangerment and threat are largely determined at the species level; elevating a rare or declining subspecies to full species level can, in theory, quickly lead to a change in its conservation status and the protections afforded to it.

A number of phylogeographic studies have focused on species in the Neotropics. Among the first was Eberhard and Bermingham's (2004) study of the Yellow-headed Amazon species complex, which consists of three closely related taxa variously described as subspecies of *Amazona ochrocephala* or as the full species *A. ochrocephala*, *A. oratrix*, and *A. auropalliata*. Using mitochondrial and nuclear DNA sequence data, the authors found distinct differences between the previously identified subspecies *A. o. oratrix* and *A. o. auropalliata*, which occur in Mesoamerica (Mexico and Central America), and between those lineages and the South American (sub)species, *A. ochrocephala*. In contrast, within South America they found less genetic structure in *A. ochrocephala* sampled across the Amazon Basin, as well as evidence of introgression between this species and a closely related congener, *A. aestiva* (Eberhard and Bermingham 2004). Another study conducted at the same scale was Wenner et al. (2012), an investigation of cryptic speciation in the Southern Mealy Amazon (*A. farinosa*). These authors also used genetic data from both mitochondrial and nuclear DNA to identify deep divisions between Central and South American subspecies and more recent divergences between pairs of subspecies found within each of these regions. This study led to the reclassification of *A. f. guatemalae* and *A. f. farinosa* to full species status. Similarly, Masello et al. (2011), in their study of the Burrowing Parrot (*Cyanoliseus patagonus*), used mitochondrial DNA sequences from several loci to examine the degree of differentiation among four subspecies, three found east of the Andes, ranging from northern to central Argentina, and one west of the Andes in central Chile. They found strong differentiation across the Andes and evidence of genetic structure between two of the subspecies to the east of the Andes, while the third subspecies appeared to consist of a long-standing hybrid zone between the other two.

Studies of two other parrot species examined variation across the Amazon Basin and neighboring areas of central South America. In both cases, these studies found contrasting patterns in maternally inherited mitochondrial DNA sequences vs. biparentally inherited nuclear microsatellite markers. In the Turquoise-fronted Amazon (*A. aestiva*), mitochondrial DNA sequences detected distinct variation among two subspecies found in the northeastern and southwestern portions of the species' range, while nuclear microsatellites showed only weak differentiation across the entire range (Leite et al. 2008; Caparroz et al. 2009b). Likewise, a study of the Blue-and-yellow Macaw (*Ara ararauna*) in the same geographic region detected structure between eastern and western populations using mitochondrial DNA sequences, while the nuclear microsatellite data did not reveal any structure (Caparroz et al. 2009a).

Several studies have examined the degree of differentiation between disjunct populations of species distributed across either true islands or isolated islands of suitable habitat. A series of studies has examined genetic structure in the highly endangered Hyacinth Macaw (*Anodorhynchus hyacinthinus*) of Brazil. The first used nuclear and mitochondrial DNA markers to test for genetic differentiation between two populations located 100 km apart in the Pantanal region and a third disjunct population in the Gerais region (Faria et al. 2008). The authors found no evidence of differentiation between the two populations in the Pantanal but, using nuclear microsatellites, found significant differences between those populations and the disjunct population in Gerais. A more recent study expanded the scale of sampling within the Pantanal region and added sampling from a third disjunct region in eastern Amazonia (Presti et al. 2015). This work confirmed the previous finding of significant differentiation between the Gerais and Pantanal regions but found lower differentiation between the disjunct Gerais and eastern Amazonia regions. The authors also detected significant differentiation between populations in the north and south of the Pantanal region, located about 700 km apart, a result that was unexpected, given the lack of an obvious barrier to dispersal between populations within the Pantanal. Overall, the level of genetic diversity exhibited by Hyacinth Macaws was considered low (average observed heterozygosity = 0.35); a related study found that levels of genetic diversity in this species were lower than those found in two other macaw species (Faria and Miyaki 2006). Surprisingly, there was no suggestion in these data that the three disjunct populations represented either true or incipient species, perhaps either because of relatively recent isolation of the disjunct populations or because of ongoing gene flow. In contrast, Russello et al. (2010), in a study of the Cuban Amazon (*Amazona leucocephala*) species complex, did detect evidence for cryptic diversity among island populations within the Bahamas and across the Greater Antilles. Using both mitochondrial DNA and nuclear microsatellite data, the authors found evidence for the genetic distinctiveness of the previously identified subspecies found on the islands of Cuba, Grand Cayman, and Cayman Brac, and also showed previously unrecognized differences among populations on three different islands in the Bahamas.

There have been fewer such studies outside the Neotropics, but those that have been conducted typically show similar patterns to those seen in the Americas. A study of the declining Eastern Ground Parrot (*Pezoporus wallicus*) in Australia found deep genetic divisions between eastern and western populations, which are found on different sides of the continent and are currently separated by more than 1,500 km of inhospitable habitat (Murphy et al. 2011). A recent study by Jackson et al. (2015a) examined the evolutionary relationships among extant and extinct species in the genus *Psittacula* from Madagascar and surrounding islands in the western Indian Ocean. They found that most of the populations endemic to various islands had diverged to the level of distinct species, indicating low levels of gene flow after initial colonization. Interestingly, the phylogenetic diversity of this region initially decreased with European colonization and subsequent extinctions but was partially restored by another human-mediated influence, namely the introduction of the invasive Rose-ringed Parakeet (Jackson et al. 2015a). Finally, a recent study of the Red-tailed Black Cockatoo (*Calyptorhynchus banksii*) is one of the first to combine the power of genome-wide sampling of single nucleotide polymorphisms (SNPs) with the expanded sampling provided by museum specimens to examine phylogeographic questions (Ewart et al. 2019). The authors examined variation in samples from across this species' continent-wide range in Australia and detected robust geographic structure among different regions that only partially corresponded to established subspecies boundaries.

Four general patterns have emerged from these studies of phylogeography. The first is that populations are often structured over large geographic scales, such as between Central American and South American populations, among islands in the Caribbean Sea or Indian Ocean, or on opposite sides of the continents of South America or Australia. Such divisions often correspond to contemporary barriers to dispersal such as mountains, oceans, or inhospitable habitats, or to past barriers created by Pleistocene climate shifts and sea level changes. The second general pattern to emerge is that these divisions often correspond to subspecies boundaries previously identified using plumage and morphology. Examples include studies of the Southern Mealy Amazon and the Turquoise-fronted Amazon, which identified genetic differences between some preexisting subspecies (Caparroz et al. 2009b; Wenner et al. 2012). In many cases, confirmation of these evolutionary differences has directly impacted the conservation status of the taxa in question. In other cases, more complex patterns are identified, such as where secondary hybridization and introgression have reduced historically present genetic differences in the Burrowing Parrot (Masello et al. 2011) or the *Amazona ochrocephala/A. aestiva* species complex (Eberhard and Bermingham 2004; Leite et al. 2008). More rarely, such studies identify true cryptic diversity among populations that have not previously been identified as genetically distinct; as a prime example, Russello et al. (2010) found consistent differences among populations on three different islands in the Bahamas, one of which had already become extinct at the time of the study. The third general pattern is that less structure is apparent between populations sampled at smaller scales than species-wide, suggesting that dispersal-mediated gene flow is relatively strong in parrots, a point that is reinforced by studies in the next section. The fourth general pattern is that structure is more likely to be detected with some types of genetic markers than others. In particular, mitochondrial DNA, with its maternal transmission, has a smaller effective population size than biparentally inherited and recombining nuclear autosomal markers such as microsatellites, which can lead to a greater degree of structuring at mitochondrial DNA. This structure can be further compounded if males tend to disperse and females are more philopatric, leading to less mixing of mitochondrial DNA haplotypes. Both factors may have been at play in studies of the Blue-and-yellow Macaw and Turquoise-fronted Amazon that found more structure using mitochondrial DNA sequences than microsatellite alleles (Leite et al. 2008; Caparroz et al. 2009a, 2009b). These results have interesting implications for naturalized populations of parrots. In particular, the suggestion that dispersal-mediated gene flow is high in many species reinforces the idea that parrots are candidates for the invasive bridgehead effect described above, in which initial founding populations give rise to many dispersed naturalized populations.

Population Structure

While the distinction between studies of phylogeography and population structure is somewhat arbitrary, there have been a number of studies in parrots that have focused on investigating the degree of genetic structure among populations on a smaller scale than the geographical range. Typically, these studies focus on inferring gene flow between populations that are hypothesized to be isolated either behaviorally or through habitat fragmentation. Among the first of such studies were two by Wright et al. (2001, 2005) that employed both mitochondrial DNA sequences and nuclear microsatellites to examine genetic structure among populations of the Yellow-naped Amazon (*Amazona auropalliata*) in Costa Rica distributed over a 100 km span. The primary focus of these studies was testing whether behavioral differences represented by two vocal dialects in learned contact calls corresponded to differences at neutral genetic markers. Both types of markers found no evidence of genetic differences across dialect boundaries and instead indicated genetic panmixia among populations at this scale (Wright and Wilkinson 2001; Wright et al. 2005).

A number of studies have examined the effects of human-induced habitat fragmentation on gene flow and population genetic structure. A study of the threatened Red-tailed Amazon (*A. brasiliensis*) found no evidence of greater DNA fingerprint band sharing within than between two populations in Brazil (Caparroz et al. 2006). This result suggests that there is no genetic differentiation between these two populations, separated by 100 km, despite extensive fragmentation of this species' coastal Atlantic Forest habitat (Caparroz et al. 2006). A study of the nomadic and critically endangered Swift Parrot (*Lathamus discolor*) examined the degree of population structure among several populations nesting on different parts of Tasmania and neighboring offshore islands (Stojanovic et al. 2018). Population genetic analyses based on microsatellite genotypic data suggested panmixia of these populations despite extensive habitat fragmentation; while the lack of inbreeding might be viewed as a positive for species conservation, the authors point out that their results emphasize the need to treat the entire species as a single conservation unit in population viability models (Stojanovic et al. 2018). Another study from Australia examined genetic variation among populations of two species collectively known as white-tailed black cockatoos, Baudin's Black Cockatoo (*Calyptorhynchus baudinii*) and Carnaby's Black Cockatoo (*C. latirostris*), in the southwestern portion of mainland Australia (White et al. 2014). These two species have undergone dramatic declines as this region has experienced extensive habitat fragmentation due to clearing for intensive agriculture. The authors used both contemporary and historical samples and a large panel of microsatellites to assess whether population isolation had led to genetic differentiation. They found clear genetic differences between populations to the east and the west of a large wheat belt; interestingly, this geographic differentiation was much stronger than that separating samples from the two putative species. The authors suggested that this geographic differentiation was driven primarily by the loss of allelic diversity in the contemporary eastern population after its isolation from the western population (White et al. 2014). Finally, a study by Raisin et al. (2012) examined changes in genetic differentiation among populations of the Echo Parakeet (*Psittacula eques*) prior to and following extensive translocations undertaken to maximize population growth of this highly endangered island endemic. They found that prior to translocations, the population most geographically isolated by extensive habitat fragmentation was the most genetically distinct from other populations. Perhaps unsurprisingly, this difference disappeared after individuals were translocated into this population. The authors suggested that loss of genetic structure was a beneficial, albeit unintended, outcome of conservation efforts and that the new genetic panmixia likely mimicked the uniform genetic structure of the species hypothesized to exist prior to human-mediated habitat loss.

To date, few papers have examined population structures of parrots in relatively undisturbed settings. One exception is a study by Olah et al. (2017) of Scarlet Macaw (*Ara macao*) populations in the lowlands of the Peruvian Amazon. They used microsatellite markers to examine variation among samples collected at clay licks and nest sites from three populations along an 80 km stretch of the Tambopata River. They found weak but statistically detectable differentiation among the three sites, driven primarily by the higher-elevation site that was separated from the others by a mountain ridge.

In summary, these studies suggest that, under undisturbed conditions, gene flow among populations tends to be high in parrots with an absence of marked population structure. When human actions lead to habitat fragmentation, however, gene flow can be reduced, leading to genetic differentiation among isolated populations. It should be noted, however, that most of these studies have focused on larger-bodied parrots with fairly general habitat preferences; it would be interesting to conduct similar studies in smaller-bodied parrots or habitat specialists, both of which might be expected to disperse less widely. These predictions could also be examined in naturalized species, which range in size from small lovebirds to large Amazon parrots.

GENETIC DIVERSITY WITHIN POPULATIONS

A handful of studies have examined the patterns of genetic variation within one or a few populations of parrots. Without exception, these have all focused on endangered species in which a substantial portion of the population is maintained in captivity; the primary questions were: How much genetic diversity remains in captive population(s), and how can it be maximized via translocations or managed breeding? One of the earliest studies to use microsatellites to characterize within-population genetic variation was conducted for the endangered St. Vincent Amazon (*Amazona guildingii*), in which the authors found moderate levels of variation and detected two subgroups characterized by a high and low degree of relatedness, respectively (Russello and Amato 2004). This information led directly to recommendations of specific breeding pairs for maximizing genetic diversity and minimizing inbreeding in this captive population of high conservation value (Russello and Amato 2004). Another study, which collected microsatellite and mitochondrial DNA data from two captive populations of the Cuban Amazon, found that the Zapata Swamp breeding population contained two distinct lineages, one of which was also found in the other breeding population located in Managua (Milián-García et al. 2015). The authors hypothesized that the unique lineage in Zapata Swamp might represent individuals unintentionally introduced from the Cayman Brac subspecies of the Cuban Amazon. As the population overall had moderately high levels of heterozygosity and low levels of inbreeding, the authors suggested that these two lineages should be managed separately until further data could be collected to better characterize their original sources. In contrast, a study of the Kakapo (*Strigops habroptila*), a highly endangered island endemic species of New Zealand, found high levels of inbreeding and low heterozygosity, as measured by microsatellites (White et al. 2015). Notably, lower levels of heterozygosity in females were significantly associated with lower rates of hatching and fecundity, suggesting that the genetic bottleneck experienced by this species had led to inbreeding depression (White et al. 2015).

While the number of genetic studies within populations is small, they do suggest that populations that undergo severe bottlenecks have the potential to experience deleterious effects from inbreeding. This result is quite relevant to invasive or naturalized populations, as many would be expected to undergo genetic bottlenecks due to founding events involving just a few breeding individuals.

GENETICS OF INVASION SUCCESS: THE CASE OF THE MONK PARAKEET

The Monk Parakeet has emerged as a model species in avian invasion genetics. Genetic studies have led to insights about where invasive Monk Parakeets originated, how populations were established in climatically distinct parts of the world, and how such populations may spread within the invasive range. Pinpointing source populations, a crucial step when asking questions about the ecology and evolution of invasive species, has been a primary focus of this research. Identifying invasive Monk Parakeets' original sources of genetic and phenotypic variation is an undertaking that has employed broad-scale sampling across the native and invasive ranges and has tracked recent innovations in genetic tools.

Source populations for invasive Monk Parakeets were first assessed with mitochondrial DNA by Russello et al. (2008), who characterized patterns of genetic similarity and structure among invasive populations in the United States (US) and native populations in South America. Parakeets in the US were most genetically similar to those in northern Argentina and Uruguay, a narrow region of this species' native range that has been tapped heavily for the global pet trade. This study highlighted one of the first stages of the Monk Parakeet invasion pathway, demonstrating that dispersal out of the native range was indeed human-mediated and associated with the pet trade (Russello et al. 2008). A subsequent study, relying on both mitochondrial DNA and nuclear microsatellites, and additional sampling in the native and invasive ranges, provided more evidence for northern Argentina and Uruguay as a source region (Fig. 4.2a–e) (Edelaar et al. 2015). Parakeets sampled from pet shops in Spain, representing an intermediate step in the invasion process between initial dispersal out of the native range and population establishment in the invasive range, were genetically similar to this source region (Edelaar et al. 2015), further supporting the role of the global pet trade in Monk Parakeet invasions (Fig. 4.2d, e).

These studies have led to important insights about potential source populations and Monk Parakeets' invasion pathways. Some limitations remain, however, that are being addressed in ongoing research to more closely evaluate origins of invasive populations. First, while both studies identified northern Argentina and Uruguay as a likely source region for invasive Monk Parakeets, sampling so far has been insufficient to identify South American source populations at a finer geographic scale. The Convention on International Trade in Endangered Species of Wild Fauna and Flora (CITES) documents Uruguay as the primary legal exporter of Monk Parakeets from their native range (CITES 1973). Indeed, from 1975 to 2015, 97% of Monk Parakeets that were legally imported to Mexico originated in Uruguay (Hobson et al. 2017). However, previous genetic sampling of Monk Parakeets in Uruguay has been sparse (Edelaar et al. 2015; Russello et al. 2008). Second, recent innovations in sequencing technology provide the means to sample genomes to a much greater degree than ever before, returning thousands of SNPs across many individuals. Ongoing research is focused on extending geographic sampling to Uruguay and broadening the sampling of individual genomes with restriction site-associated DNA sequencing (RADseq) to shed additional light on source populations for invasive Monk Parakeets (Smith-Vidaurre, Russello, and Wright, unpubl. data).

Genetic studies with Monk Parakeets have also identified patterns of population genetic structure that provide clues about population establishment. Evolutionary theory predicts that species arriving in new habitats are vulnerable

Monk Parakeets (*Myiopsitta monacus*) using a windmill as a nesting site outside Colonia del Sacramento, Uruguay, July 2017. Photo by Tania Molina.

GENETICS OF INVASIVE PARROT POPULATIONS

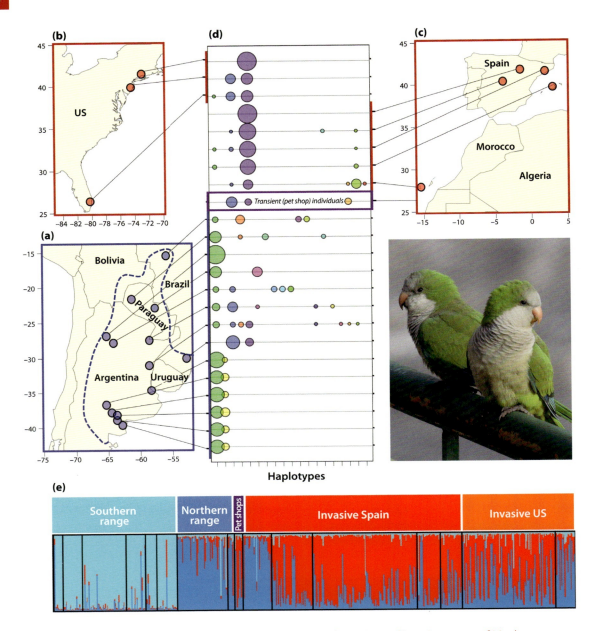

Figure 4.2. Distribution of mitochondrial DNA variation across the native and invasive ranges of Monk Parakeets (*Myiopsitta monachus*). (a). Distribution of sampled populations across the entire native range (indicated by the blue dashed line). (b). Locations of sampled populations in the United States. (c). Locations of sampled populations in Spain. Wild-caught birds sampled in pet shops (a stage between uptake and potential introduction) do not have a location. (d). Haplotype frequencies in each population, with the size of each bubble being proportional to the number of individuals in a population with that particular haplotype. (e). Bayesian clustering plot based on microsatellite genotypic data collected in the northern and southern native range in South America and the invasive range in Spain and the US. Modified from figs. 2 and 4 of (Edelaar et al. 2015). Monk Parakeet image courtesy of Jennifer J. Uehling.

to genetic bottlenecks (Dlugosch and Parker 2008). The impact of such bottlenecks depends on the size and genetic variation of the founding population and the length of time the population remains small (Nei et al. 1975). Stronger bottlenecks can increase inbreeding, potentially inhibiting population resilience to novel selection pressures, such as disease or changing climates, and are generally thought to impede establishment after introduction (Dlugosch and Parker 2008; Bock et al. 2015). All genetic studies comparing invasive and native populations of Monk Parakeets to date have found evidence of genetic bottlenecks associated with invasion (Russello et al. 2008; Edelaar et al. 2015). Edelaar et al. (2015) identified decreasing genetic variation in concert with increasing latitude, a pattern repeated across invasive populations in both the US and Spain. This pattern suggests either increasingly strong genetic bottlenecks with increasing latitude or, alternatively, parallel adaptation to shared urban and climatic selection pressures (Edelaar et al. 2015). Some invasive populations show evidence of weaker genetic bottlenecks, such as those in the Canary Islands, which could be due to many founders originating from a single genetically diverse source population, admixture among multiple source populations after dispersal from the native range, and/or weaker selection pressures in that part of the invasive range (Edelaar et al. 2015). Research is ongoing to better characterize genetic bottlenecks and the possibility of adaptation to parallel selection pressures. Despite exhibiting genetic bottlenecks, invasive Monk Parakeet populations in the US and Spain show no signs of a decline that might be attributable to inbreeding, with the possible exception of the Hyde Park population in Chicago (Pruett-Jones et al. 2012). Taken together, these findings suggest that founder effects and/or genetic bottlenecks do not prohibit Monk Parakeet population establishment, as has been observed in other invasive species (Dlugosch and Parker 2008). Future work should extend the assessment of genetic bottlenecks and multiple source populations to other regions of the invasive range to better understand population establishment in this invasive parrot.

Additional processes that could influence invasive population establishment and spread remain understudied in Monk Parakeets. For instance, we know little about gene flow among invasive populations over local or broader geographic scales, which could ameliorate the negative effects of genetic bottlenecks by increasing genetic diversity (Dlugosch and Parker 2008). Most studies to date have conducted sampling at a broad geographic scale, limiting our ability to ask whether Monk Parakeet establishment has been influenced by gene flow among neighboring invasive populations, or whether populations in parts of the invasive range (e.g., Florida) have spread by splitting off from a single successful population (i.e., invasive bridgehead effect). In that regard, Gonçalves da Silva et al. (2010) used species-specific microsatellites (Russello et al. 2007) to reconstruct a dispersal distance estimate of nearly 100 km for invasive Monk Parakeets in Florida. While this estimate is high, it suggests that invasive Monk Parakeets could establish new populations within 100 km of an existing population.

Second, we do not know whether introduction through the global pet trade has led to sex-ratio biases in invasive populations, which could impact establishment by altering population growth and/or patterns of genetic diversity. Finally, we know little about reproductive strategies in the invasive range that could influence population establishment. For instance, extra-pair copulation can alter the genetic diversity of a female's brood and is known to be common across birds (Jennions and Petrie 2000). Extra-pair copulation could work in tandem with standing genetic variation of an invasive population and sex-ratio biases to impact establishment. Monk Parakeets nest colonially and communally, a strategy that provides many opportunities for extra-pair copulation. A study of native Monk Parakeets in Córdoba, Argentina, reported significant levels of extra-pair paternity within broods, suggesting that Monk Parakeet nesting strategies are associated with copulation outside the social pair bond (Martínez et al. 2013). However, whether this reproductive strategy impacts invasive Monk Parakeet establishment success remains an open question.

A promising approach to assess the establishment and spread of invasive Monk Parakeet populations is the reconstruction of invasion pathways. Such reconstructed pathways, based on genetic data, are hypotheses about the invasion process that complement previous studies of population genetic patterns. Invasion pathways encompass estimates of historic and recent population sizes, bottleneck duration, and historic or recent admixture. Multiple hypotheses can be tested per invasive population, allowing for the simultaneous evaluation of hypotheses with different source populations and demographic processes. Recently developed analytical tools facilitate hypothesis testing using mitochondrial DNA or microsatellite markers (Estoup and Guillemaud 2010) as well as highly dimensional SNP markers that provide finer resolution of individual genomes (Raynal et al. 2016; Smith et al. 2017). Research currently ongoing with Monk Parakeets is employing SNPs to identify source populations and potential demographic and selection processes underlying invasive population establishment (Smith-Vidaurre, Russello, and Wright, unpubl. data). Accounting for demographic processes that impact population establishment is an important baseline from which to ask whether establishment has also been influenced by adaptation to new selection pressures. Future research should extend these methods to more populations across the invasive range, and evaluate patterns of spread by employing finer-scale geographic sampling and methods, such as the genetic directionality index (Peter and Slatkin 2013), to identify whether new invasive populations are founded by long- or short-range dispersal from existing populations (invasive bridgehead effect) vs. independent colonization by escaped or released birds from the pet trade.

Exciting and challenging directions for future research with Monk Parakeets include assessing gene flow and genetic similarity among invasive populations at local and broader geographic scales, extending such genetic studies to more areas across the invasive range, and testing hypotheses about adaptation to new selection pressures, which will shed light on candidate molecular markers and networks under selection following invasion. A recent study with native Monk Parakeets identified high heritability of morphological traits, suggesting that some traits, such as tarsus length and bill width, could be sensitive to new selection pressures in the invasive range (Martínez et al. 2018). Cold tolerance may serve as an important selection pressure underlying invasive population establishment and spread (Edelaar et al. 2015), as suggested for Rose-ringed Parakeets (Jackson et al. 2015b). Gene networks or other molecular markers associated with thermal tolerance are promising candidates for future research. As Monk Parakeets are recent invaders that have experienced relatively few generations of natural selection, endeavors to study evolutionary responses in invasive populations should consider not only changes in SNP allele frequencies as signatures of selection, but also changes in molecular markers sensitive to environmental change over short timescales, such as transposable elements and DNA methylation (Rey et al. 2016; Stapley et al. 2015).

FUTURE DIRECTIONS

Advances in genomic technologies now allow us to mine the genome at an unprecedented level to enhance our understanding of fundamental questions in ecology and evolutionary biology, including those related to invasion success. Although parrots in general have low environmental and economic impacts relative to the most invasive species on the planet (e.g., Kudzu, *Pueraria montana*; Asian Long-horned Beetle, *Anoplophora glabripennis*; Asian Carp, *Cyprinus carpio*; Zebra Mussel, *Dreissena polymorpha*), widespread species such as the Monk Parakeet and the Rose-ringed Parakeet may still provide opportunities for testing hypotheses related to invasion success, as well as opportunities to evaluate genomics, epigenomics, and genome editing as tools for more effective management.

Since the seminal volume edited by Baker and Stebbins (1965), there has been much debate over the primary sources of genetic variation important for invasion success: preexisting standing variation, new beneficial mutations, and/or introgression by way of hybridization of naturally distinct lineages (reviewed in Bock et al. 2015). Invasive species, with their contrast between "baseline" source populations, and "experimental" invasive populations, offer exciting opportunities to employ genomic approaches like population-wide reduced representation (e.g., RADseq; Baird et al. 2008) or whole genome sequencing (e.g., Therkildsen and Palumbi 2017) to determine the different sources of genetic variation at work. A host of studies have already begun to harness genomic scans for detecting signatures of selection, differentiation, and hybridization for identifying genes and linked regions associated with invasiveness (e.g., Prentis et al. 2008; Willoughby et al. 2018). These same genome-wide approaches also hold promise for improving our understanding of genetic patterns in native-range populations and how these change with increasing rarity, population fragmentation, or environmental changes. As invasive populations also face these processes, genomic studies in the native range can provide a baseline for understanding population history and informing invasive-species management (Sjodin et al. 2020a, 2020b).

In addition to heritable genetic variation, there is increasing recognition that epigenetics can affect ecologically important traits, which may also play a role in invasion success (Hawes et al. 2018). Epigenetics is generally defined as the study of changes in organisms caused by modification of gene expression rather than alteration of the genetic code itself. In the context of invasive species, organisms that can successfully colonize and establish areas outside their native range have been generally considered to have a high stress tolerance (Serafini et al. 2011; Zerebecki and Sorte 2011; Al Hassan et al. 2016), which may be associated with underlying epigenetic processes. For example, a recent study of corals found an association between levels of DNA methylation, the most commonly studied epigenetic change, and a measure of ocean acidification that is predicted to increase under various climate-change scenarios (Putnam et al. 2016). Understanding the relative roles of genetic and epigenetic variation in invasion success will continue to be a fundamental question moving forward (Hawes et al. 2018).

Lastly, there has been much discussion in the literature and the popular press on the promise and pitfalls of emerging genome-editing technologies; in terms of immediate deployment outside of the laboratory, invasive species (and disease vectors) have featured prominently (Burt 2003; Dearden et al. 2018). One such genome-editing technique known as Clustered Regularly Interspaced Short Palindromic Repeat (CRISPR), combined with CRISPR-associated protein 9 (Cas9), has been developed as a driving mechanism that increases the chances that the targeted genetic change will be inherited by offspring ("gene drive") (Esvelt et al. 2014; Gantz et al. 2015). In the context of invasive species, application of this technology has been actively considered for disrupting the sex determination process to bias offspring toward a single sex as a means for heavily skewing sex ratios in order to promote negative population growth and, ultimately, extirpation (Deredec et al. 2008; Alphey 2014; Esvelt et al. 2014; Gantz et al. 2015). While there is considerable scope for debate on the ethical implications and risks involved in deploying such technologies, the strong impact some invasive species have on their surrounding environments suggests it would be wise to at least investigate their potential as management tools.

ACKNOWLEDGMENTS

The authors thank members of the Wright and Russello labs for discussion of these ideas.

REFERENCES

Al Hassan, M., Chaura, J., López-Gresa, M. P., Borsai, O., Daniso, E., Donat-Torres, M. P., Mayoral, O., Vicente, O., and Boscaiu, M. 2016. Native-invasive plants vs. halophytes in Mediterranean salt marshes: Stress tolerance mechanisms in two related species. *Frontiers in Plant Science* 7:473.

Alphey, L. 2014. Genetic control of mosquitos. *Annual Review of Entomology* 59:205–224.

Baird, N. A., Etter, P. D., Atwood, T. S., Currey, M. C., Shiver, A. L., Lewis, Z. A., Selker, E. U., Cresko, W. A., and Johnson, E. A. 2008. Rapid SNP discovery and genetic mapping using sequenced RAD markers. *PLoS One* 3:e3376.

Baker, H. G., and Stebbins, G. L., eds. 1965. *The Genetics of Colonizing Species: Proceedings of the International Union of Biological Sciences Symposia on General Biology*. New York: Academic Press.

Barker, B. S., Cocio, J. E., Anderson, S. R., Braasch, J. E., Cang, F. A., Gillette, H. D., and Dlugosch, K. M. 2018. Potential limits to the benefits of admixture during biological invasion. *Molecular Ecology* 28:100–113.

Barrett, S. C. 2015. Foundations of invasion genetics: The Baker and Stebbins legacy. *Molecular Ecology* 24:927–1941.

Barrett, S. C., and Husband, B. C. 1990. The genetics of plant migration and colonization. In *Plant Population Genetics, Breeding, and Genetic Resources*, ed. A.H.D. Brown, M. T. Clegg, A. L. Kahler, and B. S. Weir, 254–277. Sunderland, MA: Sinauer Associates.

Bertelsmeier, C., and Keller, L. 2018. Bridgehead effects and role of adaptive evolution in invasive populations. *Trends in Ecology & Evolution* 33:527–534.

Blackburn, T. M., and Duncan, R. P. 2001. Determinants of establishment success in introduced birds. *Nature* 414:195.

Bock, D. G., Caseys, C., Cousens, R. D., Hahn, M. A., Heredia, S. M., Hübner, S., Turner, K. G., Whitney, K. D., and Rieseberg, L. H. 2015. What we still don't know about invasion genetics. *Molecular Ecology* 24:2277–2297.

Burt, A. 2003. Site-specific selfish genes as tools for the control and genetic engineering of natural populations. *Proceedings of the Royal Society of London B: Biological Sciences* 270:921–928.

Caparroz, R., Martuscelli, P., Scherer-Neto, P., Miyaki, C., and Wajntal, A. 2006. Genetic variability in the red-tailed amazon (*Amazona brasiliensis*, Psittaciformes) assessed by DNA fingerprinting. *Revista Brasileira de Ornitologia* 14:15–19.

Caparroz, R., Miyaki, C. Y., and Baker, A. J. 2009a. Contrasting phylogeographic patterns in mitochondrial DNA and microsatellites: Evidence of female philopatry and male-biased gene flow among regional populations of the blue-and-yellow macaw (Psittaciformes: *Ara ararauna*) in Brazil. *Auk* 126:359–370.

Caparroz, R., Seixas, G.H.F., Berkunsky, I., and Collevatti, R. G. 2009b. The role of demography and climatic events in shaping the phylogeography of *Amazona aestiva* (Psittaciformes, Aves) and definition of management units for conservation. *Diversity and Distributions* 15:459–468.

Carlton, J. T. 1996. Pattern, process, and prediction in marine invasion ecology. *Biological Conservation* 78:97–106.

Cassey, P., Blackburn, T. M., Russell, G. J., Jones, K. E., and Lockwood, J. L. 2004. Influences on the transport and establishment of exotic bird species: An analysis of the parrots (Psittaciformes) of the world. *Global Change Biology* 10:417–426.

CITES. 1973. In. 27 U.S.T. 1087, T.I.A.S. No. 8249, 993 U.N.T.S. 243, ELR Stat. 40336.

Cristescu, M. E. 2015. Genetic reconstructions of invasion history. *Molecular Ecology* 24:2212–2225.

DAISIE. 2009. *Psittacula krameri*. In *Handbook of Alien Species in Europe*. Dordrecht, Netherlands: Springer.

Darling, J. A., Bagley, M. J., Roman, J., Tepolt, C. K., and Geller, J. B. 2008. Genetic patterns across multiple introductions of the globally invasive crab genus *Carcinus*. *Molecular Ecology* 17:4992–5007.

Dearden, P. K., Gemmell, N. J., Mercier, O. R., Lester, P. J., Scott, M. J., Newcomb, R. D., Buckley, T. R., Jacobs, J.M.E., Goldson, S. G., and Penman, D. R. 2018. The potential for the use of gene drives for pest control in New Zealand: A perspective. *Journal of the Royal Society of New Zealand* 48:225–244.

Deredec, A., Burt, A., and Godfray, C. 2008. The population genetics of using homing endonuclease genes (HEGs) in vector and pest management. *Genetics* 17:2013–2026.

Dlugosch, K., and Parker, I. 2008. Founding events in species invasions: Genetic variation, adaptive evolution, and the role of multiple introductions. *Molecular Ecology* 17:431–449.

Eberhard, J. R., and Bermingham, E. 2004. Phylogeny and biogeography of the *Amazona ochrocephala* (Aves: Psittacidae) complex. *Auk* 121:318–332.

Edelaar, P., Roques, S., Hobson, E. A., Gonçalves da Silva, A., Avery, M., Russello, M. A., Senar, J. C., Wright, T. F., Carrete, M., and Tella, J. L. 2015. Shared genetic diversity across the global invasive range of the monk parakeet suggests a common restricted geographic origin and the possibility of convergent selection. *Molecular Ecology* 24:2164–2176.

Ehrlich, P. 1989. Attributes of invaders and the invading processes: Vertebrates. In *Biological Invasions: A Global Perspective*, ed. J. A. Drake et al., 315–328. Chicester, UK: Wiley.

Estoup, A., and Guillemaud, T. 2010. Reconstructing routes of invasion using genetic data: Why, how and so what? *Molecular Ecology* 19:4113–4130.

Esvelt, K. M., Smidler, A. L., Catteruccia, F., and Church, G. M. 2014. Emerging technology: Concerning RNA-guided gene drives for the alteration of wild populations. *eLife* 3:e03401.

Ewart, K. M., Johnson, R. N., Ogden, R., Joseph, L., Frankham, G. J., and Lo, N. 2019. Museum specimens provide reliable SNP data for population genomic analysis of a widely distributed but threatened cockatoo species. *Molecular Ecology Resources* 19:1578–1592.

Faria, P. J., Guedes, N.M.R., Yamashita, C., Martuscelli, P., and Miyaki, C. Y. 2008. Genetic variation and population structure of the endangered hyacinth macaw (*Anodorhynchus hyacinthinus*): Implications for conservation. *Biodiversity and Conservation* 17:765–779.

Faria, P. J., and Miyaki, C. Y. 2006. Molecular markers for population genetic analyses in the family Psittacidae (Psittaciformes, Aves). *Genetics and Molecular Biology* 29:231–240.

Ficetola, G. F., Bonin, A., and Miaud, C. 2008. Population genetics reveals origin and number of founders in a biological invasion. *Molecular Ecology* 17:773–782.

REFERENCES

Gantz, V. M., Jasinskiene, N., Tatarenkova, O., Fazekas, A., Macias, V. M., Bier, E., and James, A. A. 2015. Highly efficient Cas9-mediated gene drive for population modification of the malaria vector mosquito *Anopheles stephensi*. *Proceedings of the National Academy of Sciences (USA)* 112:E6736–E6743.

Gonçalves da Silva, A., Eberhard, J. R., Wright, T. F., Avery, M. L., and Russello, M. A. 2010. Genetic evidence for high propagule pressure and long-distance dispersal in monk parakeet (*Myiopsitta monachus*) invasive populations. *Molecular Ecology* 19:3336–3350.

Grapputo, A., Boman, S., Lindstroem, L., Lyytinen, A., and Mappes, J. 2005. The voyage of an invasive species across continents: Genetic diversity of North American and European Colorado potato beetle populations. *Molecular Ecology* 14:4207–4219.

Hawes, N. A., Fidler, A. E., Tremblay, L. A., Pochon, X., Dunphy, B. J., and Smith, K. F. 2018. Understanding the role of DNA methylation in successful biological invasions: A review. *Biological Invasions* 20:2285–2300.

Hobson, E. A., Smith-Vidaurre, G., and Salinas-Melgoza, A. 2017. History of nonnative monk parakeets in Mexico. *PLoS One* 12:e0184771.

Jackson, H., Jones, C. G., Agapow, P. M., Tatayah, V., and Groombridge, J. J. 2015a. Micro-evolutionary diversification among Indian Ocean parrots: Temporal and spatial changes in phylogenetic diversity as a consequence of extinction and invasion. *Ibis* 157:496–510.

Jackson, H., Strubbe, D., Tollington, S., Prys-Jones, R., Matthysen, E., and Groombridge, J. J. 2015b. Ancestral origins and invasion pathways in a globally invasive bird correlate with climate and influences from bird trade. *Molecular Ecology* 24: 4269–4285.

Jennions, M. D., and Petrie, M. 2000. Why do females mate multiply? A review of the genetic benefits. *Biological Reviews* 75:21–64.

Joseph, L. 2014. Perspectives from parrots on biological invasions. In *Invasion Biology and Ecological Theory: Insights from a Continent in Transformation*, ed. H. Prins and I. Gordon, 58–82. Cambridge: Cambridge Univ. Press.

Keller, S., and Taylor, D. 2010. Genomic admixture increases fitness during a biological invasion. *Journal of Evolutionary Biology* 23:1720–1731.

Kolbe, J. J., Glor, R. E., Schettino, L. R., Lara, A. C., Larson, A., and Losos, J. B. 2004. Genetic variation increases during biological invasion by a Cuban lizard. *Nature* 431:177.

Leite, K.C.E., Seixas, G.H.F., Berkunsky, I., Collevatti, R. G., and Caparroz, R. 2008. Population genetic structure of the blue-fronted amazon (*Amazona aestiva*, Psittacidae: Aves) based on nuclear microsatellite loci: Implications for conservation. *Genetics and Molecular Research* 7:819–829.

Lever, C. 2005. *Naturalised Birds of the World*. London: T. & A.D. Poyser.

Lockwood, J. L., Cassey, P., and Blackburn, T. 2005. The role of propagule pressure in explaining species invasions. *Trends in Ecology & Evolution* 20:223–228.

Lombaert, E., Guillemaud, T., Cornuet, J. M., Malausa, T., Facon, B., and Estoup, A. 2010. Bridgehead effect in the worldwide invasion of the biocontrol harlequin ladybird. *PLoS One* 5:e9743.

Mack, R. N., Simberloff, D., Lonsdale W. M., Evans, H., Clout, M., and Bazzaz, F. A. 2000. Biotic invasions: Causes, epidemiology, global consequences, and control. *Ecological Applications* 10:689–710.

Martínez, J. J., de Aranzamendi, M. C., and Bucher, E. H. 2018. Quantitative genetics in the monk parakeet (*Myiopsitta monachus*) from central Argentina: Estimation of heritability and maternal effects on external morphological traits. *PLoS One* 13:e0201823.

Martínez, J. J., de Aranzamendi, M. C., Masello, J. F., and Bucher, E. H. 2013. Genetic evidence of extra-pair paternity and intraspecific brood parasitism in the monk parakeet. *Frontiers in Zoology* 10:68.

Masello, J. F., Quillfeldt, P., Munimanda, G. K., Klauke, N., Segelbacher, G., Schaefer, H. M., Failla, M., Cortes, M., and Moodley, Y. 2011. The high Andes, gene flow and a stable hybrid zone shape the genetic structure of a wide-ranging South American parrot. *Frontiers in Zoology* 8:16.

Milián-García, Y., Jensen, E. L., Madsen, J., Alonso, S. A., Rodríguez, A. S., López, G. E., and Russello, M. A. 2015. Founded: Genetic reconstruction of lineage diversity and kinship informs *ex situ* conservation of Cuban amazon parrots (*Amazona leucocephala*). *Journal of Heredity* 106:573–579.

Miller, N., Estoup, A., Toepfer, S., Bourguet, D., Lapchin, L., Derridj, S., Kim, K. S., Reynaud, P., Furlan, L., and Guillemaud, T. 2005. Multiple transatlantic introductions of the western corn rootworm. *Science* 310:992.

Murphy, S. A., Joseph, L., Burbidge, A. H., and Austin, J. 2011. A cryptic and critically endangered species revealed by mitochondrial DNA analyses: The western ground parrot. *Conservation Genetics* 12:595–600.

Nei, M., Maruyama, T., and Chakraborty, R. 1975. The bottleneck effect and genetic variability in populations. *Evolution* 29:1–10.

Olah, G., Smith, A. L., Asner, G. P., Brightsmith, D. J., Heinsohn, R. G., and Peakall, R. 2017. Exploring dispersal barriers using landscape genetic resistance modelling in scarlet macaws of the Peruvian Amazon. *Landscape Ecology* 32:445–456.

Peter, B. M., and Slatkin, M. 2013. Detecting range expansions from genetic data. *Evolution* 67:3274–3289.

Prentis, P. J., Wilson, J. R., Dormontt, E. E., Richardson, D. M., and Lowe, A. J. 2008. Adaptive evolution in invasive species. *Trends in Plant Science* 13:288–294.

Presti, F. T., Guedes, N.M.R., Antas, P.T.Z., and Miyaki, C. Y. 2015. Population genetic structure in hyacinth macaws (*Anodorhynchus hyacinthinus*) and identification of the probable origin of confiscated individuals. *Journal of Heredity* 106:491–502.

Pruett-Jones, S., Appelt, C. W., Sarfaty, A., Van Vossen, B., Leibold, M. A., and Minor, E. S. 2012. Urban parakeets in Northern Illinois: A 40-year perspective. *Urban Ecosystems* 15:709–719.

Putnam, H. M., Davidson, J. M., and Gates, R. D. 2016. Ocean acidification influences host DNA methylation and phenotypic plasticity in environmentally susceptible corals. *Evolutionary Applications* 9:1165–1178.

Raisin, C., Frantz, A. C., Kundu, S., Greenwood, A. G., Jones, C. G., Zuel, N., and Groombridge, J. J. 2012. Genetic consequences of intensive conservation management for the Mauritius parakeet. *Conservation Genetics* 13:707–715.

Raynal, L., Marin, J. M., Pudlo, P., Ribatet, M., Robert, C. P., and Estoup, A. 2016. ABC random forests for Bayesian parameter inference. *ArXiv*:1605.05537.

Rey, O., Danchin, E., Mirouze, M., Loot, C., and Blanchet, S. 2016. Adaptation to global change: A transposable element–epigenetics perspective. *Trends in Ecology & Evolution* 31:514–526.

Russello, M., and Amato, G. 2004. *Ex situ* population management in the absence of pedigree information. *Molecular Ecology* 13:2829–2840.

Russello, M. A., Avery, M. L., and Wright, T. F. 2008. Genetic evidence links invasive monk parakeet populations in the United States to the international pet trade. *BMC Evolutionary Biology* 8:217.

Russello, M. A., Saranathan, V., Buhrman-Deever, S., Eberhard, J., and Caccone, A. 2007. Characterization of polymorphic microsatellite loci for the invasive monk parakeet (*Myiopsitta monachus*). *Molecular Ecology Notes* 7:990–992.

Russello, M., Stahala, C., Lalonde, D., Schmidt, K., and Amato, G. 2010. Cryptic diversity and conservation units in the Bahama parrot. *Conservation Genetics* 11:1809–1821.

Sakai, A. K., Allendorf, F. W., Holt, J. S., Lodge, D. M., Molofsky, J., With, K. A., Baughman, S., Cabin, R. J., Cohen, J. E., and Ellstrand, N. C. 2001. The population biology of invasive species. *Annual Review of Ecology and Systematics* 32:305–332.

Schrey, A. W., Grispo, M., Awad, M., Cook, M. B., McCoy, E. D., Mushinsky, H. R., Albayrak, T., Bensch, S., Burke, T., and Butler, L. K. 2011. Broad-scale latitudinal patterns of genetic diversity among native European and introduced house sparrow (*Passer domesticus*) populations. *Molecular Ecology* 20:1133–1143.

Serafini, L., Hann, J. B., Kültz, D., and Tomanek, L. 2011. The proteomic response of sea squirts (genus *Ciona*) to acute heat stress: A global perspective on the thermal stability of proteins. *Comparative Biochemistry and Physiology Part D: Genomics and Proteomics* 6:322–334.

Simberloff, D. 2009. The role of propagule pressure in biological invasions. *Annual Review of Ecology, Evolution, and Systematics* 40:81–102.

Sjodin, B.M.F., Irvine, R. L., Ford, A. T., Howald, G. R., and Russello, M. A. 2020a. *Rattus* population genomics across the Haida Gwaii archipelago provides a framework for guiding invasive species management. *Evolutionary Applications* 13:889–904.

Sjodin, B.M.F., Irvine, R. L., and Russello, M. 2020b. RapidRat: Development, validation and application of a genotyping-by-sequencing panel for rapid biosecurity and invasive species management. *PLoS One* 15:e0234694.

Smith, M. L., Ruffley, M., Espíndola, A., Tank, D. C., Sullivan, J., and Carstens, B. C. 2017. Demographic model selection using random forests and the site frequency spectrum. *Molecular Ecology* 26:4562–4573.

Stapley, J., Santure, A. W., and Dennis, S. R. 2015. Transposable elements as agents of rapid adaptation may explain the genetic paradox of invasive species. *Molecular Ecology* 24:2241–2252.

Stojanovic, D., Olah, G., Webb, M., Peakall, R., and Heinsohn, R. 2018. Genetic evidence confirms severe extinction risk for critically endangered swift parrots: Implications for conservation management. *Animal Conservation* 21:313–323.

Strubbe, D., and Matthysen, E. 2009. Establishment success of invasive ring-necked and monk parakeets in Europe. *Journal of Biogeography* 36:2264–2278.

Therkildsen, N. O., and Palumbi, S. R. 2017. Practical low-coverage genomewide sequencing of hundreds of individually barcoded samples for population and evolutionary genomics in nonmodel species. *Molecular Ecology Resources* 17:194–208.

Wenner, T. J., Russello, M. A., and Wright, T. F. 2012. Cryptic species in a Neotropical parrot: Genetic variation within the *Amazona farinosa* species complex and its conservation implications. *Conservation Genetics* 13:1427–1432.

White, K. L., Eason, D. K., Jamieson, I. G., and Robertson, B. C. 2015. Evidence of inbreeding depression in the critically endangered parrot, the kakapo. *Animal Conservation* 18:341–347.

White, N. E., Bunce, M., Mawson, P. R., Dawson, R., Saunders, D. A., and Allentoft, M. E. 2014. Identifying conservation units after large-scale land clearing: A spatio-temporal molecular survey of endangered white-tailed black cockatoos (*Calyptorhynchus* spp.). *Diversity and Distributions* 20:1208–1220.

Willoughby, J. R., Harder, A. M., Tennessen, J. A., Scribner, K. T., and Christie, M. R. 2018. Rapid genetic adaptation to a novel environment despite a genome-wide reduction in genetic diversity. *Molecular Ecology* 27:4041–4051.

Wright, T. F., Rodriguez, A. M., and Fleischer, R. C. 2005. Vocal dialect, sex-biased dispersal and microsatellite population structure in the parrot *Amazona auropalliata*. *Molecular Ecology* 14:1197–1205.

Wright, T. F., and Wilkinson, G. S. 2001. Population genetic structure and vocal dialects in an Amazon parrot. *Proceedings of the Royal Society of London B: Biological Sciences* 268:609–616.

Zerebecki, R. A., and Sorte, C. J. 2011. Temperature tolerance and stress proteins as mechanisms of invasive species success. *PLoS One* 6:e14806.

5

NATURALIZED PARROTS: CONSERVATION AND RESEARCH OPPORTUNITIES

Simon Kiacz and **Donald J. Brightsmith**

INTRODUCTION

Upwards of 60 of the world's 398 species of psittacines have established naturalized populations outside their normal ranges (Cassey et al. 2004; Runde et al. 2007; Menchetti and Mori 2014). These populations vary in size from a few breeding pairs to thousands of individuals (Minor et al. 2012; Uehling et al. 2019). Much has been written about the real and potential negative ecological, economic, and social impacts caused by these naturalized populations (Pimentel et al. 2005; Sax and Gaines 2008; Menchetti and Mori 2014; in this vol.: Crowley, chap. 3; Mori and Menchetti, chap. 6; Bucher, chap. 8). These negative impacts, coupled with the rate at which some of these populations have spread and grown, has led many to refer to these populations as invasive (Gonçalves da Silva et al. 2010; Newson et al. 2011; Russell and Blackburn 2017). However, most literature on naturalized parrots has made little effort to describe and quantify the real or potential benefits of these populations. While most literature on naturalized parrots has focused on negative impacts, other authors have noted that in order to define a species as invasive, one must quantify the overall impact, both negative and positive, that these naturalized populations have on an ecosystem (Kueffer and Hadorn 2008; Goodenough 2010; Kumschick and Nentwig 2010; Schlaepfer et al. 2011; Strubbe et al. 2011; Brightsmith and Kiacz, chap. 9 this vol.). While some benefits may be trivial and others more significant, a balanced review of these populations needs to look critically at both the positives and negatives. In this chapter, we will analyze the real and potential benefits of naturalized parrot populations, with a focus on conservation, research opportunities, and societal impacts. Table 5.1 summarizes the benefits of naturalized populations of parrots that we discuss in detail below.

NATURALIZED POPULATIONS AS SOURCES OF PARROTS FOR CONSERVATION

The IUCN (International Union for Conservation of Nature) Red List data show that 55% (~227 species) of all parrot species are in decline, and roughly 28% of all parrot species are considered threatened (IUCN 2019). Of the ~227 species in decline, at least 35 have naturalized populations (Runde et al. 2007; Menchetti and Mori 2014; IUCN 2019). In some cases, such as those of the Red-crowned Amazon (*Amazona viridigenalis*) and the Yellow-crested Cockatoo (*Cacatua sulphurea*), the species are listed as endangered, and the naturalized populations are, or may soon be, larger than the native populations (Runde et al. 2007; Sullivan et al. 2009; Gibson and Yong 2017). Many of these native populations are continuing to decrease, and intensive management has yet to begin, signaling that these populations are likely years away from recovery (Berkunsky et al. 2017).

TABLE 5.1

Challenges facing native and captive populations of parrots, and the benefits of naturalized populations when considering them for conservation initiatives.

CHALLENGES FACING NATIVE AND/OR CAPTIVE PARROT POPULATIONS	BENEFITS OF NATURALIZED POPULATIONS
Populations decreasing in the wild due to habitat loss	Naturalized populations almost entirely use urban areas, which are expanding, representing potential areas of growth for backup populations
Populations decreasing in the wild due to poaching for the pet trade	Naturalized populations of nonthreatened parrots could be used as stock for the pet trade, possibly decreasing demand from native flocks
Loss of genetic variation in native populations due to stochastic events and genetic bottlenecking	Naturalized populations provide genetic reservoirs, free from the selection favoring captivity and from many of the stochastic pressures acting on native populations
Parrots held in captivity or in the wild can be susceptible to disease because they are held in close quarters or are in areas that lack easy access to management and veterinary care	Naturalized populations are free-living and not constrained in unsanitary conditions, and urban populations are accessible to biologists and veterinarians for monitoring and treatment
Rewilding techniques of captive populations are difficult, time consuming, and costly	Naturalized populations are able to forage and avoid humans and other predators, limiting any necessary rewilding before release
More parrot natural history studies are needed	Accessible urban parrot populations allow researchers and citizen scientists to easily study breeding, foraging, and nesting biology as well as other topics
Designing and implementing new research techniques or equipment for parrots can be risky and expensive	Naturalized populations are useful for development of tracking techniques and new equipment, and could reduce the risks and cost of testing new techniques on threatened or inaccessible populations
Some parrot populations are considered invasive (Monk Parakeets and Rose-ringed Parakeets) and cause economic or ecological damage	Study of invasive populations can inform invasion biology and can also help develop techniques for control, which could be useful in case naturalized populations develop invasive tendencies

For threatened species, naturalized populations could act as backup populations or population reservoirs (Menchetti and Mori 2014). Species with one or few populations are at higher risk of extinction (Boyd et al. 2017), so additional populations, naturalized or otherwise, can help reduce the overall risk of loss. Especially useful would be naturalized populations in countries outside the species' native range, because different social and political systems can play important roles in how native or naturalized fauna are perceived and/or protected (Dallimer and Strange 2015), and parrot populations restricted to only one country are at a higher risk of endangerment (Olah et al. 2016). Naturalized population reservoirs could serve as tools for conservationists, biologists, and land managers by allowing them to address threats in native ranges without the threat of losing the entire species (Gibson and Yong 2017). Although current

Red-crowned Amazon (*Amazona viridigenalis*) in courtship display. Honolulu, Hawaii, US, March 2020. Photo by Nicholas P. Kalodimos.

naturalized populations of threatened parrots are not being actively managed to mitigate losses within their native range, they are still serving the important role of a backup population and genetic reservoir, including populations of Red-crowned Amazons and Yellow-crested Cockatoos.

Since many of the conservation benefits of naturalized parrots revolve around their potential usefulness in translocation programs, it is worth exploring the pros and cons of the usual methods. Translocations have allowed managers to establish new populations, bring species back from the brink of extinction, and maintain species' genetic viability for a wide variety of birds, including the California Condor (*Gymnogyps californianus*), Puerto Rican Amazon (*A. vittata*), Hawaiian Crow or 'Alalā (*Corvus hawaiiensis*), Nene (*Branta sandvicensis*), and Black Robin (*Petroica traversi*) (Reed and Merton 1991; Kuehler et al. 1994; Black 1995; Snyder and Snyder 2000; Butchart et al. 2006). If wildlife managers working with threatened parrots seek to create new wild populations or reinforce existing ones, they first need a source of birds. Birds are usually translocated directly from wild populations, obtained from wildlife confiscations, or bred in captivity (Snyder et al. 1994; Sanz and Grajal 1998; Oehler et al. 2001; Plair et al. 2008; Lopes et al. 2018). When confiscated or captive individuals are not available, capturing birds from the wild is often needed, but justifying the take of an already threatened species from the wild to put into a captive-breeding or translocation program is often difficult, expensive, challenged legally, and logistically complicated (Nielsen 2006; Kalmar et al. 2010). For species deemed in need of *ex situ* conservation, utilizing individuals from naturalized populations instead of native populations can remove or reduce many of these potential problems. Although naturalized parrots have not been utilized in this manner, this is a potentially valuable conservation benefit (Marchetti and Engstrom 2016).

The source populations for translocations are normally held in captivity, either briefly when individuals are translocated directly from wild populations (Sisson et al. 2017) or permanently, as in many cases of captive breeding (Snyder et al. 1996; Heinrichs et al. 2019). But naturalized populations also maintain animals for the long term and could provide individuals useful for direct conservation actions. By definition, naturalized populations are breeding at a rate sufficient to maintain or grow the population (Colautti and MacIsaac 2004), and this is accomplished usually without direct human involvement. Comparatively, captive breeding can be difficult, and while practices have advanced tremendously in the past decades and have succeeded in helping a number of plants and animals from perishing, captive breeding is not without its drawbacks (Yamamoto et al. 1989; Snyder et al. 1996; Comizzoli and Holt 2019). Having the correct environmental cues, maintaining proper health (i.e., exercise and diet), providing the proper breeding environment, and pairing birds correctly are all critical to maximizing successful breeding in captivity (Kalmar et al. 2010). In the sections that follow, we will provide a comparison between utilizing captive-breeding or naturalized populations as a source of parrots for conservation.

Costs and Time

Utilizing a naturalized population in place of captive breeding could save money. Costs for captive-breeding programs can exceed $1 million per year, as is the case for Black Stilts (*Himantopus novaezelandiae*) in New Zealand (Moran et al. 2005) and California Condors in the United States (US) (Snyder et al. 1996). Yearly costs for a captive population of about 170 Puerto Rican Amazons are over $700,000, which accounts for about 57% of the total costs of the endangered species program for that species (USFWS 2009), and Kakapo (*Strigops habroptila*) recovery programs in New Zealand are in a similar situation (Moran et al. 2005). Initial funding for these efforts can be difficult to obtain, and many agencies, private and governmental, must work together to obtain the necessary funding and permits required (Garnett et al. 2018). Captive programs also compete for funding that could otherwise go to the conservation of wild (or naturalized) populations (Snyder et al. 1997).

Utilizing a naturalized population as a source for translocations also has associated costs, for items such as trapping teams, quarantine, veterinary care, disease testing, etc., but given the much shorter time in captivity, the overall cost of translocations should be much less than the cost of captive breeding.

Another limited resource that influences this debate is time. Legal restrictions, issuing of permits for capture, obtaining initial breeding stock, and housing threatened species can slow down captive conservation efforts (Marchetti and Engstrom 2016). Practitioners and institutions also need time to build aviaries, get initial breeding stock, and learn how to breed captive individuals (Snyder et al. 1996). This is complicated by the fact that many parrot species have long generation times and take multiple years to reach sexual maturity (Young et al. 2012), and even after reaching maturity, many parrots have small clutch sizes and may not breed every year.

Some delays may occur when utilizing naturalized populations for translocation, such as the time it takes to obtain funding and permits for capture. Public relations and the societal implications of removing birds from urban areas need to be taken into consideration as well, as some members of the public may have a sense of ownership over birds that use their feeders or nest in their yards (Crowley et al. 2019).

Some costs would be the same, whether utilizing a naturalized population or a captive-breeding facility as a source for birds for conservation actions. These include costs for conserving or rehabilitating release sites, capture and release permits, health checks, release of birds, and protection of the release site and birds after release. However, all indications are that naturalized populations can maintain populations and produce cohorts for use in conservation in a more efficient manner than captive breeding.

Genetic Issues

The usually small populations held in captivity can suffer from negative genetic effects (Snyder et al. 1996; Robert 2009) that can lead to health problems for individuals and reduced population growth. Furthermore, populations of organisms kept in captivity may experience unexpected selection pressures. Many times, individuals that are better adapted to life in a captive environment survive and reproduce better, passing on heritable traits such as neophobia, boldness, tameness, and sociability (Boissy 1995; McDougall et al. 2006; Faure and Mills 2013; Grandin and Deesing 2013). Unfortunately, these adaptations to captivity may be detrimental to life in the wild (McDougall et al. 2006; Carrete and Tella 2015; Shier 2016).

Genetic issues can also arise in naturalized populations, although a continued supply of new escapees may help ameliorate this (Kolbe et al. 2004; Simberloff 2009; Gonçalves da Silva et al. 2010). Naturalized parrots may also evolve or develop behaviors ideal for life in urban ecosystems but unsuitable for life in their native habitat. For example, birds may become reliant on food sources uncommon in their native environment, such as bird feeders or non-native flora. These birds would need to undergo at least some rewilding or a soft release during translocations.

Hybridization

In areas where congeners have become naturalized, individuals may hybridize readily (see summary by Mori and Menchetti, chap. 6 this vol.). Naturalized *Amazona* parrots are known to hybridize in California (Mabb 1997a), and naturalized Lilac-crowned Amazons (*A. finschi*) are hybridizing with Red-crowned Amazons in Texas (Kiacz and Brightsmith, unpubl. data). The population of Red-crowned Amazons in Texas is believed to be a combination of native birds, naturalized birds that escaped captivity, and their offspring (Neck 1986; Garrett 1997; Mabb 1997b; Uehling et al. 2019; USFWS 2019; Enkerlin-Hoeflich and Hogan 2020). Both Lilac-crowned and Red-crowned Amazons are endangered within their native ranges, which adds another layer of complexity to management of the naturalized populations (IUCN 2019). Corruption of the gene pools for these species could reduce the value of these naturalized populations for use in future translocations or captive breeding programs (Rocha and Bergallo 2012; Mori et al. 2017). Conversely, captive breeding programs have total control over breeding individuals, although genetic testing and organization can be costly.

Disease

Parrots are known to carry a variety of contagious diseases, including psittacine beak and feather disease (PBFD), exotic Newcastle disease (END), avian chlamydiosis (AC), parrot bornavirus, and polyomavirus (Lever 2005; Harrison and Lightfoot 2006; Raidal and Peters 2018). Some of these diseases are of major conservation concern for wild parrots (PBFD) (Ortiz-Catedral et al. 2009), while others can impact a variety of avian taxa (END and AC) and even humans (AC) (Smith et al. 2011). Of note, different diseases affect parrot species differently, resulting in some species becoming healthy carriers of diseases that are highly virulent in others (Payne et al. 2011). All parrot populations are at risk from these diseases, including captive, naturalized, and wild populations, so disease control is a vital part of all parrot management, including translocations or captive breeding (Doak et al. 2013; Raidal and Peters 2018).

Captive parrots are susceptible to disease from a variety of sources, including the founding stock of the population, the mixing of species in large facilities, and the addition of new individuals from a variety of sources (the wild, other breeding facilities, pet owners, etc.). In captive facilities, birds are usually kept in close quarters, fed from the same food sources, and handled by the same keepers, which can cause rapid spread of disease if infectious agents enter the facility. Common precautions to reduce this risk include maintaining single species in isolated facilities, high levels of biosecurity, regular disease testing, vaccination, and strict quarantine and testing of new stock—all costly procedures (Australian Dept. of the Environment and Heritage 2006; Doak et al. 2013; Raidal and Peters 2018).

Diseases are unlikely to spread as quickly among naturalized populations as in captive breeding facilities due to lower densities of birds; however, where large communal roosts or use of bird feeders are common, this may not be the case (Bradley and Altizer 2007; Robb et al. 2008). When taking individuals from a naturalized population, it is imperative to screen for diseases before housing them with other birds or releasing them into the wild. Screening for disease can be expensive and time consuming, but it is a necessary procedure for both captive and naturalized birds and has been done with positive results (Collazo et al. 2003; Brightsmith et al. 2005).

Rewilding and Release Preparation

Many conservation efforts utilizing captive-reared individuals have had very high losses post release: 59% mortality of Puerto Rican Amazons in 2002 (White et al. 2005), 96% mortality of captive-reared Thick-billed Parrots (*Rhynchopsitta pachyrhyncha*) after just two months in the wilds of Arizona (Snyder et al. 1994), and failure to establish a breeding population of Orange-bellied Parrots (*Neophema chrysogaster*), even after 423 birds were released in Australia over a 10-year period (Orange-Bellied Parrot Recovery Team 2006). In order to reduce these losses, many projects engage in rewilding of captive-bred birds in preparation for release. However, this activity is time consuming and costly and not always fully effective (Snyder et al. 1994; White et al. 2005; Stojanovic et al. 2017). Unlike captive-bred populations, naturalized populations already have the ability to find food, avoid predators (including humans), and breed in the wild. This eliminates the need for most prerelease training and should greatly increase post-release survival (Carrete and Tella 2015). Geographic and habitat similarity between the birds' naturalized and native ranges would be of concern in these situations, with individuals from peripatric naturalized populations presumably better equipped for translocation into their native ranges (Jeschke and Strayer 2008), while naturalized birds from habitats dissimilar to their native range would need at least some prerelease training to help them identify local food sources and novel predation risks.

NATURALIZED PARROTS FOR CONSERVATION: SUMMARY

Problems with captive breeding are not new and have been an issue for conservationists and wildlife managers for decades (Snyder et al. 1996). Nevertheless, over the last few decades, the breeding of rare species in captivity has been successful enough that it's no longer fringe science but a reliable method of species conservation (Griffiths and Pavajeau 2008). The things that haven't changed over this time are the costs and time associated with breeding animals in captivity. Currently, ~227 species of parrots have declining populations, and 37 of those are listed as in need of *ex situ* conservation (IUCN 2019). Of those 37, six have at least one naturalized population: Yellow-crested Cockatoo, Salmon-crested Cockatoo (*Cacatua moluccensis*), Grey Parrot (*Psittacus erithacus*), Red-crowned Amazon, Yellow-headed Amazon (*Amazona oratrix*), and Kuhl's Lorikeet (*Vini kuhlii*). Although no naturalized population is currently being utilized for conservation relative to its native populations, all six of these high-priority species have conservation programs within their native ranges that could benefit from the managed use of their naturalized populations. Additionally, 137 parrot species have declining populations and are listed as "near threatened" or "vulnerable" (IUCN 2019), meaning that over the coming decades, many more species are likely to be in need of *ex situ* conservation, and naturalized populations of those species could become very useful (Berkunsky et al. 2017). Although not completely free of costs or disadvantages, naturalized populations may be a viable alternative to captive breeding for long-term maintenance of parrots of conservation interest. Managers would still need to monitor for hybridization and disease, and genetic testing should be done in most cases, but utilizing naturalized populations for translocation should still be less expensive than captive breeding and provide individuals better prepared for life in the wild.

OPPORTUNITIES FOR RESEARCH ON NATURALIZED PARROTS

Parrots are a greatly understudied portion of the world's avifauna (Collar 1998), and there are critical gaps in our scientific knowledge of parrots as it relates to conservation of threatened and endangered species (Marsden and Royle 2015; Renton et al. 2015). Even basic knowledge of psittacine natural history is still lacking for most parrot species worldwide. Naturalized populations can allow researchers to study the basic natural history traits important for conservation purposes (Marchetti and Engstrom 2016; Uehling et al. 2019). Naturalized parrots also represent "natural experiments" that can be used to test adaptation of species to urban habitats as well as scientific hypotheses at spatial and temporal scales too large for controlled experiments. Research on naturalized parrots could further our understanding of autecology, ecosystem function, and biodiversity in ways that cannot be achieved easily with wild populations (Blackburn and Duncan 2001; Richardson and Pyšek 2008).

Research on naturalized bird populations has been conducted at least since the early 1900s, when Joseph Grinnell studied House Sparrows (*Passer domesticus*) in Death Valley, California (Grinnell 1919). Since then, researchers have used naturalized birds to study a broad array of topics, including life history traits, genetics, adaptability, invasion biology, and disease (Moles et al. 2008; Gonçalves da Silva et al. 2010; Martin-Albarracin et al. 2015; Blanvillain et al. 2017). In this section, we will review some current research topics involving naturalized populations (including parrots and other birds) and highlight some potential avenues for future work with naturalized parrots.

Basic Natural History

Knowledge of basic parrot ecology and natural history is critical to help us conserve threatened species (Collar 1998; Masello and Quillfeldt 2002; Marsden and Royle 2015). Naturalized populations of parrots can be studied to gain information on their life history traits, such as longevity, fecundity, number of eggs per brood, number of broods per season, nesting, dispersal, etc. (Leech et al. 2008; Simberloff 2009; Marchetti and Engstrom 2016; Blanvillain et al. 2017; Uehling et al. 2019). This is obviously most useful if the naturalized species itself is threatened, but it is also useful if congeners or other species that share similar traits are imperiled and lacking this general information.

Genetic Issues

Naturalized populations are ideal models with which to study the relationships of genetic bottlenecks and founding effects on allele frequency, heterozygosity, and polymorphic loci (Kolbe et al. 2004; Simberloff 2009; Gonçalves da Silva et al. 2010). Baker and Moeed (1987) found that introduced populations of Common Mynas (*Acridotheres tristis*) had an 18% loss of alleles and lower heterozygosity than did native populations. The largest loss of diversity in mynas was in naturalized populations that had the smallest number of founders (Baker and Moeed 1987). Other researchers have since shown similar losses in genetic diversity among naturalized or introduced populations (Fleischer et al. 1991; Jamieson 2011). However, Simberloff (2009) points out that not all naturalized species suffer from genetic impoverishment if the propagule pressure is high, and Kolbe et al. (2004) were able to show this in naturalized populations of lizards, which actually have higher genetic variation than populations of their native counterparts. Additionally, Gonçalves da Silva et al. (2010) showed little loss of genetic variation in naturalized populations of Monk Parakeets (*Myiopsitta monachus*) in the US, pointing to continued propagule pressure as the likely explanation. Barring high propagule pressure, genetic drift and fixation events could prove detrimental to a threatened species, so understanding the relationships between population numbers, genetic bottlenecks, and their possible effects is crucial to the design of effective translocations and captive breeding projects.

Naturalized populations also provide the opportunity to study adaptation and phenotypic change due to novel selection pressures as a result of living in novel ecosystems (Ross 1983; Baker and Moeed 1987; Cabe 1998; Suarez and Tsutsui 2008; Jackson et al. 2015). This type of work can also shed light on how organisms successfully invade new ecosystems and how they may respond in the face of changing climatic regimes and urbanization.

Genetic studies can also pinpoint source populations of naturalized populations, guiding where to focus poaching and trade control efforts (Kirk et al. 2013; Perdereau et al. 2013; Jackson et al. 2015). This genetic information can also help researchers understand the value of a naturalized population, since individuals with varying degrees of introgression or of a certain subspecies may be deemed undesirable for translocation projects (Amato 1995; Sanz and Grajal 1998).

Disease

Parrots and their associated diseases are moved at a global scale through the pet trade (Smith et al. 2009; Symes et al., chap. 17 this vol.). Testing for disease, such as PBFD, on captive and wild individuals in native and naturalized ranges allows researchers to understand the spread of these infections and the susceptibility of populations and can guide conservation, management, and national policy (Ha et al. 2007; Fogell et al. 2018). Disease also has implications for translocation and breeding success (Tollington et al. 2013; Tollington et al. 2015) and, as such, is of major concern to conservationists and aviculturists alike (Harkins et al. 2014).

Vaccine testing and disease prevalence studies could be done using naturalized populations that have been deemed invasive and are suitable for removal (Kirkpatrick et al. 2011). Capture of these individuals and use in research that could potentially help their species or relatives in aviculture or in the wild could be a meaningful way to justify take. By using individuals from less critical, naturalized populations instead of pulling from native populations, we can limit the loss of native birds while gaining insight on possible methods to cure or prevent diseases.

Naturalized populations can also offer insight to potential environmental contamination. Naturalized *Aratinga* parakeets in California have been found to be affected by bromethalin toxicosis (Van Sant et al. 2019). Bromethalin is a common ingredient in rat poison; the source of this toxin in the environment is currently unknown but under investigation because of the studies performed by Van Sant and collaborators.

DEVELOPMENT OF RESEARCH TECHNIQUES

Naturalized populations can be useful when designing and implementing new research techniques or equipment. In the early 1990s, researchers tested multiple types of very high frequency (VHF) transmitters for fit and wear on naturalized *Amazona* parrots in Puerto Rico before using them on critically endangered Puerto Rican Amazons (Meyers 1996). Naturalized Rose-ringed Parakeets were used to study hand-rearing and soft-release methodology of critically endangered Echo Parakeets (*Psittacula eques*) on Mauritius Island (Jones et al. 1998). Rose-ringed Parakeets were also used as surrogate fosters for raising some of the initial breeding crop of Echo Parakeets. Due to the possibility of future hybridization, no Echo Parakeets raised by Rose-ringed Parakeets were released into the wild but were instead kept and used for captive breeding and fostering of Echo Parakeets (Jones et al. 1998).

INVASION BIOLOGY

Parrots' repeated successes and failures to become established in different areas set up opportunities for researchers to test theories on how species successfully colonize new environments. Studies on naturalized parrots can look at the four major aspects of invasion biology: transport, introduction, establishment, and spread (Case 1996; Williamson 1996; Duncan et al. 2003; Martin-Albarracin et al. 2015), and in some cases the decline to extirpation. Two naturalized avian populations, Budgerigars (*Melopsittacus undulatus*) in Florida and Common Mynas in Vancouver, Canada, have gone through all four stages as well as extirpation in just a few decades. Roughly 7,000 Budgerigars and 20,000 Common Mynas populated their respective naturalized ranges at their peaks, and both, without clear explanation, declined in number to virtually zero, and neither species is currently established where it was once common (Long 1981; Colautti and MacIsaac 2004; Butler 2005). These events took place before scientists were actively engaged in invasion biology, but by studying the events that led to introduction, establishment, spread, decline, and extirpation of these populations, conservationists and managers could be better prepared to contain or manage naturalized populations.

Since naturalization events (especially introduction and establishment) mirror similar events that must take place during translocations, understanding the main factors in successful establishment of non-native species can be useful as a resource of ideology and theory for better understanding translocations (Royle and Donner, chap. 2 this vol.). Many studies place propagule pressure as the most significant factor correlated with establishment success (Veltman et al. 1996; Williamson 1996; Duncan et al. 2001; Brook 2004; Lockwood et al. 2005; Duncan and Forsyth 2006; Simberloff 2009). For example, Brook (2004) showed that number of introduction attempts and total number of individuals released predicted the successful establishment of almost 90% of 77 avian introduction cases in Australasia. Likewise, Veltman et al. (1996) showed that the highest correlate with naturalization success for 27 bird species in New Zealand was propagule pressure. By modeling the historical and current global trade of parrots, or the highest percentage of households with parrots as pets, researchers could predict areas with higher chances of establishment of non-native parrots, informing management and trade decisions at a worldwide scale (Uehling et al. 2019).

Species distribution models (SDMs) can greatly help our understanding of where an organism can successfully invade (Peterson 2003; Yackulic et al. 2015). By developing models of naturalized parrot ranges, we can understand which regions across the globe could be successful points of introduction (Peterson 2003; Strubbe and Matthysen 2009). Modeling of these ranges can take into account urbanization and climate change for better predictive abilities (Jeschke and Strayer 2008; Strubbe and Matthysen 2009; Bellard et al. 2013). For most species, matching the naturalized environment (including climate) to that of their native environment can increase successful establishment (Daehler and Strong 1993; Thuiller et al. 2005; Strubbe and Matthysen 2009). However, this is not always the case, as with habitat generalists (Monk Parakeets, Rose-ringed Parakeets), which can survive in a wide range of habitats (Cassey et al. 2004; Strubbe and Matthysen 2009). Additionally, climate change is predicted to create range shifts for many species both in native and non-native ranges (Bellard et al. 2013). These shifts include range shrinkage and enlargement and can cause extirpation of currently established non-natives or create new habitats for species to exploit.

SOCIAL IMPACTS OF NATURALIZED PARROTS

Although there are in-depth reviews of the societal impacts of naturalized parrot populations in this book (see Crowley, chap. 3, and Bucher, chap. 8), we provide a brief review of the conservation-related benefits of interactions between naturalized parrots and people. Fifty-five percent of humans live in urban areas, and as that figure continues to increase, human engagement with nature continues to decline (Turner et al. 2004; Soga and Gaston 2016). This increasing disconnect from nature is termed "extinction of experience" (Pyle 1978; Miller 2005) or "nature deficit disorder" (Louv 2008). Fortunately, most naturalized parrots occur in areas of high human activity, which offers opportunities for people to have interactions with groups of these sociable and noticeable birds. Research has shown that exposure to natural vs. urban environments can reduce blood pressure (Hartig et al. 2003) and increase happiness (MacKerron and Mourato 2013; Shanahan et al. 2015) as well as change how the human population interacts with and views nature (Bixler et al. 2002; Miller 2005). Participation in recreational activities, including birdwatching, has a positive influence on pro-environmental behavior, including contribution of money to conservation organizations (Nord et al. 1998). In order to increase support from the general public for biodiversity conservation, we urgently need to increase people's interactions with the natural world (Miller 2005). We argue here that naturalized parrots afford us with an opportunity to help connect urban human populations with nature.

Naturalized parrots have high visibility and great charisma, making them a natural attraction for many people (Avery et al. 2006; Crowley et al. 2019). Monk Parakeets that regularly nest on utility poles and palm trees in urban landscapes offer humans an opportunity to observe colorful "wild" birds courting, singing, and socializing in areas where the native fauna has been drastically reduced (Burger and Gochfeld 2009). This opportunity is not unique to just a few cases—Monk Parakeets, Red-masked Parakeets (*Psittacara erythrogenys*), and Blue-and-yellow Macaws (*Ara ararauna*) have garnered enough public support in cities like Brooklyn, New York; San Francisco, California; and Caracas, Venezuela, respectively, to thwart attempts at removal. In Texas, local municipalities have passed laws that protect native and naturalized parrots from poaching and restrict destruction of trees used for roosting or nesting (Brownsville 1992; Weslaco 1999; Harlingen 2011; McAllen 2014). This type of local support is important—public backing, perception, and education are critical to successful conservation initiatives, as shown in island nations with endemic parrots (Christian 1993; Christian et al. 1996; White et al. 2011). Dunn et al. (2006) argue that even non-native species, like naturalized parrots, can lead the way for conservation-minded management and local involvement in urban areas. In addition, many, if not most naturalized parrot populations have few negative environmental impacts (Brightsmith and Kiacz, chap. 9 this vol.). As a result, many naturalized parrot populations can act as ideal

Blue-and-yellow Macaws (*Ara ararauna*) flying over the University of Miami. This population has suffered from poaching and is greatly reduced in size. Miami, Florida, US, 2011. Photo by Roelant Jonker.

flagships to help lead a reconnection with nature and encourage local participation in conservation-minded recreation.

Naturalized parrots may also provide some small-scale economic opportunities to local communities, which in turn build support for the parrots' populations (Seymour 2013). At a local park in Brownsville, Texas, an average of six birdwatchers per evening are present year-round to watch as flocks of native and naturalized parrots come to roost, making them a local tourist attraction (Kiacz and Brightsmith, unpubl. data). Festivals for birds and birdwatchers are gaining popularity throughout the Americas and Europe, and in some cases, parrots can be a major draw for participation. At the annual Rio Grande Valley Birding Festival in Texas, nearly 120 participants pay to ride through neighborhoods in search of the native and naturalized urban parrots. Similarly, the San Diego Bird Festival fills its Parrot Party Bus with participants hoping to catch a glimpse of some of the 13 naturalized parrot species in southern California. Proceeds from the Rio Grande Valley festival have been used to support parrot research and local conservation initiatives, and the San Diego festival supports local wildlife rehabilitation and educational programs through the San Diego Audubon Society. Events such as these are likely to grow in popularity and should be embraced by local communities as ways to gain benefits from naturalized parrots.

Although negative economic and social impacts from naturalized parrots do exist, these impacts are mostly insignificant and caused by only two species (Monk and Rose-ringed Parakeets), and the general public largely sees the birds as affable and charismatic additions to their daily lives (Crowley, chap. 3 this vol.). The fact that many people rally around "their" parrots to protect them from extirpation is evidence that there is positive value that can balance the negatives in many cases. Events such as those mentioned above suggest that parrots have already started the process of bringing nature to the people and may be helping to reduce nature deficit disorder and increase support for conservation initiatives, including research on parrots in native and naturalized habitats.

CONCLUSIONS

In this chapter, we examined actual and potential benefits of naturalized populations of parrots. Some of these benefits have real-world conservation value—naturalized parrots can be used as cheap, low-impact sources for captive breeding programs and are especially useful if they are one of the six naturalized species in need of *ex situ* conservation. These populations are further useful for translocations, as rewilding them should be less expensive, and they are likely better equipped for problems they will face living in the wild. However, when using naturalized populations, biologists lose their ability to control hybridization and disease, and societal connections with the parrots can hamper capture efforts.

Naturalized parrots may also have value as research systems to help quantify life history parameters of parrots, advance our understanding of invasion biology, and develop new conservation management techniques for use in captive, native, and naturalized populations. Study of these populations is necessary if we want to fully understand the parrots' impacts on ecosystems, so we can rationally define and manage these populations.

The social impacts naturalized parrots make in some urban areas, while hard to quantify, may not be trivial, based on the actions local people take to protect the birds. It remains to be seen, however, if these parrots can get people outdoors and reconnect them with nature in urban environments at a large scale, and if this can translate into more concern for nature and actions that benefit conservation. From an economic perspective, naturalized parrots may generate some tourism revenue, but at a global level, the money generated is certainly less than the damage caused (Brightsmith and Kiacz, chap. 9 this vol.). However, individual populations that do not cause economic damage could have a net positive contribution.

All these impacts, negative, neutral, or positive, deserve to be analyzed and taken into consideration when determining how these populations are to be defined and managed. Most documented negative impacts have been caused by specific populations of Rose-ringed Parakeets and Monk Parakeets. Therefore, analysis of each naturalized population of each parrot species should stand on its own merit and not be lumped with other species or populations, since they are all distinctive. The impacts each population has depend on life history strategies as well as the ecosystems it inhabits, and each arouses a different perspective in the society with which it coinhabits. All these factors make naturalized parrots an interesting and worthy study system for researchers across the globe.

REFERENCES

Amato, G. 1995. Report on molecular genetic assessment of subspecies status for *Amazona barbadensis*. Unpublished report to the Wildlife Conservation Society, New York.

Australian Dept. of the Environment and Heritage. 2006. *Hygiene Protocols for the Prevention and Control of Diseases (Particularly Beak and Feather Disease) in Australian Birds*. N.p.: Commonwealth of Australia.

Avery, M., Lindsay, J., Newman, J., Pruett-Jones, S., and Tillman, E. 2006. Reducing monk parakeet impacts to electric utility facilities in south Florida. *Advances in Vertebrate Pest Management* 4:125–136.

Baker, A. J., and Moeed, A. 1987. Rapid genetic differentiation and founder effect in colonizing populations of common mynas (*Acridotheres tristis*). *Evolution* 41:525–538.

Bellard, C., Thuiller, W., Leroy, B., Genovesi, P., Bakkenes, M., and Courchamp, F. 2013. Will climate change promote future invasions? *Global Change Biology* 19:3740–3748.

Berkunsky, I., Quillfeldt, P., Brightsmith, D. J., Abbud, M. C., Aguilar, J.M.R.E., Alemán-Zelaya, U., Aramburú, R. M., Arce Arias, A., Balas McNab, R., Balsby, T.J.S., et al. 2017. Current threats faced by Neotropical parrot populations. *Biological Conservation* 214:278–287.

Bixler, R. D., Floyd, M. F., and Hammitt, W. E. 2002. Environmental socialization: Quantitative tests of the childhood play hypothesis. *Environment and Behavior* 34:795–818.

Black, J. M. 1995. The Nene *Branta sandvicensis* recovery initiative: Research against extinction. *Ibis* 137:S153–S160.

Blackburn, T. M., and Duncan, R. P. 2001. Establishment patterns of exotic birds are constrained by non-random patterns in introduction. *Journal of Biogeography* 28:927–939.

Blanvillain, C., Ghestemme, T., Withers, T., and O'Brien, M. 2017. Breeding biology of the critically endangered Tahiti monarch *Pomarea nigra*, a bird with a low productivity. *Bird Conservation International* 28:606–619.

Boissy, A. 1995. Fear and fearfulness in animals. *Quarterly Review of Biology* 70:165–191.

Boyd, C., DeMaster, D. P., Waples, R. S., Ward, E. J., and Taylor, B. L. 2017. Consistent extinction risk assessment under the U.S. Endangered Species Act. *Conservation Letters* 10:328–336.

Bradley, C. A., and Altizer, S. 2007. Urbanization and the ecology of wildlife diseases. *Trends in Ecology & Evolution* 22:95–102.

Brightsmith, D., Hilburn, J., del Campo, A., Boyd, J., Frisius, M., Frisius, R., Janik, D., and Guillen, F. 2005. The use of hand-raised psittacines for reintroduction: A case study of scarlet macaws (*Ara macao*) in Peru and Costa Rica. *Biological Conservation* 121:465–472.

Brook, B. W. 2004. Australasian bird invasions: Accidents of history? *Ornithological Science* 3:33–42.

Brownsville, 1992. Art. 4: Wild Birds, Sec. 7-75, Ordinance 92-1249, City of Brownsville, TX.

Burger, J., and Gochfeld, M. 2009. Exotic monk parakeets (*Myiopsitta monachus*) in New Jersey: Nest site selection, rebuilding following removal, and their urban wildlife appeal. *Urban Ecosystems* 12:185–196.

Butchart, S.H.M., Stattersfield, A. J., and Collar, N. J. 2006. How many bird extinctions have we prevented? *Oryx* 40:266–288.

Butler, C. J. 2005. Feral parrots in the continental United States and United Kingdom: Past, present, and future. *Journal of Avian Medicine and Surgery* 19:142–149.

Cabe, P. R. 1998. The effects of founding bottlenecks on genetic variation in the European starling (*Sturnus vulgaris*) in North America. *Heredity* 80:519.

Carrete, M., and Tella, J. L. 2015. Rapid loss of antipredatory behaviour in captive-bred birds is linked to current avian invasions. *Scientific Reports* 5:18274.

Case, T. J. 1996. Global patterns in the establishment and distribution of exotic birds. *Biological Conservation* 78:69–96.

Cassey, P., Blackburn, T. M., Jones, K. E., and Lockwood, J. L. 2004. Mistakes in the analysis of exotic species establishment: Source pool designation and correlates of introduction success among parrots (Aves: Psittaciformes) of the world. *Journal of Biogeography* 31:277–284.

Christian, C. S. 1993. The challenge of parrot conservation in St Vincent and the Grenadines. *Journal of Biogeography* 20:463–469.

Christian, C. S., Potts, T., Burnett, G., and Lacher, T., Jr. 1996. Parrot conservation and ecotourism in the Windward Islands. *Journal of Biogeography* 23:387–393.

Colautti, R. I., and MacIsaac, H. J. 2004. A neutral terminology to define "invasive" species. *Diversity and Distributions* 10:135–141.

Collar, N. 1998. Information and ignorance concerning the world's parrots: An index for twenty-first century research and conservation. *Papageienkunde* 2:201–235.

REFERENCES

Collazo, J. A., White, T. H., Jr., Vilella, F. J., and Guerrero, S. A. 2003. Survival of captive-reared Hispaniolan parrots released in Parque Nacional del Este, Dominican Republic. *Condor* 105:198–207.

Comizzoli, P., and Holt, W. V. 2019. Breakthroughs and new horizons in reproductive biology of rare and endangered animal species. *Biology of Reproduction* 101:514–525.

Crowley, S. L., Hinchliffe, S., and McDonald, R. A. 2019. The parakeet protectors: Understanding opposition to introduced species management. *Journal of Environmental Management* 229:120–132.

Daehler, C. C., and Strong, D. R. 1993. Prediction and biological invasions. *Trends in Ecology & Evolution* 8:380.

Dallimer, M., and Strange, N. 2015. Why socio-political borders and boundaries matter in conservation. *Trends in Ecology & Evolution* 30:132–139.

Doak, D. F., Bakker, V. J., and Vickers, W. 2013. Using population viability criteria to assess strategies to minimize disease threats for an endangered carnivore. *Conservation Biology* 27:303–314.

Duncan, R. P., Blackburn, T. M., and Sol, D. 2003. The ecology of bird introductions. *Annual Review of Ecology, Evolution, and Systematics* 34:71–98.

Duncan, R. P., Bomford, M., Forsyth, D. M., and Conibear, L. 2001. High predictability in introduction outcomes and the geographical range size of introduced Australian birds: A role for climate. *Journal of Animal Ecology* 70:621–632.

Duncan, R. P., and Forsyth, D. M. 2006. Modelling population persistence on islands: Mammal introductions in the New Zealand archipelago. *Proceedings of the Royal Society of London B: Biological Sciences* 273:2969–2975.

Dunn, R. R., Gavin, M. C., Sanchez, M. C., and Solomon, J. N. 2006. The pigeon paradox: Dependence of global conservation on urban nature. *Conservation Biology* 20:1814–1816.

Enkerlin-Hoeflich, E. C., and K. M. Hogan. 2020. Red-crowned parrot (*Amazona viridigenalis*), (v.1.0). In *The Birds of North America*, ed. A. F. Poole and F. B. Gill. Ithaca, NY: Cornell Lab of Ornithology.

Faure, J. M., and Mills, A. D. 2013. Improving the adaptability of animals by selection. In *Genetics and the Behavior of Domestic Animals*, 2nd ed., ed. T. Grandin and M. J. Deesing: chap. 8. Cambridge, MA: Academic Press.

Fleischer, R. C., Conant, S., and Morin, M. P. 1991. Genetic variation in native and translocated populations of the Laysan finch (*Telespiza cantans*). *Heredity* 66:125–130.

Fogell, D. J., Martin, R. O., Bunbury, N., Lawson, B., Sells, J., McKeand, A. M., Tatayah, V., Trung, C. T., and Groombridge, J. J. 2018. Trade and conservation implications of new beak and feather disease virus detection in native and introduced parrots. *Conservation Biology* 32:1325–1335.

Garnett, S., Woinarski, J., Lindenmayer, D., and Latch, P. 2018. *Recovering Australian Threatened Species: A Book of Hope*. Clayton South, Victoria: CSIRO Publishing.

Garrett, K. L. 1997. Population status and distribution of naturalized parrots in southern California. *Western Birds* 28:181–195.

Gibson, L., and Yong, D. L. 2017. Saving two birds with one stone: Solving the quandary of introduced, threatened species. *Frontiers in Ecology and the Environment* 15:35–41.

Gonçalves da Silva, A. G., Eberhard, J. R., Wright, T. F., Avery, M. L., and Russello, M. A. 2010. Genetic evidence for high propagule pressure and long-distance dispersal in monk parakeet (*Myiopsitta monachus*) invasive populations. *Molecular Ecology* 19:3336–3350.

Goodenough, A. 2010. Are the ecological impacts of alien species misrepresented? A review of the "native good, alien bad" philosophy. *Community Ecology* 11:13–21.

Grandin, T., and Deesing, M. J., eds. 2013. *Genetics and the Behavior of Domestic Animals*. 2nd ed. Cambridge, MA: Academic Press.

Griffiths, R. A., and Pavajeau, L. 2008. Captive breeding, reintroduction, and the conservation of amphibians. *Conservation Biology* 22:852–861.

Grinnell, J. 1919. The English sparrow has arrived in Death Valley: An experiment in nature. *American Naturalist* 53:468–472.

Ha, H., Anderson, I., Alley, M., Springett, B., and Gartrell, B. 2007. The prevalence of beak and feather disease virus infection in wild populations of parrots and cockatoos in New Zealand. *New Zealand Veterinary Journal* 55:235–238.

Harkins, G. W., Martin, D. P., Christoffels, A., and Varsani, A. 2014. Towards inferring the global movement of beak and feather disease virus. *Virology* 450:24–33.

Harlingen, 2011. Sec. 1, Chap. 90, Sec. 90.03, Ordinance 11-55, City of Harlingen, TX.

Harrison, G. L., and Lightfoot, T. L., eds. 2006. *Clinical Avian Medicine*. Palm Beach, FL: Spix Publishing.

Hartig, T., Evans, G. W., Jamner, L. D., Davis, D. S., and Gärling, T. 2003. Tracking restoration in natural and urban field settings. *Journal of Environmental Psychology* 23:109–123.

Heinrichs, J. A., McKinnon, D. T., Aldridge, C. L., and Moehrenschlager, A. 2019. Optimizing the use of endangered species in multi-population collection, captive breeding and release programs. *Global Ecology and Conservation* 17:e00558.

IUCN. 2019. The IUCN Red List of Threatened Species (v.2019-1). https://www.iucnredlist.org.

Jackson, H., Strubbe, D., Tollington, S., Prys-Jones, R., Matthysen, E., and Groombridge, J. J. 2015. Ancestral origins and invasion pathways in a globally invasive bird correlate with climate and influences from bird trade. *Molecular Ecology* 24:4269–4285.

Jamieson, I. G. 2011. Founder effects, inbreeding, and loss of genetic diversity in four avian reintroduction programs. *Conservation Biology* 25:115–123.

Jeschke, J. M., and Strayer, D. L. 2008. Usefulness of bioclimatic models for studying climate change and invasive species. *Annals of the New York Academy of Sciences* 1134:1–24.

Jones, C., Swinnerton, K., Thorsen, M., and Greenwood, A. 1998. The biology and conservation of the echo parakeet *Psittacula eques* of Mauritius. In *Proceedings of IV International Parrot Convention, Tenerife* (n.p.), 110–123.

Kalmar, I. D., Janssens, G. P., and Moons, C. P. 2010. Guidelines and ethical considerations for housing and management of psittacine birds used in research. *Institute for Laboratory Animal Research Journal* 51:409–423.

Kirk, H., Dorn, S., and Mazzi, D. 2013. Worldwide population genetic structure of the oriental fruit moth (*Grapholita molesta*), a globally invasive pest. *BMC Ecology* 13:12.

Kirkpatrick, J. F., Lyda, R. O., and Frank, K. M. 2011. Contraceptive vaccines for wildlife: A review. *American Journal of Reproductive Immunology* 66:40–50.

Kolbe, J. J., Glor, R. E., Schettino, L. R., Lara, A. C., Larson, A., and Losos, J. B. 2004. Genetic variation increases during biological invasion by a Cuban lizard. *Nature* 431:177.

Kueffer, C., and Hadorn, G. H. 2008. How to achieve effectiveness in problem-oriented landscape research: The example of research on biotic invasions. *Living Reviews in Landscape Research* 2.

Kuehler, C., Kuhn, M., McIlraith, B., and Campbell, G. 1994. Artificial incubation and hand-rearing of 'Alala (*Corvus hawaiiensis*) eggs removed from the wild. *Zoo Biology* 13:257–266.

Kumschick, S., Nentwig, W. 2010. Some alien birds have as severe an impact as the most effectual alien mammals in Europe. *Biological Conservation* 143:2757–2762.

Leech, T. J., Gormley, A. M., and Seddon, P. J. 2008. Estimating the minimum viable population size of kaka (*Nestor meridionalis*), a potential surrogate species in New Zealand lowland forest. *Biological Conservation* 141:681–691.

Lever, C. 2005. *Naturalised Birds of the World*. London: A & C Black.

Lockwood, J. L., Cassey, P., and Blackburn, T. 2005. The role of propagule pressure in explaining species invasions. *Trends in Ecology & Evolution* 20:223–228.

Long, J. L. 1981. *Introduced Birds of the World: The Worldwide History, Distribution and Influence of Birds Introduced to New Environments*. New York: Universe Books.

Lopes, A.R.S., Rocha, M. S., Mesquita, W. U., Drumond, T., Ferreira, N. F., Camargos, R.A.L., Vilela, D.A.R., and Azevedo, C. S. 2018. Translocation and post-release monitoring of captive-raised blue-fronted amazons *Amazona aestiva*. *Acta Ornithologica* 53:37–48.

Louv, R. 2008. *Last Child in the Woods: Saving Our Children from Nature-Deficit Disorder*. Chapel Hill, NC: Algonquin Books.

Mabb, K. T. 1997a. Nesting behavior of *Amazona* parrots and rose-ringed parakeets in the San Gabriel Valley, California. *Western Birds* 28:209–217.

Mabb, K. T. 1997b. Roosting behavior of naturalized parrots in the San Gabriel Valley, California. *Western Birds* 28:202–208.

MacKerron, G., and Mourato, S. 2013. Happiness is greater in natural environments. *Global Environmental Change* 23:992–1000.

Marchetti, M. P., and Engstrom, T. 2016. The conservation paradox of endangered and invasive species. *Conservation Biology* 30:434–437.

Marsden, S. J., and Royle, K. 2015. Abundance and abundance change in the world's parrots. *Ibis* 157:219–229.

Martin-Albarracin, V. L., Amico, G. C., Simberloff, D., and Nuñez, M. A. 2015. Impact of non-native birds on native ecosystems: A global analysis. *PLoS One* 10:e0143070.

Masello, J. F., and Quillfeldt, P. 2002. Chick growth and breeding success of the burrowing parrot. *Condor* 104:574–586.

McAllen, 2014. Art. 5: Wild Birds, Sec. 14-114–115, Ordinance 2014-06, City of McAllen, TX.

McDougall, P. T., Réale, D., and Sol, D., Reader, S. M. 2006. Wildlife conservation and animal temperament: Causes and consequences of evolutionary change for captive, reintroduced, and wild populations. *Animal Conservation* 9:39–48.

Menchetti, M., and Mori, E. 2014. Worldwide impact of alien parrots (Aves Psittaciformes) on native biodiversity and environment: A review. *Ethology Ecology & Evolution* 26:172–194.

Meyers, J. M. 1996. Evaluation of 3 radio transmitters and collar designs for *Amazona*. *Wildlife Society Bulletin* 15–20.

Miller, J. R. 2005. Biodiversity conservation and the extinction of experience. *Trends in Ecology & Evolution* 20:430–434.

Minor, E. S., Appelt, C. W., Grabiner, S., Ward, L., Moreno, A., and Pruett-Jones, S. 2012. Distribution of exotic monk parakeets across an urban landscape. *Urban Ecosystems* 15:979–991.

Moles, A. T., Gruber, M. A., and Bonser, S. P. 2008. A new framework for predicting invasive plant species. *Journal of Ecology* 96:13–17.

Moran, E. M., Cullen, R., and Hughey, K.F.D. 2005. Financing threatened species management: The costs of single species programmes and the budget constraint. Paper presented at 2005 NZARES (New Zealand Agricultural and Resource Economics Society) Conference, Nelson, NZ.

Mori, E., Grandi, G., Menchetti, M., Tella, J., Jackson, H., Reino, L., van Kleunen, A., Figueira, R., and Ancillotto, L. 2017. Worldwide distribution of non-native Amazon parrots and temporal trends of their global trade. *Animal Biodiversity and Conservation* 40:49–62.

Neck, R. W. 1986. Expansion of red-crowned parrot, *Amazona viridigenalis*, into southern Texas and changes in agricultural practices in northern Mexico. *Bulletin of the Texas Ornithological Society* 19:6–12.

Newson, S. E., Johnston, A., Parrott, D., and Leech, D. I. 2011. Evaluating the population-level impact of an invasive species, ring-necked parakeet *Psittacula krameri*, on native avifauna. *Ibis* 153:509–516.

Nielsen, J. 2006. *Condor: To the Brink and Back; The Life and Times of One Giant Bird*. New York: HarperCollins.

REFERENCES

Nord, M., Luloff, A., and Bridger, J. C. 1998. The association of forest recreation with environmentalism. *Environment and Behavior* 30:235–246.

Oehler, D. A., Boodoo, D., Plair, B., Kuchinski, K., Campbell, M., Lutchmedial, G., Ramsubage, S., Maruska, E. J., and Malowski, S. 2001. Translocation of blue and gold macaw *Ara ararauna* into its historical range on Trinidad. *Bird Conservation International* 11:129–141.

Olah, G., Butchart, S.H.M., Symes, A., Guzmán, I. M., Cunningham, R., Brightsmith, D. J., and Heinsohn, R. 2016. Ecological and socio-economic factors affecting extinction risk in parrots. *Biodiversity and Conservation* 25:205–223.

Orange-Bellied Parrot Recovery Team. 2006. National recovery plan for the orange-bellied parrot (*Neophema chrysogaster*). East Melbourne, Victoria: Dept. of Primary Industries and Water.

Ortiz-Catedral, L., McInnes, K., Hauber, M. E., and Brunton, D. H. 2009. First report of beak and feather disease virus (BFDV) in wild red-fronted parakeets (*Cyanoramphus novaezelandiae*) in New Zealand. *Emu* 109:244–247.

Payne, S., Shivaprasad, H., Mirhosseini, N., Gray, P., Hoppes, S., Weissenböck, H., and Tizard, I. 2011. Unusual and severe lesions of proventricular dilatation disease in cockatiels (*Nymphicus hollandicus*) acting as healthy carriers of avian bornavirus (ABV) and subsequently infected with a virulent strain of ABV. *Avian Pathology* 40:15–22.

Perdereau, E., Bagnères, A. G., Bankhead-Dronnet, S., Dupont, S., Zimmermann, M., Vargo, E., and Dedeine, F. 2013. Global genetic analysis reveals the putative native source of the invasive termite, *Reticulitermes flavipes*, in France. *Molecular Ecology* 22:1105–1119.

Peterson, A. T. 2003. Predicting the geography of species' invasions via ecological niche modeling. *Quarterly Review of Biology* 78:419–433.

Pimentel, D., Zuniga, R., and Morrison, D. 2005. Update on the environmental and economic costs associated with alien-invasive species in the United States. *Ecological Economics* 52:273–288.

Plair, B. L., Kuchinski, K., Ryan, J., Warren, S., Pilgrim, K., Boodoo, D., Ramsubage, S., Ramadhar, A., Lal, M., and Rampaul, B. 2008. Behavioral monitoring of blue-and-yellow macaws (*Ara ararauna*) reintroduced to the Nariva Swamp, Trinidad. *Ornitologia Neotropical* 19:113–122.

Pyle, R. M. 1978. The extinction of experience. *Horticulture* 56:64–67.

Raidal, S. R., and Peters, A. 2018. Psittacine beak and feather disease: Ecology and implications for conservation. *Emu* 118:80–93.

Reed, C., and Merton, D. 1991. Behavioural manipulation of endangered New Zealand birds as an aid towards species recovery. *Acta XX Congressus Internationalis Ornithologici* 4:2514–2522.

Renton, K., Salinas-Melgoza, A., De Labra-Hernández, M. A., and de la Parra-Martínez, S. M. 2015. Resource requirements of parrots: Nest site selectivity and dietary plasticity of Psittaciformes. *Journal of Ornithology* 156:73–90.

Richardson, D. M., and Pyšek, P. 2008. Fifty years of invasion ecology: The legacy of Charles Elton. *Diversity and Distributions* 14:161–168.

Robb, G. N., McDonald, R. A., Chamberlain, D. E., and Bearhop, S. 2008. Food for thought: Supplementary feeding as a driver of ecological change in avian populations. *Frontiers in Ecology and the Environment* 6:476–484.

Robert, A. 2009. Captive breeding genetics and reintroduction success. *Biological Conservation* 142:2915–2922.

Rocha, C.F.D., and Bergallo, H. G. 2012. When invasive exotic populations are threatened with extinction. *Biodiversity and Conservation* 21:3729–3730.

Ross, H. A. 1983. Genetic differentiation of starling (*Sturnus vulgaris*: Aves) populations in New Zealand and Great Britain. *Journal of Zoology* 201:351–362.

Runde, D. E., Pitt, W. C., and Foster, J. 2007. Population ecology and some potential impacts of emerging populations of exotic parrots. In *Managing Vertebrate Invasive Species: Proceedings of an International Symposium*, ed. G. W. Witmer, W. C. Pitt, and K. A. Fagerstone, 338–360. Fort Collins, CO: USDA/APHIS Wildlife Services, National Wildlife Research Center.

Russell, J. C., and Blackburn, T. M. 2017. The rise of invasive species denialism. *Trends in Ecology & Evolution* 32:3–6.

Sanz, V., and Grajal, A. 1998. Successful reintroduction of captive-raised yellow-shouldered amazon parrots on Margarita Island, Venezuela. *Conservation Biology* 12:430–441.

Sax, D. F., and Gaines, S. D. 2008. Species invasions and extinction: The future of native biodiversity on islands. *Proceedings of the National Academy of Sciences (USA)* 105:11490–11497.

Schlaepfer, M. A., Sax, D. F., and Olden, J. D. 2011. The potential conservation value of non-native species. *Conservation Biology* 25:428–437.

Seymour, M. 2013. "Support your local invasive species": Animal protection rhetoric and nonnative species. *Society & Animals* 21:54–73.

Shanahan, D. F., Fuller, R. A., Bush, R., Lin, B. B., and Gaston, K. J. 2015. The health benefits of urban nature: How much do we need? *BioScience* 65:476–485.

Shier, D. 2016. Manipulating animal behavior to ensure reintroduction success. In *Conservation Behavior: Applying Behavioral Ecology to Wildlife Conservation and Management*, ed. O. Berger-Tal and D. Saltz, 275–304. Cambridge: Cambridge Univ. Press.

Simberloff, D. 2009. The role of propagule pressure in biological invasions. *Annual Review of Ecology, Evolution, and Systematics* 40:81–102.

Sisson, D. C., Terhune, I., Theron, M., Palmer, W. E., and Thackston, R. 2017. Contributions of translocation to northern bobwhite population recovery. *National Quail Symposium Proceedings* 8:151–159.

Smith, K., Acevedo-Whitehouse, K., and Pedersen, A. 2009. The role of infectious diseases in biological conservation. *Animal Conservation* 12:1–12.

Smith, K. A., Campbell, C. T., Murphy, J., Stobierski, M. G., and Tengelsen, L. A. 2011. Compendium of measures to control *Chlamydophila psittaci* infection among humans (psittacosis) and pet birds (avian chlamydiosis), 2010 National Association of State Public Health Veterinarians (NASPHV). *Journal of Exotic Pet Medicine* 20:32–45.

Snyder, N., and Snyder, H. 2000. *The California Condor: A Saga of Natural History and Conservation*. San Diego, CA: Academic Press.

Snyder, N. F., Derrickson, S. R., Beissinger, S. R., Wiley, J. W., Smith, T. B., Toone, W. D., and Miller, B. 1996. Limitations of captive breeding in endangered species recovery. *Conservation Biology* 10:338–348.

Snyder, N. F., Derrickson, S. R., Beissinger, S. R., Wiley, J. W., Smith, T. B., Toone, W. D., and Miller, B. 1997. Limitations of captive breeding: Reply to Gippoliti and Carpaneto. *Conservation Biology* 11:808–810.

Snyder, N. F., Koenig, S. E., Koschmann, J., Snyder, H. A., and Johnson, T. B. 1994. Thick-billed parrot releases in Arizona. *Condor* 96:845–862.

Soga, M., and Gaston, K. J. 2016. Extinction of experience: The loss of human–nature interactions. *Frontiers in Ecology and the Environment* 14:94–101.

Stojanovic, D., Alves, F., Cook, H., Crates, R., Heinsohn, R., Peters, A., Rayner, L., Troy, S. N., and Webb, M. H. 2017. Further knowledge and urgent action required to save orange-bellied parrots from extinction. *Emu* 118:126–134.

Strubbe, D., and Matthysen, E. 2009. Establishment success of invasive ring-necked and monk parakeets in Europe. *Journal of Biogeography* 36:2264–2278.

Strubbe, D., Shwartz, A., and Chiron, F. 2011. Concerns regarding the scientific evidence informing impact risk assessment and management recommendations for invasive birds. *Biological Conservation* 144:2112–2118.

Suarez, A. V., and Tsutsui, N. D. 2008. The evolutionary consequences of biological invasions. *Molecular Ecology* 17:351–360.

Sullivan, B. L., Wood, C. L., Iliff, M. J., Bonney, R. E., Fink, D., and Kelling, S. 2009. eBird: A citizen-based bird observation network in the biological sciences. *Biological Conservation* 2282–2292.

Thuiller, W., Richardson, D. M., Pyšek, P., Midgley, G. F., Hughes, G. O., and Rouget, M. 2005. Niche-based modelling as a tool for predicting the risk of alien plant invasions at a global scale. *Global Change Biology* 11:2234–2250.

Tollington, S., Greenwood, A., Jones, C. G., Hoeck, P., Chowrimootoo, A., Smith, D., Richards, H., Tatayah, V., and Groombridge, J. J. 2015. Detailed monitoring of a small but recovering population reveals sublethal effects of disease and unexpected interactions with supplemental feeding. *Journal of Animal Ecology* 84:969–977.

Tollington, S., Jones, C. G., Greenwood, A., Tatayah, V., Raisin, C., Burke, T., Dawson, D. A., and Groombridge, J. J. 2013. Long-term, fine-scale temporal patterns of genetic diversity in the restored Mauritius parakeet reveal genetic impacts of management and associated demographic effects on reintroduction programmes. *Biological Conservation* 161:28–38.

Turner, W. R., Nakamura, T., and Dinetti, M. 2004. Global urbanization and the separation of humans from nature. *BioScience*. 54:585–590.

Uehling, J. J., Tallant, J., and Pruett-Jones, S. 2019. Status of naturalized parrots in the United States. *Journal of Ornithology* 160:907–921.

USFWS. 2009. *Recovery Plan for the Puerto Rican Parrot (Amazona vittata)*. Atlanta, GA: US Fish and Wildlife Service.

USFWS. 2019. Species status assessment report for red-crowned parrot (*Amazona viridigenalis*) (v.3.0.). US Fish and Wildlife Service, Albuquerque, NM.

Van Sant, F., Hassan, S. M., Reavill, D., McManamon, R., Howerth, E. W., Seguel, M., Bauer, R., Loftis, K. M., Gregory, C. R., Ciembor, P. G., and Ritchie, B. W. 2019. Evidence of bromethalin toxicosis in feral San Francisco "Telegraph Hill" conures. *PLoS One* 14:e0213248.

Veltman, C. J., Nee, S., and Crawley, M. J. 1996. Correlates of introduction success in exotic New Zealand birds. *American Naturalist* 147:542–557.

Weslaco, 1999. Art. 3, Sec. 22-221, in Code 1969, Sec. 6-117, Ordinance 99-14, Sec. 1, 9-7-1999, City of Wesalco, TX.

White, T. H., Jr., Camacho, A. J., Bloom, T., Diéguez, P. L., and Sellares, R. 2011. Human perceptions regarding endangered species conservation: A case study of Saona Island, Dominican Republic. *Latin American Journal of Conservation* 2:18–29.

White, T. H., Jr., Collazo, J. A., and Vilella, F. J. 2005. Survival of captive-reared Puerto Rican parrots released in the Caribbean National Forest. *Condor* 107:424–432.

Williamson, M. 1996. *Biological Invasions*. Heidelberg: Germany: Springer Science & Business Media.

Yackulic, C. B., Nichols, J. D., Reid, J., and Der, R. 2015. To predict the niche, model colonization and extinction. *Ecology* 96:16–23.

Yamamoto, J., Shields, K., Millam, J., Roudybush, T., and Grau, C. 1989. Reproductive activity of force-paired cockatiels (*Nymphicus hollandicus*). *Auk* 106:86–93.

Young, A. M., Hobson, E. A., Lackey, L. B., and Wright, T. F. 2012. Survival on the ark: Life-history trends in captive parrots. *Animal Conservation* 15:28–43.

6

THE ECOLOGICAL IMPACTS OF INTRODUCED PARROTS

Emiliano Mori and Mattia Menchetti

INTRODUCTION

Introduced and invasive species are negatively influencing native species at an alarming and increasing rate. Invasive species may alter the structure and the function of ecosystems through hybridization, competition, predation, parasitism, and disease. More than 60 parrot species have established at least one breeding population outside their natural extent of occurrence, thus making Psittaciformes the most introduced bird order throughout the world. Impacts of introduced parrots are known mainly for the two most widespread species, the Rose-ringed Parakeet (*Psittacula krameri*) and the Monk Parakeet (*Myiopsitta monachus*). These impacts may vary across different study sites and population densities. Hybridization occurs in both tropical native areas, where similar species are naturally present, and novel areas where species have been introduced. Behavioral interference by introduced parrots with a number of native species (birds, mammals, and insects) sharing the same breeding or feeding areas has been reported. Other types of impacts by introduced parakeets include economic effects (e.g., to agriculture and electrical infrastructures) and human well-being. In this chapter, we summarize the ecological impacts of introduced parrots, updating and improving the review by Menchetti and Mori (2014) and attempting to draw conclusions from a worldwide perspective.

THE GLOBAL INTRODUCTION OF PARROTS

Biological invasions are currently known to shape and modify biodiversity in native communities by means of ecological impacts, health consequences, and economic costs (Mack et al. 2000; Kumschick and Nentwig 2010; Mazza et al. 2014). European Regulation 1143/2014, concerning introduced and invasive species, lists only five bird species among the 49 species on its "blacklist" (cf. Nentwig et al. 2018): Ruddy Duck (*Oxyura jamaicensis*), which is hybridizing with the native White-headed Duck (*Oxyura leucocephala*); Egyptian Goose (*Alopochen aegyptiaca*), an agricultural pest; African Sacred Ibis (*Threskiornis aethiopicus*), which competes with native wading birds and hybridizes with the native African Spoonbill (*Platalea alba*); Indian House Crow (*Corvus splendens*), a predator of native small mammals and birds; and the Common Myna (*Acridotheres tristis*), which competes with native passerines and is reported to be an agricultural pest. Parrots are the most common bird pets (Drews 2001; Cassey et al. 2004; Dyer et al. 2017; Mori et al. 2017b, 2020a), undoubtedly because of their vividly colored plumages and great synanthropy (Menchetti and Mori 2014). Records and numbers of introduced populations have increased since the late 1970s and are still growing (Mori et al. 2013b; Pârâu et al. 2016; Hobson et al. 2017; Mori et al. 2017b; Falcón and Tremblay 2018; Per 2018; Mori et al. 2020a). Populations of introduced parrots have been recorded in at least 47 countries, on all continents excluding Antarctica (Ancillotto et al. 2015; Menchetti et al. 2016; iNaturalist Alien Parrots

Observatory 2019; see Pruett-Jones, chap. 19 this vol., for updated numbers).

ECOLOGICAL AND BEHAVIORAL IMPACTS

Competitive Exclusion

Interspecific competition can include either or both competitive exclusion and behavioral interference. Most parrots rely on cavities for nesting (Cramp 1985; Diamond and Ross 2019). Unlike woodpeckers (Piciformes), which, as primary cavity nesters, directly excavate their nesting cavities, parakeets, as secondary cavity nesters, usually rely upon already existing cavities, sometimes adapting them, e.g., by enlarging their entrances. Competitive exclusion between parrots and woodpeckers may therefore be dependent on the abundance of cavities (Keijl 2001; Menchetti and Mori 2014; Şahin and Arslangündoğdu 2019). Where cavities are a limiting resource (e.g., in urban areas or managed parks), introduced parrots may outcompete native secondary cavity users of comparable body size, including bats and dormice using cavities for resting or roosting (Menchetti and Mori 2014; Diamond and Ross 2019).

To date, no study has confirmed the impact of introduced parakeets on any other species throughout the whole of their range of overlap. Therefore, impacts by parrots are dependent on the geographic context, resource availability, and the complexity of the native community. Competitive exclusion has been observed by the Rose-ringed Parakeet against a number of native cavity-dependent species. The success of the Rose-ringed Parakeet is due, in part, to the fact that it starts breeding early in the season. Therefore, the parakeets may secure the best nest sites before native species begin their breeding process. Strubbe et al. (2010) predicted moderate competition between Rose-ringed Parakeets and the native Eurasian Nuthatch (*Sitta europaea*) in Belgium, putting perhaps one-third of the population of the native species at risk. Conversely, Newson et al. (2011) did not find any significant evidence for competition between Rose-ringed Parakeets and nuthatches in the United Kingdom (UK); however, they do not rule out that competitive exclusion may occur where availability of nest cavities is limiting.

Since 1998, the Rose-ringed Parakeet has become one of the most common bird species in Israel (Roll et al. 2008). Before the parakeet introduction, the Eurasian Hoopoe (*Upupa epops*) was a very common and successful species in Israel, as reported by long-term monitoring programs (Yosef et al. 2016). After the establishment of the parakeets, breeding success of the hoopoe dramatically declined. Yosef et al. (2016) claimed that the parakeets were the main factor responsible for hoopoe breeding failures. Rose-ringed Parakeets occupy trunk holes earlier in the season than native hoopoes, and the parakeets can displace hoopoes from occupied holes. In Italy, a comparison of nesting-site occupation before and after the Rose-ringed Parakeet invasion suggested a potential competition with the Eurasian Scops Owl (*Otus scops*). Nesting sites of scops owls overlap with those selected by Rose-ringed Parakeets, and after the parakeets became established the owls were forced to use suboptimal breeding areas (Mori et al. 2017a). Rose-ringed Parakeets are similary forcing Common Swifts (*Apus apus*) to use suboptimal nesting areas in northern Italy (Grandi et al. 2018).

In Seville, Spain, where one of the world's largest colonies of the native Lesser Kestrel (*Falco naumanni*) breeds (Tella et al. 1996), Rose-ringed Parakeets have been observed aggressively attacking the kestrel (Hernández-Brito et al. 2014). In the same city, an urban park hosts the largest European colony of the Greater Noctule Bat (*Nyctalus lasiopterus*), a rare cavity-breeding species (Popa-Lisseanu et al. 2008). Since 2003, the population of Rose-ringed Parakeets has increased in this park, displacing female bats from roosting places, even killing them, and triggering a remarkable population decline of this rare species (Hernández-Brito et al. 2014, 2018). Ivanova and Symes (2019) showed a remarkable decline in native populations of Laughing Doves (*Spilopelia senegalensis*) and Southern Fiscals (*Lanius collaris*) in South Africa coinciding with an increase in Rose-ringed Parakeets.

ECOLOGICAL AND BEHAVIORAL IMPACTS

Competition for nesting sites is particularly concerning on islands, where endemic species may be threatened by the introduction of the Rose-ringed Parakeet. Particularly, in Mauritius (Indian Ocean), the native Echo Parakeet (*Psittacula eques*) may suffer competition with introduced and spreading Rose-ringed Parakeets (Tatayah et al. 2007; Jackson et al. 2015). Displacements by the parakeets have also been recorded with the native Seychelles Black Parrot (*Coracopsis barklyi*) in Mahè Island (Seychelles, Indian Ocean), the Norfolk Parakeet (*Cyanoramphus cookii*) in Norfolk Island (Oceania), and Carnaby's Black Cockatoo (*Calyptorhynchus latirostris*) in southwestern Australia (Lever 1994, 2005; Chapman 2005; Jackson et al. 2015; Menchetti et al. 2016).

Behavioral Interference

Behavioral interference can be either direct or indirect. Hole-nesting parrots strongly defend cavities they use to breed, sometimes displacing and killing other species (Menchetti and Mori 2014). Fatal attacks against cavity-roosting bats have been observed after the establishment of new parrot colonies or at the start of their spread (Menchetti et al. 2016). For instance, Menchetti et al. (2014) observed a Rose-ringed Parakeet killing and partially consuming a male Lesser Noctule Bat (*Nyctalus leisleri*) in a central Italian site where no breeding population of the parakeet occurred. In northern Italy, four Savi's Pipistrelles (*Hypsugo savii*) are known to have been killed by Rose-ringed Parakeets in early spring 2019.

In the Netherlands, several Common Noctule Bats (*Nyctalus noctula*) were reportedly found under parakeet nests (Haarsma and Van der Graaf 2013). A dead Gould's Wattled Bat (*Chalinolobus gouldii*) was observed under the nest of a Coconut Lorikeet (*Trichoglossus haematodus*), in Perth, Australia, where the lorikeet is invasive (Menchetti and Mori 2014). Gebhardt (1996) reported that Rose-ringed Parakeets displaced a Geoffroy's Bat (*Myotis emarginatus*)—and also two dormouse species (*Glis glis* and *Eliomys quercinus*)—from tree holes in Germany. Most parrot and bat species use tree cavities both for breeding and for roosting, therefore suggesting that behavioral interference may be more frequent than observed or may occur directly within cavities.

In Israel, Rose-ringed Parakeets have often been observed attacking another invasive bird, the Common Myna, around nesting sites (Orchan et al. 2012). Attacks by Rose-ringed Parakeets against Little Owls (*Athene noctua*) around nesting cavities have been observed in Italy and in Algeria, where an individual of this nocturnal raptor was killed by a small group of Rose-ringed Parakeets (Menchetti and Mori 2014). In one case, in Germany, Rose-ringed Parakeets displaced a breeding couple of Common Starlings (*Sturnus vulgaris*) from a nest (Czajka et al. 2011).

Rose-ringed Parakeet (*Psittacula krameri*) attacking a Greater Noctule Bat (*Nyctalus lasiopterus*). The bat can be seen clinging to the tree, just above the bird's right wingtip. This bat was later found dead at the base of the tree. Seville, Spain, May 2016. Photo by Dailos Hernández-Brito.

Introduced parrots have also been recorded attacking and in some cases killing arboreal mammals, such as squirrels (Clergeau et al. 2009; Mori et al. 2013a). In France and Turkey, groups of Rose-ringed Parakeets were observed attacking and killing Red Squirrels (*Sciurus vulgaris*) (Menchetti et al. 2016; Çalışkan 2018). This aggression is likely the result of the fact that the Eurasian Red Squirrel is a known nest predator and has been observed attacking parakeet chicks in Rome (Mori et al. 2013a). Similarly, the introduced Eastern Gray Squirrel (*Sciurus carolinensis*) is reported to be a predator of young Rose-ringed Parakeets in the UK (Shwartz et al. 2008). A total of 21 aggressions by Rose-ringed Parakeets against Black Rats (*Rattus rattus*) have been reported around parakeet nests in Seville and Tenerife (Canary Islands). In the majority of attacks the rats escaped, but two were killed (Hernández-Brito et al. 2015).

Parrot attacks may be due to competition for food resources or may be territorial defense behavior (Covas et al. 2017). In 2013, in Seville, harassment and noise intimidation by nesting Rose-ringed Parakeets were recorded against Spotless Starlings (*Sturnus unicolor*), Rock Doves (*Columba livia* var. *domestica*), House Sparrows (*Passer domesticus*), Western Jackdaws (*Coloeus monedula*), Eurasian Collared Doves (*Streptopelia decaocto*), Eurasian Blue Tits (*Cyanistes caeruleus*), Black Kites (*Milvus migrans*), Great Tits (*Parus major*), and Booted Eagles (*Hieraaetus pennatus*) (Hernández-Brito et al. 2014). Similarly, aggressive interactions were also observed against three other introduced parrot species, the Senegal Parrot (*Poicephalus senegalus*), the Blue-crowned Parakeet (*Thectocercus acuticaudatus*) and the Monk Parakeet (Hernández-Brito et al. 2014).

The presence of Rose-ringed Parakeets significantly increased vigilance among native birds at bird feeders (Peck et al. 2014). Particularly, Le Louarn et al. (2016) found that the most affected species at feeders in northern France is the Common Starling. In turn, breeding Rose-ringed Parakeets are rarely displaced by native species, although this does occasionally occur (Czajka et al. 2011).

Around their colonial nests, introduced Monk Parakeets in the United States (US) may harass and kill Blue Jays (*Cyanocitta cristata*), American Robins (*Turdus migratorius*), and House Sparrows (MacGregor-Fors et al. 2011). In Belgium, this parakeet has been observed harrassing Western Jackdaws and Hooded Crows (*Corvus cornix*) at nesting sites (Dangoisse 2009). Similar observations have been made in central Italy (Menchetti and Mori 2014; Di Santo et al. 2017). Monk Parakeets share feeding sites with the Rock Dove, Common Wood Pigeon (*Columba palumbus*), blackbirds, and other native birds, without competing with them, throughout the introduced range (Dangoisse 2009; Zocchi et al. 2009; Appelt et al. 2016). In Spain, Monk Parakeets occasionally displaced blackbirds from feeding areas (Batllori and Nos 1985). Breeding Turquoise-fronted Amazons (*Amazona aestiva*) have been frequently observed attacking Black Rats, Common Starlings and Western Jackdaws in Italy; and American Crows (*Corvus brachyrhynchos*) in California (Mori et al. 2017b).

At feeding sites in Western Australia, introduced Coconut Lorikeets show aggressive behavior toward Red Wattlebirds (*Anthochaera carunculata*), honeyeaters (Meliphagidae), and corellas (*Cacatua* spp.). During the breeding season, this lorikeet displaces from nesting sites native Galahs (*Eolophus roseicapilla*), Australian Ringnecks (*Barnardius zonarius*), Carnaby's Black Cockatoos, and Laughing Kookaburras (*Dacelo novaeguineae*) (Chapman 2005). Lamont (1996) reported that introduced lorikeets may also kill the nestlings of Australian Ringnecks before taking over the nest. Nest displacement of native African Palm Swifts (*Cypsiurus parvus*) by introduced Fischer's Lovebirds (*Agapornis fischeri*), Yellow-collared Lovebirds (*A. personatus*), and their hybrids occurs in Kenya (Lever 1994; Mori et al. 2020a), but no data on swift population trends are available (Menchetti and Mori 2014). The Red-breasted Parakeets (*Psittacula alexandri*) introduced to Singapore feed mainly on unripe fruits, thus competing with the native Yellow-vented Bulbul (*Pycnonotus goiavier*), which eats only ripe ones (Neo 2012). In the same area (central Singapore), Red-breasted Parakeets and Tanimbar Corellas (*Cacatua goffiniana*) may affect breeding success of the rare Oriental Pied Hornbill (*Anthracoceros albirostris*) (Peh 2010).

Competition in Singapore between introduced parrots and native Common Flamebacks (*Dinopium javanense*) and Oriental Dollarbirds (*Eurystomus orientalis*) has also been suggested (Neo 2012). Fischer's Lovebirds and House Sparrows may compete in southern France (Dubois 2007). Additionally, Runde et al. (2007) suggested that introduced Mitred Parakeets (*Psittacara mitratus*) may displace native marine birds from their nests in Hawaii.

Mobbing and Harassment

In Singapore, introduced Tanimbar Corellas have been sporadically observed mobbing native Pink-necked Green Pigeons (*Treron vernans*) (Neo 2012). Single Rose-ringed Parakeet individuals have been observed harrasing a Peregrine Falcon (*Falco peregrinus*), a Black Stork (*Ciconia nigra*), a Eurasian Wryneck (*Jynx torquilla*), and a European Green Woodpecker (*Picus viridis*) in Rome (Mori and Menchetti, unpubl. data). Flocks of Rose-ringed Parakeets may occasionally harass larger species; a group of about 60 parakeets was observed mobbing a Booted Eagle in Seville (Hernández-Brito et al. 2014). Single mobbing events or harassments in the surroundings of nesting sites by Rose-ringed Parakeets targeted Hooded Crows, Rock Doves, and Stock Doves (*Columba oenas*) (Menchetti et al. 2016). In Turkey, Per (2018) and Şahin and Arslangündoğdu (2019) reported harassments by Rose-ringed Parakeets of Yellow-legged Gulls (*Larus michahellis*), Hooded Crows, jackdaws, starlings, nuthatches, Syrian Woodpeckers (*Dendrocopos syriacus*), and Caucasian Squirrels (*Sciurus anomalus*).

OTHER IMPACTS ON FAUNA

Hybridization

Hybridization is common among congeneric species of parrots and may naturally occur in overlapping native ranges (cf. Forshaw 2010). While hybridization between different lovebird species does not alter the individual behavior or invasive success (Mori et al. 2020a), in other cases, hybrids may be better invaders

Hybrid of Salmon-crested Cockatoo (*Cacatua moluccensis*) × White Cockatoo (*C. alba*). Hybrids between these species show crest coloration intermediate between the parent species. Honolulu, Hawaii, US, August 2017. Photo by Nicholas P. Kalodimos.

than either of the pure species. For instance, hybrids of Rose-ringed Parakeet × Alexandrine Parakeet (*Psittacula eupatria*) are larger and more aggressive than parental Rose-ringed Parakeets (Krause 2004; Ancillotto et al. 2015; Postigo 2016).

Some introduced parrots are endangered in the native range; therefore, self-sustaining introduced populations may represent a source of individuals for reintroduction events (Menchetti and Mori 2014; Kiacz and Brightsmith, chap. 5 this vol.). However, hybridization with other introduced parrot species may represent a problem for the genetic integrity of threatened species. An example may be represented in Germany by the introduced Yellow-headed Amazon (*Amazona oratrix*), which is endangered in its native range, hybridizing and producing fertile hybrids with the congeneric Turquoise-fronted Amazon (Martens et al. 2013).

Facilitation of Success of Other Introduced Species

The presence of introduced parakeets in an area may help facilitate the establishment success of other non-native species. Ancillotto et al. (2015) stated that the presence of previously established Rose-ringed Parakeets contributes to niche expansion and invasion success of the congeneric Alexandrine Parakeet. In Israel, enlargement of trunk cavities by the Rose-ringed Parakeets may increase the breeding success of the invasive Common Myna by creating more suitable nesting sites (Orchan et al. 2012). The Common Myna is spreading in Israel and affecting the breeding success of many native hole-nesting species (Roll et al. 2008). Colonial nests of Monk Parakeet are used by many other species, including the Italian Sparrow (*Passer italiae*) in its native range (Moltoni 1945) and the White Stork (*Ciconia ciconia*) in Italy (G. Grandi, pers. comm.) and in Spain (Hernández-Brito et al. 2020). In contrast to the White Stork, the introduced African Sacred Ibis in northern Italy uses the nests of Monk Parakeets as a source of material to build its nests elsewhere, sometimes partially destroying the parakeet nest (Castiglioni et al. 2015). The nesting association between Monk Parakeets and White Storks in Spain has been hypothesized to provide the parakeets with protection from potential predators (Hernández-Brito et al. 2020).

Parasite-Mediated Competition

Parasites may have a substantial impact on the success and impact of biological invasions. When a species establishes new populations outside its natural range, it also introduces parasites to the new range (i.e., spillover) and it is subjected to the infection by local parasites (i.e., spillback) (Kelly et al. 2009; Romeo et al. 2014). Two species may thus interact with one another by means of parasites and diseases (Price et al. 1986; McInnes et al. 2013). To date, ectoparasites of introduced parakeets have been examined in Italian and Spanish populations of naturalized parrots (Mori et al. 2015, 2019; Ancillotto et al. 2018). Among introduced species, the Indian lice *Neopsittaconirmus lybartota* and *Echinophilopterus tota* were observed on Rose-ringed Parakeets in Italy; the South American louse *Paragoniocotes fulvofasciatus* and the biting mite *Ornithonyssus bursa* occurred on Monk Parakeets in Italy and Spain. Conversely, several species of native parasite arthropods were detected on parrots introduced to Italy: the squirrel flea (*Tarsopsylla octodecimdentata*), the pigeon soft tick (*Argas reflexus*; responsible for Lyme disease transmission), the chicken mite (*Dermanyssus gallinae*), the buzzard louse (*Laemobothrion maximum*), and the flat fly (*Ornithomyia avicularia*) on Rose-ringed Parakeets; the pigeon louse (*Columbicola columbae*) and the flat flies *Crataerina pallida* and *Ornithophila metallica* on Monk Parakeets (Mori et al. 2015; Ancillotto et al. 2018). Thus, introduced parakeets may be prone to spillback from coexisting native species with which they interact. Accordingly, in Barcelona, the mite *O. bursa*, introduced with Monk Parakeets, was often recorded on native Rock Doves; conversely, the pigeon louse was rare on Monk Parakeets (Mori et al. 2019).

Introduced parakeets are reservoirs of a high number of bacterial and viral diseases, and therefore, they may affect the fitness of native wild species (Kalodimos 2013; Menchetti and Mori 2014; Mori et al. 2018; Pisanu et al. 2018). Among those, *Chlamydophila psittaci* is reported to be harmful for raptors, including the Bald Eagle (*Haliaeetus leucocephalus*) and the Osprey (*Pandion haliaetus*) (Avery and Shiels 2018).

The role of introduced parrots in spreading avian influenza H5N1 (cf. Fletcher and Askew 2007), potentially harmful to native birds, has been dismissed (D. Brightsmith, pers. comm., 2019). Exotic Newcastle disease, caused by *Paramyxovirus* and carried by Monk Parakeets, increases mortality in wild bird communities (Nelson et al. 1952; Mori et al. 2018). The psittacine beak and feather disease virus (*Circovirus*) has been detected in Rose-ringed Parakeets (Gerlach 1994; Sa et al. 2014; Fogell et al. 2018; Morinha et al. 2020) and is one of the worst parrot pathologies worldwide. Native to Oceania, this disease is particularly concerning when introduced parrots come into contact with autochthonous parrots (Ha et al. 2007), such as the Seychelles Black Parrot (Jackson et al. 2015). Bites by the mite *O. bursa*, introduced with Monk Parakeets in Italy and Spain, may provoke dermatitis, asthma, irritations, and skin rashes in humans and native animal species (Mori et al. 2018).

PREDATORS OF INTRODUCED PARROTS

Pithon and Dytham (1999) reported that in the United Kingdom, both the Eurasian Sparrowhawk (*Accipiter nisus*) and the Northern Goshawk (*A. gentilis*) regularly prey upon Rose-ringed Parakeets, and Harris (2015) observed that Peregrine Falcons and Eurasian Hobbies (*Falco subbuteo*) occasionally do so. Black Rats were responsible for the extinction of the Monk Parakeet colony in Milan in the 1940s (Scortecci 1953; Menchetti and Mori 2014). Eastern Gray Squirrels and Red Squirrels are reported to be occasional predators of Rose-ringed Parakeet chicks (Shwartz et al. 2008; Mori et al. 2013a). Domestic cats have been reported to be effective predators of introduced parakeets; for example, Budgerigars (*Melopsittacus undulatus*) are widely preyed upon by feral cats in Florida (Menchetti and Mori 2014). Remains of a Budgerigar (feathers) have also been recorded in a pellet of a Tawny Owl (*Strix aluco*) in London (Harris 2015). Corvids, such as Western Jackdaw and Carrion Crow (*Corvus corone*) are reported to be predators of parrot chicks in Belgium and Italy (Weiserbs and Jacob 1999). The Rose-ringed Parakeet represented nearly 10% of the total volume of the diet of urban Long-Eared Owls (*Asio otus*) in Follonica, Italy (Mori et al. 2020b).

Human persecution against introduced parakeets is not common, but it has occurred throughout the world. In Italy, two populations of Monk Parakeet have been extirpated by poachers, one in central and one in northern Italy (Mori et al. 2014). Furthermore, a small population of Rose-ringed Parakeets established in the Borromean Islands (Isola Madre, Lake Maggiore, northwestern Italy) was eradicated by the human residents in 1990 (Brichetti and Fracasso 2006). In Israel, where Rose-ringed Parakeets represent an agricultural pest, individuals are commonly killed. In addition to these examples, there are many others in which populations of native or naturalized parrots have been reduced or controlled by humans. Bucher (chap. 8 this vol.) discussed widespread control programs of Monk Parakeets and other species in South America and Pruett-Jones (chap. 19 this vol.) gives other examples of control of naturalized parrots.

IMPACTS ON PLANTS

Apart from the damage parrots cause to cultivated plants (Senar et al. 2016; Mentil et al. 2018; Shiels et al. 2018; Rehman et al. 2019), Tella et al. (2015) recently reported clear evidence of the importance of parrots as potential seed dispersers, a feature that has been long overlooked (Menchetti and Mori 2014). While data on seed dispersion by parrots in their native range are increasingly available (Blanco et al. 2016), there is very little in the literature on the impacts exerted by introduced parakeets on flora and vegetation of local environments (Shiels et al. 2018). Runde et al. (2007) suggested that, given its feeding habits, the invasive Mitred Parakeet may help spread the invasive Velvet Tree (*Miconia calvescens*) in the Hawaiian archipelago. Conversely, if introduced parrots totally remove the seed coat, they may increase the deterioration of the seed, therefore reducing its germination success (Howe 1977; Galetti 1993; Norconk et al. 1998). In turn, parrots may limit the reproductive success of endemic plants in native environments. Consumption of flowers and blossoms by Rose-ringed and Monk Parakeets may affect plant reproduction success.

The folivorous diet of introduced Sulphur-crested Cockatoos (*Cacatua galerita*) in New Zealand is currently altering forest dynamics, increasing scrub richness, and triggering the population decline of some tree species (Styche 2008; Menchetti and Mori 2014). Fletcher and Askew (2007) suggested that colonial roosts of Rose-ringed Parakeets may affect the vegetal composition on the ground because of the chemical alteration produced by droppings, but this hypothesis has yet to be verified (Menchetti and Mori 2014).

Monk Parakeets may strip small branches of trees and other plants, both natural and cultivated (e.g., *Vitis vinifera*), to build their nests (Menchetti and Mori 2014).

CONCLUSIONS AND MANAGEMENT IMPLICATIONS

This summary reveals that knowledge on impacts by introduced parrots has increased since the first assessment review (Menchetti and Mori 2014; see Tables 6.1 and 6.2). For Europe, data obtained through ParrotNet, an EU COST (European Cooperation in Science and Technology) Action, allowed us also to draw the impact assessment for the two most widespread parrot species, the Rose-ringed Parakeet and the Monk Parakeet, outside their natural extent of occurrence (Turbé et al. 2017).

Impacts of introduced parrots are context-dependent and vary highly across geography, local resource availability, and the native bird community. Therefore, it is still hard to draw general conclusions and management actions. From a general perspective, native hole-nesting species and cavity-roosting bat species are most affected, and they deserve specific attention, particularly at the start of the invasion process (Hernández-Brito et al. 2014; Menchetti et al. 2014). Conversely, the main gap remains in analyses of effects on the flora (both native and invasive) of the introduced range, which have not been quantitatively assessed yet in any country. Apart from ecological impacts, Menchetti and Mori (2014) reported that introduced parakeets also affect human economy, by impacting agriculture and electrical infrastructures (Reed et al. 2014) and wellness, through noise pollution, disease, and parasite transmission (Avery and Shiels 2018; Mori et al. 2020c).

Prevention and early detection of new invasions are reported to be the most cost-effective strategies (Vall-llosera et al. 2016), which may be obtained also through the help of popular online internet resources (e.g., biodiversity platforms and social networks). Online projects are effective tools for parakeet passive surveillance (Vall-llosera et al. 2016). Once parrots are established, attempts to eradicate or numerically control them may trigger controversy and conflicts (Crowley et al. 2019). Furthermore, eradication and numerical control may be ineffective or, at least, unsustainably expensive (Crowley et al. 2016; Vall-llosera et al. 2016). Where damage to human activity is minimal, parakeets are widely appreciated. For instance, in New Zealand, introduced Eastern Rosellas (*Platycercus eximius*) and Crimson Rosellas (*P. elegans*), both native to Australia, are reported to be efficient predators of a destructive fly species, *Calliphora laemica* (Dipera: Calliphoridae), a perpetrator of fly-strike on domestic sheep (Lever 1994), and therefore are likely appreciated by local shepherds. After educational campaigns, assessment of local behavioral ecology (see Le Louarn et al. 2017; Le Louarn et al. 2018; Postigo et al. 2019) and social impacts may make management actions including eradication and numerical control more democratic and, in turn, much more effective (Crowley et al. 2016).

ACKNOWLEDGMENTS

We would like to thank all the people who helped us in data/paper collection and analysis: Leonardo Ancillotto, Donald Brightsmith, Silvia Giuntini, Gioele Grandi, Lorenzo Mentil, and all the other ParrotNet participants. Jessica Peruzzo, Giuliano Petreri, Mario Monfrini, Alberto Calveri, Milos Di Gregorio, Marco Vicariotto, and Luciano Giacomino kindly provided us with important documentary photos.

TABLE 6.1

Summary of interactions between bird and mammal species and the Rose-ringed Parakeet in its introduced range (Sources: Menchetti and Mori 2014; Menchetti et al. 2016; White et al. 2019; this study). Positive effects refer to those where a species depredates the parakeets; neutral effects refer to harassments or mobbing events by the parakeets; negative effects refer to population/individual effects (i.e., killings, population reduction). Species are listed in alphabetical order by common name.

CLASS	SPECIES	EFFECT ON SPECIES
Experimental Studies:		
Aves	Eurasian Hoopoe (*Upupa epops*)	Negative
Aves	Eurasian Nuthatch (*Sitta europaea*)	Negative
Aves	Eurasian Scops Owl (*Otus scops*)	Negative
Aves	Laughing Dove (*Spilopelia senegalensis*)	Negative
Aves	Long-eared Owl (*Asio otis*)	Negative
Aves	Southern Fiscal (*Lanius collaris*)	Negative
Mammalia	Greater Noctule Bat (*Nyctalus lasiopterus*)	Negative
Observational Studies and Occasional Observations:		
Aves	Black Kite (*Milvus migrans*)	Neutral
Aves	Black Stork (*Ciconia nigra*)	Neutral
Aves	Booted Eagle (*Aquila pennata*)	Neutral
Aves	Carnaby's Black Cockatoo (*Calyptorhynchus latirostris*)	Negative
Aves	Common Blackbird (*Turdus merula*)	Negative
Aves	Common Myna (*Acridotheres tristis*)	Neutral
Aves	Common Starling (*Sturnus vulgaris*)	Negative
Aves	Echo Parakeet (*Psittacula eques*)	Negative
Aves	Eurasian Blue Tit (*Cyanistes caeruleus*)	Negative
Aves	Eurasian Tree Sparrow (*Passer montanus*)	Negative
Aves	Eurasian Wryneck (*Jynx torquilla*)	Neutral
Aves	European Green Woodpecker (*Picus viridis*)	Negative
Aves	Great Tit (*Parus major*)	Neutral
Aves	Grey Heron (*Ardea cinerea*)	Neutral
Aves	Hooded Crow (*Corvus cornix*)	Negative and Positive
Aves	House Sparrow (*Passer domesticus*)	Neutral
Aves	Italian Sparrow (*Passer italiae*)	Neutral
Aves	Lesser Kestrel (*Falco naumanni*)	Negative
Aves	Lesser Vasa Parrot (*Coracopsis nigra*)	Negative
Aves	Little Owl (*Athene noctua*)	Negative
Aves	Monk Parakeet (*Myiopsitta monachus*)	Neutral
Aves	Norfolk Parakeet (*Cyanoramphus cookii*)	Negative
Aves	Peregrine Falcon (*Falco peregrinus*)	Positive
Aves	Pink-necked Green Pigeon (*Treron vernans*)	Negative
Aves	Rock Dove (*Columba livia*)	Negative

continued overleaf

THE ECOLOGICAL IMPACTS OF INTRODUCED PARROTS

TABLE 6.1 continued

CLASS	SPECIES	EFFECT ON SPECIES
Aves	Spotless Starling (*Sturnus unicolor*)	Negative
Aves	Stock Dove (*Columba oenas*)	Negative
Aves	Syrian Woodpecker (*Dendrocopos syriacus*)	Negative
Aves	Western Jackdaw (*Coloeus monedula*)	Negative and Positive
Aves	Yellow-legged Gull (*Larus michahellis*)	Neutral
Mammalia	Black Rat (*Rattus rattus*)	Positive
Mammalia	Caucasian Squirrel (*Sciurus anomalus*)	Negative
Mammalia	Common Noctule Bat (*Nyctalus nyctala*)	Negative
Mammalia	Eastern Gray Squirrel (*Sciurus carolinensis*)	Positive
Mammalia	Geoffroy's Bat (*Myotis emarginatus*)	Negative
Mammalia	Lesser Noctule Bat (*Nyctalus leisleri*)	Negative
Mammalia	Red Squirrel (*Sciurus vulgaris*)	Negative and Positive
Mammalia	Savi's Pipistrelle (*Hypsugo savii*)	Negative

TABLE 6.2

Summary of documented behavioral interactions between native species and introduced parrots, except for Rose-ringed Parakeet (Sources: Menchetti and Mori 2014; White et al. 2019; this study). Species are listed in alphabetical order by common name, and only common names are listed.

OTHER SPECIES	Budgerigar	Coconut Lorikeet	Crimson Rosella	Fischer's Lovebird	Little Corella	Long-billed Corella	Mitred Parakeet	Monk Parakeet	Red-breasted Parakeet	Sulphur-crested Cockatoo	Superb Parrot	Tanimbar Corella	Turquoise-fronted Amazon	Yellow-collared Lovebird
Birds														
African Palm Swift				X										X
American Crow													X	
American Robin								X						
Black Noddy							X							
Blue Jay								X						
Common Blackbird								X						
Common Flameback										X		X		
Common Kestrel								X						
Common Myna		X												
Common Starling													X	
Honeyeater spp.		X												

TABLES

OTHER SPECIES	Budgerigar	Coconut Lorikeet	Crimson Rosella	Fischer's Lovebird	Little Corella	Long-billed Corella	Mitred Parakeet	Monk Parakeet	Red-breasted Parakeet	Sulphur-crested Cockatoo	Superb Parrot	Tanimbar Corella	Turquoise-fronted Amazon	Yellow-collared Lovebird
Hooded Crow								X						
House Sparrow	X			X				X						
Italian Sparrow								X						X
Laughing Kookaburra		X												
Little Wattlebird		X												
Mourning Dove	X													
New Holland Honeyeater		X												
Oriental Dollarbird									X				X	
Oriental Pied Hornbill													X	
Petrel spp.							X							
Pied Currawong		X												
Pink-necked Green Pigeon													X	
Purple Martin	X													
Red Wattlebird		X												
Rock Dove								X						
Shearwater spp.							X							
Western Jackdaw													X	
Yellow-vented Bulbul									X					
Other species of native parrots		X	X		X	X						X		
Mammals														
Black Rat	X													
Gould's Wattled Rat			X											
Red Squirrel											X			
Native Flora							X	X	X	X				

THE ECOLOGICAL IMPACTS OF INTRODUCED PARROTS

REFERENCES

Ancillotto, L., Strubbe, D., Menchetti, M., and Mori, E. 2015. An overlooked invader? Ecological niche, invasion success and range dynamics of the Alexandrine parakeet in the invaded range. *Biological Invasions* 18:583–595.

Ancillotto, L., Studer, V., Howard, T., Smith, V. S., McAlister, E., Beccaloni, J., Manzia, F., Renzopaoli, F., Bosso, L., Russo, D., and Mori, E. 2018. Environmental drivers of parasite load and species richness in introduced parakeets in an urban landscape. *Parasitology Research* 117:3591–3599.

Appelt, C. W., Ward, L. C., Bender, C., Fasenella, J., Van Vossen, B. J., and Knight, L. 2016. Examining potential relationships between exotic monk parakeets (*Myiopsitta monachus*) and avian communities in an urban environment. *Wilson Journal of Ornithology* 128:556–566.

Avery, M. L., and Shiels, A. B. 2018. Monk and rose-ringed parakeets. In *Ecology and Management of Terrestrial Vertebrate Invasive Species in the United States*, ed. W. C. Pitt, J. C. Beasley, and G. W. Witmer. Boca Raton, FL: CRC Press.

Batllori, X., and Nos, R. 1985. Presencia de la cotorrita gris (*Myiopsitta monachus*) y de la cotorrita de collar (*Psittacula krameri*) en el Area Metropolitana de Barcelona. *Miscellanea Zoologica* 9:407–411.

Blanco, G., Bravo, C., Pacifico, E. C., Chamorro, D., Speziale, K. L., Lambertucci, S. A., Hiraldo, F., and Tella, J. L. 2016. Internal seed dispersal by parrots: An overview of neglected mutualism. *PeerJ* 4:e1688.

Brichetti, P., and Fracasso, G. 2006. *Ornitologia Italiana*. Vol. 3: *Stercoraridae–Caprimulgidae*. Bologna, Italy: Oasi Alberto Perdisa Editions.

Çalışkan, O. 2018. Rose-ringed parakeets (*Psittacula krameri*) and geographical evaluation of habitats in Turkey. *International Journal of Geography and Geography Education* 38:279–294.

Cassey, P., Blackburn, T. M., Russel, G. J., Jones, K. E., and Lockwood, J. L. 2004. Influences on the transport and establishment of exotic bird species: An analysis of the parrots (Psittaciformes) of the world. *Global Change Biology* 10:417–426.

Castiglioni, R., Azzola, C., Capelli, F., and Biancardi, C. 2015. Aspetti ecologici e riproduttivi del parrocchetto monaco (*Myiopsitta monachus* Boddaert, 1783) in una colonia in provincia di Bergamo. *XVIII Congresso Italiano di Ornitologia*, 17. Caramanico Terme (Pescara), 17–20 Sept. 2015.

Chapman, T. 2005. The Status and Impact of the Rainbow Lorikeet (*Trichoglossus haematodus moluccanus*) in South-west Western Australia. Perth, Western Australia: Dept. of Agriculture and Food.

Clergeau, P., Vergnes, A., Delanoue, R. 2009. La perruche à collier *Psittacula krameri* introduite en Ile-de-France: Distribution et régime alimentaire. *Alauda* 77:121–132.

Covas, L., Senar, J. C., Roquè, L., and Quesada, J. 2017. Records of fatal attacks by rose-ringed parakeets *Psittacula krameri* on native avifauna. *Revista Catalana d'Ornitologia* 33:45–49.

Cramp, S. 1985. *Handbook of the Birds of Europe, the Middle East and North Africa: The Birds of the Western Palearctic*. Vol. 4: *Terns to Woodpeckers*. New York: Oxford Univ. Press.

Crowley, S., Hinchliffe, S., and McDonald, R.A. 2016. Invasive species management will benefit from social impact assessment. *Journal of Applied Ecology* 54:351–357.

Crowley, S. L., Hinchliffe, S., and McDonald, R. A. 2019. The parakeet protectors: Understanding opposition to introduced species management. *Journal of Environmental Management* 229:120–132.

Czajka, C., Braun, M., and Wink, M. 2011. Resource use by non-native ring-necked parakeets (*Psittacula krameri*) and native starlings (*Sturnus vulgaris*) in central Europe. *Open Ornithology Journal* 4:17–22.

Dangoisse, G. 2009. Étude de la population de Conures veuves (*Myiopsitta monachus*) de Bruxelles-Capitale. *Aves* 46:57–69.

Diamond, J. M., and Ross, M. S. 2019. Exotic parrots breeding in urban tree cavities: Nesting requirements, geographic distribution, and potential impacts on cavity nesting birds in southeast Florida. *Avian Research* 10:39.

Di Santo, M., Battisti, C., and Bologna, M. A. 2017. Interspecific interactions in nesting and feeding urban sites among introduced monk parakeet (*Myiopsitta monachus*) and syntopic bird species. *Ethology Ecology & Evolution* 29:138–148.

Drews, C. 2001. Wild animals and other pets kept in Costa Rican households: Incidence, species and numbers. *Society & Animals* 9:107–126.

Dubois, P. J. 2007. Les oiseux allochtones en France: Status et interactions avec les espèces indigènes. *Ornithos* 14:329–364.

Dyer, E. E., Cassey, P., Redding, D. W., Collen, B., Franks, V., Gaston, K., Jones, K. E., Kark, S., Orme, C.D.L., and Blackburn, T. M. 2017. The global distribution and drivers of introduced bird species richness. *PLoS Biology* 15:e2000942.

Falcón, W., and Tremblay, R. L. 2018. From the cage to the wild: Introductions of Psittaciformes to Puerto Rico. *PeerJ* 6:e5669.

Fletcher, M., and Askew, N. 2007. *Review of the Status, Ecology and Likely Future Spread of Parakeets in England*. York, UK: Central Science Laboratory.

Fogell, D. J., Martin, R. O., Bunbury, N., Lawson, B., Sells, J., McKeand, A. M., Tatayah, V., Tien Trung, C., and Groombridge, J. J. 2018. Trade and conservation implications of new beak and feather disease virus detection in native and introduced parrots. *Conservation Biology* 32:1325–1335.

Forshaw, J. M. 2010. *Parrots of the World*. London: Helm Field Guides, Christopher Helm Editions.

Galetti, M. 1993. Diet of the scaly-headed parrot (*Pionus maximiliani*) in a semideciduous forest in southeastern Brazil. *Biotropica* 25 419–425.

Gebhardt, H. 1996. Ecological and economics consequences of introductions of exotic wildlife (birds and mammals) in Germany. *Wildlife Biology* 2:205–211.

REFERENCES

Gerlach, H. 1994. Viruses. In *Avian Medicine Principles and Application*, ed. B. W. Ritchie, 862–948. 1st ed. Lake Worth, FL: Wingers Publishing.

Grandi, G., Menchetti, M., and Mori, E. 2018. Vertical segregation by ring-necked parakeets *Psittacula krameri* in northern Italy. *Urban Ecosystems* 21:1011–1017.

Ha, H. J., Anderson, I. L., Alley, M. R., Springett, B. P., and Gartrell, B. D. 2007. The prevalence of beak and feather disease virus infection in wild populations of parrots and cockatoos in New Zealand. *New Zealand Veterinary Journal* 55:235–238.

Haarsma, A. J., and Van der Graaf, C. 2013. Halsbandparkieten een bedreiging voor Rosse vleermuizen. *De Levende Natuur* 114:10–13.

Harris, A. 2015. Predation of Rose-ringed parakeets by raptors and owls in Inner London. *British Birds* 108:349–353.

Hernández-Brito, D., Blanco, G., Tella, J. L., and Carrete, M. 2020. A protective nesting association with native species counteracts biotic resistance for the spread of an invasive parakeet from urban into rural habitats. *Frontiers in Zoology* 17:13.

Hernández-Brito, D., Carrete, M., Ibáñez, C., Juste, J., and Tella, J. L. (2018) Nest-site competition and killing by invasive parakeets cause the decline of a threatened bat population. *Royal Society Open Science* 5:172477.

Hernández-Brito, D., Carrete, M., Popa-Lisseanu, A. G., Ibáñez, C., and Tella, J. L. 2014. Crowding in the city: Losing and winning competitors of an invasive bird. *PLoS One* 9:e100593.

Hernández-Brito, D., Luna, A., Carrete, M., and Tella, J. L. 2015. Introduced rose-ringed parakeets (*Psittacula krameri*) attack black rats (*Rattus rattus*) sometimes resulting in death. *Hystrix, the Italian Journal of Mammalogy* 25:121–123.

Hobson, E. A., Smith-Vidaurre, G., and Salinas-Melgoza, A. 2017. History of nonnative monk parakeets in Mexico. *PLoS One* 12:e0184771.

Howe, H. F., 1977. Bird activity and seed dispersal of a tropical wet forest tree. *Ecology* 58:539–550.

iNaturalist Alien Parrots Observatory. 2019. Accessed Dec. 2019. https://www.inaturalist.org/projects/alien-parrots-observatory.

Ivanova, I. M., and Symes, C. T. 2019. Invasion of *Psittacula krameri* in Gauteng, South Africa: Are other birds impacted? *Biodiversity and Conservation* 28:3633–3656.

Jackson, H., Jones, C. G., Agapow, P. M., Tatayah, V., and Groombridge, J. J. 2015 Micro-evolutionary diversification among Indian Ocean parrots: Temporal and spatial changes in phylogenetic diversity as a consequence of extinction and invasion. *Ibis* 157:496–510.

Kalodimos, N. P. 2013. First account of a nesting population of monk parakeets, *Myiopsitta monachus*, with nodule-shaped bill lesions in Katehaki, Athens, Greece. *Bird Populations* 12:1–6.

Keijl, G. O. 2001. Halsbandparkieten *Psittacula krameri* in Amsterdam, 1976–2000. *Limosa* 74:29–31.

Kelly, D. W., Paterson, R. A., Townsend, C. R., Poulin, R., and Tompkins, D. M. 2009. Parasite spillback: A neglected concept in invasion ecology? *Ecology* 90:2047–2056.

Krause, T. 2004. F1-und F2-Hybriden zwischen Alexandersittich *Psittacula eupatria* und Halsbandsittich *P. krameri* im Volksgarten in Dusseldorf. *Charadrius* 40:7–12.

Kumschick, S., and Nentwig, W. 2010. Some introduced birds have as severe an impact as the most effectual introduced mammals in Europe. *Biological Conservation* 143:2757–2762.

Lamont, D. A. 1996. The changing status of the rainbow lorikeet, *Trichoglossus haematodus* (Linnaeus 1771), in southwestern Australia: Its potential for range extension. MS thesis, Univ. of New England, Armadale, Australia.

Le Louarn, M., Clergeau, P., Briche, E., and Deschamps-Cottin, M. 2017. "Kill two birds with one stone": Urban tree species classification using bi-temporal pléiades images to study nesting preferences of an invasive bird. *Remote Sensing* 9:916.

Le Louarn, M., Clergeau, P., Strubbe, D., and Deschamps-Cottin, M. 2018. Dynamic species distribution models reveal spatiotemporal habitat shifts in native range-expanding versus non-native invasive birds in an urban area. *Journal of Avian Biology* 2018:e01527.

Le Louarn, M., Couillens, B., Deschamps-Cottin, M., and Clergeau, P. 2016. Interference competition between an invasive parakeet and native bird species at feeding sites. *Journal of Ethology* 34:291–298.

Lever, C. 1994. *Naturalized Animals: The Ecology of Successfully Introduced Species*. London: Poyser Natural History.

Lever, C. 2005. *Naturalised Birds of the World*. London: T. & A. D. Poyser.

MacGregor-Fors, I., Calderón-Parra, R., Meléndez-Herrada, A., López-López, S., and Schondube, J. E. 2011. Pretty, but dangerous! Records of non-native monk parakeets (*Myiopsitta monachus*) in Mexico. *Revista Mexicana de Biodiversidad* 82:1053–1056.

Mack, R. N., Simberloff, D., Lonsdale, W. M., Evans, H., Clout, M., and Bazzaz, F. A. 2000. Biotic invasions: Causes, epidemiology, global consequences, and control. *Ecological Appications* 10:689–710.

Martens, J., Hoppe, D., and Woog, F. 2013. Diet and feeding behaviour of naturalised Amazon parrots in a European city. *Ardea* 101:71–76.

Mazza, G., Tricarico, E., Genovesi, P., and Gherardi, F. 2014. Biological invaders are threats to human health: An overview. *Ethology Ecology & Evolution* 26:112–129.

McInnes, C. J., Coulter, L., Dagleish, M., Deane, D., Gilray, J., Percival, A., Willoughby, K., Scantlebury, M., Marks, N., Graham, D., et al. 2013. The emergence of squirrel pox in Ireland. *Animal Conservation* 16:51–59.

Menchetti, M., and Mori, E. 2014. Worldwide impact of introduced parrots (Aves Psittaciformes) on native biodiversity and environment: A review. *Ethology Ecology & Evolution* 26:172–194.

Menchetti, M., Mori, E., and Angelici, F. M. 2016. Effects of the recent world invasion by ring-necked parakeets *Psittacula krameri*. In *Problematic Wildlife: A Cross-Disciplinary Approach*, ed. F. M. Angelici, 253–266. Switzerland and New York: Springer International Publishing.

Menchetti, M., Scalera, R., and Mori, E. 2014. First record of a possibly overlooked impact by introduced parrots on a bat (*Nyctalus leisleri*). *Hystrix, the Italian Journal of Mammalogy* 25:61–62.

Mentil, L., Battisti, L., and Carpaneto, G. M. 2018. The impact of *Psittacula krameri* (Scopoli, 1769) on orchards: First quantitative evidence for southern Europe. *Belgian Journal of Zoology* 148:129–134.

Moltoni, E. 1945. Pappagalli in libertà nei giardini pubblici di Milano e loro nidificazione in colonia in associazione con il passero. *Rivista Italiana di Ornitologia* 15:98–106.

Mori, E., Ancillotto, L., Groombridge, J., Howard, T., Smith, V. S., and Menchetti, M. 2015. Macroparasites of introduced parakeets in Italy: A possible role for parasite-mediated competition. *Parasitology Research* 114:3277–3281.

Mori, E., Ancillotto, L., Menchetti, M., Romeo, C., and Ferrari, N. 2013a. Italian red squirrels and introduced parakeets: Victims or perpetrators? *Hystrix, the Italian Journal of Mammalogy* 24:195–196.

Mori, E., Ancillotto, L., Menchetti, M., and Strubbe, D. 2017a. "The early bird catches the nest": Possible competition between scops owls and ring-necked parakeets. *Animal Conservation* 20:463–470.

Mori, E., Cardador, L., Reino, L., White, R. L., Hernández-Brito, D., Le Louarn, M., Mentil, L., Edelaar, P., Pârâu, L. G., Nikolov, B. P., and Menchetti, M. 2020a. Lovebirds in the air: Trade patterns, establishment success and niche shift of *Agapornis* parrots within their non-native range. *Biological Invasions* 22:421–435.

Mori, E., Di Febbraro, M., Foresta, M., Melis, P., Romanazzi, E., Notari, A., and Boggiano, F. 2013b. Assessment of the current distribution of free-living parrots and parakeets (Aves: Psittaciformes) in Italy: A synthesis of published data and new records. *Italian Journal of Zoology* 80:158–167.

Mori, E., Grandi, G., Menchetti, M., Tella, J. L., Jackson, H. A., Reino, L., van Kleunen, A., Figueira, R., and Ancillotto, L. 2017b. Worldwide distribution of non-native Amazon parrots and temporal trends of their global trade. *Animal Biodiversity and Conservation* 40:49–62.

Mori, E., Malfatti, L., Le Louarn, M., Hernández-Brito, D., ten Cate, B., Ricci, M., and Menchetti, M. 2020b. "Some like it alien": Predation on invasive ring-necked parakeets by the long-eared owl in an urban area. *Animal Biodiversity and Conservation* 43:151–158.

Mori, E., Meini, S., Strubbe, D., Ancillotto, L., Sposimo, P., and Menchetti, M. 2018. Do introduced free-ranging birds affect human health? A global summary of known zoonoses. In *Invasive Species and Human Health*, ed. G. Mazza and E. Tricarico, 120–129. New York: CABI International Edition.

Mori, E., Monaco, A., Sposimo, P., and Genovesi, P. 2014. Low establishment success of introduced non-passerine birds in a central Italy wetland. *Italian Journal of Zoology* 81:593–598.

Mori, E., Onorati, G., and Giuntini, S. 2020c. Loud callings limit human tolerance towards invasive parakeets in urban areas. *Urban Ecosystems* 23:755–760.

Mori, E., Sala, J. P., Fattorini, N., Menchetti, M., Montalvo, T., and Senar, J. C. 2019. Ectoparasite sharing among native and invasive birds in a metropolitan area. *Parasitology Research* 118:399–409.

Morinha, F., Carrete, M., Tella, J. L., and Blanco, G. 2020. High prevalence of novel beak and feather disease virus in sympatric invasive parakeets introduced to Spain from Asia and South America. *Diversity* 12:192.

Nelson, C. B., Pomeroy, B. S., Schrall, K., Park, W. E., and Lindeman, R. J. 1952. An outbreak of conjunctivitis due to Newcastle disease virus (NDV) occurring in poultry workers. *American Journal of Public Health and the Nation Health* 42:672–678.

Nentwig, W., Bacher, S., Kumschick, S., Pyšek, P., and Vilà, M. 2018. More than "100 worst" introduced species in Europe. *Biological Invasions* 20:1611–1621.

Neo, M. L. 2012. A review of three introduced parrots in Singapore. *Nature in Singapore* 5:241–248.

Newson, S. E., Johnston, A., Parrott, D., and Leech, D. I. 2011. Evaluating the population-level impact of an invasive species, ring-necked parakeet *Psittacula krameri*, on native avifauna. *Ibis* 153:509–516.

Norconk, M. A., Grafton, B. W., Conklin-Brittain, N. L. 1998. Seed dispersal by neotropical seed predators. *American Journal of Primatology* 45:103–126.

Orchan, Y., Chiron, F., Shwarts, A., and Kark, S. 2012. The complex interaction among multiple invasive bird species in a cavity-nesting community. *Biological Invasions* 15:429–445.

Pârâu, L. G., Strubbe, D., Mori, E., Menchetti, M., Ancillotto, L., van Kleunen, A., White, R. L., Hernández-Brito, D., Le Louarn, M., Clergeau, P., et al. 2016. Rose-ringed parakeet *Psittacula krameri* populations and numbers in Europe: A complete overview. *Open Ornithology Journal* 9:1–13.

Peck, H. L., Pringle, H. E., Marshall, H. H., Owens, I.P.F., and Lord, A. M. 2014. Experimental evidence of impacts of an invasive parakeet on foraging behavior of native birds. *Behavioral Ecology* 25:582–590.

Peh, K.S.H. 2010. Invasive species in Southeast Asia: The knowledge so far. *Biodiversity and Conservation* 19:1083–1099.

Per, E. 2018. The spread of the rose-ringed parakeet, *Psittacula krameri*, in Turkey between 1975 and 2015 (Aves: Psittacidae). *Zoology in the Middle East* 64:297–303.

Pisanu, B., Laroucau, K., Aaziz, R., Vorimore, F., Le Gros, A., Chapuis, J. L., and Clergeau, P. 2018. *Chlamidya avium* detection from a ring-necked parakeet (*Psittacula krameri*) in France. *Journal of Exotic Pet Medicine* 27:68–74.

Pithon, J. A., and Dytham, C. 1999. Breeding performance of Ring-necked parakeets *Psittacula krameri* in small introduced populations in southeast England. *Bird Study* 46:342–347.

Popa-Lisseanu, A. G., Bontadina, F., Mora, O., and Ibáñez, C. 2008. Highly structured fission–fusion societies in an aerial hawking, carnivorous bat. *Animal Behaviour* 75:471–482.

Postigo, J. L. 2016. New records of invasive parakeet hybrids in Spain: A great opportunity to apply the rapid response mechanism. *European Journal of Ecology* 2:19–22.

REFERENCES

Postigo, J. L., Strubbe, D., Mori, E., Ancillotto, L., Carneiro, I., Latsoudis, P., Menchetti, M., Pârâu, L. G., Parrott, D., Reino, L., Weiserbs, A., and Senar, J. C. 2019. Mediterranean versus Atlantic monk parakeets *Myiopsitta monachus*: Towards differentiated management at the European scale. *Pest Management Science* 75:915–922.

Price, P. W., Westoby, M., Rice, B., Atsatt, P. R., Fritz, R. S., Thompson, J. N., and Mobley, K. 1986. Parasite mediation in ecological interactions. *Annual Review of Ecology and Systematics* 17:487–505.

Reed, J. E., McCleery, R. A., Silvy, N. J., Smeins, F. E., and Brightsmith, D. J. 2014. Monk parakeet nest site selection of electric utility structures in Texas. *Landscape and Urban Planning* 129:65–72.

Rehman, Q.S.U., Ali, K. W., Ali, W.S.M., Waqar, M., Muhammad, N., Abdul, S., and Ullah, K. A. 2019. Damage impact of vertebrate pests on different crops and stored food items. *GSC Biological and Pharmaceutical Sciences* 6:16–20.

Roll, U., Dayan, T., and Simberloff, D. 2008. Non-indigenous terrestrial vertebrates in Israel and adjacent areas. *Biological Invasions* 10:659–672.

Romeo, C., Wauters, L. A., Ferrari, N., Lanfranchi, P., Martinoli, A., Pisanu, B., Preatoni, D. G., and Saino, N. 2014. Macroparasite fauna of introduced grey squirrels (*Sciurus carolinensis*): Composition, variability and implications for native species. *PLoS One* 9:e88002.

Runde, D. E., Pitt, W. C., and Foster, J. 2007. Population ecology and some potential impacts of emerging populations of exotic parrots. In *Managing Vertebrate Invasive Species: Proceedings of an International Symposium*, ed. G. W. Witmer, W. C. Pitt, and K. A. Fagerstone, 338–360. Fort Collins, CO: USDA/APHIS Wildlife Services, National Wildlife Research Center.

Sa, R.C.C., Cunningham, A. A., Dagleish, M. P., Wheelhouse, N., Pocknell, A., Borel, N., Peck, H., and Lawson, B. 2014. Psittacine beak and feather disease in a free-living ring-necked parakeet (*Psittacula krameri*) in Great Britain. *European Journal of Wildlife Research* 60:395–398.

Şahin, D., and Arslangündoğdu, Z. 2019. Breeding status and nest characteristics of rose-ringed (*Psittacula krameri*) and Alexandrine parakeets (*Psittacula eupatria*) in Istanbul's city parks. *Applied Ecology and Environmental Research* 17:2461–2471.

Scortecci, G. 1953. Monaco (*Myiopsitta monachus*). In *Animali: Come sono, dove vivono, come vivono*. Milan: Labor Editors.

Senar, J. C., Domènech, J., Arroyo, L., Torre, I., and Gordo, O. 2016. An evaluation of monk parakeet damage to crops in the metropolitan area of Barcelona. *Animal Biodiversity and Conservation* 39:141–145.

Shiels, A. B., Bukoski, W. P., and Siers, S. R. 2018. Diets of Kauai's invasive rose-ringed parakeet (*Psittacula krameri*): Evidence of seed predation and dispersal in a human-altered landscape. *Biological Invasions* 20:1449–1457.

Shwartz, A., Strubbe, D., Butler, C. J., Matthysen, E., and Kark, S. 2008. The effect of enemy-release and climate conditions on invasive birds: A regional test using the ring-necked parakeet (*Psittacula krameri*) as a case study. *Diversity and Distributions* 15:310–318.

Strubbe, D., Matthysen, E., and Graham, C. H. 2010. Assessing the potential impact of invasive ring-necked parakeets *Psittacula krameri* on native nuthatches *Sitta europeae* in Belgium. *Journal of Applied Ecology* 47:549–557.

Styche, A. 2008. Distribution and behavioural ecology of the sulphur-crested cockatoo (*Cacatua galerita* L.) in New Zealand. MS diss., Victoria Univ. of Wellington, New Zealand.

Tatayah, R.V.V., Malham, J., and Haverson, P. 2007. The use of copper strips to exclude invasive African giant land-snails *Achatina* spp. from echo parakeet *Psittacula eques* nest cavities, Black River Gorges National Park, Mauritius. *Conservation Evidence* 4:6–8.

Tella, J. L., Baños-Villalba, A., Hernández-Brito, D., Rojas, A., Pacífico, E., Díaz-Luque, J. A., Carrete, M., Blanco, G., and Hiraldo, F. 2015. Parrots as overlooked seed dispersers. *Frontiers in Ecology and the Environment* 13:338–339.

Tella, J. L., Hiraldo, F., Donazar, J. A., and Negro, J. J. 1996. Costs and benefits of urban nesting in the Lesser Kestrel. In *Raptors in Human Landscapes: Adaptations to Built and Cultivated Environments*, ed. D. Bird, D. Varland, J. J. Negro, 53–60. London: Academic Press.

Turbé, A., Strubbe, D., Mori, E., Carrete, M., Chiron, F., Clergeau, P., Gonzalez-Moreno, P., Le Louarn, M., Luna, A., Menchetti, M., et al. 2017. Assessing the assessments: Evaluation of four impact assessment protocols for invasive introduced species. *Diversity and Distributions* 23:297–307.

Vall-llosera, M., Woolnough, A. P., Anderson, D., and Cassey, P. 2016. Improved surveillance for early detection of a potential invasive species: The introduced ring-necked parakeet *Psittacula krameri* in Australia. *Biological Invasion* 19:1273–1284.

Weiserbs, A., and Jacob, J. P. 1999. Etude de la population de perriche jeune-veuve *Myiopsitta monachus* à Bruxelles. *Aves* 36:207–223.

White, R., Strubbe, D., Dallimer, M., Davies, Z. G., Davis, A.J.S., Edelaar, P., Groombridge, J., Jackson, H. A., Menchetti, M., Mori, E., et al. 2019. Assessing the ecological and societal impacts of alien parrots in Europe using a transparent and inclusive evidence-mapping scheme. *Neobiota* 48:45-69.

Yosef, R., Zduniak, P., and Żmihorski, M. 2016. Invasive ring-necked parakeet negatively affects indigenous Eurasian hoopoe. *Annales Zoologici Fennici* 53:281–287.

Zocchi, A., Battisti, C., and Santoro, R. 2009. Note sul pappagallo monaco, *Myiopsitta monachus*, a Roma (Villa Pamphili). *Rivista Italiana di Ornitologia* 78:135–137.

DECISION-MAKING MODELS AND MANAGEMENT OF THE MONK PARAKEET

Juan Carlos Senar, Michael Conroy, and Tomás Montalvo

INTRODUCTION

Species in the order Psittaciformes have long attracted human interest because of their brilliant coloring, intelligence, sociality, and ability to vocalize. Because of this charisma, parrots and parakeets have been highly appreciated as pets for generations. Parrots were, for instance, among the first goods traded by American indigenous people to European explorers (Collar and Juniper 1992). With the rise of economic prosperity in North America and Europe in the 1960s, the parrot trade became a highly profitable activity, and thousands of birds entered the pet market in both regions (FAO 2011). In the 1970s, the trading of wild birds was immense. In 1975 alone, an estimated 7.5 million wild birds were traded. According to a survey carried out in 1988 in the United States (US), an estimated 5.2 million residences (6% of all residences) were home to approximately 13.9 million exotic birds, of which 43% were parakeets (Thomsen and Mulliken 1992). By the 1990s, the international bird trade was estimated at 5 million specimens per year (FAO 2011). This trend is even more evident in Monk Parakeets (*Myiopsitta monachus*); unlimited exploitation of this species was allowed based on its status as a pest in its native countries (Bucher 1992). Additionally, because the species was abundant, its sale price was low, which increased trade even more (Bull 1973). As a consequence, millions of wild-caught Monk Parakeets were transported from their native range to homes across the globe (Edelaar et al. 2015).

The next chapter in this story involves a number of mostly accidental escapes or small-scale releases (as with other species), which resulted in the establishment of new feral populations. These populations have increased exponentially in size since the 1970s, in both Europe and the US (van Bael and Pruett-Jones 1996; Avery et al. 2002; Domènech et al. 2003; Pruett-Jones et al. 2005; Conroy and Senar 2009; Postigo et al. 2019), and as a consequence, negative impacts related to these birds have also risen. Brightsmith and Kiacz (chap. 9 this vol.) and Mori and Menchetti (chap. 6 this vol.) review the most important potential damage caused by feral parakeets. In this chapter, we evaluate methods to mitigate these negative impacts. Additionally, we develop population dynamics and decision-making models in order to provide the best strategy for the control of these populations.

A DEMOGRAPHIC MODEL TO EXAMINE CONTROL METHODS ON POPULATION PARAMETERS

Science-based conservation and management of pest species requires the estimation of key population dynamics parameters, modeling, and decision-making analyses (Williams et al. 2002). Models are basic to articulating the biological response to different biological and control scenarios and to predicting biological response to management (Williams et al. 2002; Conroy and

Carroll 2009; Giunchi et al. 2012). In this section, we review the population dynamics parameters of the Monk Parakeet and develop a population dynamics model based on these parameters, which allows the prediction of the behavior of Monk Parakeet populations in the coming years under different management scenarios.

Survival Rate

Survival estimations of Monk Parakeets from Barcelona, Spain, within the invasive range of the species in Europe, yielded an adult survival rate of 80% (Conroy and Senar 2009). In the bird's native range in Argentina, annual survival of first-year birds was estimated at 61% and that of adults at 81% (Bucher et al. 1991). Values from the invasive and original range are therefore very similar.

Breeding Parameters

Data from South America indicate that only a proportion (37%–60%) of the adult Monk Parakeets typically breed in any one year (Bucher et al. 1991). In the invasive range in Barcelona, approximately 50% of juveniles that fledged the previous year bred as one-year-old birds (Senar et al. 2019). In the native range, breeding by yearling birds seems to be extremely rare (Bucher et al. 1991; Martín and Bucher 1993). The percentage of pairs engaged in second broods, either true second broods or replacement broods, was far greater in the invasive range than in the native range (56% ± 9.93% vs. 15%) (Navarro et al. 1992; Senar et al. 2019).

Mean clutch size for the first brood in Barcelona over five years was estimated at 5.0 ± 1.90 eggs, with a maximum clutch size of 16 eggs (Senar et al. 2019). Average clutch size in Florida was estimated at 5.2 ± 0.91 eggs (Avery et al. 2008, 2012). Clutch size in invasive areas in Brazil was 5.5 ± 1.6 eggs (Viana et al. 2016). Average clutch size for the second brood in Barcelona was significantly smaller than first clutch size (3.6 ± 1.63 eggs) (Senar et al. 2019). Clutch size within the native range in South America was not significantly different from that in invasive areas, for both first and second clutches (1st: 5.8 ± 0.90; 2nd: 4.0 ± 1.88) (Navarro et al. 1992; Peris and Aramburú 1995; Eberhard 1998).

Fledging success (i.e., productivity) in Barcelona for the first brood was estimated to be 3.3 ± 2.10 chicks per pair (see Table 7.1). The maximum number of fledglings produced by a pair from a single breeding attempt was 11. Fledging success from second broods in Barcelona was significantly lower (1.5 ± 1.61 chicks) than that of first broods (Senar et al. 2019). Fledging success during the first brood in the invasive range was substantially larger than in the native range (1.6 ± 0.53) (Navarro et al. 1992; Peris and Aramburú 1995). The lower fledging success in the native compared to the invasive range could be due to the greater incidence of nest predation in the native range (Navarro et al. 1992).

Population Growth

Population growth rate (r) is a key variable to define population dynamics, summarizing trends in population size over time. Values greater than 0 indicate that the population size is increasing, and smaller than 0 that it is decreasing (Sibly and Hone 2002; Conroy and Carroll 2009). The time interval for a population to double in size is also very illustrative of the trends in the population; it can be computed with $t = \ln(2)/r$, where r is the population growth rate.

Invasive Monk Parakeet populations display a typically exponential growth rate. Overall population growth rate in Spain from 1997 to 2002 was estimated at 0.20, with a doubling time of 3.5 years (Muñoz 2003). More recent estimations focusing on Mediterranean populations and including data up to 2015 indicate a population growth rate of 0.23, with a population doubling time of 3 years (Postigo et al. 2019). The growth rate in Italy is similar (0.20), with a population doubling time of 3.5 years, but values are considerably larger in Greece, where the population growth rate is 0.42 and population doubling time is just 1.7 years (Postigo et al. 2019). In North America, the values are similarly directed. The average annual rate of population growth in the US from 1971 to 1995 was 0.15, yielding a population doubling time of 4.8 years (van Bael and Pruett-Jones 1996). Overall population growth of Monk Parakeets in the US also fitted an exponential model of population growth (Pruett-Jones and Tarvin 1998). Similar patterns appear when we focus on more local scales. Population growth rate in Barcelona

TABLE 7.1

Summary of population parameter values used in the population growth model of Monk Parakeets. Data on mean number of offspring produced by each female (F), 1st and 2nd broods, has been modified from Senar et al. (2019) by adding data from 2019, which provides a better fit of the model.

PARAMETERS	VALUE	REFERENCE
Survival juveniles (annual rate)	0.61	Bucher et al. (1991)
Survival adults (annual rate)	0.81	Conroy and Senar (2009)
Longevity (years)	13	Senar, pers.obs.
Age of 1st reproduction	1	Senar et al. (2019)
% individuals that breed at age 1	0.50	Senar et al. (2019)
% individuals that breed at age 2	0.70	Pruett-Jones et al. (2007)
% individuals that breed at age 3	0.73	Pruett-Jones et al. (2007)
Age of F/M when breeding ceases	8	Pruett-Jones et al. (2007)
Maximum # offspring produced/year	11	Senar et al. (2019)
Mean # offspring produced by each F 1st brood	2.9	Senar et al. (2019)
SD in # offspring produced by each F 1st	1.28	Senar et al. (2019)
Mean # offspring produced by each F 2nd brood	1.2	Senar et al. (2019)
SD in # offspring produced by each F 2nd	0.3	Senar et al. (2019)
% 2nd broods per year	0.56	Senar et al. (2019)
Sex ratio at birth	1:1	Pruett-Jones et al. (2007)
Density dependence in reproduction	no	Pruett-Jones et al. (2007)
% adult F that breed each year	0.70	Pruett-Jones et al. (2007)

from 1975 to 2015 was estimated at 0.19, with a population doubling time of 3.6 years. Population growth also fitted to an exponential growth pattern (Fig. 7.1). The population growth rate in Madrid was estimated at 0.31, with a population doubling time of 2.3 years (Molina et al. 2016). The population growth rate in Málaga, Spain, from 1994 to 2017 was estimated at 0.15, with a population doubling time of 4.7 years (Postigo and Senar 2017).

Although all the European populations with a representative historical record are experiencing exponential growth, the Mediterranean countries are experiencing a higher exponential growth, greater spread rate, and more rapid colonization of new municipalities than Atlantic countries (Postigo et al. 2019). It is also important to note that agricultural populations in Israel grew faster ($r = 0.42$) than urban populations ($r = 0.21$) (Postigo et al. 2017), which stresses the risk that urban populations can spread into agricultural areas.

OUTLINE OF THE MODEL

We created a life-cycle model that simulates the growth of a long-lived bird, such as the Monk Parakeet, based on the population dynamics parameters previously described (Table 7.1), and with additional parameters reflecting the impact of control efforts on specific model parameters, and thus on population growth (Fig. 7.2). We assumed, similar to other studies, that the decisions at our disposal mostly included reducing survival of adults and juveniles and reducing productivity.

The model was initialized at $t = 0$ by specifying an initial population size (user-specified) and an initial age distribution (age_0) produced by running the model with our parameter values for Monk Parakeets (Appendix 7.1) but without controls until a stable age distribution was achieved.

OUTLINE OF THE MODEL

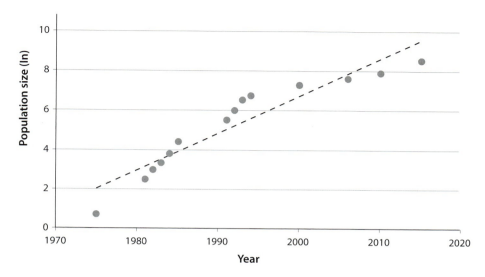

Figure 7.1. Population growth of the Barcelona Monk Parakeet population from 1975, when the first individuals were detected, to 2015. The growth fits an exponential function, with a slope (growth rate) of 0.19 (r^2 = 0.88, $p < 0.001$). Note that growth rate (slope = 0.31; doubling time = 2.2 years) from 1975 to 1995 (r^2 = 0.99, $p < 0.001$) was higher than from 2000 on (slope = 0.08; r^2 = 0.96, $p < 0.001$), which suggests slowing growth from 2000 on (doubling time = 8.4 years).

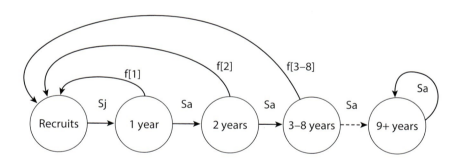

Figure 7.2. Life-cycle model used to simulate the growth of Monk Parakeet populations, based on population-dynamics parameters previously described in the chapter, and with additional parameters reflecting the impact of control efforts on specific model parameters and thus on population growth. *Sj* refers to survival rate of juveniles, and *Sa* refers to survival rate of adult birds (>1 year). The parameter *f* refers to fecundity, as number of fledged chicks, for the 1st, 2nd, and additional years of life of the female.

$Nm[0] = Nf[0] = {}^{N_0}\!/_2 \times age_0$

where $Nm[0]$ and $Nf[0]$ are vectors representing distribution in the nine age classes.

All parameter values (age-specific survival and recruitment parameters, initial age distribution and abundance) can be modified by other users as discussed below.

Recruitment of Individuals into the Breeding Population

Our model structure assumes that first-year, second-year, and third- through eighth-year birds have nonzero and different probabilities of breeding. In addition, we assume that first and second broods differ with respect to fecundity (average clutch size), with realized clutch sizes

$f_{1st\ brood}[t]$ and $f_{2nd\ brood}[t]$ drawn each year from a log-normal distribution with the specified means and a standard deviation based on field data in Barcelona, with the latter adjusted for proportion of females with second broods.

$$f.first[t] = \exp(\text{Normal}(\mu.first, \sigma.first)),$$

$$f.second[t] = \exp(\text{Normal}(\mu.second, \sigma.second)) \times p.second.brood, \ t = 1, \ldots, sim.years,$$

$$\mu.first = \log\left(\frac{mean.f.first.brood^2}{\sqrt{sd.f.first^2\ mean.f.first.brood^2}}\right),$$

$$\mu.second = \log\left(\frac{mean.f.second.brood^2}{\sqrt{sd.f.first^2 + mean.f.second.brood^2}}\right),$$

$$\sigma.first = \sqrt{\log(1 + sd.f.first.brood^2 + mean.f.first.brood^2)},$$

$$\sigma.second = \sqrt{\log(1 + sd.f.second.brood^2 + mean.f.second.brood^2)}$$

Recruitment in year *t* was then simulated as the sum of Poisson outcomes over the nine potential breeding ages:

$$R[t] = \sum_{i=1}^{9} \text{Pois}(Nf[i,t] \times p[i] \times (f_{1st\ brood}[t] + f_{2nd\ brood}[t]) \times p.breed),$$

where $p[i]$, $i = 1,\ldots,9$ are probabilities of breeding for each age class, and *p.breed* is the overall probability that a female breeds.

$$P_{1 \times 9} = [(p.breed.1 \quad p.breed.2 \quad p.breed.3 + \ldots p.breed.3 \quad 0)].$$

From this value, the control of recruits (if any) is included by

$$R[t] = R[t] - nest.rem \times eff.rem.nest - poking \times eff.poking,$$

where *nest.rem* and *poking* are, respectively, hours of nest removal and egg poking, and *eff.rem.nest* and *eff.poking* are efficiencies in terms of reductions in recruitment per hour of effort. To preclude negative recruitment values, the maximum of the above result and zero was taken as the value of *R[t]*.

Survival Process

Survival was taken as a binomial stochastic process, given assumed constant annual survival probabilities of S_j for recruits to one year, and S_a for birds one year and older. Thus, survival of recruits to one year of age was modeled as:

$$Nm[1,t+1] = \text{Binom}(R[t]/2, S_j),$$

$$Nf[1,t+1] = \text{Binom}(R[t]/2, S_j)$$

with the value $R[t]/2$ rounded to the nearest integer, and assuming a 50:50 sex ratio at hatching.

Before modeling the survival process of the remaining age classes to the next year, we applied controls (removal of adults), if any, as follows:

ad.removed = cannon × eff.cannon + trapping × eff.trapping + shooting × eff.shooting

where *cannon* (cannon netting), *trapping*, and *shooting* are the time efforts (hours) engaged in each method, multiplied by the respective efficiencies (birds removed per hour) of each. This quantity was then allocated to age and sex classes using the year t proportion in each group:

$N[t] = N[t] -$ ad.removed $\times c[t]$,

$c[t] = [Nm[t] \quad Nf[t]]/N[t]$,

$$N[t] = \sum_{i=1}^{9} Nm[i,t] + \sum_{i=1}^{9} Nf[i,t].$$

As with post-control recruitment, the elements of the resulting vector $N[t]$ were rounded to integer values and a lower limit of zero imposed on each. Survival was then modeled as:

$Nm[i+1, t+1] = Binom(Nm[i,t], S_a)$,

$Nf[i+1, t+1] = Binom(Nf[i,t], S_a)$, $i = 1,...,7$.

Finally, survival to the ninth (post-breeding) age class was modeled as:

$Nm[9, t+1] = Binom(Nm[8,t], S_a) + Binom(Nm[9,t], S_a)$

$Nf[9, t+1] = Binom(Nf[8,t], S_a) + Binom(Nf[9,t], S_a)$

that is, survivors to age class nine-plus includes survivors of animals that were age eight or nine-plus the previous year.

The effectiveness of removal efforts usually declines in response to reduced population density, because the animals become increasingly difficult to locate and are likely to be more wary of humans. Thus, actual culling effort may be greater than depicted by our models (Brook et al. 2003). To take that into account, and similar to Brook et al. (2003), we assumed that the same number of labor hours should be dedicated to the control program, to compensate for the effect that the absolute number of parakeets culled will decline as population density is suppressed.

We coded the simulation model in R v.3.5.1 (R Core Team 2018) and created an HTML interface using the Shiny package that allows users to select model inputs graphically. The model can also be run as stand-alone R code. Code and documentation are available via web links provided in Appendix 7.1.

PERFORMANCE OF THE MODEL

We ran the model assuming the population dynamics parameters summarized in Table 7.1 and an initial population size of 50 birds, which is approximately the number of Monk Parakeets estimated in Barcelona in 1984. According to the model, the population was predicted to grow to approximately 5,277 birds by 2015, with a population growth rate of 0.15 (Fig. 7.3). This is very close to the 5,078 birds found in the survey of 2015 and to the estimated growth rate of 0.19 computed for Barcelona (Fig. 7.1) and 0.20 for the whole of Spain (Muñoz 2003). This makes us confident that both the parameters in the model and the model itself are closely reflecting the population dynamics of Monk Parakeets in Spain and, by extension, in most invaded areas in Europe and North America (see "Population Growth," above).

Previous population dynamics models on the population of Monk Parakeets in the US did not fit true population-growth data, as it appeared to grow faster than predicted based on life history data obtained for the species in Argentina (Pruett-Jones et al. 2007). We now know that the breeding potential of the species in invaded areas is far greater than in native areas (Senar et al. 2019), and our modeling shows that by using these new parameters, we can more accurately describe the fate of invading populations. This will allow us to accurately model, in the next section, the response of Monk Parakeets to different management scenarios.

DECISION-MAKING MODELS AND MANAGEMENT OF THE MONK PARAKEET

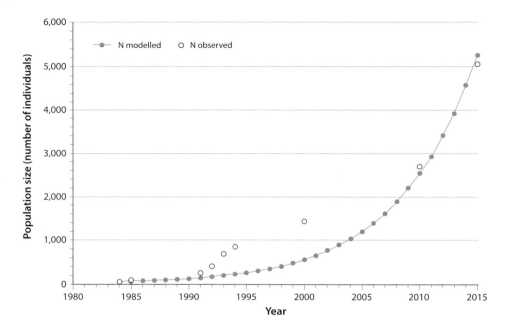

Figure 7.3. Population growth of Monk Parakeets in Barcelona from 1984 to 2015 (filled circles) according to the life-cycle model that simulates the growth of the population based on the population dynamics parameters previously described (Table 7.1). We provide for comparison the observed population size of Monk Parakeets (open circles) in Barcelona according to survey data (see Fig. 7.1).

ALTERNATIVE CONTROL MEASURES: ADVANTAGES, DISADVANTAGES, AND IMPACTS

Local vs. Regional Actions

The aim of wildlife managers may sometimes be to avoid the presence of Monk Parakeets in specific areas or to prevent placement of their nests on specific structures. This can clearly be the case when the species damages human utilities. However, visual scare devices such as models of owls, rubber snakes, taxidermic Monk Parakeet effigies, scare-eye balloons, handheld red lasers, and loud noises are mostly ineffective in deterring Monk Parakeets (Avery et al. 2002; Avery and Shiels 2018). Modifying certain structural components of transmission lines or substations can make utility structures less suitable for breeding, reducing the frequency of nesting (Avery et al. 2002; Reed et al. 2014).

Similarly, in order to reduce crop damage, some suggestions for local management of agricultural areas have been provided, such as increasing plant density, synchronizing planting times of different crops within the region, moving the harvest date forward to decrease exposure of crops to foraging birds, or planting non-preferred crops on the borders of the fields (Canavelli et al. 2012, 2014). However, we believe that such local actions simply shift the problem from one location to another, and although they may be efficient in some specific situations, this is not a solution to the problem. For this reason, the rest of the review focuses on global approaches that reduce the problem at a larger scale.

Culling: Captures at the Nest

The capture of adult Monk Parakeets can be done at the nest during the night using a long-handled net to cover the nest opening and to catch the birds as they fly out of the nest (Martella and Navarro 1987; Avery and Shiels 2018). When applied at transmission lines in Florida, the method was quite effective, with a capture success of 51% of individuals flushing from the nest (Tillman et al. 2004). The method, however, is quite expensive, with an estimated cost per nest of $1,000 (Avery et al. 2002). An additional problem of capturing birds at the nest is that since Monk Parakeets nest colonially, the disturbance at the first nest causes other birds from nearby nests to leave to avoid capture (Avery and Shiels 2018). This is also the case in urban areas, where the distress calls of the first individuals captured alert other individuals in the colony (Esteban 2016). Additionally, in urban areas, street lighting allows parakeets to easily detect people or a net approaching the nest. This makes this method largely inapplicable in most Monk Parakeet management scenarios. A promising alternative is the use of spring traps at the entrances of nest chambers closed by remote control at night, but this has not been fully tested, and no data on costs and efficiency are available.

Culling: Captures by Traps or Nets

Trapping around the nests has been suggested as a control alternative (Fitzwater 1988). Two main kinds of traps have been described: (1) passive traps, which the birds enter by themselves, attracted by decoys or food, and are unable to exit, so that no presence of managers is needed; and (2) active traps, in which the manager must activate doors to capture the birds.

Modified crow drop-in traps have been successful in trapping Rose-ringed Parakeets (*Psittacula krameri*) in Pakistan, with an estimated average efficiency of 7.2 birds/trap day (Bashir 1979). This trap has also been successful in Germany (R. Jonker, pers. comm.). However, this trap has proven to be totally ineffective for capturing Monk Parakeets, even when using live decoys (Avery et al. 2002; Tillman et al. 2004). Funnel traps have been useful for trapping Australian Ringnecks (*Barnardius zonarius*) in Australia (Morgan and McNeely 2000). However, the capture rate of the funnel trap with live decoys was less successful with Monk Parakeets (14 birds over 127 trap days), because the birds were easily able to exit the traps (Tillman et al. 2004). Some attempts to use funnel traps in Barcelona showed that, although Rose-ringed Parakeets were easily captured, Monk Parakeets quickly learned to exit the traps (J. C. Senar, pers. obs.). We have also tried traps with one-way swinging doors designed for pigeons (Krebs 1974), but Monk Parakeets soon learn not to enter the trap when it is active and even learn to exit the traps by pulling the doors. We have also tried cage traps, with doors that close when the birds enter the trap, which have been useful for trapping Eurasian Magpies (*Pica pica*) (Díaz-Ruiz et al. 2010). This technique proved successful during the first trapping day, but the birds quickly became trap shy, and they learned to open the doors of the cage (Esteban 2016).

Remotely triggered traps mounted on platform feeders have been used successfully to capture Monk Parakeets attracted by bait (Avery and Shiels 2018), and despite the considerable time commitment, active trapping seem to be more efficient than passive trapping (Tillman et al. 2004). In Barcelona over the past 18 years, we have been using a modified Yunick trap (Yunick 1971), trapping about 1,500 individuals and obtaining >2,000 recaptures. Translating this into a measure of efficiency, we estimate that with an active trap we can capture 1.6 Monk Parakeets per hour, at a cost of €30 per hour (Table 7.2). When modeling the fate of the population of Barcelona in the next 10 years, with a trapping effort of 300 hours per year and a starting population size of about 5,000 birds (as recorded in 2015), we were able to reduce the population growth rate from 0.15 to 0.10, but the population still doubled in size in 6.9 years (Table 7.3). The method is therefore very inefficient compared to other methods (Fig. 7.4). An additional technical problem was the number of different traps that needed to be set. Because the home range of Monk Parakeets is quite small, generally about 80 ha (Shields 1974; Faus et al. 2010), any given trap normally only captured birds in a radius of 350 m (Senar and Carrillo-Ortiz 2007), with capture efficiency reduced over time as a result of the birds avoiding the traps (Tillman et al. 2004).

Netting parakeets with mist nets in agricultural settings has been largely unsuccessful because the birds become accustomed to the nets, and the operations are costly and time consuming (Schwab and Gwynn 1992; Avery and Shiels 2018). In Barcelona, we have used gas-propelled cannon nets (or clap nets) without much success.

DECISION-MAKING MODELS AND MANAGEMENT OF THE MONK PARAKEET

TABLE 7.2

Efficiency (number of birds captured per hour) and cost (€ to capture one bird and € to capture for one hour) of the main methods used to manage Monk Parakeet populations. Capturing birds with a trap has the additional cost of the time to set a trap and bait it over months; thus, the actual cost of trapping is higher than estimated here and depends on the number of traps used, etc.

METHOD	CAPTURE RATE (# BIRDS/HOUR)	COST/BIRD	COST/HOUR	DATA FROM	YEARS
Trapping	1.6	~$36 (€30)	~$36 (€30)	Barcelona	2004–18
Cannon netting	0.7	~$345 (€291)	~$298 (€198)	Barcelona	2016–17
Shooting	5.2	~$13 (€11)	~$65 (€55)	Zaragoza	2015–16
Removing nests	8.8	~$27 (€23)	~$242 (€204)	Barcelona	2015–17
Poking eggs	10.3	~$24 (€19)	~$231 (€195)	Zaragoza	2013–15

TABLE 7.3

Variation in population increase rate (r) based on different control methods. The model assumes an initial population size of 5,000 birds, an effort of 300 hours, and a period of control of 10 years.

METHOD	R	FINAL POPULATION SIZE (10 YEARS)
No action	0.15	22,682
Trapping	0.10	13,914
Cannon netting	0.14	19,359
Shooting	−0.69	0
Nest removal	−0.15	1,130
Egg poking	−0.18	861

The method captures 0.7 Monk Parakeets per hour of effort, at a cost of €291 per bird and €198 per hour, which is very expensive (Table 7.2). The reason for this high cost is that, as with traps, the area has to be pre-baited for 10–15 days to attract birds to the area, which requires significant time and personnel. When modeling the fate of the population of Barcelona over the next 10 years, with a standard effort of 300 hours per year, we could reduce population growth rate slightly, from 0.15 to 0.14, but the population doubling time was still an unacceptable five years (Table 7.3). For this reason, we do not advise the use of these methods, as they have a negligible effect on the population.

TABLE 7.4

Variation in population increase rate (r) and final population size after 10 years of control by shooting according to different culling efforts (hours). We assume an initial population size of 5,000 individuals. We also provide the expected number of years to achieve eradication (if possible), the number of individuals that should be culled during the 1st year, and the percentage of individuals this number implies in relation to the total number of individuals in the population.

SHOOTING EFFORT (H)	R	FINAL POPULATION SIZE (10 YRS)	YEARS TO ERADICATE POPULATION	INDIVIDUALS CULLED 1ST YEAR	% INDIVIDUALS IN POPULATION
0	0.15	22,682			
100	0.10	12,529		540	11
200	−0.07	1,827		1,080	22
300	−0.69	0	9	1,620	32
400	−1.38	0	6	2,160	43
500	−1.32	0	4	2,700	54

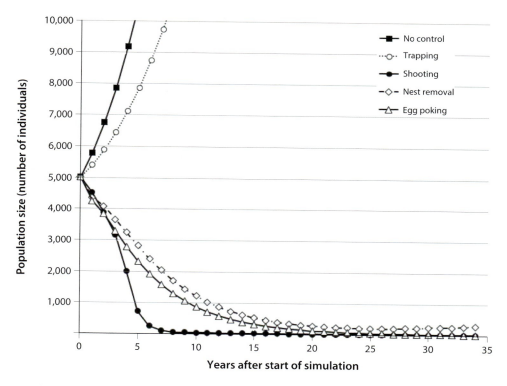

Figure 7.4. Modeling of Monk Parakeet populations across years according to various possible control methods. The simulation was begun assuming an initial population size of 5,000 individuals. The x-axis shows the projected years after the start of the simulation.

Shooting

Shooting has been a common method to control parakeets in many areas (Fitzwater 1988; Orueta 2007; Avery and Shiels 2018). Although some reports state that shooting did not effectively reduce populations or suppress population growth in the Monk Parakeet's native range (Petersen and Grasso 2010), the method was reported to be successful in Uruguay (Bruggers et al. 1998). In 1981–82, the Uruguayan government carried out a lethal control campaign over an area of >500,000 km². This involved eight people for a year that killed approximately 250,000 birds at a cost about $150,000 ($0.60 per bird) (Linz et al. 2015). The method was also effectively used in Mallorca (Orueta 2007) and the US (Neidermyer and Hickey 1977; Avery and Shiels 2018). The most recent and clear example of success of shooting comes from Zaragoza, Spain, where the city council successfully eliminated the 1,500 Monk Parakeets in the urban area within two years (2015–16) (Esteban 2016). In Mallorca, the method also allowed the removal of most of the 500 individuals recorded in 2011 (Víctor Colomer, pers. comm.). When modeling the fate of the population of Barcelona over the next 10 years, with a standard effort of 300 hours of shooting per year, we get a negative population growth rate that moves from 0.15 (with no action) to -0.69. In this way, the population is eradicated in less than 10 years (Tables 7.3 and 7.4, Fig. 7.4). We can also see that the population growth rate becomes negative, and hence the population decreases, when effort results in the culling of >25%–30% of the individuals in the population (Table 7.4).

Shooting is clearly the most effective method to reduce Monk Parakeet populations. Intensive shooting has also proven to be effective in the control of the Australian Ringneck (Morgan and McNeely 2000). We therefore strongly advise this method to effectively control Monk Parakeets, in both rural and urban habitats.

Toxic Substances

Nest poisoning in Monk Parakeets can potentially reduce the population size of the species (Bucher 1992). Since 1980, agencies in Argentina and Uruguay have managed parakeet populations by smearing a mixture of grease and a toxic insecticide (e.g., carbofuran) around the nest openings. The birds die from ingesting the toxicant as they preen the paste from their feathers (Linz et al. 2015). However, later monitoring showed that the populations soon recovered, forcing costly new campaigns (Linz et al. 2015). Spraying nests with an endrin solution has also been used with some success in Uruguay and Argentina (Mott 1973). Other lethal control alternatives include chemical control, such as DRC-1339 (Starlicide). Approved by the US Environmental Protection Agency, this toxicant is used to control starlings and other problem bird species, but its efficacy on Monk Parakeets is not established (Newman et al. 2008). Despite the effectiveness of the toxins, all these methods are generally unacceptable because of the negative impacts on nontarget species, which occur either when they use empty parakeet nests or consume carcasses of dead parakeets (Bruggers et al. 1998; Linz et al. 2015). The method is also expensive and time consuming (Newman et al. 2008).

Biological Control

Another possible lethal approach to population control is the selective application of an endemic protozoan parasite. *Sarcocystis falcatula* is a protozoan parasite that cycles between Virginia Opossums (*Didelphis virginiana*) and several bird species, and although it is apparently not harmful to native bird species, it is lethal to psittacines. However, effective doses have yet to be developed, and it is still necessary to evaluate a selective delivery procedure so that only Monk Parakeets will be affected by field application of this control method (Avery et al. 2002). Additionally, more recent work has found the Monk Parakeet resistant to this protozoan parasite (Newman et al. 2008). Thus, the method should be further evaluated, especially when used in invasive areas where it could affect native species.

Employees of Florida Power and Light prepare to remove a large, compound nest of Monk Parakeets (*Myiopsitta monachus*) on a transmission tower. Fort Lauderdale, Florida, US, November 2002. Photo by James Lindsay.

Monk Parakeets (*Myiopsitta monachus*) can chew through perch guards on electrical structures, as shown here. These guards were installed to manage where raptors, particularly vultures, perch on transmission structures. The guards were designed to minimize guano contamination of the conductors, which can cause electrical flashovers, equipment damage, and micro-outages. Dade County, Florida, US, December 2003. Photo by James Lindsay.

REDUCTION OF REPRODUCTIVE SUCCESS

Removing Nests

An apparently easy way to reduce Monk Parakeet breeding success is to remove the nests while the birds are breeding. Nest removal from electrical utility structures was estimated in the US to cost $415 to $1,500 per nest (Avery et al. 2008). In Barcelona, we estimated the average cost to be €204 per hour, with a cost of €23 per Monk Parakeet removed; the effectiveness was 8.8 Monk Parakeets removed per hour (Table 7.2). However, these are not adult birds but chicks, and with this method, breeding success is reduced. When modeling the fate of the 5,000 birds of the population of Barcelona over the next 10 years, with a standard effort of 300 hours of nest removal per year, we observed a reduction in population increase of -0.15, with a population size after 10 years of 1,130 birds (Table 7.3, Fig. 7.4).

Besides being quite expensive because it requires a cherry-picker crane and specialized personnel, nest removal has other drawbacks. Monk Parakeets are highly persistent in returning to build nests on the same supports despite nest removal (Burger and Gochfeld 2009), and several studies have shown that removed nests are rebuilt within a few weeks (Bucher and Martin 1987; Monzón 1997; Avery et al. 2002; Avery and Shiels 2018), sometimes starting on the first day after removal (Burger and Gochfeld 2009). Nest removal can also induce birds to disperse to begin new nesting colonies (Avery et al. 2002). Additionally, since many chambers have been abandoned and are empty, we end up directing a significant effort in the removal of nests that are not in use. In Argentina, about 70% of nests are not used (Eberhard 1998). In Barcelona, this number is about 50% (Domènech et al. 2003). Because of these drawbacks, we do not believe this control method is a good option for reducing Monk Parakeet population size.

Removing, Poking, and Oiling Eggs

Removing eggs, or poking or oiling them, is another way to reduce breeding success. Egg poking or oiling is better than egg removal, because the parakeets continue incubating the eggs and do not start replacement clutches. The method is obviously cheaper than removing nests, and based on data from Barcelona, it may cost about €195 per hour (€19 per egg removed) (Table 7.2). The method is slightly more effective than removing nests, in that it allows for the elimination of about 10.3 eggs per hour. When modeling the fate of the population of Barcelona in the next 10 years, with a standard effort of 300 hours of egg poking per year, we observed a reduction in population increase of -0.18, but population size after 10 years was still at 861 birds (Table 7.3, Fig. 7.4). This low level of effectiveness was in fact observed in Zaragoza, where 10,500 eggs were poked from 2006 to 2014, without any reduction in population size (Fig. 7.5) (Esteban 2016). Another drawback of the method is that much effort may be wasted in visiting nests that are either abandoned (as in nest removal) or in which egg laying has not yet started. The time window during which egg poking is feasible in invasive areas lasts about two to three months (most egg laying occurs during March, and incubation lasts about one additional month). In a population like that in Barcelona, with an estimated 3,500 nesting chambers in 2015, 1,750 (50%) of which were occupied (Pascual et al. 2015), poking all the eggs would require a minimum 900 hours of work, assuming no revisits. This would require three teams working simultaneously during the two- to three-month window, and about 40 years to achieve population eradication (Table 7.5). For all these reasons, we do not advise this method to control Monk Parakeets.

Contraception

Another way to reduce breeding success is to use contraceptives. DiazaCon, as the active ingredient, in a dose less than 50 mg/kg body weight, has been used successfully to reduce Monk Parakeet breeding success (Yoder et al. 2007; Avery and Shiels 2018). DiazaCon is a promising contraceptive tool, since it needs to be consumed for only 5–10 days to affect reproduction for the length of a breeding season

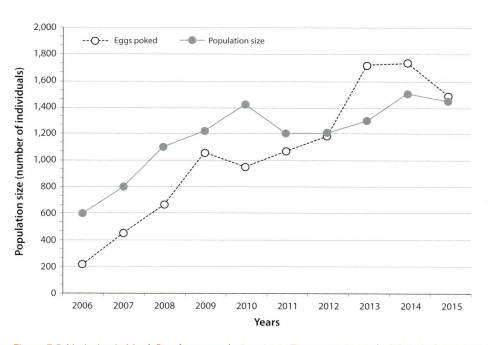

Figure 7.5. Variation in Monk Parakeet population size in Zaragoza, Spain (solid circles), in relation to the number of eggs poked each year (open circles). Data from Esteban (2016 and pers. comm.).

TABLE 7.5

Variation in population increase rate (r) and final population size of Monk Parakeets after 10 years of control by egg poking based on different culling efforts (hours). We assume an initial population size of 5,000 individuals. We also provide the expected number of years to achieve eradication (if possible), the number of eggs that should be poked during the 1st year, and the percentage of eggs and nests (chambers) this number implies in relation to the total number of eggs and nests in the population.

EGG POKING EFFORT (H)	R	FINAL POPULATION SIZE (10 YRS)	YEARS TO ERADICATE POPULATION	EGGS POKED/YEAR	NESTS VISITED/YEAR	% NESTS VISITED
0	0.15	22,682		0		
100	0.09	11,623		1,030	206	15
200	−0.10	2,430	>50	2,060	412	29
300	−0.18	826	43	3,090	618	44
400	−0.20	665	42	4,120	824	59
500	−0.21	621	41	5,150	1,030	74
600	−0.21	617	41	6,180	1,236	88
700	−0.21	613	41	7,210	1,442	103

(Yoder et al. 2007). The use of DiazaCon caused a clutch-size reduction close to 60% (1.6 vs. 3.9 eggs per clutch) and a 100% reduction in hatching success (Yoder et al. 2007; Avery et al. 2008). Potential constraints to using DiazaCon are its possible effect on nontarget native species (Avery et al. 2008; Petersen and Grasso 2010) and the fact that it must be continually reapplied to food items for effectiveness (Petersen and Grasso 2010). DiazaCon may also be associated with some adverse health effects and mortality in Monk Parakeets (Yoder et al. 2007). Thus, fertility-control drugs should be tested for their effects on population dynamics of the target species rather than just on reproductive output (Bomford and O'Brien 1992). Given that contraception is, from a population dynamics and effectiveness perspective, very similar to egg poking, and given that it has additional drawbacks related to its implementation, we do not advise this method to reduce the spread of the Monk Parakeet.

LIMITING FACTORS

An alternative method to culling and impeding reproduction may be to act on the ecological factors that limit population growth. Used to control feral pigeons (*Columba livia domestica*), for example, this method has allowed important reductions in population size by reducing food supply, educating the general public about food provisioning, and closing building holes needed by pigeons to breed (Haag-Wackernagel 1993, 1995; Giunchi et al. 2012; Stock and Haag-Wackernagel 2016; Senar et al. 2017). As with pigeons, the two most likely limiting factors in Monk Parakeets are nest-tree availability and food (Bull 1973; Bruggers et al. 1998; Rodríguez-Pastor et al. 2012). Although tree removal is obviously not a solution, it has been suggested that specific trees be selected in urban planning (Romero et al. 2015) or that selective pruning be applied to preferred trees (Volpe and Aramburú 2011) to reduce the availability of suitable nest substrate. However, we think this is quite unfeasible. Additionally, Monk Parakeets show great flexibility regarding nesting requirements and can use any kind of structure, even artificial ones, for building nests (Avery and Shiels 2018).

Limiting food availability may be more effective than removing tress. The diet of Monk Parakeets is highly supplemented by food provided by humans, either in feeders in backyard gardens (Bull 1973; Neidermyer and Hickey 1977; Hyman and Pruett-Jones 1995; South and Pruett-Jones

2000; Burger and Gochfeld 2009) or by people providing food directly as bread or seeds (Weiserbs and Jacob 1999; Carrillo-Ortiz 2009). Stable isotopic analyses have shown that food of human origin may account for up to 42% of total ingested food (Borray-Escalante et al. 2020). In fact, abundance and distribution of Monk Parakeets in Barcelona have been highly associated with the presence of people over 65 years of age, the demographic that accounts for most food provisioning to parakeets (Rodríguez-Pastor et al. 2012). As a consequence, it may be very important to establish educational programs to convince the public that controlling this exotic species is a high-priority conservation issue and that they should not feed these birds (MacGregor-Fors et al. 2011; Rodríguez-Pastor et al. 2012; Borray-Escalante et al. 2020). Specific monitoring should be carried out to evaluate whether this approach has a real effect on population size or whether the birds alternatively shift their diet to other food sources.

UNDERSTANDING THE DECISION PROBLEM

A decision model requires three components: (1) a quantitative statement of the objective, (2) delineation of decision alternatives, and (3) a model relating the decision alternatives or controls to the objective. The task is then to determine the combination of decisions that best meet the resource objective, taking into account biological and economic constraints (Conroy and Senar 2009). Here, we develop a population dynamics model that, taking into account the biological parameters of the population and the efficiency of the different methods, allows us to forecast the effect that the different control methods would have on the population.

The Optimization Model

We used a simulation-optimization approach to find quasi-optimal combinations of control levels to achieve desired reductions in abundance while minimizing costs. We computed monetary costs of combinations by means of a cost function that incorporated as inputs the specified number of hours of effort for each control and costs per hour specified as a parameter value. Total costs were then summed over years of control for a specified combination of hours of control of each type and years of control. We selected up to five levels of each control and number of years of control, for up to 15,625 combinations. For each combination, we simulated population growth under the control regime and calculated total costs to achieve the specified level of reduction and averaged these costs over replicated simulations. Finally, we compared averaged costs for scenarios that achieved the reduction goal and selected the scenario as "optimal" that minimized total cost. We coded the optimization and ranking functions using R v. 3.5.1, referencing the simulation-control model described above. The optimization function is run from stand-alone R code rather than an interface. The code to run the optimization and summarize the results is available in Appendix 7.1.

Results from the Optimization Decision Model

Population models were simulated for 10 years, under the assumption that this reflects a reasonable duration for an actual control program, similar to the approach of Pruett-Jones et al. (2007) and Brook et al. (2003). We focused on methods for which we had estimations of efficiency and costs (Table 7.2). Optimization analysis provides the best method (or combination of methods) to control the population, within a time period, additionally providing the estimated cost.

Analyses showed that for very small population sizes (around 100 birds), trapping would be feasible, provided the birds are so highly concentrated in an area that we would not need to set more than one trap. In this case, the population could be eradicated in one year (Table 7.6). For larger populations, the best method is clearly shooting, and the time needed for eradication would increase to seven years (Table 7.6). For very large population sizes (10,000-plus birds), shooting still would be the best method, but time needed for eradication would increase to 14 years (Table 7.6).

TABLE 7.6

Results from the simulation-optimization approach to find the best combinations of control levels to achieve Monk Parakeet eradication while minimizing costs. We provide data for different starting population sizes. Years refers to the number of years to achieve eradication according to that population size; otherwise the numbers refer to the number of hours to be invested per year for each main control method, and the estimated total cost of the control (based on estimations from Table 7.2).

STARTING POPULATION	YEARS	NEST REMOVAL	POKING	CANNON NET	TRAPPING	SHOOTING	TOTAL COST
100	1	0	0	0	125	0	~$5,400 (€4,500)
500	1	0	0	0	0	125	~$8,225 (€6,875)
1,000	1	0	0	0	0	333	~$22,000 (€18,315)
5,000	7	0	0	0	0	333	~$154,000 (€128,205)
10,000	14	0	0	0	0	500	~$460,600 (€385,000)

GENERAL DISCUSSION

Adequacy of the Population Dynamics Model

With other invasive species, the usefulness of population models in predicting outcomes and assessing cost-effectiveness of control strategies is limited by the lack of detailed demographic information, which constrains the development of realistic models (Govindarajulu et al. 2005). This has been true of the Monk Parakeet for many years. Because the Monk Parakeet was previously thought to make only seasonal breeding attempts, and juveniles in their first year were believed not to breed, the species had been regarded as lacking the characteristics of an efficient bird pest (Linz et al. 2015). However, this optimistic prediction has not not proved true in naturalized populations (Postigo et al. 2019). Previous population dynamics models on the population of Monk Parakeets in the US did not fit the population-growth data, in that it appeared to be growing faster than predicted based on life history data obtained for the species in Argentina (Pruett-Jones et al. 2007). Clearly, the problem was that accurate demographic information from invaded areas was lacking. Detailed data from Europe has shown that Monk Parakeets in fact have a high breeding capacity (Senar et al. 2019), and our model including these parameters was highly successful in describing the population growth of Monk Parakeets in Barcelona in the past 30 years. This has allowed us to accurately model different control scenarios. We now briefly discuss these scenarios.

Analysis of Different Control Scenarios

It has been suggested that, in general, lethal management of granivorous bird populations has shortcomings, such as low cost-effectiveness, challenging logistics, public resistance, and potential environmental risks, especially to nontarget birds (Linz et al. 2015). For example, in spite of the removal of more than 60,000 Australian Ringnecks by trapping, no regional-scale reduction in the parrot's numbers or the extent of damage was observed in southwestern Western Australia (Morgan and McNeely 2000). This generally led to the view that the effort likely necessary to reduce population growth of Monk Parakeets was probably impractical (Pruett-Jones et al. 2007). Nevertheless, the success of programs carried out by the US Fish and Wildlife Service, the Zaragoza city council, and the Balearic Islands government indicates that the species can be effectively controlled through the eradication of birds at their colonies (Neidermyer and Hickey 1977; van Bael and Pruett-Jones 1996; Conroy and Senar 2009; Esteban 2016; Postigo et al. 2019). Our modeling work supports this view.

Having established that control of these populations is possible and necessary, the aim now should be to determine the best method to attain this control. From a social point of view, farmers generally prefer reproductive and lethal control of Monk Parakeets (Canavelli et al. 2013). Nevertheless, in urban areas, management can encounter opposition from the public, and this should be taken into account (Crowley et al. 2018). From a technical point of view, however, the results are very clear: both our current modeling approach and previous work support the view that culling of adult birds is the most efficient method (Pruett-Jones et al. 2007; Conroy and Senar 2009). We will now discuss the key results from a broader perspective.

Reproductive output can be lowered by destroying eggs, providing chemosterilants, or destroying nests (Pruett-Jones et al. 2007). However, both our models and those previously published clearly show that none of these approaches achieves population extinction within a reasonable period of time (Pruett-Jones et al. 2007; Conroy and Senar 2009). Additionally, egg-control operations or nest removal may be difficult because of the high synchronicity in breeding phenology of the species and the need to apply control within the last two weeks of the breeding period to optimize the effort and to reduce time left to the parakeets for rebreeding (Conroy and Senar 2009). Hence, and in accordance with our models, reducing recruitment and breeding is not a good strategy.

Previous work has shown that the most rapid decrease in population growth can be achieved by concentrating on decreasing survival rates (Conroy and Senar 2009). This is also in line with more general work that has shown that culling leads to a more rapid reduction in density than do methods impeding breeding (Barlow et al. 1997). Our new models stress that, in general, shooting is the most effective and cost-efficient method to cull Monk Parakeet populations. As previously stated, all instances of successful eradication of Monk Parakeet populations have used this method (Postigo et al. 2019). Additionally, if properly carried out, shooting is considered one of the most humane methods to eliminate pest birds (Sharp 2012). The optimal allocation of removal effort is to assign the bulk of removal to the winter period, when reproductive recruitment is minimal (Conroy and Senar 2009), although from a practical point of view, culling the individuals in the periods in which they are most linked to their nests may be more efficient. Data also suggest that to be successful, culling efforts should remove >25%–30% of the population per year. This is in line with work on feral pigeons (Kautz and Malecki 1990) and House Crows (*Corvus splendens*; Brook et al. 2003). Clearly, unless culling effort is sufficiently intense, survival and nest success may increase as the density of birds is reduced, compensating for any culling effort (Brook et al. 2003).

We should also stress that it is likely that populations, if not eliminated, would rapidly expand following cessation of control (Conroy and Senar 2009). As a consequence, eradication is a better strategy that simply controlling the populations to a given minimum population size. Eradication programs should also be coordinated among adjoining local authorities, to avoid migration from non-culled to culled localities. The large-scale eradication attempts for House Crows in Aden, Yemen, in the late 1980s, in which over 240,000 birds were culled during a two-year period, was unsuccessful because of a lack of comparable effort in neighboring cities and areas from which migrating individuals quickly replaced culling losses (Jennings 1992).

FINAL CONSIDERATION

For decisions requiring a rapid course of action, the amount of scientific research available is often minimal, but delaying action can make control more difficult and expensive (Simberloff 2003). To do nothing will lead to further damage from a spreading population that will become more difficult to control in the future (Chapman and Massam 2007). As stated elsewhere, given the rate of increase and spread of the Monk Parakeet into newly established populations, the potential damage that this can cause, and the designation of the Monk Parakeet as an exotic invasive species, social considerations should not prevent relevant governmental bodies from pursuing efforts to control the species (Conroy and Senar 2009).

APPENDIX 7.1

Parameter values for the demographic model and computation of effects and costs of control measures.

The R code to perform simulations and optimizations is available as a Zip archive at: https://press.princeton.edu/books/hardcover/9780691204413/naturalized-parrots-of-the-world. Running the code requires a current version of R (v.3.5.1 or later; https://www.r-project.org). R Studio (https://www.rstudio.com/products/rstudio/download/) is optional but helpful. The R packages Shiny and Progress must be installed from R or R Studio.

DEMOGRAPHIC PARAMETERS

PARAMETER	DESCRIPTION	VALUE
age.last.breed	Last age (years) that breeding occurs	8
initial.age	Initial age distribution	(0.337,0.228,0.146,0.098,0.064,0.0422,0.030,0.019,0.037)
initial.pop	Initial abundance (all ages and sexes)	User defined (e.g., 1,000)
mean.f.1st.brood	Mean fecundity per female: 1st brood	2.9
mean.f.2nd.brood	Mean fecundity per female: 2nd brood	1.2
p.breed.1	Proportion 1-year-old females breeding	0.50
p.breed.2	Proportion 2-year-old females breeding	0.7
p.breed.3pl	Proportion 3+-year-old females breeding	0.73
p.2nd.brood	Proportion of females with 2nd broods	0.56
S_a	Annual survival probability of adults (1+ years)	0.81
sd.f.1st.brood	sd clutch size per female: 1st brood	1.28
sd.f.1st.brood	sd clutch size per female: 2nd brood	0.3
sim.years	Number of years to simulate population growth	User-specified (e.g., 20)
S_j	Annual survival probability of juveniles (0–1 year)	0.61

CONTROL PARAMETERS

PARAMETER	DESCRIPTION	VALUE
birds.per.trap	Capacity of individual traps	200
cost.bait.per.trap	Annual cost of bait per trap	1,200
cost.can	Cost per hour of cannon netting	198
cost.nest	Cost per hour of nest removing	204
cost.per.trap	Cost per hour of trapping	300
cost.poke	Cost per hour of egg poking	195
cost.shoot	Cost per hour of shooting	55
cost.trap	Cost per hour of trap operation	30
cost.units	Currency units of costs	euro
eff.cannon	Per hour removal of adults from cannon netting	0.7
eff.poking	Per hour removal of recruits from egg poking	10.3
eff.rem.nest	Per hour removal of recruits from nest removal	8.8
eff.shooting	Per hour removal of adults from shooting	5.1
eff.trapping	Per hour removal of adults from trapping	1.6

REFERENCES

Avery, M. L., Greiner, E. C., Lindsay, J. R., Newman, J. R., and Pruett-Jones, S. 2002. Monk parakeet management at electric utility facilities in south Florida. In *Proceedings of the 20th Vertebrate Pest Conference*, ed. R. M. Timm and R. H. Schmidt, 140–145. Davis: Univ. of California.

Avery, M. L., and Shiels, A. B. 2018. Monk and rose-ringed parakeets. In *Ecology and Management of Terrestrial Vertebrate Invasive Species in the United States*, ed. W. C. Pitt, J. C. Beasley, and G. W. Witmer, 333–357. Boca Raton, FL: CRC Press.

Avery, M. L., Tillman, E. A., Keacher, K. L., Arnett, J. E., and Lundy, K. J. 2012. Biology of invasive monk parakeets in south Florida. *Wilson Journal of Ornithology* 124:581–588.

Avery, M. L., Yoder, C. A., and Tillman, E. A. 2008. Diazacon inhibits reproduction in invasive monk parakeet populations. *Journal of Wildlife Management* 72:1449–1452.

Barlow, N. D., Kean, J. M., and Briggs, C. J. 1997. Modelling the relative efficacy of culling and sterilisation for controlling populations. *Wildlife Research* 24:129–141.

Bashir, E. A. 1979. A new "parotrap" adapted from the MAC trap for capturing live parakeets in the field. *Proceedings of the Bird Control Seminar* 8:167–171.

Bomford, M., and O'Brien, P. 1992. A role for fertility control in wildlife management in Australia? In *Proceedings of the 15th Vertebrate Pest Conference*, ed. J. E. Borrecco and R. E. Marsh, 344–347. Davis: Univ. of California.

Borray-Escalante, N. A., Mazzoni, D., Ortega-Segalerva, A., Arroyo, L., Morera-Pujol, V., González-Solís, J., and Senar, J. C. 2020. Diet assessments as a tool to control invasive species: Comparison between monk and rose-ringed parakeets with stable isotopes. *Journal of Urban Ecology* 6.

Brook, B. W., Sodhi, N. S., Soh, M.C.K., and Lim, H. C. 2003. Abundance and projected control of invasive house crows in Singapore. *Journal of Wildlife Management* 67:808.

Bruggers, R. L., Rodriguez, E., and Zaccagnini, M. E. 1998. Planning for bird pest problem resolution: A case study. *International Biodeterioration and Biodegradation* 42:173–184.

Bucher, E. H. 1992. Neotropical parrots as agricultural pests. In *New World Parrots in Crisis: Solutions from Conservation Biology*, ed. S. R. Beissinger and N.F.R. Snyder, 201–219. Washington, DC: Smithsonian Institution Press.

Bucher, E. H., and Martín, L. F. 1987. Los nidos de cotorras (*Myiopsitta monachus*) como causa de problemas en lineas de transmision electrica. *Vida Silvestre Neotropical* 1:50–51.

Bucher, E. H., Martín, L. F., Martella, M. B., and Navarro, J. L. 1991. Social behaviour and population dynamics of the monk parakeet. In *Acta XX International Ornithological Congress*, ed. B. Bell, J. Cossee, J. Flux, B. Heather, R. Hitchmough, C. Robertson, and M. J. Williams, 681–689. Christchurch, New Zealand: Ornithological Trust Board, Wellington.

Bull, J. 1973. Exotic birds in the New York City area. *Wilson Bulletin* 85:501–505.

Burger, J., and Gochfeld, M. 2009. Exotic monk parakeets (*Myiopsitta monachus*) in New Jersey: Nest site selection, rebuilding following removal, and their urban wildlife appeal. *Urban Ecosystems* 12:185–196.

Canavelli, S. B., Aramburú, R. M., and Zaccagnini, M. E. 2012. Aspectos a considerar para disminuir los conflictos originados por los daños de la cotorra (*Myiopsitta monachus*) en cultivos agrícolas. *El Hornero* 27:89–101.

Canavelli, S. B., Branch, L. C., Cavallero, P., González, C., and Zaccagnini, M. E. 2014. Multi-level analysis of bird abundance and damage to crop fields. *Agriculture, Ecosystems and Environment* 197:128–136.

Canavelli, S. B., Swisher, M. E., and Branch, L. C. 2013. Factors related to farmers' preferences to decrease monk parakeet damage to crops. *Human Dimensions of Wildlife* 18:124–137.

Carrillo-Ortiz, J. 2009. Dinámica de poblaciones de la cotorra de pecho gris (*Myiopsitta monachus*) en la ciudad de Barcelona. PhD diss., Univ. of Barcelona, Spain.

Chapman, T., and Massam, M. 2007. Rainbow lorikeet. *Pestnote* 200:1–4.

Collar, N. J., and Juniper, A. T. 1992. Dimensions and causes of the parrot conservation crisis. In *New World Parrots in Crisis: Solutions from Conservation Biology*, ed. S. R. Beissinger and N.F.R. Snyder, 1–24. Washington, DC: Smithsonian Institution Press.

Conroy, M. J., and Carroll, J. P. 2009. *Quantitative Conservation of Vertebrates*. Oxford, UK: Wiley-Blackwell.

Conroy, M. J., and Senar, J. C. 2009. Integration of demographic analyses and decision modeling in support of management of invasive monk parakeets, an urban and agricultural pest. *Environmental and Ecological Statistics* 3:491–510.

Crowley, S. L., Hinchliffe, S., and McDonald, R. A. 2018. The parakeet protectors: Understanding opposition to introduced species management. *Journal of Environmental Management* 229:120–132.

Díaz-Ruiz, F., García, J. T., Pérez-Rodríguez, L., and Ferreras, P. 2010. Experimental evaluation of live cage-traps for black-billed magpies *Pica pica* management in Spain. *European Journal Wildlife Research* 56:239–248.

Domènech, J., Carrillo-Ortiz, J., and Senar, J. C. 2003. Population size of the monk parakeet *Myiopsitta monachus* in Catalonia. *Revista Catalana d'Ornitologia* 20:1–9.

Eberhard, J. R. 1998. Breeding biology of the monk parakeet. *Wilson Bulletin* 110:463–473.

Edelaar, P., Roques, S., Hobson, E. A., Gonçalves da Silva, A., Avery, M. L., Russello, M. A., Senar, J. C., Wright, T. F., Carrete, M., and Tella, J. L. 2015. Shared genetic diversity across the global invasive range of the monk parakeet suggests a common restricted geographic origin and the possibility of convergent selection. *Molecular Ecology* 24:2164–2176.

Esteban, A. 2016. *Control de la especie cotorra argentina (Myiopsitta monachus) en Zaragoza*. Zaragoza, Spain: Zaragoza City Council.

FAO. 2011. *International Trade in Wild Birds, and Related Bird Movements, in Latin America and the Caribbean*. Animal Production and Health Paper no. 166. Rome: Food and Agriculture Organization of the United Nations.

Faus, J., Ortega, A., Arroyo, L., and Senar, J. C. 2010. *Determinació de l'àrea de deambulació de la cotorra de pit gris a la ciutat de Barcelona*. Barcelona: Museu de Ciències Naturals de Barcelona.

Fitzwater, W. D. 1988. Solutions to urban bird problems. In *Proceedings of the 13th Vertebrate Pest Conference*, ed. A. C. Crabb and R. E. Marsh, 254–259. Davis: Univ. of California.

Giunchi, D., Albores-Barajas, Y. V., Baldaccini, N. E., Vanni, L., and Soldatini, C. 2012. Feral pigeons: Problems, dynamics and control methods. In *Integrated Pest Management and Pest Control: Current and Future Tactics*, ed. S. Soloneski, 215–240. Rijeka, Croatia: InTech.

Govindarajulu, P., Altwegg, R., and Anholt, B. R. 2005. Matrix model investigation of invasive species control: Bullfrogs on Vancouver Island. *Ecological Applications* 15:2161–2170.

Haag-Wackernagel, D. 1993. Street pigeons in Basel. *Nature* 361:200.

Haag-Wackernagel, D. 1995. Regulation of the street pigeon in Basel. *Wildlife Society Bulletin* 23:256–260.

Hyman, J., and Pruett-Jones, S. 1995. Natural history of the monk parakeet in Hyde Park, Chicago. *Wilson Bulletin* 107:510–517.

Jennings, M. C. 1992. The house crow *Corvus splendens* in Aden (Yemen) and an attempt at its control. *Sandgrouse* 14:27–33.

Kautz, J. E., and Malecki, R. A. 1990. Effects of harvest on feral rock dove survival, nest success, and population size. *Fish and Wildlife Technical Report* 31:1–16.

Krebs, L. B. 1974. Feral pigeon control. In *Proceedings of the 6th Vertebrate Pest Conference*, ed. W. V. Johnson, 257–262. Lincoln: Univ. of Nebraska.

Linz, G. M., Bucher, E. H., Canavelli, S. B., Rodriguez, E., and Avery, M. L. 2015. Limitations of population suppression for protecting crops from bird depredation: A review. *Crop Protection* 76:46–52.

MacGregor-Fors, I., Calderón-Parra, R., Meléndez-Herrada, A., López-López, S., and Schondube, J. E. 2011. Pretty, but dangerous! Records of non-native monk parakeets (*Myiopsitta monachus*) in Mexico. *Revista Mexicana de Biodiversidad* 82:1053–1056.

Martella, M. B., and Navarro, J. L. 1987. Metodo para la captura de cotorras (*Myiopsitta monachus*) en sus nidos. *Vida Silvestre Neotropical* 1:52–53.

Martín, L. F., and Bucher, E. H. 1993. Natal dispersal and first breeding age in monk parakeets. *Auk* 110:930–933.

Molina, B., Postigo, J. L., Román-Muñoz, A., and Del Moral, J. C. 2016. *La Cotorra argentina en España: Población reproductora en 2015 y método de censo*. Madrid: SEO/BirdLife.

Monzón, G. 1997. Problemática de la presencia de cotorras en la ciudad de Barcelona. In *Jornadas sobre el control de estorninos y otras aves gregarias*, 37–42. Huesca, Spain: Concejalía de Medio Ambiente.

Morgan, B., and McNeely, S. 2000. Control of the Australian ringneck parrot by trapping in south-west Western Australia. *TreeNote* 34:1–4.

Mott, D. F. 1973. Monk parakeet damage to crops in Uruguay and its control. In *Bird Control Seminars Proceedings*: 79–81. Lincoln: Univ. of Nebraska.

Muñoz, A. R. 2003. Cotorra Argentina *Myiopsitta monachus*. In *Atlas de las aves reproductoras de España*, ed. R. Martí and J. C. Del Moral, 638–639. Madrid: Dirección General de Conservación de la Naturaleza, Sociedad Española de Ornitología.

Navarro, J. L., Martella, M. B., and Bucher, E. H. 1992. Breeding season and productivity of monk parakeets in Cordoba, Argentina. *Wilson Bulletin* 104:413–424.

Neidermyer, W. J., and Hickey, J. J. 1977. The monk paraket in the United States 1970–75. *American Birds* 31:273–278.

Newman, J. R., Newman, C. M., Lindsay, J. R., Merchant, B., Avery, M. L., and Pruett-Jones, S. 2008. Monk parakeets: An expanding problem on power lines and other electrical utility structures. In *8th International Symposium on Environmental Concerns in Rights-of-Way Management*, ed. J. W. Goodrich-Mahoney, L. P. Abrahamson, J. L. Ballard, and S. M. Tikalsky, 343–354. New York: Elsevier Science.

Orueta, J. F. 2007. *Vertebrados invasores: Problemática ambiental y gestión de sus poblaciones*. Madrid: Organismo Autónomo Parques Nacionales.

Pascual, J., Carrillo-Ortiz, J., and Senar, J. C. 2015. *Cens de la població de cotorreta de pit gris a la ciutat de Barcelona 2015*. Barcelona: Museu de Ciències Naturals de Barcelona.

Peris, S. J., and Aramburú, R. M. 1995. Reproductive phenology and breeding success of the monk parakeet (*Myiopsitta monachus*) in Argentina. *Studies on Neotropical Fauna and Environment* 30:115–119.

Petersen, C. L., and Grasso, F. W. 2010. Seasonal changes in nest maintenance behavior of monk parakeets (*Myiopsitta monachus*). *Integrative and Comparative Biology* 50:E281-E281.

Postigo, J. L., and Senar, J. C. 2017. *Informe diagnóstico sobre la situación de las cotorras invasoras en el municipio de Málaga*. Málaga, Spain: Ayuntamiento de Málaga.

Postigo, J. L., Shwartz, A., Strubbe, D., and Muñoz, A. R. 2017. Unrelenting spread of the alien monk parakeet *Myiopsitta monachus* in Israel: Is it time to sound the alarm? *Pest Management Science* 73:349–353.

Postigo, J. L., Strubbe, D., Mori, E., Ancillotto, L., Carneiro, I., Latsoudis, P., Menchetti, M., Pârâu, L. G., Parrott, D., Reino, L., Weiserbs, A., and Senar, J. C. 2019. Mediterranean versus Atlantic monk parakeets *Myiopsitta monachus*: Towards differentiated management at the European scale. *Pest Management Science* 75:915–922.

Pruett-Jones, S., Newman, J. R., Newman, C. M., Avery, M. L., and Lindsay, J. R. 2007. Population viability analysis of monk parakeets in the United States and examination of alternative management strategies. *Human-Wildlife Conflicts* 1:35–44.

Pruett-Jones, S., Newman, J. R., Newman, C. M., and Lindsay, J. R. 2005. Population growth of monk parakeets in Florida. *Florida Field Naturalist* 33:1–14.

Pruett-Jones, S., and Tarvin, K. A. 1998. Monk parakeets in the United States: Population growth and regional patterns of distribution. In *Proceedings of the 18th Vertebrate Pest Conference*, ed. R. O. Baker and A. C. Crabb, 55–58. Davis: Univ. of California.

R Core Team. 2018. R: A language and environment for statistical computing. R Foundation for Statistical Computing, Vienna.

Reed, J. E., McCleery, R. A., Silvy, N. J., Smeins, F. E., and Brightsmith, D. J. 2014. Monk parakeet nest-site selection of electric utility structures in Texas. *Landscape and Urban Planning* 129:65–72.

Rodríguez-Pastor, R., Senar, J. C., Ortega, A., Faus, J., Uribe, F., and Montalvo, T. 2012. Distribution patterns of invasive monk parakeets (*Myiopsitta monachus*) in an urban habitat. *Animal Biodiversity and Conservation* 35:107–117.

Romero, I. P., Codesido, M., and Bilenca, D. N. 2015. Nest building by monk parakeets *Myiopsitta monachus* in urban parks in Buenos Aires, Argentina: Are tree species used randomly? *Ardeola* 62:323–333.

Schwab, D. J., and Gwynn, T. M., III. 1992. Monk parakeets nesting in Newport News, Virginia. *Raven* 63:34.

Senar, J. C., and Carrillo-Ortiz, J. 2007. *Dinàmica de poblacions de la cotorra de pit gris* (Myiopsitta monachus) *i recomanacions per al seu control*. Barcelona: Museu de Ciències Naturals de Barcelona.

Senar, J. C., Carrillo-Ortiz, J., Ortega-Segalerva, A., Dawson-Pell, F.S.E., Pascual, J., Arroyo, L., Mazzoni, D., Montalvo, T., and Hatchwell, B. J. 2019. The reproductive capacity of monk parakeets *Myiopsitta monachus* is higher in their invasive range. *Bird Study* 66:136–140.

Senar, J. C., Montalvo, T., Pascual, J., and Peracho, V. 2017. Reducing the availability of food to control feral pigeons: Changes in population size and composition. *Pest Management Science* 73:313–317.

Sharp, T. 2012. Shooting of pest birds: Standard operating procedure. PestSMART. Accessed 20 Oct. 2020. https://pestsmart.org.au/toolkit-resource/shooting-of-pest-birds/.

Shields, W. M. 1974. Use of native plants by monk parakeets in New Jersey. *Wilson Bulletin* 86:172–173.

Sibly, R. M., and Hone, J. 2002. Population growth rate and its determinants: An overview. *Philosophical Transactions of the Royal Society of London B: Biological Sciences* 357:1153–1170.

Simberloff, D. 2003. How much information on population biology is needed to manage introduced species? *Conservation Biology* 17:83–92.

South, J. M., and Pruett-Jones, S. 2000. Patterns of flock size, diet, and vigilance of naturalized monk parakeets in Hyde Park, Chicago. *Condor* 102:848–854.

Stock, B., and Haag-Wackernagel, D. 2016. Food shortage affects reproduction of feral pigeons *Columba livia* at rearing of nestlings. *Ibis* 158:776–783.

Thomsen, J. B., and Mulliken, T. A. 1992. Trade in Neotropical psittacines and its conservation implications. In *New World Parrots in Crisis: Solutions from Conservation Biology*, ed. S. R. Beissinger and N.F.R. Snyder, 221–239. Washington, DC: Smithsonian Intitution Press.

Tillman, E. A., Genchi, A. C., Lindsay, J. R., Newman, J. R., and Avery, M. L. 2004. Evaluation of trapping to reduce monk parakeet populations at electric utility facilities. In *Proceedings of the 21st Vertebrate Pest Conference*, ed. R. M. Timm and W. P. Gorenzel, 126–129. Davis: Univ. of California.

van Bael, S., and Pruett-Jones, S. 1996. Exponential population growth of monk parakeets in the United States. *Wilson Bulletin* 108:584–588.

Viana, I. R., Strubbe, D., and Zocche, J. J. 2016. Monk parakeet invasion success: A role for nest thermoregulation and bactericidal potential of plant nest material? *Biological Invasions* 18:1305–1315.

Volpe, N. L., and Aramburú, R. M. 2011. Nesting preferences of the monk parakeet (*Myiopsitta monachus*) in an urban area of Argentina. *Ornitologia Neotropical* 22:111–119.

Weiserbs, A., and Jacob, J. P. 1999. Etude de la population de Perriche jeune-veuve *Myiopsitta monachus* à Bruxelles. *Aves* 36:209–223.

Williams, B. K., Nichols, J. D., and Conroy, M. J. 2002. *Analysis and Management of Animal Populations: Modeling, Estimation, and Decision Making*. New York: Academic Press.

Yoder, C. A., Avery, M. L., Keacher, K. L., and Tillman, E. A. 2007. Use of DiazaCon™ as a reproductive inhibitor for monk parakeets (*Myiopsitta monachus*). *Wildlife Research* 34:8–13.

Yunick, R. P. 1971. A platform trap. *EBBA News* 34:122–125.

8

MANAGEMENT OF HUMAN-PARROT CONFLICTS: THE SOUTH AMERICAN EXPERIENCE

Enrique H. Bucher

INTRODUCTION

In South America, parrots were considered agricultural pests by the native peoples well before the arrival of the European conquerors. For example, the Peruvian Incas protected corn crops by posting guards equipped with several implements to scare the birds (see illustration) (Guaman Poma de Ayala 1614). After the European conquest, arriving settlers converted large portions of the continent to agriculture. This process was particularly widespread in Argentina and Uruguay, where conflicts with birds and agriculture became widespread and significant by the end of the 19th century. Damaged crops included mainly sunflower, corn, and cereals in the temperate, southern part of the continent and several fruit species in the subtropical and tropical regions (Bucher 1992). The Monk Parakeet (*Myiopsitta monachus*) became one of the most significant bird pests, rapidly invading the Pampas grasslands region shortly after the area was converted to agriculture at the end of the 19th century (Pergolani de Costa 1953; Bucher and Aramburú 2014). Besides damaging crops, the species affected human structures, particularly windmills and utility (electrical and telegraph) poles, with their large communal nests (Martin and Bucher 1987). During the early 20th century, the continued expansion of agriculture and the associated increase in bird damage problems led government agencies to create the legal status of "pest bird" (*aves plaga*), which allowed large-scale investments in control campaigns (Pergolani de Costa 1953). Later, the same status was applied to other parrots in the Argentine territory. The information obtained from these large-scale

Figura I
(Guaman Poma 1993 [1614]: 705)
Indios • labrador, arariua, parian, trabaja

To protect their ripening crops from parrots, the Inca of South America installed guards specially equipped to scare the birds. From F. Guaman Poma de Ayala, *Primera Nueva Crónica y Buen Gobierno* (1614). Image from Danish Royal Library (http://www5.kb.dk/permalink/2006/poma/info/en/frontpage.htm).

MANAGEMENT OF HUMAN-PARROT CONFLICTS: THE SOUTH AMERICAN EXPERIENCE

campaigns provided significant insight into the effectiveness of different control techniques and also—and probably most important—the opportunity to learn from failures.

This chapter attempts to summarize the South American experience in parrot control and provides some alternatives for the control of invasive parrot species in other continents. The following sections are included: (1) detailed case analyses of large-scale control campaigns, including their characteristics, complexities, and results; (2) a summary analysis of the lessons learned from parrot control in South America; and (3) an outline of criteria and alternatives for designing and implementing parrot-control strategies.

CASE ANALYSES

This section includes three cases of parrot-control problems from South America, concerning the Monk Parakeet, Burrowing Parrot (*Cyanoliseus patagonus*), and Turquoise-fronted Amazon (*Amazona aestiva*). These three species have been considered pests in their countries of origin, and as such subjected to large-scale control campaigns, and also have been exported in large numbers to Europe and North America for the pet trade.

The Monk Parakeet in the Pampas of Argentina and Uruguay

The Monk Parakeet is a paradigmatic example of a pest species in South America. No doubt, this is the parrot species that has been the target of the most intense and widespread control efforts in the whole continent. Here, the cases of efforts in Argentina and Uruguay are presented, including their strategies, implementation approaches, and results.

ARGENTINA: A HISTORICAL OVERVIEW

At the beginning of the 20th century, the Monk Parakeet's range expanded across the Pampas grasslands, following agriculture development and the introduction of *Eucalyptus*, a highly preferred nesting tree (Bucher and Aramburú 2014). As early as in 1901, the Argentine government delivered a law dealing with the problem of vertebrates damaging crops in rural areas. The regulation named the Monk Parakeet and obliged farmers to eliminate this pest from their properties. The recommended procedure for this species included shooting adults and burning the compound nests with fire-starting arrows (Pergolani de Costa 1953).

As the species continued expanding across agricultural lands, in 1947 government agencies decided to subsidize control campaigns by paying farmers for parakeet legs as an incentive for achieving control at the regional scale. From 1958 to 1960, bounties were paid on 427,206 pairs of parakeet legs in just one Argentine province (Buenos Aires) (Spreyer and Bucher 1998; Linz et al. 2015).

Later during the second half of the 20th century, the Argentine and Uruguayan governments introduced changes in the technology of massive control campaigns. Retribution for parakeet legs was discontinued, and the new approved control method was based on spraying nests with toxic insecticides (Bucher 1992). This practice was hazardous for the personnel involved in the applications, given the high human toxicity of the pesticides and the difficulty (and cost) of reaching the nests, usually built in the upper third of high eucalyptus trees (20–30 m range). Accordingly, new legislation was passed stating that control operations should be performed only by authorized, certified enterprises, paid for by the property owner.

Since the 1980s, a further improvement of this technique was introduced in Argentina, consisting of smearing a mixture of grease and a toxic insecticide (e.g., carbofuran) around the nest openings (Linz et al. 2015). This new method was very effective and required lesser amounts of insecticides, making it also less hazardous for the staff involved in the operation. On the negative side, nontarget wildlife such as foxes and scavenger birds were affected when feeding on dead parakeets that fell from the nest. Unfortunately, and despite the massive scale of the government-organized control campaigns, the wildlife agencies kept no records regarding the effects of this strategy, such as the number of

Monk Parakeets killed, operation costs, or—more important—crop damage reduction in the treated areas (Bucher 1992).

At present, this practice continues in some provinces of Argentina, particularly Buenos Aires, although there is growing disagreement on its effectiveness and economic justification. Sections of the population that are in favor of control include farming communities and companies involved in the parakeet-control measures; those opposing sectors are mostly in the conservation and academic areas (Canavelli et al. 2013).

An additional control measure was added during the 1980s: the promotion of massive exports of Monk Parakeets for the international pet trade by the governments of Argentina and Uruguay.

Uruguay: Assessing the Effectiveness of Lethal Control

In general terms, Uruguay followed a similar approach to that of Argentina regarding Monk Parakeet control, based mostly on lethal-control campaigns using nest poisoning (Linz et al. 2015). In terms of monitoring the success of control, an advance was made in 1981–83. During that period, the Uruguayan government and local farmers, with the financial support of the UN Food and Agriculture Organization (FAO), implemented an experimental, large-scale control campaign in a vast agricultural area in the western area of the country (Soriano Department, 9,008 km^2). Fortunately, detailed records were kept, which provided valuable information on the effectiveness of the control operation (Rodriguez et al. 2013).

The objective was to reduce crop damage (mostly to sunflowers; see photograph) through poisoning of the parakeets' nests by smearing grease mixed with a carbamate insecticide, covering the whole area simultaneously, and repeating the same procedure one year later to prevent reinvasion. Each farm in the department was visited twice between 1987 and 1989. During the first visit, the number of nests was recorded, and all the nests were treated. On the second visit, active nests were recorded and treated again. Unfortunately, no information was collected about adverse side effects on nontarget species. A total of 114 farms were inspected, and 4,841 nests were treated in the initial operation. During the second visit, about 17 months later, occupied nests were found in all farms, indicating that no eradication was achieved at the farm level. In that round, 2,203 active nests (45% of the initial number) were found, including old and newly built nests. Given that nests were on average smaller in the second check, a better

Damage by Monk Parakeets (*Myiopsitta monachus*) to corn and sunflower in Argentina. Photos by Enrique H. Bucher.

indication of the parakeet population change was obtained by adding up the number of chambers in each nest. This number decreased from 11,957 to 3,525 between checks (29% of the initial value) (Rodriguez et al. 2013).

In summary, even if a substantial reduction of the initial parakeet population was achieved, the campaign results were far from complete elimination, and the remaining surviving population was still large enough to cause agricultural damage. Unfortunately, the effect of the second treatment was not assessed one year later. It is disappointing that a parallel crop-damage assessment was not included in the project. Consequently, an open question remains about whether the control campaign resulted in a decrease in crop damage, and if so, how the savings compared with the large-scale operation cost.

The Burrowing Parrot in Patagonia

The Burrowing Parrot is a colonial parrot native to Argentina and parts of Chile. In Argentina, it occurs from the Andean slopes in the northwest of the country to the Patagonian steppes in the south. In Argentina, the Burrowing Parrot has long been considered an agriculture pest, affecting cereals (mainly corn, sunflower, barley, and wheat) as well as fruits (almonds, apples, walnuts, and grapes). The pest status was decided basically as a response to farmers' claims and was not supported by accurate, methodologically reliable assessments (Sanchez et al. 2016).

Most claims of crop damage came from the region around the large parrot colonies at El Cóndor, in the Río Negro province, which comprises about 71% of the species' total population. Until the end of the 1990s, most local farmers claimed damage to wheat, in an area marginally suitable for cultivation because of climate and soil limitations (Sanchez et al. 2016). These local adverse factors frequently led farmers to abandon their crops unharvested, leaving standing plants with grains available to parrots and other birds for an extended period (Bucher and Bedano 1976). At that time, the governments of the provinces of Buenos Aires and Río Negro attempted population control by spraying the colonies with toxic insecticides and even blowing them up with dynamite (Sanchez et al. 2016).

At the end of the 1990s, a combination of weather changes (an increase in rainfall) and development of irrigation projects in the Buenos Aires and Río Negro regions allowed agricultural expansion and cultivation of corn. Since then, local farmers have claimed that corn is the crop most affected by the Burrowing Parrot. In response to these claims, a detailed assessment of the magnitude of parrot damage to irrigated corn plantation was performed (Sanchez et al. 2016). Results showed that damage to irrigated corn crops was economically insignificant, affecting 0.1%–0.4% of the corn harvest. Moreover, even if large numbers of parrots were seen in the irrigated corn fields, they were not feeding on the standing crop but on corn grain spilled on the ground. This wasted grain represented a loss attributable to harvest machines, with no connection to the presence of parrots.

In summary, the study concluded that there is no need for control of parrots as crop pests in northeastern Patagonia, and these results also provided further support to the view that parrot damage is often exaggerated and overstated (Bucher 1992).

The Pet Trade as a Means of Control of Monk Parakeet and Turquoise-fronted Amazon

During the 1980s, the volume of birds traded in the international bird pet trade exploded in numbers. Parrots represented a large portion of the total of birds exported from South America, as reported by the Convention on International Trade in Endangered Species of Wild Fauna and Flora (CITES 1991). In most cases, pest status in country of origin was the primary justification for requesting export–import permits (Thomsen and Mulliken 1992). We consider here the process that led to the justification of the parrot trade as a complementary population-control alternative for Monk Parakeets and Turquoise-fronted Amazons, in order to better understand: (1) the effect, if any, of exports on the native populations, and (2) the possible effect of exports on crop damage.

With regard to Monk Parakeets, there is no evidence that exports reduced local populations. Based on CITES data, a total of 636,648 individuals were exported from Argentina and Uruguay between 1980 and 2010, and yet there was no observable reduction in local populations. From a different perspective, Monk Parakeet exports resulted in what may be seen as a unique case in history of a massive intercontinental dispersal of a well-known pest species mediated (even if unintentionally) by a human-managed trade operation (this vol.: Cardador et al., chap 1; Calzada Preston et al., chap. 11; Carrete et al., chap. 15).

The Turquoise-fronted Amazon was once one of the most common parrots in drier habitats in the Chaco Province of Argentina. This species was allowed to be exported from Argentina, at a time when it was clear that the population was already declining due to intense deforestation in the Chaco and the growing pet trade (Bucher and Martella 1988). The justification for approving export permits was that the species had been given the legal status of pest species by the provinces of Tucumán and Salta, because of the damage caused to citrus plantations (CITES 1991). No consideration was given to the lack of supporting scientific information or to the fact that the parrot was included in Appendix 2 of CITES, which lists species that are not necessarily threatened with extinction but may become so unless trade is strictly controlled. On the contrary, the pet-trade agreement encouraged the Argentine government to establish high annual export quotas. As a result, about 900,000 Turquoise-fronted Amazons were exported from Argentina between 1982 and 1988 (Thomsen and Mulliken 1992).

To counter the lack of reliable information on the damage caused by the Turquoise-fronted Amazon, researchers from Argentina's Universidad Nacional de Córdoba developed a detailed study, growing orange, grapefruit, and lemon, in a citrus plantation in northeastern Tucumán province, from May to September 1990 (Navarro et al. 1991). That study showed that damage caused by the Turquoise-fronted Amazon was very low (1.0% of all produced fruits were damaged). Orange was the most affected crop (2.0%), whereas damage to lemon, the main citrus crop produced in the region, was 0.3%, and grapefruit showed no damage. These figures were considered economically irrelevant, even by orchard owners. The authors concluded that, despite the claims of citrus growers, the Turquoise-fronted Amazon was not a significant problem for the studied orchard and presumably for most of the citrus orchards surrounding the study area. Moreover, results showed not only that the Turquoise-fronted Amazon was not causing significant damage to orchards but also that nestling capture was instead generating a severe impact on the already endangered Chaco forest. Collectors trying to extract parrot nestlings from nest cavities were damaging and even felling trees when the nest hole was high and out of the reach. Given the large number of parrots exported, this practice had a massive impact on the preferred nesting tree, Quebracho Blanco (*Aspidosperma quebracho-blanco*), a highly valued, dominant tree of the Chaco forest (Berkunsky et al. 2012; Bucher 1997; Bucher and Martella 1988; Bucher et al. 1992).

LESSONS LEARNED

Analysis of the experience gained through a long history of bird control actions and related research in South America provides interesting and useful insight, applicable to the management of conflicts generated by invasive parrots worldwide. This section summarizes the key aspects that characterize parrots as management problems, based on the existing literature and the author's personal experience.

Parrots as Peculiar Pest Birds

One of the most difficult challenges wildlife managers face consists in dealing with a pest species that is harmful but at the same time attractive. This situation, frequent in parrot-human conflicts, on many occasions leads to a management dilemma between public preference and the real magnitude of the problem. Accordingly, understanding stakeholders' attitudes and the underlying factors supporting their points of view becomes essential (Brightsmith and Kiacz, chap. 9 this vol.). Research and extension actions need to be proactive to avoid public controversies about management decisions (Canavelli et al. 2013).

Bird-Caused Crop Damage Tends to Be Exaggerated

Farmers tend to overstate bird damage worldwide (as previously mentioned in the cases of the Burrowing Parrot and the Turquoise-fronted Amazon) (Bucher 1992). This tendency is related to the high visibility of parrots. Indeed, large and noisy flocks of colorful birds over the crops scare farmers, who tend to correlate the number of birds to the magnitude of the expected crop damage, without considering a more accurate damage assessment. On the few occasions when crop damage has been assessed following strict statistical procedures, the damage proved to be lower than that estimated by farmers (see previous "Case Analyses" section). Unfortunately, in both agricultural and urban situations, damage evaluation is time consuming and expensive, and sometimes requires a complicated sampling design.

Furthermore, opposed and extreme attitudes to parrot conflicts may become a political issue. This scenario is likely to develop in cases when control is funded and implemented by government agencies, involving complaints about effectiveness, environmental concerns, biodiversity conservation, and parrot affection and care (Bucher 1992). Moreover, vested economic interests may play an influential role, as in the case of the international pet trade, described previously.

Lethal Control as Preferred Option

In South America, authorities confronted with bird pest situations have been inclined almost invariably to use massive control campaigns based on lethal control as the first (and usually the only) alternative (Linz et al. 2015). The accumulated experience indicates that lethal control alone has not been capable of reducing population levels and therefore of mitigating crop damage by birds (Linz et al. 2015). Despite these limitations, lethal control methods (both legal and illegal) are still used (though at a decreasing rate), in some cases with the support of farmers, hunters, and government agencies. Consequently, understanding the factors involved in this contradictory social response appears to be very important.

Crop Damage by Parrots Is Usually Irregularly Distributed

When investigating parrots as agricultural pests, it is often found that a few plots may be severely damaged, whereas many others are left slightly damaged or untouched (Bucher 1992), a pattern that differs from those of other crop pests such as insects or bacteria. For example, at a regional level, Monk Parakeet damage is not considered economically significant in Argentina, although locally, damage may be serious on some occasions (Linz et al. 2015). The same may apply regarding parrot control in cities, particularly in the case of the Monk Parakeet. This landscape pattern implies that the benefit-to-cost ratio of control actions at a large scale tends to be very low, which discourages the implementation of large-scale control campaigns (see the Monk Parakeet discussion in the previous section). Under these circumstances, damage compensation insurance becomes worth considering as a management alternative.

ARE PARROTS "WEAK" BIRD PESTS?

Parrots lack a variety of characteristics typical of most bird pest species, which may limit their potential to resist population control measures, particularly in terms of low distributional opportunism and reproductive flexibility.

Distributional Opportunism

Defined as the ability of birds to displace according to the unpredictable distribution of short-lived resources, distributional opportunism in most parrots appears limited to the nonbreeding season, given their marked nest-site fidelity (Bucher 1992). Several species, including the Burrowing Parrot and the Turquoise-fronted Amazon, are partly nomadic outside the breeding season, whereas the Monk Parakeet remains close to the nesting area year-round (Martin and Bucher 1993).

However, evidence has accumulated indicating that parrot dispersal in invasive areas depended not only on the species' natural behavior—human dispersal no doubt has also played a very important role. For example, Muñoz and Real (2006) showed that the Monk Parakeet's spread in Spain was much faster and more scattered than could be expected via the natural neighborhood diffusion pattern, by which individuals expand from their established territories without leaving gaps in between. They concluded that two processes, neighborhood diffusion and jump dispersal mediated by artificial transportation and escapes, were occurring simultaneously. In the United States, Gonçalves da Silva et al. (2010) found genetic evidence of rapid dispersal of Monk Parakeets of up to 100 km/year, one order of magnitude higher than in Argentina. Again, the most plausible explanation in this case is high propagule pressure from large numbers of parakeets dispersed by the pet trade (Bucher and Aramburú 2014).

Breeding Opportunism

In general, parrots have very specific breeding requirements that limit their ability to adapt to changing conditions (Beissinger and Snyder 1992). Many species nest in tree cavities, which are usually associated with mature forests. Cliff-nesting species, like the Burrowing Parrot, require suitable escarpments according to their specific height and texture requirements. The Monk Parakeet is the only parrot that builds its own communal nests and therefore appears more flexible about breeding sites, although the large size of its nests limits the range of usable supporting trees or structures. Parrots usually have a single, seasonally fixed breeding season and are nesting-site faithful. Both factors constrain the range and pattern of their annual movements and therefore reduce their mobility for exploiting new resources.

Population Dynamics

Most parrots tend to delay their first breeding to about two to three years of age (Beissinger and Snyder 1992). Delayed maturity may retard population growth in newly invaded areas or in populations that are recovering from lethal control. The combination of delayed maturity and a fixed breeding season makes it difficult for parrots to increase rapidly in response to abundant but ephemeral resources such as crops. These two parameters are insufficient, however, for estimating the population behavior, given that another two additional key parameters, survival and the number of nonbreeding adults in the population, are also needed. Both parameters are highly variable and difficult to estimate (Bucher et al. 1991; Sandercock et al. 2000).

It is also important to bear in mind that parrots' population parameters may change in invasive areas with respect to the original population, due, for example, to the effect of density-dependent factors. This possibility is supported by the finding that the reproductive capacity (productivity) of Monk Parakeets is higher in their invasive range (e.g., Barcelona, Spain) than in the native area (Argentina) (Senar et al. 2019). Whether the observed differences derive from environmental conditions (e.g., food availability) or respond to physiological and/or genetic processes remains an open (and very challenging) question.

DEVELOPING INTEGRATED CONTROL STRATEGIES

An ideal strategy for parrot control should prioritize methods and tactics that are the most permanent, least expensive, and least environmentally damaging over those that are repetitive, expensive, and damaging to the environment (Bucher 1992). Based on this principle, this section includes guiding concepts that summarize the experience gained in South America, as described in this chapter. Management options are presented in the form of decision-making alternatives.

Is Control Action Necessary?

Justification for control actions should be carefully considered. Damage (in both urban and agricultural situations) should be evaluated in the most objective and precise way possible. The social perception and acceptance of the need for control should also be assessed carefully in advance of any action in the field. Once control actions have started, managers should continually evaluate both the control campaign and the achieved results carefully, and also be prepared to stop or change planned activities if needed.

For example, during the campaign to control Monk Parakeets and other pest birds in Argentina, managers often became reluctant to stop the control campaigns, even when there were clear indications of their lack of success. They usually argued, "Such large investment should not be discontinued now, when just a little bit of extra effort and investment will bring us success." This kind of misjudgment is known as the "Concorde fallacy," a metaphor related to the supersonic airplane built in France and England in the 1970s. Despite early evidence of its economic unviability, both governments decided to complete the project based on what had already been spent—and the national pride attached to the project. The Concorde fallacy should always be kept in mind when launching parrot-control campaigns.

Eradication or Damage Control?

Eradication is an extreme management action that should be thoroughly considered and justified, and it should be implemented in accordance with legal regulations. Additionally, as far as possible, it should be conducted taking into account the views of all stakeholders and with public support. In principle, eradication appears more feasible with parrots than with other pest birds, given their ecological and population characteristics (as mentioned previously). However, eradication appears achievable only for small populations, ideally in isolated areas, where reinvasion is less likely. In any case, the risk of reinvasion is always present, either by natural or human-induced mechanisms, such as accidental or voluntary pet release.

Damage control aims at reducing the negative impact caused by the problem species, ideally with minimal cost and negative side effects. If, after careful assessment, control is decided upon, the design of an integrated strategy should follow, which requires adequate scientific base information. In this sense, the recent important advances regarding distribution and population trends of the Monk Parakeet in Europe provide a very much needed starting point for any regional management plan (Postigo et al. 2017, 2019).

CONTROL MANAGEMENT ALTERNATIVES

A list of possible control tactics worth being considered is presented below. As mentioned previously, the ultimate goal should be to reduce damage, giving priority to single, definitive actions, instead of implanting repetitive procedures (see also Bucher 1992; Avery and Shiels 2018).

Habitat Change

Alteration of feeding and nesting conditions may make a habitat unsuitable for the target species. For example, in the case of the Monk Parakeet, making bird feeders inaccessible during winter in cities with cold, snowy winters may decrease food availability below a critical level. In the case of human-made structures, changes to their design may make them unsuitable for supporting Monk Parakeet nests (Avery et al. 2006).

Barriers

Physical barriers may isolate affected crops or buildings from parrots. Examples include covering vineyards with nets, and redesigning electricity-transport towers to prevent Monk Parakeets from building communal roosts affecting power lines.

Crop Substitution

When possible, replacing susceptible crops with others less affected by parrots but similarly profitable may provide a one-step definitive solution. This alternative is particularly feasible in sites where the affected crops are cultivated in marginally suitable areas, making productivity low and production costs high. A variant of this approach is the use of resistant varieties of susceptible crops. For example, new varieties of sunflower resistant to damage by Monk Parakeets and other birds are now widely used in Argentina.

Repellents

Although repellents are frequently advertised, the available experience indicates that auditory repellents are not effective, while taste repellents have been tested with only limited results (Mason and Clark 1995). In the case of chemical repellents, one important negative factor is the risk of contamination of the seeds and fruits for human consumption.

Compensation Programs

Insurance is worth considering when parrot damage is restricted to a few sites in a wide area, a situation that makes regional population control unviable. On the negative side, implementing insurance compensation may require a significant investment in damage evaluation.

Lethal Control

Today, lethal control is considered a last alternative; it is in many cases ineffective, with undesirable side effects and often meets growing opposition from the public (Canavelli et al. 2013; Linz et al. 2015). This method may be considered for treating small areas with an intolerable number of parrots, such as parks, gardens, buildings, or even small neighborhoods. In all cases, special care has to be taken when manipulating parrots and their nests during control operations, given the risk of disease (psittacosis) transmission (Chereau et al. 2018).

At the regional scale, on the contrary, lethal control does not seem feasible in most cases, not so much due to the lack of adequate capability but fundamentally because of the low benefit-to-cost ratio resulting from the high economic and operational demands, as well as the long-term persistence required, as shown in the case of the large-scale control campaign attempted in Uruguay.

FINAL COMMENT

The cumulative experience in South America indicates that invasive parrots are unlikely to become a critical management problem and can be managed adequately on a local level if the alternative of learning to accept them as new members of the local fauna is accepted. In addition, as human population growth increases and the magnitude of international transport of wildlife continues unabated, the global intercontinental mixing of species and faunal communities seems inevitable.

REFERENCES

Avery, M. L., Lindsay, J. R., Newman, J. R., Pruett-Jones, S., and Tillman, E. A. 2006. Reducing monk parakeet impacts to electric utility facilities in south Florida. In *Advances in Vertebrate Pest Management*, ed. C. J. Feare, and D. P. Cowan, 4:125–136. Furth, Germany: Filander Verlag.

Avery, M. L., and Shiels, A. B. 2018. Monk and rose-ringed parakeets. In *Ecology and Management of Terrestrial Vertebrate Invasive Species in the United States*, ed. W. C. Pitt, J. C. Beasley, and G. W. Witmer, 333–357. Boca Raton, FL: CRC Press.

Beissinger, S. R., and Snyder, N. F. 1992. *New World Parrots in Crisis: Solutions from Conservation Biology*. Washington, DC: Smithsonian Institution Press.

Berkunsky, I. I., Ruggera, R., Aramburú, R., and Reboreda, J. 2012. Principales amenazas para la conservación del loro hablador (*Amazona aestiva*) en la región del Impenetrable, Argentina. *Hornero* 27:39–41.

Bucher, E. H., 1992. Neotropical parrots as agricultural pests. In *New World Parrots in Crisis: Solutions from Conservation Biology*, ed. S. R. Beissinger and N.F.R. Snyder, 201–219. Washington, DC: Smithsonian Institution Press.

Bucher, E. H. 1997. Situación actual y prioridades para la conservación del loro hablador (*Amazona aestiva*) en la Argentina. *Naturaleza y Conservación (Aves Argentinas)* 1:23.

Bucher, E. H., and Aramburú, R. M. 2014. Land-use changes and monk parakeet expansion in the Pampas grasslands of Argentina. *Journal of Biogeography* 41:1160–1170.

Bucher, E. H., and Bedano, P. 1976. Bird damage problems in Argentina. *International Studies on Sparrows* 9:3–16.

Bucher, E. H., and Martella, M. B. 1988. Preliminary report on the current status of *Amazona aestiva* in the western Chaco, Argentina. *Parrotletter* 1:9–10.

Bucher, E. H., Martin, H., Martella, M. B., and Navarro, J. 1991. Social behaviour and population dynamics of the monk parakeet. In *Acta XX International Ornithological Congress*, ed. B. Bell, R. Cossee, J. Flux, B. Heather, R. Hitchmough, C. Robertson, and M. J. Williams, 681–689. Christchurch, New Zealand: Ornithological Trust Board, Wellington.

Bucher, E. H., Saravia, C., Miglietta, S., and Zaccagnini, M. E. 1992. Status and management of the Blue-fronted Amazon parrot in Argentina. *PsittaScene* 4:3–6.

Canavelli, S. B., Swisher, M. E., and Branch, L. C. 2013. Factors related to farmers' preferences to decrease monk parakeet damage to crops. *Human Dimensions of Wildlife* 18:124–137.

Chereau, F., Rehn, M., Pini, A., Kühlmann-Berenzon, S., Ydring, E., and Ringberg, H. 2018. Wild and domestic bird faeces likely source of psittacosis transmission: A case control study in Sweden, 2014–2016. *Zoonoses and Public Health* 65:1–8.

CITES. 1991. Psittaciformes: Quotas for Argentina. *Traffic Bulletin* 12:22.

Gonçalvez da Silva, A., Eberhard, J. R., Wright, T. F., Avery, M. F., and Rusello, M. A. 2010. Genetic evidence for high propagule pressure and long-distance dispersal in monk parakeet (*Myiopsitta monachus*) invasive populations. *Molecular Ecology* 19:3336–3350.

Guaman Poma de Ayala, F. 1614 [1993]. *Primera Nueva Crónica y Buen Gobierno*. Lima, Peru: Fondo de Cultura Económica.

Linz, G., Bucher, E. H., Canavelli, S. B., Rodriguez, E. N., and Avery, M. L. 2015. Limitations of population suppression for protecting crops from bird depredation: A review. *Crop Protection* 76:46–52.

Martín, L. F., and Bucher, E. H. 1987. Los nidos de cotorras (*Myiopsitta monachus*) como causa de problemas en líneas de transmisión eléctrica. *Vida Silvestre Neotropical* 1:50–51.

Martín, L. F., and Bucher, E. H. 1993. Natal dispersal and first breeding age in monk parakeets. *Auk* 110:930–933.

Mason, J. R., and Clark, L. 1995. Avian repellents: Options, modes of action, and economic considerations. National Wildlife Research Center Repellents Conference. USDA National Wildlife Research Center Symposia. Lincoln: Univ. of Nebraska.

Muñoz, A., and Real, R. 2006. Assessing the potential range expansion of the exotic monk parakeet in Spain. *Diversity and Distributions* 12:656–665.

Navarro, J. L., Martella, M. B., and Chediack, A. 1991. Analysis of Blue-fronted Amazon damage to a citrus orchard in Tucumán, Argentina. *Agriscientia* 8:75–78.

Pergolani de Costa, M. J. 1953. La lucha contra las cotorras en la República Argentina. Instituto de Sanidad Vegetal, Ministerio de Agricultura y Ganadería de la Nación Argentina, ser. A, 9 (53):1–32.

Postigo, J., Shwartz, A., Strubbe, D., and Muñoz, A. 2017. Unrelenting spread of the alien monk parakeet *Myiopsitta monachus* in Israel. Is it time to sound the alarm? *Pest Management Science* 73:349–353.

Postigo, J. L., Strubbe, D., Mori, E., Ancillotto, L., Carneiro, I., Latsoudis, P., Menchetti, M., Pârâu, L. G., Parrott, D., Reino, L., Weiserbs, A., and Senar, J. C. 2019. Mediterranean versus Atlantic monk parakeets *Myiopsitta monachus*: Towards differentiated management at the European scale. *Pest Management Science* 75:915–922.

Rodriguez, E., Aramburú, R. B., and Bucher, E. H. 2013. Recovery of monk parakeet populations after massive lethal control in agricultural areas of Uruguay: Lessons learnt. Workshop: Neotropical psittacines as agricultural pests: Building capacity to manage conflicts between people and parrots. *19th Regional Meeting of the Society for the Conservation and Study of Caribbean Birds*, 27–31 July 2013, St. George's Univ., Grenada.

Sanchez, R., Ballari, S. A., Bucher, E. H., and Masello, J. F. 2016. Foraging by burrowing parrots has little impact on agricultural crops in northeastern Patagonia, Argentina. *International Journal of Pest Management* 62:326–335.

Sandercock, B. K., Beissinger, S. R., Stoleson, S. H., Melland, R. R., and Hughes, C. R. 2000. Survival rates of a neotropical parrot: Implications for latitudinal comparisons of avian demography. *Ecology* 81:1351–1370.

Senar, J. C., Carrillo-Ortiz, J. G., Ortega-Segalerva, A., Dawson Pell, F.S.E., Pascual, J., Arroyo, L., Mazzoni, D., Montalvo, T., and Hatchwell, B. J. 2019. The reproductive capacity of monk parakeets *Myiopsitta monachus* is higher in their invasive range. *Bird Study*, 66:136–140.

Spreyer, M. F., and Bucher, E. H. 1998. Monk parakeet *Myiopsitta monachus*. In *The Birds of North America*, ed. A. Poole. Ithaca, NY: Cornell Lab of Ornithology.

Thomsen, J. B., and Mulliken, T. A. 1992. Trade in neotropical psittacines and its conservation implications. In *New World Parrots in Crisis: Solutions from Conservation Biology*, ed. S. R. Beissinger and N.F.R. Snyder, 221–239. Washington, DC.: Smithsonian Institution Press.

ARE NATURALIZED PARROTS PRIORITY INVASIVE SPECIES?

Donald J. Brightsmith and **Simon Kiacz**

INTRODUCTION

Non-native species are considered one of the leading threats to global biodiversity (Wilcove et al. 1998; Gurevitch and Padilla 2004; Salo et al. 2007; Simberloff et al. 2013; Olah et al. 2016), but this contention has been hotly debated (Davis et al. 2011; Chew 2015; Dueñas et al. 2018). Some scientists and sectors of society are questioning the traditional view that naturalized species in general, including naturalized parrots, should be eradicated (Burger and Gochfeld 2009; Davis et al. 2011; Valéry et al. 2013). Currently this debate is raging in the pages of esteemed scientific journals, in government agencies, and in the streets of cities like London and Madrid (Seymour 2013; Briggs 2017; Crowley et al. 2017, 2019; Russell and Blackburn 2017; Ayuntamiento de Madrid 2019).

Much of this debate is among groups that view exotic species issues from different perspectives (Schlaepfer et al. 2011). Many stakeholders see non-native species through the lens of the precautionary principle, which calls for the rapid elimination of exotic species before they have the chance to become established and cause ecological, economic, or social harm (Kriebel et al. 2001; Polkanov and Keeling 2001; Chapman 2005; Simberloff 2005; Simberloff et al. 2013). At the other end of the extreme are groups that purport that we should not worry so much about the native or exotic origins of species but instead embrace the fact that exotic species represent our new ecological reality and try to control only the most harmful species, be they native or exotic (Carroll 2011; Davis et al. 2011; Schlaepfer et al. 2011; Davis and Chew 2017). Between these two extremes is a broad array of stakeholders who recognize that non-native species vary greatly in their real and potential harm, and that harm should often be mitigated through management actions (Williams et al. 2010; Strubbe et al. 2011; Kumschick et al. 2012; Tassin et al. 2017). Many environmental managers are de facto mitigators, as they must make difficult decisions and allocate their limited resources to target the species that are causing the most harm to their local environments (Kumschick and Nentwig 2010; Kuebbing and Simberloff 2015).

At the global level, the largest single driver of invasive species management is the Convention on Biological Diversity (CBD 1992). The CBD is an international treaty with 196 participants (195 countries plus the European Union), which has as one of its main goals the conservation of global biodiversity. The convention recognizes that non-native species are a threat to global biodiversity and calls on the participating nations to "prevent the introduction of, control or eradicate those alien species which threaten ecosystems, habitats or species" (CBD 1992). It projected that, "by 2020, invasive alien species and pathways are identified and prioritized, priority species are controlled or eradicated, and measures are in place to manage pathways to prevent their introduction and establishment" (CBD 2010). The CBD defines "invasive alien species" as any species whose introduction and/or spread outside its native range threatens biological diversity (CBD 2002). By distinguishing invasive alien species as those that damage the environment and by calling for management of *priority* invasive species, the approach used by the CBD most closely aligns with the mitigation perspective. The text of the CBD is globally important, as it drives the formation of national level legislation in nearly 200 nations. However, the CBD leaves it up to each nation to determine which species it considers to be priority invasive species. Once nations declare a species to be a priority invasive, the nation is obliged by the CBD to engage in control or eradication. However, for those non-native species that are not considered priority invasives, control and management are lower priority and less likely to be conducted. For this reason, the determination of whether or not a species is a priority invasive has important repercussions (Fig. 9.1).

In the determination of invasive status, the CBD and many other regulatory bodies consider only the biological harm caused by a species (CBD 2002; Cassinello 2018; Crowley et al. 2019). However, Goodenough (2010) rightly points out that "consideration of the entire spectrum of impacts is a necessary prerequisite in formulating objective and justifiable policies and management initiatives." Recognizing this, a variety of authors propose that determination of invasive status should include ecological, economic, and social harm and benefits when assessing a species' invasive status (Kueffer and Hadorn 2008; Schlaepfer et al. 2011; Strubbe et al. 2011; Kumschick et al. 2012). However, in order to operationalize any of these definitions, governments and other stakeholders must decide upon which species are, in fact, invasive and which of those are priorities for control (Postigo et al. 2019).

In this chapter we present and evaluate the scientifically documented and likely impacts of naturalized parrots from a mitigation perspective. Our goal is not to engage in "scientific denialism" of the impacts of naturalized parrots (Russell and Blackburn 2017) but to objectively evaluate when and where naturalized parrots have been documented as causing ecological, economic, or social harm or benefit. By doing this, we hope to advance scientific understanding of the circumstances under which naturalized parrots do and do not have important positive and negative impacts. At the end of the chapter, we return to

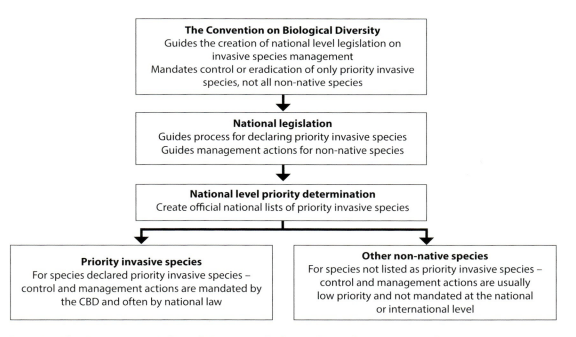

Figure 9.1. The Convention on Biological Diversity is the largest driver of management of non-native species worldwide. Its text drives the development of national-level legislation and mandates that participating parties (195 nations and the EU but not the US) control or eradicate priority invasive species. At the national level, the key step in this process is determining which species are designated as priority invasive species, as this has important implications for how or if non-native species are managed.

our central question—"Are naturalized parrots priority invasive species?"—so that we can provide a tool for stakeholders at all levels who must determine whether, when, and where to manage naturalized parrots.

ECOLOGICAL IMPACTS

Much has been written about the ecological harm caused by naturalized parrot populations, and the reviews fall into two general types. Some focus exclusively on naturalized parrots from a precautionary perspective, and for this reason provide a mixture of potential impacts, anecdotal observations, and scientific documentation of ecological impacts (Polkanov and Keeling 2001; Menchetti and Mori 2014; Menchetti et al. 2016; Vall-llosera et al. 2017; Mori and Menchetti, chap. 6 this vol.; Senar et al., chap. 7 this vol.). Other reviews have been written from more of a mitigation-based perspective and present mostly impacts backed up with statistically significant research findings (Strubbe et al. 2011; Martin-Albarracin et al. 2015; White et al. 2019). These two types of reviews paint very different pictures of the overall impacts of naturalized parrots. The

TABLE 9.1

Confirmed and probably negative ecological impacts caused by naturalized parrots.

IMPACT TYPE	PROBABLE	CONFIRMED
Competitive exclusion	Reduction in nesting success in Italian Sparrows, Common Swifts, Lesser Kestrels, and Eurasian Scops Owls from competition with Rose-ringed Parakeets (Hernández-Brito et al. 2014; Mori et al. 2017a; Grandi et al. 2018)	Eurasian Nuthatches in Belgium may be suffering a ~10% decline in the region due to competition with Rose-ringed Parakeets (Strubbe and Matthysen 2007, 2009; Strubbe et al. 2010)
	Echo Parakeet decline exacerbated by competition for cavities with Rose-ringed Parakeets (Jones 1980)	Decline in Greater Noctule Bats due to competition with Rose-ringed Parakeets (Hernández-Brito et al. 2018)
	Interactions with individuals commonly reported but no population-level impacts detected (see review by Mori and Menchetti, chap. 6 this vol.)	
Disease transmission	PBFD in Echo Parakeets may have come from naturalized Rose-ringed Parakeets (Kundu et al. 2012, Fogell et al. 2018)	Monk Parakeets have spread mites to Rock Doves, but no population-level impacts detected (Mori et al. 2019; Mori and Menchetti, chap. 6 this vol.)
Hybridization with native species	Not reported	Red-lored and Lilac-crowned Amazons have hybridized with Red-crowned Parrots in a mixed native/naturalized population in Texas (Kiacz and Brightsmith, unpubl.)
Herbivory	Only small localized damage reported (Shields et al 1974; Styche 2000; Fletcher and Askew 2007)	Not reported
Facilitation of non-native species	The presence of Rose-ringed Parakeets may aid the establishment of Alexandrine Parakeets (Ancillotto et al. 2016)	In Israel, Rose-ringed Parakeets may facilitate breeding by the invasive Common Mynas by opening cavities too small for them to use normally (Orchan et al. 2013)

former is written mostly to support the use of precautionary management to control species before they become problems, and the latter is written more to support decisions to control species based on reducing known impacts (see Crowley et al. 2019 for a discussion of precautionary and mitigation management).

Like the works of other authors, our review focuses on six major classes of ecological impacts relevant to naturalized parrot populations: competition, transmission of diseases, hybridization, interactions with other non-native species, herbivory and impact on ecosystems, and ecological benefits (Martin-Albarracin et al. 2015; White et al. 2019). For each topic, we critically review the literature and summarize the nature and magnitude of the documented and highly likely ecological impacts of naturalized parrot populations in order to provide a mitigation-style review of the impacts of naturalized parrots (Table 9.1).

Competition

Competition is one of the most commonly studied impacts of introduced birds on native species (Martin-Albarracin et al. 2015; White et al. 2019). Competition between naturalized parrots and native species is most likely to occur for two principal resources: tree cavities and food. All naturalized parrots, except for the Monk Parakeet (*Myiopsitta monachus*), use cavities for nesting, and tree cavities for bird nesting are a limited resource (Newton 1994; Marsden and Jones 1997; Orchan et al. 2013). There are many anecdotal accounts of naturalized parrots fighting over tree cavities with native birds, but there are relatively few instances in which competitive exclusion by naturalized parrots may be significantly reducing populations of other species. Here we highlight the most important of those impacts in order to set the stage for further discussions on their relative magnitude.

Eurasian Nuthatches (*Sitta europaea*) in Belgium are reported to be suffering a ~10% decline in the region due to competition with Rose-ringed Parakeets (*Psittacula krameri*) (Strubbe and Matthysen 2007, 2009; Strubbe et al. 2010). However, in the United Kingdom (UK) such impacts are apparently absent (Newson et al. 2011). In Italy, the spatial distribution of the cavity-nesting Eurasian Scops Owl (*Otus scops*) shifted after invasion by Rose-ringed Parakeets, presumably to reduce nest competition, but owl population declines were not detected (Mori et al. 2017a). In holes in brick towers in Italy, an increase in Rose-ringed Parakeets corresponded with a decrease in nesting Italian Sparrows (*Passer italiae*), a reduction in nesting height of Common Swifts (*Apus apus*), and an increase in fallen swift fledglings, suggesting a causative relationship (Grandi et al. 2018). Yosef et al. (2016) suggest that competition between native Eurasian Hoopoes (*Upupa epops*) and naturalized Rose-ringed Parakeets in desert date-palm plantations in Israel led to precipitous declines in hoopoes. However, the data from that study suggest local hoopoe populations were declining before the arrival of the parakeets, and in only one of two plots did the decline increase after the arrival of the parakeets. In Spain, competition for cavities from Rose-ringed Parakeets has caused an 80% reduction in the largest European population of the Greater Noctule Bat (*Nyctalus lasiopterus*) (Hernández-Brito et al. 2018). The parakeets may also be competing with Lesser Kestrels (*Falco naumanni*) for nest cavities (Hernández-Brito et al. 2014). In Mauritius, one of the likely factors contributing to the decline of the endangered Echo Parakeet (*Psittacula eques*) was competition for cavities with an introduced congener, the Rose-ringed Parakeet (Jones 1980).

Besides competition for nest sites, competition for food resources between naturalized parrots and native species has been repeatedly suggested or anecdotally documented (Batllori and Nos 1985; Lever 2005; Neo 2012; Menchetti and Mori 2014). Studies show that the presence of parakeets at bird feeders reduces feeding by native birds (Peck et al. 2014; Le Louarn et al. 2016). However, there is no evidence to date that this competition has population- or ecosystem-level impacts (Menchetti and Mori 2014; Martin-Albarracin et al. 2015; Appelt et al. 2016; Menchetti et al. 2016).

Multiple authors have concluded that the Rose-ringed Parakeets in Europe and Africa may be having impacts on individual birds but very little impact at the population and ecosystem levels (Newson et al. 2011; Hernández-Brito et

Example of a large, compound nest of Monk Parakeets (*Myiopsitta monachus*) at an electrical substation. Miami-Dade County, Florida, US, April 2007. Photo by James Lindsay.

al. 2014; Martin-Albarracin et al. 2015). However, the impacts on Echo Parakeets in Mauritius and Greater Noctule Bats in Spain show that Rose-ringed Parakeets can have noteworthy impacts under some circumstances. However, apart from the cases presented here, there are few if any statistically significant population-level impacts of competition between native species and the hundreds of naturalized parrot populations globally (Martin-Albarracin et al. 2015).

Disease

Disease spread has been greatly facilitated by the global transport of goods and animals. Captive parrots and naturalized populations carry diseases capable of causing significant ecological harm. However, there have been relatively few studies that show disease spread from naturalized birds to native species (Peters et al. 2014; Martin-Albarracin et al. 2015; Fogell et al. 2018; Mori et al. 2018; Raidal and Peters 2018), but see Strubbe et al. (2011). Parrots are known carriers of a variety of avian diseases and parasites, some of which can infect humans and other bird species, including psittacine beak and feather disease (PBFD), exotic Newcastle disease (END), psittacosis (Lever 2005; Harrison and Lightfoot 2006; Menchetti and Mori 2014; Peters et al. 2014; Sa et al. 2014; Fogell et al. 2018; Raidal and Peters 2018), and parasites such as lice and mites (Mori et al. 2015; Mori et al. 2019). There is little information to suggest that naturalized parrots spread diseases to native non-psittacine birds, but Monk Parakeets have apparently spread mites to Rock Doves (*Columba livia*) (Mori et al. 2019). In addition, Millsap et al. (2004) report that Monk Parakeets made a nest in the base of an urban Bald Eagle (*Haliaeetus leucocephalus*) nest, and one of the fledgling eagles in the nest died of an infection most likely caused by *Chlamydophila psittaci*. However, in neither of these cases were any population-level impacts on native species suspected.

Recent genetic studies have looked for evidence of disease exchange between naturalized parrot populations and native parrot populations. An outbreak of PBFD among endangered Echo Parakeets on the island of Mauritius caused death in over 50% of those infected (Richards 2010; Kundu et al. 2012; BirdLife International 2019). The researchers suggest that the infection in the Echo Parakeets came from Rose-ringed Parakeets, but whether the transmission occurred among birds in the wild or while birds of both species were being held in captivity was unclear (Kundu et al. 2012; Fogell et al. 2018). Regardless, the data do suggest that the disease is being passed back and forth between both species in the wild, and that this disease spread remains a threat to the Echo Parakeet (BirdLife International 2019). In New Zealand, PBFD virus has been detected in the endemic and vulnerable Red-crowned (*Cyanoramphus novaezelandiae*) and Yellow-crowned Parakeets (*C. auriceps*) (Massaro et al. 2012; Jackson et al. 2015). PBFD virus has also been found at high rates in naturalized Crimson Rosellas (*Platycercus elegans*; 15%) and Sulphur-crested Cockatoos (*Cacatua galerita*; 28%) in New Zealand (Ha et al. 2007), although the source of the virus is unclear (Massaro et al. 2012; Jackson et al. 2015). A molecular study of PBFD from endangered Cape Parrots (*Poicephalus robustus*) in Africa suggests that the virus currently circulating in the wild population came from a recent infection, perhaps from captive birds, but there is no evidence that the infection came from naturalized parrot populations (Regnard et al. 2015). In summary, a variety of diseases in wild parrot populations likely came from parrots moved internationally as part of legal or illegal trade. However, with the exception of the Echo Parakeets in Mauritius, there is little hard evidence that known naturalized parrot populations are spreading diseases to wild parrot populations (Ha et al. 2007, 2009; Ortiz-Catedral et al. 2009; Kundu et al. 2012; Massaro et al. 2012; Menchetti and Mori 2014). While disease spread to humans and other species is a suggested potential impact of naturalized species, we can find no evidence of diseases being spread from naturalized parrot populations to humans or other mammals (Menchetti and Mori 2014; Mori et al. 2018; White et al. 2019).

Hybridization

Hybridization is a complex and difficult topic in conservation biology (Allendorf et al. 2001; Baker et al. 2014; Hamilton and Miller 2016; Kovach et al. 2016). Some authors suggest that the new genes infused to taxa through hybridization may enhance long-term survival (Hamilton and Miller 2016), while others clearly point out that interspecific hybridization can lead to extinction of parental species. There is concern that naturalized parrots could hybridize with endangered endemic species like the Puerto Rican Amazon (*Amazona vittata*) (Snyder et al. 1987). In the southern tip of Texas, there is a population of Red-crowned Amazons (*A. viridigenalis*) that is considered native (USFWS 2019) but is likely composed of a mix of native and naturalized individuals (Neck 1986; Enkerlin-Hoeflich and Hogan 1997; AOU 1998; Burgess 2001). Recent work in this population has found at least 12 nesting pairs consisting of a Red-crowned Amazon and a Red-lored (*A. autumnalis*) or Lilac-crowned Amazon (*A. finschi*), and at least 11 apparently hybrid offspring in this population of about 700 birds (Kiacz and Brightsmith, unpubl. data). Other than this one somewhat special case in Texas, there seem to be no well-documented cases of extensive hybridization between native and naturalized parrots (Wiley 1991; Menchetti and Mori 2014; Martin-Albarracin et al. 2015).

The threat of hybridization is still very real, because a variety of naturalized parrot species have been documented hybridizing. Many hybrids among naturalized *Amazona* parrots have been recorded: Turquoise-fronted (*A. aestiva*), Red-crowned, Lilac-crowned, and Yellow-crowned (*A. ochrocephala*) Amazons have been recorded forming mixed pairs, hybridizing, and producing fertile offspring in both the US and Europe (Garrett 1997; Mori et al. 2017b; USFWS 2019; Kiacz and Brightsmith, chap. 5 this vol.). Members of the genus *Psittacara* (ex *Aratinga*) also hybridize in naturalized populations (Bittner 2004; Pyle and Pyle 2017). Recently, the first hybrids of Rose-ringed Parakeet × Alexandrine Parakeet (*Psittacula eupatria*) have been recorded in Spain (Postigo 2016). Hybridization can also occur among native parrots. For example, genetically pure Chatham Parakeets (*Cyanoramphus forbesi*) are rare now due to hybridization with native Red-crowned Parakeets in New Zealand (Tompkins et al. 2006).

Interactions with Other Non-native Bird Species and Exotic Plants

Naturalized parrots, through their propensity to occupy urban habitats, often come in contact with other non-native species (Ortega-Álvarez and MacGregor-Fors 2009; Orchan et al. 2013). There are a few reports of naturalized parrots both inhibiting and facilitating the nesting of other naturalized birds (Menchetti and Mori 2014; White et al. 2019). In Israel, Rose-ringed Parakeets may facilitate breeding by the invasive Common Myna (*Acridotheres tristis*) by opening cavities too small for it to use normally (Orchan et al. 2013), and the presence of Rose-ringed Parakeets may aid the establishment of Alexandrine Parakeets (Ancillotto et al. 2016). Some authors report that naturalized Rainbow Lorikeets (*Trichoglossus moluccanus*) may reduce nesting success of Common Mynas (Menchetti and Mori 2014), but it appears that this behavior was recorded in the lorikeet's native range (Chapman 2005). Overall, based on our current state of knowledge, the interactions of naturalized parrots with other non-native bird species are unlikely to have major population or ecosystem-level impacts.

Parrots are increasingly recognized as important seed dispersers capable of influencing

Red-masked Parakeets (*Psittacara erythrogenys*) in the financial district of San Francisco, California, US. Photo by Eliya, Flickr User/CC BY 2.0 (https://creativecommons.org/licenses/by/2.0).

long-term forest dynamics (Tella et al. 2015; Baños-Villalba et al. 2017; Blanco et al. 2018). However, there is little research on naturalized parrots and their interactions with vegetation (Menchetti and Mori 2014; Mori and Menchetti, chap. 6 this vol.). Studies suggest that seeds of fleshy-fruited invasive plants consumed by Rose-ringed Parakeets have poor germination, suggesting that they are not important dispersers of these invasive plants (Thabethe et al. 2015; Shiels et al. 2018). However, these and other authors suggest that the parakeets may successfully disperse seeds of exotics by transporting them in their beaks or by occasional seeds surviving gut passage (Runde et al. 2007; VanderWerf and Kalodimos, chap. 13 this vol.). Given the limited evidence to date (Mori and Menchetti, chap. 6 this vol.) and the propensity for most naturalized parrots to inhabit mostly urban areas (see "Urban Habitats" section, below), it seems unlikely that naturalized parrots are or will become a major force in the establishment and spread of invasive exotic plants.

Herbivory and Impacts on Ecosystems

The diets of most parrots are dominated by plant reproductive parts (seeds, fruits, and flowers) (Collar 1997; Renton et al. 2015). As a result, naturalized parrots are unlikely to cause massive die-off of adult plants similar to those caused by herbivorous insects. It is not surprising, therefore, that impacts of naturalized parrots on native vegetation and ecosystem structure have rarely been reported (Menchetti and Mori 2014; Mori and Menchetti, chap. 6 this vol.). In New Zealand, a wintering group of naturalized Sulphur-crested Cockatoos clipped branches and caused localized damage to canopy trees in a native forest (Styche 2000). In Australia, Rose-ringed Parakeets strip bark and have killed some trees, which may alter local forest composition and structure (Fletcher and Askew 2007). Similarly, localized damage by Monk Parakeets to native canopy trees was observed in a park in New Jersey (Shields et al. 1974). Despite these observations, there seems to be no evidence that naturalized parrots are causing large-scale ecologically important damage to native plants and ecosystem structure.

Ecological Benefits

Naturalized parrots rarely provide the more traditional conservation benefits documented for other exotic species, such as food for native species, facilitating recovery of native species, filling niches vacated by extinct native species, or providing beneficial ecosystem services (Maris and Béchet 2010; Schlaepfer et al. 2011). However, one study suggests that native birds nesting near Rose-ringed Parakeets may benefit from protection from predators offered by the aggressive parakeets (Hernández-Brito et al. 2014). Additionally, Rose-ringed Parakeets may enlarge cavities, thus making them useful for other species such as Stock Doves (*Columba oenas*) (Czajka et al. 2011). Another study suggests that Italian Sparrows (*Passer italiae*) may benefit from nesting in Monk Parakeet nests (Di Santo et al. 2017). Introduced parrots can also serve as food for native predators (Mori and Menchetti, chap. 6 this vol.). However, none of these impacts are likely to result in significant population- or ecosystem-level benefits.

Most of the real and potential ecological benefits of naturalized parrots seem to be more indirect (Kiacz and Brightsmith, chap. 5 this vol.). Naturalized populations of threatened species may serve as source populations for captive breeding or direct translocation back to the native range. The naturalized parrots may also serve as research populations to help investigators obtain knowledge or develop techniques that can aid parrots in their native ranges or help advance our understanding of invasion biology.

In summary, most of the ecological benefits of naturalized parrots documented to date will not likely have large-scale impacts on native populations or native ecosystems. However, the potential for naturalized populations as sources for captive-breeding stock or translocation and as study systems to advance basic knowledge of parrots and invasion biology should not be discounted, and all potential benefits should be considered when weighing the potential impacts of these naturalized populations.

Ecological Impacts Summary

Our review suggests that the cumulative global impacts of naturalized parrots on native species and ecosystems has been relatively small (Table 9.1). The magnitude of the conservation benefits of naturalized parrots are also apparently quite small. Of the approximately 60 parrot species with naturalized populations, the Rose-ringed Parakeet is responsible for nearly all the well-documented and highly likely negative ecological impacts. The most serious and well-documented impacts are probably the displacement of the Greater Noctule Bat by Rose-ringed Parakeets in Seville, Spain (Hernández-Brito et al. 2014) and the competition and disease sharing with Echo Parakeets in Mauritius (Kundu et al. 2012). The competition between the nonthreatened Eurasian Nuthatch and the Rose-ringed Parakeet has been well documented, but the evidence—that the impacts are of small magnitude (0%–10%), concentrated in urban areas, and spatially variable—suggests that this will have a relatively minor impact on the nuthatches overall (Strubbe et al. 2010).

To put these data on naturalized parrots into perspective, we present the following analysis: The IUCN (International Union for Conservation of Nature) Red List includes 11,126 species of birds and 5,792 species of mammals in its current database (IUCN 2019). Of these, 4,556 birds and mammals are threatened by "biological resource use," defined as harvesting of the species itself or of species needed for its survival (e.g., poaching, logging, etc.); 4,104 by "agriculture and aquaculture"; and 1,312 by "non-native invasive species," including diseases (IUCN 2019). Our analysis suggests that globally, at most, one species of bird and one species of bat are currently being directly impacted by naturalized parrots. This represents 0.012% of the evaluated species and 0.15% of the species threatened by introduced species. It is hard for us to agree with the statement that the increase of parakeets in Europe has "begun to exert serious problems to native environments, wildlife and humans" (Menchetti et al. 2016). Instead, our analysis agrees more with the suggestion that naturalized parrots have not lived up to the dire forecasts of large-scale environmental impacts (Pruett-Jones et al. 2012).

ECONOMIC IMPACTS

For decades, authors have been concerned about the potential economic damage that naturalized parrot populations could cause (Shields et al. 1974; Williams et al. 2010). Despite this persistent worry, we still have a relatively poor idea of the economic impacts naturalized parrots cause. Economic damage attributed to naturalized parrots normally falls into one of four categories: agricultural losses, electrical infrastructure damage, building damage, and airplane collisions (Lever 2005; Williams et al. 2010). Here we examine the scientific evidence for each of these (Table 9.2).

Agricultural Damage

Damage by parrots to commercial agriculture has been widely reported (Forshaw 1989; Collar 1997; Burgio et al. 2016; White et al. 2019). However, much of this evidence may be based on anecdotes and inflated claims by stakeholders (Bomford and Sinclair 2002; Burgio et al. 2016; Sánchez et al. 2016). For parrots, most reported agricultural damage comes from areas where the parrots are native, and relatively few claims have been attributed to naturalized parrot populations (Styche 2000; Koopman and Pitt 2007; FERA 2009; Menchetti and Mori 2014; Yosef et al. 2016; Shiels et al. 2018; White et al. 2019).

Additionally, the true economic costs have rarely been estimated. In agricultural areas adjacent to Barcelona, Spain, Monk Parakeets damaged an average of 28% of the corn crops studied, while in orchards, damages ranged from 6.5% of quinces to 37% of pears (Senar et al. 2016). Of note, damage in individual plots in this study occasionally exceeded 60% for plums, pears, and corn. In the same area, these parakeets damaged an estimated ~$11,000 (€7,800) worth of tomatoes in 2001 (Conroy and Senar 2009). In Israel, a single community reportedly suffered losses of $250,000 per year to Rose-ringed Parakeets in peri-urban date farms (Yosef et al. 2016). In south Florida, Monk Parakeets feeding in small longan fruit orchards near urban areas damaged 20%–60% of the crops, resulting in losses of $15,000/ha at two

TABLE 9.2

Documented negative economic impacts caused by naturalized parakeets. Values preceded by ~ were not reported in US$ in the original publication and were converted using historical rates for the year of publication. See Table 9.3 for a discussion of how these values compare with damage caused by other naturalized species in the United States and United Kingdom.

IMPACT TYPE	DESCRIPTION	PARROT SPECIES	COST	REFERENCES
Peri-urban agriculture	Longan fruits in south Florida, US	Monk Parakeet	$15,000	Tillman et al. 2000
	Grapes, raspberries, apples, and corn in UK	Rose-ringed Parakeet	~$3300–$8200	FERA 2009
	Tomatoes near Barcelona, Spain	Monk Parakeet	~$11,000	Conroy and Senar 2009
	Crops near Barcelona (corn, quince, and pears, 6.5%–37% of crop)	Monk Parakeet	Value not quantified	Senar et al. 2016
	Corn crops in New Zealand	Sulphur-crested Cockatoo	Value not quantified	Styche 2000
	Corn crops in Hawaii, US	Rose-ringed Parakeet	Value not quantified	Koopman and Pitt 2007; Shiels et al. 2018; VanderWerf and Kalodimos, chap. 13 this vol.
	Date farms in Israel	Rose-ringed Parakeet	$250,000, as reported by landowner	Yosef et al. 2016
Electrical infrastructure	About 1,000 parakeet-caused outages and removal of 90 nests in Florida	Monk Parakeet	$585,000/year	Avery et al. 2002
	Removing parakeet nests in Florida	Monk Parakeet	$340,000–$940,000/year	Avery et al. 2008
	Removing parakeet nests in Connecticut, US	Monk Parakeet	$70,000	Silverman 2009
	New York, US: 8 fires blamed on nests	Monk Parakeet	Value not quantified	Kilgannon and Singer 2009
Building damage	Historic windmill and church damaged, UK	Rose-ringed Parakeet	~$76,000 (windmill) and ~$15,000 (church)	Williams et al. 2010
Airplane strikes	Of 98 total strikes over two years in the UK, three were caused by parakeets	Rose-ringed Parakeet	~$27,000	Williams et al. 2010

sites (Tillman et al. 2000). Of note in this study, only longan orchards were damaged, and not lychees, and the seven other naturalized parrots in the area were not reported to feed in orchards. In the UK, Rose-ringed Parakeets feeding in a small vineyard caused a reported 80% drop in wine production and a loss of ~$8,250 (£5,000) per year (FERA 2009). Also in the UK, Rose-ringed Parakeets fed on raspberries, apples, and corn at a small "pick-your-own" fruit farm and damaged at least 10%–15% of the apple crop, resulting in losses of over $3,300 (£2,000) (FERA 2009).

Williams et al. (2010) estimate that parakeets do about $15,000 (£10,000) of damage per year to commercial agriculture in the UK. This is about 0.001% of the ~$1.62 billion (£1.07 billion) in agricultural damage caused by invasive non-native species in the UK (Williams et al. 2010). In all the cases reported here, the agricultural plots impacted were immediately adjacent to large urban areas (within 4 km) with large populations of naturalized parrots. In Spain, the agricultural plots studied were spread across an area of about 1,000 ha, but in all the other cases the agricultural areas were only a few hectares in size.

It is noteworthy that even after many years of parrots' establishment in North America and Europe, areas with a long history of ornithological study, few documented cases of agricultural damage by naturalized parrots exist (Table 9.2). The review by Menchetti and Mori (2014) reports anecdotal or other evidence that 28% (17 of 60 species) of naturalized parrot species damage agricultural crops. However, 70% of naturalized parrot species are reported to be agricultural pests in their native ranges (Runde et al. 2007). Pointing out the difference between the 70% pest status in native ranges and only 28% in naturalized ranges, Menchetti and Mori (2014) imply that it is only a matter of time until greater agricultural impacts of naturalized parrots will be recorded. However, it is important to consider that there may be real ecological reasons underlying this difference between native and naturalized ranges.

Most large-scale commercial agricultural landscapes are highly altered and have little tree cover. They are dominated by relatively few crops, a small number of which are potential parrot foods, and these crops are available in huge amounts for short periods annually. Therefore, parrot populations in areas dominated by large-scale agriculture are unlikely to grow very large, due to severe seasonal food shortages, in addition to a lack of nesting substrate. In order to inflict large-scale agricultural damage, large groups of parrots would need to leave their breeding areas and move to agricultural areas to exploit these seasonally available foods. Most parrots are not truly migratory, and large-scale seasonal movements are driven mostly by local food shortages (Bjork 2000; Renton 2001; USFWS 2019). If the timing of these movements corresponds with crop ripening in agricultural lands, there is potential for conflict (Santos Neto and Gomes 2007). Fortunately, most naturalized parrot populations are restricted to urban and suburban landscapes.

Urban areas have a wide variety of native and exotic species planted to provide fruits and flowers across the seasons for aesthetic and other reasons (McKinney 2002). During times of low food availability for birds, local people may provide supplemental food in the form of bird feeders (Clergeau and Vergnes 2011; Pruett-Jones et al. 2012). As a result, there may be less local annual variation in food supply for naturalized parrot populations than in many native habitats. These lower fluctuations in food supply likely reduce the need for urban naturalized parrots to move en masse to distant agriculture areas or other sites in search of abundant food. Instead, it is more likely that the naturalized parrots will move relatively short distances from urban environments into peri-urban or suburban settings to exploit seasonally abundant preferred foods. This scenario is consistent with the case studies from the US and Europe provided above, in which relatively small groups of parakeets fed on preferred crops in agricultural areas immediately adjacent to nesting areas (Tillman et al. 2000; FERA 2009).

There are a few instances of naturalized parrots damaging crops away from urban centers (Styche 2000; Shiels et al. 2018). Styche (2000) reports that naturalized populations of Sulphur-crested Cockatoos in New Zealand were damaging maize crops in remote mountain valleys of the North Island. This naturalized population was breeding and foraging in native podocarp forest remnants over a large geographic area during most of the year. During the winter, the birds from hundreds of square kilometers formed large flocks that fed on the maize crops (Styche 2000). In this case, the birds were behaving more like native populations.

Other Economic Damage

Damage to electrical infrastructure caused by Monk Parakeet nests is commonly reported from many parts of their naturalized and native ranges (Bucher and Martin 1987; Avery et al. 2002; Minor et al. 2012). The nests cause overheating, shorts, and fires (Bucher and Martin 1987; Spreyer and Bucher 1998; Avery et al. 2002; Reed et al. 2014) that result in elevated repair costs, disruption of service, and lost revenue, and lead to costly prevention measures like removing nests. Surprisingly, the magnitude of Monk Parakeet–induced damage and control has rarely been quantified, with only the following few exceptions (Table 9.2). In New York City, an estimated 8 of 18 power-pole fires were blamed on Monk Parakeets (Kilgannon and Singer 2009), but no dollar estimate for the damage was given. Estimates that do exist include $60,000–$70,000 per year to remove nests in Connecticut (Silverman 2009); $340,000–$940,000 per year to remove nests in Florida (Avery et al. 2008), and $585,000 per year for ~1,000 parakeet-caused outages and removal of 90 nests in Florida (Avery et al. 2002). Of note, none of these reports provide information on the total cost of power grid maintenance, damage from squirrels, woodpeckers, or wind or other natural causes, or the total annual profits from the business, so stakeholders are unable to estimate the relative costs from parakeet damage. However, annual financial reports suggest that Florida Power and Light made about $871 million dollars in 2003, suggesting that the parakeets caused a drop in adjusted earnings of about 0.067%, or about 0.15 cents per share (FPL Group 2004).

Reports of Monk Parakeets nesting on electrical infrastructure come from their native range and their naturalized range throughout the US (Roscoe et al. 1973; Bucher and Martin 1987; Burger and Gochfeld 2009; Burgio et al. 2014; Reed et al. 2014). However, the parakeets' use of electrical infrastructure appears to vary geographically, as studies in and around Barcelona found that of 445 nests, only two were on electrical distribution poles, and two were on street or stadium lights (Domènech et al. 2003). Even more noteworthy, in Israel, studies of rapidly growing populations of Monk Parakeets found over 100 nests, and all were in trees, with none on power poles or other man-made structures (Postigo et al. 2017). Similarly, in Puerto Rico, there is no evidence of Monk Parakeets damaging electrical systems (Falcón and Tremblay 2018). Why parakeets do not use electrical infrastructure in some areas remains unknown, but these reports provide hope that some regions may escape parakeet damage to their electrical systems.

Damage to buildings by naturalized parrots is rarely reported (Table 9.2). Rose-ringed Parakeets and Monk Parakeets are known to damage buildings in Europe, and the partially naturalized populations of Red-crowned Amazons are known to nest in holes in buildings in southern Texas. The only reports with monetary values of the damage estimated that parakeets did ~$15,000 (£10,000) worth of damage per year (averaged over the survey period) to buildings in the UK. Single-year, one-time costs of damage included $75,600 (£50,000) worth of damage done by parakeets to a historic windmill, and ~$15,000 (£10,000) worth of damage done by parakeets removing roofing tiles off a church (Metro 2008; Williams et al. 2010).

Damage to airplanes by naturalized parrots has been recorded rarely (Menchetti and Mori 2014; Menchetti et al. 2016). At Heathrow Airport in the UK, 98 bird strikes were recorded over two years, and three of these were caused by Rose-ringed Parakeets (Williams et al. 2010). These collisions did not harm any people but resulted in an estimated annual cost of about $26,500 (£17,500).

Economic Benefits

To date, there is little to no evidence of major economic benefits from naturalized parrots. In Texas and California, naturalized-parrot-watching tours generate about $2,000 to $5,000 per year for conservation organizations, and some businesses report that local parrots attract customers and are "good for business" (Seymour 2013; Crowley, chap. 3 this vol.; Kiacz and Brightsmith, chap. 5 this vol.).

Economic Impacts Summary

The economic damage done by naturalized parrots to agriculture has been documented and must be acknowledged as a problem in some areas (Tillman et al. 2000; Senar et al. 2016; Yosef et al. 2016) but not in others (Pimentel et al. 2005; FERA 2009; Pruett-Jones et al. 2012; Falcón and Tremblay 2018). While localized impacts of tens to hundreds of thousands of dollars may occur, and mitigation management may be required, it is unlikely that naturalized parrots will have large-scale impacts on major agricultural landscapes. Only where large populations of naturalized parrots break their attachment to urban areas and spread out into large natural areas should the probabilities of large-scale impacts on agriculture increase. The economic and social impacts of Monk Parakeets on electrical infrastructure are real, and this conflict will continue in many areas. However, the use of electrical infrastructure for nesting is spatially variable, so some areas are being spared this impact.

To put the economic impacts of naturalized parrots in perspective, it is worthwhile to compare economic damage caused by naturalized parrots to that caused by other non-native species. In the UK, Williams et al. (2010) report a total impact by all invasive species of ~$2.36 billion (£1.56 billion) per year. Of this, they estimate that Monk and Rose-ringed Parakeets do ~$57,500 (£38,000) of damage per year. This represents about 0.0024% of the total caused by all invasive species combined (Williams et al. 2010) and only 0.000002% of the ~$2.3 trillion (£1.53 trillion) national gross domestic product (Thomas and Williamson 2019). In the US, Pimentel et al. (2005) report an annual cost of $120 billion per year from invasive alien species, equivalent to about 1% of the nation's $13 trillion GDP. However, naturalized parrots were not even mentioned in this work. If we sum up all the values for the US from Table 9.2, we obtain an estimate of $1.6 million per year of damage by naturalized parrots. This value suggests that parakeets are responsible for 0.0013% of the total damage caused by invasive species in the US.

TABLE 9.3

Economic impacts of naturalized terrestrial vertebrates in the United Kingdom and United States with comparison to naturalized parrots. Numbers for the UK are from Williams et al. (2010). Numbers for the US for all species except naturalized parrots are from Pimentel et al. 2005. The estimate for naturalized parrots is a sum of all quantified values for the US from Table 9.2. All economic damage by naturalized parrots in this table is from Rose-ringed and Monk Parakeets. Total is the sum of economic damage from the species presented in the table. Taxonomy follows the original sources.

UK		US	
TAXON	PERCENT OF TOTAL	TAXON	PERCENT OF TOTAL
Rabbits	66%	Rats	48%
Rats	16%	Cats	43%
Deer	8.7%	Pigeons	2.8%
House Mouse	4.5%	Feral pigs	2.0%
Eastern Gray Squirrel	3.5%	Starlings	2.0%
Mink	1.2%	Dogs	1.6%
Geese/swans	0.90%	Mongooses	0.13%
Edible Dormouse	0.091%	Brown Tree Snake	0.030%
Naturalized parrots	0.0095%	Naturalized parrots	0.0041%
Total	£401,001,000	Total	$39,389,100,000

ARE NATURALIZED PARROTS PRIORITY INVASIVE SPECIES?

Another way to put these values in perspective is to compare the economic damage by naturalized parrots directly to the damage caused only by other non-native terrestrial vertebrates (see summary in Table 9.3). In this comparison, mammals are responsible for over 99% of the recorded ~$606 million (£401 million) in damage in the UK and 95% of the $39 billion in damage in the US. By comparison, parakeets are responsible for only 0.0095% and 0.0041% of the total damage by terrestrial vertebrates in the UK and US, respectively.

SOCIAL IMPACTS

A variety of negative impacts of naturalized parrots on human quality of life have been suggested (reviewed by Crowley, chap. 3 this vol.). Aside from the economic impacts discussed above, the most commonly cited include power outages, noise, fouling with feces, and disease spread (Conroy and Senar 2009; Menchetti and Mori 2014; Menchetti et al. 2016; Crowley et al. 2019). One of the more serious social impacts from naturalized parrots may be the power outages caused by Monk Parakeets nesting on infrastructure (Avery et al. 2002). Power outages have great potential to cause anything from minor annoyance to loss of life, so these represent real impacts (Klinger et al. 2014). The most common complaint about naturalized parrots is the noise they make when breeding and roosting in residential neighborhoods (Chapman 2005; Pruett-Jones et al. 2012; Seymour 2013; Menchetti et al. 2016). In Western Australia, the Department of Agriculture has received complaints about damage to backyard fruits and fouling with feces caused by naturalized Rainbow Lorikeets, but no details are available (Chapman 2005). As mentioned above, there are no known cases of disease spread from naturalized parakeets to humans.

Juvenile Red-breasted Parakeets (*Psittacula alexandri*) at a nest in Changi, Singapore. Red-breasted Parakeets are native and widespread in Indonesia and parts of Southeast Asia, but are introduced and naturalized in Singapore, where they compete with other bird species for nest cavities. February 2019. Photo by Evan Landy.

Social Benefits

There are reports from many different areas of the world of local people forming strong attachments to naturalized parrot populations (Low 2000; Burger and Gochfeld 2009; Seymour 2013; Reed 2014; Crowley et al. 2019; Uehling et al. 2019). Nevertheless, it is hard to quantify the magnitude of the social costs and benefits of naturalized parrots and compare them objectively to ecological or economic impacts. Instead, it takes more nuanced approaches to understand what populations of naturalized species may mean to local communities (Pejchar and Mooney 2009; Maris and Béchet 2010, Reed 2014; McMillen et al. 2018). In addition, local managers may not always accept the validity of or recognize the potential importance of these techniques and the findings they produce. One way to measure the importance of naturalized parrots to local people is through the actions they take to protect them, including introducing state and local legislation to protect parrot populations, concealing the locations of nests from eradication teams, scaring birds away from nests to protect them from eradication teams, protesting utility companies that remove the birds, leading tours to see the birds, starting positive blogs and websites, promoting online petitions, working with reporters to provide bad press for control efforts, threatening lawsuits, and soliciting landowners not to allow eradication teams access to their lands (Spreyer 1994; Pruett-Jones et al. 2012; Seymour 2013; Crowley et al. 2019; Uehling et al. 2019). For a more detailed discussion of why people may be taking these actions see the review by Crowley (chap. 3 this vol.).

Social Impacts Summary

Economic considerations aside, the annoyances and hardships resulting from power outages caused by Monk Parakeets are likely the largest negative social impacts of naturalized parrots. Beyond that, and occasional noise complaints, there is little evidence that naturalized parrots have significant negative impacts on human society. However, naturalized parrots may help connect people with nature in urban environments. Regardless of what academics think may be the social impacts of these birds, the actions repeatedly taken against the removal of naturalized parrots show that local people often have strong attachments to the birds and wish to maintain them in their communities.

URBAN HABITATS AS A MITIGATING FACTOR OF ECOLOGICAL HARM

The vast majority of naturalized parrots live in urban or suburban ecosystems (Garrett 1997; Butler 2003; Mabb 2003; Runde et al. 2007; Menchetti and Mori 2014). This seems to hold true even for populations that have been established for decades (Garrett 1997; Spreyer and Bucher 1998; Strubbe and Matthysen 2007; Pruett-Jones et al. 2012; Le Louarn et al. 2018; USFWS 2019). Even more surprisingly, this may hold true in some areas that also support native parrot species, although there is less information from these populations (Chapman 2005; Falcón and Tremblay 2018), and exceptions have been documented (Jones 1980; Styche 2000; Wright and Clout 2001)

In apparent exception to this confinement to urban areas, two recent papers suggest that Monk Parakeets are moving into agricultural areas in Israel and Spain. In Israel, Postigo et al. (2017) report that of the nearly 100 nests they found, 64% were in agricultural habitats, 18% were in semi-agricultural habitats, and only 18% were in urban habitats. However, nests were classified as in agricultural or semi-agricultural areas if they were within a 1 km or 2 km radius, respectively, of any agricultural field. The maps of nesting sites from this study suggests that >90% of the nests were in urban areas and <10% were in agricultural areas (Postigo et al. 2017 Supplemental Material), bringing the habitat usage for this population more in line with the global trends for naturalized parrots. In Spain, as discussed earlier, the parakeets are significantly impacting agricultural areas, but these areas are all within about 4 km of Barcelona's suburbs, suggesting that these parakeets also remain at least partially tied to urban areas.

The confinement of nearly all naturalized parrot populations to urban and peri-urban areas is very important, because it reduces the chances of their having major impacts on large-scale commercial agriculture and greatly reduces the chances of their inflicting major harm on native species and natural ecosystems. In general, urbanization reduces native biodiversity and is a major threat to native species (McKinney 2002; Chace and Walsh 2006; McDonald et al. 2008; van Heezik et al. 2008; Ortega-Álvarez and MacGregor-Fors 2009; Aronson et al. 2014; Symes et al., chap. 17 this vol.), although urban areas can hold populations of threatened native species (Hernández-Brito et al. 2014; Ives et al. 2016; Luna et al. 2018). However, most native species, including threatened ones, that do use urban areas are more likely to occupy patches of remnant native vegetation (Mason 2000; van Heezik et al. 2008; Winchell and Doherty 2008; Ortega-Álvarez and MacGregor-Fors 2010; Aronson et al. 2014) and less likely to use the highly urbanized areas occupied by most naturalized parrots. This reduced spatial overlap should reduce competition between most native species and naturalized parrots.

In locations where naturalized parrots live outside of urban environments, the potential threats to native biodiversity increase. For example, naturalized Rose-ringed Parakeets occupying native habitats in Mauritius are implicated in the decline of Echo Parakeets through competition and disease spread (Jones 1980; Kundu et al. 2012). However, it is worth noting that moving out of urban areas does not automatically result in large ecosystem-level impacts. Naturalized Eastern Rosellas (*Platycercus eximius*) have colonized most of the North Island of New Zealand since establishment around 1910. However, researchers have failed to document any direct impacts on native species during this 100-year period (Wright and Clout 2001; Innes et al. 2010; Massaro et al. 2012; Graham et al. 2013; Jackson et al. 2015).

CONCLUSIONS: NATURALIZED PARROTS AS PRIORITY INVASIVE SPECIES

Naturalized parrots clearly provide a microcosm of the wider, ongoing debate among scientists, governments, stakeholders, and members of the public over what to do about naturalized species (Briggs 2017; Crowley et al. 2017; Davis and Chew 2017; Russell and Blackburn 2017; Crowley et al. 2019). The Convention on Biological Diversity (1992, 2002, 2010) provides an international standard on how nations should act with respect to alien species. As a result, it is worthwhile to consider the question about where and when naturalized parrots rise to the level of priority invasive species, because in these instances, most national governments have an obligation, through the CBD, to control or eradicate these populations.

It is clear from our review that many of the dire predictions about the global ecological and economic impacts of naturalized parrots have not come true (Burger and Gochfeld 2009; Pruett-Jones et al. 2012; Hernández-Brito et al. 2014) and that nearly all naturalized parrot populations have no or low-level impacts (Turbé et al. 2017; White et al. 2019). We recognize that naturalized populations of Rose-ringed Parakeets and Monk Parakeets are expanding rapidly in many areas globally, and this could result in increased impacts in the future (Menchetti et al. 2016; Senar et al. chap. 7 this vol.). However, documented ecological damage to native species and ecosystems in areas without native parrots has been almost negligible to date. In some specific cases, naturalized parrot impacts have risen to the level of conservation concern, as in the case of the displacement of Greater Noctule Bats from tree cavities by Rose-ringed Parakeets in Seville (Hernández-Brito et al. 2014). However, this is the only well-documented significant conservation impact of the approximately 60 naturalized parrot species representing hundreds of populations spread across the globe.

For managers who do not want to make their decisions only on scientifically documented negative impacts but wish to take a more precautionary approach, our interpretation of the scientific literature suggests there are two circumstances under which naturalized parrots living in areas with no native parrots could have an increased probability of ecological impacts: (1) in areas where cavities in major urban centers provide an ecologically important refuge for threatened species, naturalized parrots could compete for these cavities; and (2) where naturalized parrots are likely to survive outside urban areas, their populations could have increased ecological impacts (but such impacts remain undocumented to date). However, outside these two scenarios, the evidence suggests that nearly all naturalized parrots living in urban areas outside of nations with native parrot ranges do not rise to the level of priority invasive species that should be targeted for control.

In our review we were surprised to find that there are no published studies clearly documenting negative impacts of naturalized parrots on native parrots. However, in areas with native parrots, the opportunities for naturalized parrots to spread beyond urban areas seems higher (Jones 1980; Styche 2000; Wright and Clout 2001), and once they are outside urban areas, the opportunities for negative impacts increase.

The CBD does not mandate control of naturalized species that do not cause ecological harm, so there is more leeway at the national level when dealing with exotics that cause only economic impacts. Calculating economic costs of naturalized parrots is important because rarely should businesses or governments spend more to mitigate an economic loss than the cost of the impact itself. When social and moral concerns of local stakeholders are added to evaluations (Crowley, chap. 3 this vol.), the number of control efforts deemed viable will likely be even further reduced. However, there are circumstances under which naturalized parrot populations may be raised to the level of priority invasive species for economic reasons. Cities with heavy investments in peri-urban agriculture, regions with costly power outages, date-farming communes in desert areas, islands where parrots move out from urban areas to impact agriculture, and others may decide that the cost of the parakeets is more than they are willing to bear and may choose control or eradication (Senar et al. 2016; Yosef et al. 2016; Postigo et al. 2017; Shiels et al. 2018). Overall, the literature suggests that there are relatively few circumstances under which naturalized parrots cause large-scale economic damage, and the cost of most agricultural damage reported to date has been low. As a result, we suspect that if societies openly debate the ecological, economic, and social costs and benefits of naturalized parrot populations in urban areas, few will be raised to the level of priority invasive species.

Moving forward, we suggest that scientists continue to monitor and document the impacts—be they beneficial, neutral, or negative—of naturalized parrots and present their findings clearly and unapologetically. Only with a broad array of data from studies of all types can we perform the meta-analyses needed to predict where naturalized populations may cause important ecological or economic impacts, and where they will not. In addition, we hope that researchers will not just document impacts, but also put these impacts in perspective, so that stakeholders from all sectors of society can easily judge the relative magnitudes of these impacts compared to other naturalized species and other sources of harm. Only by using transparent and evidence-based assessments can society make informed decisions on whether or not naturalized parrots should be classified as priority invasive species.

REFERENCES

Allendorf, F. W., Leary, R. F., Spruell, P., and Wenburg, J. K. 2001. The problems with hybrids: Setting conservation guidelines. *Trends in Ecology & Evolution* 16:613–622.

Ancillotto, L., Strubbe, D., Menchetti, M., and Mori, E. 2016. An overlooked invader? Ecological niche, invasion success and range dynamics of the Alexandrine parakeet in the invaded range. *Journal of Biological Invasions* 18:583–595.

AOU. 1998. *Check-list of North American Birds*. 7th ed. Washington, DC: American Ornithologists' Union.

Appelt, C. W., Ward, L. C., Bender, C., Fasenella, J., Van Vossen, B. J., and Knight, L. 2016. Examining potential relationships between exotic monk parakeets (*Myiopsitta monachus*) and avian communities in an urban environment. *Wilson Journal of Ornithology* 128:556–566.

Aronson, M.F.J., La Sorte, F. A., Nilon, C. H., Katti, M., Goddard, M. A., Lepczyk, C. A., Warren, P. S., Williams, N.S.G., Cilliers, S., Clarkson, B., et al. 2014. A global analysis of the impacts of urbanization on bird and plant diversity reveals key anthropogenic drivers. *Proceedings of the Royal Society B: Biological Sciences* 281:20133330.

Avery, M. L., Greiner, E. C., Lindsay, J. R., Newman, J. R., and Pruett-Jones, S. 2002. Monk parakeet management at electric utility facilities in south Florida. In *Proceedings of the 20th Vertebrate Pest Conference*, ed. R. M. Timm and R. H. Schmidt, 140–145. Davis: Univ. California.

Avery, M. L., Yoder, C. A., and Tillman, E.A. 2008. Diazacon inhibits reproduction in invasive monk parakeet populations. *Journal of Wildlife Management* 72:1449–1452.

Ayuntamiento de Madrid. 2019. El censo de cotorras argentina en Madrid crece un 33% en los últimos tres años. Portal web del Ayuntamiento de Madrid. 10 Oct. 2019.

Baker, J., Harvey, K. J., and French, K. 2014. Threats from introduced birds to native birds. *Emu* 114:1–12.

Baños-Villalba, A., Blanco, G., Díaz-Luque, J. A., Dénes, F. V., Hiraldo, F., and Tella, J. L. 2017. Seed dispersal by macaws shapes the landscape of an Amazonian ecosystem. *Scientific Reports* 7:7373.

Batllori, X., and Nos, R. 1985. Presencia de la cotorrita gris (*Myiopsitta monachus*) y de la cotorrita de collar (*Psittacula krameri*) en el área metropolitana de Barcelona. *Miscelània Zoològica* 9:407–411.

BirdLife International. 2019. Echo parakeet, *Psittacula eques*. The IUCN Red List of Threatened Species 2019: e. T22685448A154065622. Accessed 10 Feb. 2019. https://dx.doi.org/10.2305/IUCN.UK.2019-3.RLTS.T22685448A154065622.en.

Bittner, M. 2004. *The Wild Parrots of Telegraph Hill: A Love Story … with Wings*. New York: Harmony Books.

Bjork, R. 2000. Reserve network design and management in moist lowland tropical forests: Habitat, site, and spatial requirements of the mealy parrot as a guide. PhD diss., Dept. of Fisheries and Wildlife, Oregon State Univ., Corvallis.

Blanco, G., Hiraldo, F., and Tella, J. L. 2018. Ecological functions of parrots: An integrative perspective from plant life cycle to ecosystem functioning. *Emu* 118:36–49.

Bomford, M., and Sinclair, R. 2002. Australian research on bird pests: Impact, management and future directions. *Emu* 102:29–45.

Briggs, J. C. 2017. Rise of invasive species denialism? A response to Russell and Blackburn. *Trends in Ecology & Evolution* 32:231–232.

Bucher, E. H., and Martín, L. F. 1987. Los nidos de cotorras (*Myiopsitta monachus*) como causa de problemas en líneas de transmisión eléctrica. *Vida Silvestre Neotropical* 1:50–51.

Burger, J., and Gochfeld, M. 2009. Exotic monk parakeets (*Myiopsitta monachus*) in New Jersey: Nest site selection, rebuilding following removal, and their urban wildlife appeal. *Urban Ecosystems* 12:185–196.

Burgess, H. H. 2001. Red-crowned parrot. In *The Texas Breeding Bird Atlas*, ed. K.L.P. Benson and K. A. Arnold. College Station and Corpus Christi, TX: Texas A&M Univ. System.

Burgio, K. R., Rubega, M. A., and Sustaita, D. 2014. Nest-building behavior of monk parakeets and insights into potential mechanisms for reducing damage to utility poles. *PeerJ* 2:e601.

Burgio, K. R., van Rees, C. B., Block, K. E., Pyle, P., Patten, M. A., Spreyer, M. F., and Bucher, E. H. 2016. Monk parakeet (*Myiopsitta monachus*) (v.3.0). In *The Birds of North America*, ed. P. G. Rodewald. Ithaca, NY: Cornell Lab of Ornithology.

Butler, C. J. 2003. Population biology of the introduced rose-ringed parakeet *Psittacula krameri* in the UK. PhD diss., Oxford Univ., UK.

Carroll, S. P. 2011. Conciliation biology: The eco-evolutionary management of permanently invaded biotic systems. *Evolutionary Applications* 4:184–199.

Cassinello, J. 2018. Misconception and mismanagement of invasive species: The paradoxical case of an alien ungulate in Spain. *Conservation Letters* 11:e12440.

CBD. 1992. The United Nations Convention on Biological Diversity. Repr. in *International Legal Materials* 31: 818.

CBD. 2002. Report of the sixth meeting of the conference of the parties to the Convention on Biological Diversity. Decision VI/23. UNEP. 27 May 2002. https://www.cbd.int/doc/meetings/cop/cop-06/official/cop-06-20-en.pdf.

CBD. 2010. COP 10 Decision X/2: Strategic plan for biodiversity 2011–2020. Convention on Biodiversity. Accessed 15 Mar. 2019. https://www.cbd.int/decision/cop/?id=12268.

Chace, J. F., and Walsh, J. J. 2006. Urban effects on native avifauna: A review. *Landscape and Urban Planning* 74:46–69.

Chapman, T. 2005. *The Status and Impact of the Rainbow Lorikeet (*Trichoglossus haematodus moluccanus*) in South-west Western Australia*. Perth, Western Australia: Dept. of Agriculture and Food.

Chew, M. K. 2015. Ecologists, environmentalists, experts, and the invasion of the second greatest threat. *International Review of Environmental History* 1:7–40.

Clergeau, P., and Vergnes, A. 2011. Bird feeders may sustain feral rose-ringed parakeets (*Psittacula krameri*) in temperate Europe. *Journal of Wildlife Biology* 17:248–252.

Collar, N. J. 1997. Family Psittacidae. In *Handbook of the Birds of the World*, ed. J. del Hoyo, A. Elliott, and J. Sargatal, 280–479. Barcelona: Lynx Edicions.

Conroy, M. J., and Senar, J. C. 2009. Integration of demographic analyses and decision modeling in support of management of invasive monk parakeets, an urban and agricultural pest. In *Modeling Demographic Processes in Marked Populations*, ed. D. L. Thomson, E. G. Cooch, and M. J. Conroy, 491–510. Boston, MA: Springer US.

Crowley, S. L., Hinchliffe, S., and McDonald, R. A. 2019. The parakeet protectors: Understanding opposition to introduced species management. *Journal of Environmental Management* 229:120–132.

Crowley, S. L., Hinchliffe, S., Redpath, S. M., and McDonald, R. A. 2017. Disagreement about invasive species does not equate to denialism: A response to Russell and Blackburn. *Trends in Ecology & Evolution* 32:228–229.

Czajka, C., Braun, M. P., and Wink, M. 2011. Resource use by non-native ring-necked parakeets (*Psittacula krameri*) and native starlings (*Sturnus vulgaris*) in Central Europe. *Open Ornithology Journal* 4:17–22.

REFERENCES

Davis, M. A., and Chew, M. K. 2017. "The denialists are coming!" Well, not exactly: A response to Russell and Blackburn. *Trends in Ecology & Evolution* 32:229–230.

Davis, M. A., Chew, M. K., Hobbs, R. J., Lugo, A. E., Ewel, J. J., Vermeij, G. J., Brown, J. H., Rosenzweig, M. L., Gardener, M. R., Carroll, S. P., et al. 2011. Don't judge species on their origins. *Nature* 474:153.

Di Santo, M., Battisti, C., and Bologna, M. A. 2017. Interspecific interactions in nesting and feeding urban sites among introduced monk parakeet (*Myiopsitta monachus*) and syntopic bird species. *Ethology Ecology & Evolution* 29:138–148.

Domènech, J., Carrillo, J., and Senar, J. C. 2003. Population size of the monk parakeet (*Myiopsitta monachus*) in Catalonia. *Revista Catalana d'Ornitologia* 20:1–9.

Dueñas, M.-A., Ruffhead, H. J., Wakefield, N. H., Roberts, P. D., Hemming, D. J., Diaz-Soltero, H. 2018. The role played by invasive species in interactions with endangered and threatened species in the United States: A systematic review. *Biodiversity and Conservation* 27:3171–3183.

Enkerlin-Hoeflich, E. C., and Hogan, K. M. 1997. Red-crowned parrot. In *The Birds of North America*, ed. A. Poole and F. Gill. Washington, DC: Academy of Natural Sciences and the American Ornithologists' Union.

Falcón, W., and Tremblay, R. L. 2018. From the cage to the wild: Introductions of Psittaciformes to Puerto Rico. *PeerJ* 6:e5669.

Fletcher, M., and Askew, N. 2007. *Review of the Status, Ecology and Likely Future Spread of Parakeets in England*. York, UK: Central Science Laboratory.

FPL Group. 2004. *Restoring Power, Restoring Lives: FPL Group 2004 Annual Review*. http://www.investor.nexteraenergy.com/~/media/Files/N/NEE-IR/reports-and-fillings/annual-reports/2004fplgroupannualreport.pdf.

Fogell, D. J., Martin, R. O., Bunbury, N., Lawson, B., Sells, J., McKeand, A. M., Tatayah, V., Trung, C. T., and Groombridge J. J. 2018. Trade and conservation implications of new beak and feather disease virus detection in native and introduced parrots. *Conservation Biology* 32:1325–1335.

FERA. 2009. *Rose-ringed Parakeets in England: A Scoping Study of Potential Damage to Agricultural Interests and Management Measures*. York: UK: Food and Environment Research Agency.

Forshaw, J. M. 1989. *Parrots of the World*. 3rd ed. Melbourne, Australia: Landsdowne Editions.

Garrett, K. L. 1997. Population status and distribution of naturalized parrots in southern California. *Western Birds* 28:181–195.

Goodenough, A. 2010. Are the ecological impacts of alien species misrepresented? A review of the "native good, alien bad" philosophy. *Community Ecology* 11:13–21.

Graham, M., Veitch, D., Aguilar, G., and Galbraith, M. 2013. Monitoring terrestrial bird populations on Tiritiri Matangi Island, Hauraki Gulf, New Zealand, 1987–2010. *New Zealand Journal of Ecology* 37:359–369.

Grandi, G., Menchetti, M., and Mori, E. 2018. Vertical segregation by breeding ring-necked parakeets *Psittacula krameri* in northern Italy. *Journal of Urban Ecosystems* 21:1011–1017.

Gurevitch, J., and Padilla, D. K. 2004. Are invasive species a major cause of extinctions? *Trends in Ecology & Evolution* 19:470–474.

Ha, H. J., Alley, M. R., Cahill, J. I., Howe, L., and Gartrell, B. D. 2009. The prevalence of psittacine beak and feather disease virus infection in native parrots in New Zealand. *New Zealand Veterinary Journal* 57:50–52.

Ha, H., Anderson, I., Alley, M., Springett, B., and Gartrell, B. 2007. The prevalence of beak and feather disease virus infection in wild populations of parrots and cockatoos in New Zealand. *New Zealand Veterinary Journal* 55:235–238.

Hamilton, J. A., and Miller, J. M. 2016. Adaptive introgression as a resource for management and genetic conservation in a changing climate. *Conservation Biology* 30:33–41.

Harrison, G. L., and Lightfoot, T. L., eds. 2006. *Clinical Avian Medicine*. Palm Beach, FL: Spix Publishing.

Hernández-Brito, D., Carrete, M., Ibáñez, C., Juste, J., and Tella, J. L. 2018. Nest-site competition and killing by invasive parakeets cause the decline of a threatened bat population. *Royal Society Open Science* 5:172477.

Hernández-Brito, D., Carrete, M., Popa-Lisseanu, A. G., Ibáñez, C., and Tella, J. L. 2014. Crowding in the city: Losing and winning competitors of an invasive bird. *PLoS One* 9:e100593.

Innes, J., Kelly, D., Overton, J. M., and Gillies, C. 2010. Predation and other factors currently limiting New Zealand forest birds. *New Zealand Journal of Ecology* 34:86.

IUCN. 2019. The IUCN Red List of Threatened Species (v.2019-1). https://www.iucnredlist.org.

Ives, C. D., Lentini, P. E., Threlfall, C. G., Ikin, K., Shanahan, D. F., Garrard, G. E., Bekessy, S. A., Fuller, R. A., Mumaw, L., Rayner, L., et al. 2016. Cities are hotspots for threatened species. *Global Ecology and Biogeography* 25:117–126.

Jackson, B., Varsani, A., Holyoake, C., Jakob-Hoff, R., Robertson, I., McInnes, K., Empson, R., Gray, R., Nakagawa, K., and Warren, K. 2015. Emerging infectious disease or evidence of endemicity? A multi-season study of beak and feather disease virus in wild red-crowned parakeets (*Cyanoramphus novaezelandiae*). *Archives of Virology* 160:2283–2292.

Jones, C. G. 1980. Parrot on the way to extinction. *Oryx* 15:350–354.

Kilgannon, C., and Singer, J. E. 2009. Fighting real parrots with a fake owl. *New York Times*, 17 Apr. 2009.

Klinger, C., Landeg, O., and Murray, V. 2014. Power outages, extreme events and health: A systematic review of the literature from 2011–2012. *PLoS Currents* 6:ecurrents.dis.04eb01dc05e73dd1377e-1305a1310e1379edde1673.

Koopman, M. E., and Pitt, W. C. 2007. Crop diversification leads to diverse bird problems in Hawaiian agriculture. *Human-Wildlife Conflicts* 1:235–243.

Kovach, R. P., Luikart, G., Lowe, W. H., Boyer, M. C., and Muhlfeld, C. C. 2016. Risk and efficacy of human-enabled interspecific hybridization for climate-change adaptation: Response to Hamilton and Miller (2016). *Conservation Biology* 30:428–430.

Kriebel, D., Tickner, J., Epstein, P., Lemons, J., Levins, R., Loechler, E. L., Quinn, M., Rudel, R., Schettler, T., and Stoto, M. 2001. The precautionary principle in environmental science. *Environmental Health Perspectives* 109:871–876.

Kuebbing, S. E., and Simberloff, D. 2015. Missing the bandwagon: Nonnative species impacts still concern managers. *NeoBiota* 25:73.

Kueffer, C., and Hadorn, G. H. 2008. How to achieve effectiveness in problem-oriented landscape research: The example of research on biotic invasions. *Living Reviews in Landscape Research* 2.

Kumschick, S., Bacher, S., Dawson, W., Heikkilä, J., Sendek, A., Pluess, T., Robinson, T. B., and Ingolf, K. 2012. A conceptual framework for prioritization of invasive alien species for management according to their impact. *NeoBiota* 15:69–100.

Kumschick, S., and Nentwig, W. 2010. Some alien birds have as severe an impact as the most effectual alien mammals in Europe. *Biological Conservation* 143:2757–2762.

Kundu, S., Faulkes, C. G., Greenwood, A. G., Jones, C. G., Kaiser, P., Lyne, O. D., Black, S. A., Chowrimootoo, A., and Groombridge, J. 2012. Tracking viral evolution during a disease outbreak: The rapid and complete selective sweep of a circovirus in the endangered echo parakeet. *Journal of Virology* 86:5221–5229.

Le Louarn, M., Clergeau, P., Strubbe, D., and Deschamps-Cottin, M. 2018. Dynamic species distribution models reveal spatiotemporal habitat shifts in native range-expanding versus non-native invasive birds in an urban area. *Journal of Avian Biology* 49:jav-01527.

Le Louarn, M., Couillens, B., Deschamps-Cottin, M., and Clergeau, P. 2016. Interference competition between an invasive parakeet and native bird species at feeding sites. *Journal of Ethology* 34:291–298.

Lever, C. 2005. *Naturalised Birds of the World*. Oxford: Oxford Univ. Press.

Low, R. 2000. The rainbow lorikeet: A modern witch hunt. *PsittaScene* 12:14–15.

Luna, A., Romero-Vidal, P., Hiraldo, F., and Tella, J. L. 2018. Cities may save some threatened species but not their ecological functions. *PeerJ* 6:e4908.

Mabb, K. T. 2003. Naturalized parrot roost flock characteristics and habitat utilization in a suburban area of Los Angeles County, California. MS thesis, California State Polytechnic Univ., Pomona.

Maris, V., and Béchet, A. 2010. From adaptive management to adjustive management: A pragmatic account of biodiversity values. *Conservation Biology* 24:966–973.

Marsden, S. J., and Jones, M. L. 1997. The nesting requirements of the parrots and hornbill of Sumba, Indonesia. *Biological Conservation* 82:279–287.

Martin-Albarracin, V. L., Amico, G. C., Simberloff, D., and Nuñez, M. A. 2015. Impact of non-native birds on native ecosystems: A global analysis. *PLoS One* 10:e0143070.

Mason, C. F. 2000. Thrushes now largely restricted to the built environment in eastern England. *Diversity and Distributions* 6:189–194.

Massaro, M., Ortiz-Catedral, L., Julian, L., Galbraith, J. A., Kurenbach, B., Kearvell, J., Kemp, J., van Hal, J., Elkington, S., Taylor, G., et al. 2012. Molecular characterisation of beak and feather disease virus (BFDV) in New Zealand and its implications for managing an infectious disease. *Archives of Virology* 157:1651–1663.

McDonald, R. I., Kareiva, P., and Forman, R.T.T. 2008. The implications of current and future urbanization for global protected areas and biodiversity conservation. *Biological Conservation* 141:1695–1703.

McKinney, M. L. 2002. Urbanization, biodiversity, and conservation: The impacts of urbanization on native species are poorly studied, but educating a highly urbanized human population about these impacts can greatly improve species conservation in all ecosystems. *Bioscience* 52:883–890.

McMillen, H., Campbell, L. K., and Svendsen, E. S. 2018. Weighing values and risks of beloved invasive species: The case of the survivor tree and conflict management in urban green infrastructure. *Urban Forestry & Urban Greening* 40:44–52.

Menchetti, M., and Mori, E. 2014. Worldwide impact of alien parrots (Aves Psittaciformes) on native biodiversity and environment: A review. *Ethology Ecology & Evolution* 26:172–194.

Menchetti, M., Mori, E., and Angelici, F. M. 2016. Effects of the recent world invasion by ring-necked parakeets *Psittacula krameri*. In *Problematic Wildlife: A Cross-Disciplinary Approach*, ed. F. M. Angelici, 253–266. Switzerland and New York: Springer International Publishing.

Metro. 2008. Parakeets destroying precious windmill. Metro.co.uk. 19 Oct. http://www.metro.co.uk/news/362878-parakeets-destroying-precious-windmill.

Millsap, B., Breen, T., McConnell, E., Steffer, T., Phillips, L., Douglass, N., and Taylor, S. 2004. Comparative fecundity and survival of bald eagles fledged from suburban and rural natal areas in Florida. *Journal of Wildlife Management* 68:1018–1031.

Minor, E. S., Appelt, C. W., Grabiner, S., Ward, L., Moreno, A., and Pruett-Jones. S. 2012. Distribution of exotic monk parakeets across an urban landscape. *Urban Ecosystems* 15:979–991.

Mori, E., Ancillotto, L., Groombridge, J., Howard, T., Smith, V. S., and Menchetti, M. 2015. Macroparasites of introduced parakeets in Italy: A possible role for parasite-mediated competition. *Parasitology Research* 114:3277–3281.

Mori, E., Ancillotto, L., Menchetti, M., and Strubbe, D. 2017a. "The early bird catches the nest": Possible competition between scops owls and ring-necked parakeets. *Animal Conservation* 20:463–470.

Mori, E., Grandi, G., Menchetti, M., Tella, J., Jackson, H., Reino, L., van Kleunen, A., Figueira, R., and Ancillotto, L. 2017b. Worldwide distribution of non-native Amazon parrots and temporal trends of their global trade. *Animal Biodiversity and Conservation* 40:49–62.

REFERENCES

Mori, E., Meini, S., Strubbe, D., Ancillotto, L., Sposimo, P., and Menchetti, M. 2018. Do alien free-ranging birds affect human health? A global summary of known zoonoses. In *Invasive Species and Human Health*, ed. G. Mazza and E. Tricarico, 120–129. Wallingford, UK: CAB International.

Mori, E., Sala, J. P., Fattorini, N., Menchetti, M., Montalvo, T., and Senar, J. C. 2019. Ectoparasite sharing among native and invasive birds in a metropolitan area. *Parasitology Research* 118:399–409.

Neck, R. W. 1986. Expansion of red-crowned parrot, *Amazona viridigenalis*, into southern Texas and changes in agricultural practices in northern Mexico. *Bulletin of the Texas Ornithological Society* 19:6–12.

Neo, M. L. 2012. A review of three alien parrots in Singapore. *Journal of Nature in Singapore* 5:241–248.

Newson, S. E., Johnston, A., Parrott, D., and Leech, D. I. 2011. Evaluating the population-level impact of an invasive species, ring-necked parakeet *Psittacula krameri*, on native avifauna. *Ibis* 153:509–516.

Newton, I. 1994. Experiments on the limitation of bird breeding densities: A review. *Ibis* 136:397–411.

Olah, G., Butchart, S.H.M., Symes, A., Guzmán, I. M., Cunningham, R., Brightsmith, D. J., and Heinsohn, R. 2016. Ecological and socio-economic factors affecting extinction risk in parrots. *Biodiversity and Conservation* 25:205–223.

Orchan, Y., Chiron, F., Shwartz, A., and Kark, S. 2013. The complex interaction network among multiple invasive bird species in a cavity-nesting community. *Biological Invasions* 15:429–445.

Ortega-Álvarez, R., and MacGregor-Fors, I. 2009. Living in the big city: Effects of urban land-use on bird community structure, diversity, and composition. *Landscape and Urban Planning* 90:189–195.

Ortega-Álvarez, R., and MacGregor-Fors, I. 2010. What matters most? Relative effect of urban habitat traits and hazards on urban park birds. *Ornitologia Neotropical* 21:519–533.

Ortiz-Catedral, L., McInnes, K., Hauber, M. E., and Brunton, D. H. 2009. First report of beak and feather disease virus (BFDV) in wild red-fronted parakeets (*Cyanoramphus novaezelandiae*) in New Zealand. *Emu* 109:244–247.

Peck, H. L., Marshall, H. H., Owens, I.P.F., Pringle, H. E., and Lord, A. M. 2014. Experimental evidence of impacts of an invasive parakeet on foraging behavior of native birds. *Behavioral Ecology* 25:582–590.

Pejchar, L., and Mooney, H. A. 2009. Invasive species, ecosystem services and human well-being. *Trends in Ecology & Evolution* 24:497–504.

Peters, A., Patterson, E. I., Baker, B.G.B., Holdsworth, M., Sarker, S., Ghoroshi, S. A., and Raidal, S. R. 2014. Evidence of psittacine beak and feather disease virus spillover into wild critically endangered orange-bellied parrots (*Neophema chrysogaster*). *Journal of Wildlife Diseases* 50:288–296.

Pimentel, D., Zuniga, R., and Morrison, D. 2005. Update on the environmental and economic costs associated with alien-invasive species in the United States. *Ecological Economics* 52:273–288.

Polkanov, A., and Keeling, P. 2001. Pest psittacenes: The rainbow lorikeet in New Zealand; an update. *Eclectus* 11:6–9.

Postigo, J. L. 2016. New records of invasive parakeet hybrids in Spain: A great opportunity to apply the rapid response mechanism. *European Journal of Ecology* 2:19.

Postigo, J. L., Shwartz, A., Strubbe, D., and Muñoz, A. R. 2017. Unrelenting spread of the alien monk parakeet *Myiopsitta monachus* in Israel: Is it time to sound the alarm? *Pest Management Science* 73:349–353.

Postigo, J. L., Strubbe, D., Mori, E., Ancillotto, L., Carneiro, I., Latsoudis, P., Menchetti, M., Pârâu, M.L.G., Parrott, D., Reino, L., Weiserbs, A., and Senar, J. C. 2019. Mediterranean versus Atlantic monk parakeets *Myiopsitta monachus*: Towards differentiated management at the European scale. *Pest Management Science* 75:915–922.

Pruett-Jones, S., Appelt, C. W., Sarfaty, A., Van Vossen, B., Leibold, M. A., and Minor, E. S. 2012. Urban parakeets in northern Illinois: A 40-year perspective. *Urban Ecosystems* 15:709–719.

Pyle, R. L., and Pyle, P. 2017. *The Birds of the Hawaiian Islands: Occurrence, History, Distribution, and Status* (v.2, 1 Jan.). Honolulu: B. P. Bishop Museum. http://hbs.bishopmuseum.org/birds/rlp-monograph.

Raidal, S. R., and Peters, A. 2018. Psittacine beak and feather disease: Ecology and implications for conservation. *Emu* 118:80–93.

Reed, J. E. 2014. Spatial ecology of and public attitudes toward monk parakeets nesting on electrical utility structures in Dallas and Tarrant Counties, Texas. PhD diss., Texas A & M Univ.

Reed, J. E., McCleery, R., Silvy, N. J., Smeins, F. E., and Brightsmith, D. J. 2014. Monk parakeet nest-site selection of electric utility structures in Texas. *Landscaping and Urban Planning* 129:65–72.

Regnard, G. L., Boyes, R. S., Martin, R. O., Hitzeroth, I. I., and Rybicki, E. P. 2015. Beak and feather disease viruses circulating in Cape parrots (*Poicephalus robustus*) in South Africa. *Archives of Virology* 160:47–54.

Renton, K. 2001. Lilac-crowned parrot diet and food resource availability: Resource tracking by a parrot seed predator. *Condor* 103:62–69.

Renton, K., Salinas-Melgoza, A., De Labra-Hernández, M. Á., and de la Parra-Martínez, S. M. 2015. Resource requirements of parrots: Nest site selectivity and dietary plasticity of Psittaciformes. *Journal of Ornithology* 156:73–90.

Richards, H. 2010. The 500 mark: A landmark season. *PsittaScene* 22:6–10.

Roscoe, D., Zeh, J., Stone, W., Brown, L., and Renkavinsky, J. 1973. Observations of the monk parakeet in New York State. *New York Fish and Game Journal* 20:170–173.

Runde, D. E., Pitt, W. C., and Foster, J. 2007. Population ecology and some potential impacts of emerging populations of exotic parrots. In *Managing Vertebrate Invasive Species: Proceedings of an International Symposium*, ed. G. W. Witmer, W. C. Pitt, and K. A. Fagerstone, 338–360. Fort Collins, CO: USDA/APHIS Wildlife Services, National Wildlife Research Center.

Russell, J. C., and Blackburn, T. M. 2017. The rise of invasive species denialism. *Trends in Ecology & Evolution* 32:3–6.

Sa, R.C.C., Cunningham, A. A., Dagleish, M. P., Wheelhouse, N., Pocknell, A., Borel, N., Peck, H., and Lawson, B. 2014. Psittacine beak and feather disease in a free-living ring-necked parakeet (*Psittacula krameri*) in Great Britain. *European Journal of Wildlife Research* 60:395–398.

Salo, P., Korpimäki, E., Banks, P. B., Nordström, M., and Dickman, C. R. 2007. Alien predators are more dangerous than native predators to prey populations. *Proceedings of the Royal Society B: Biological Sciences* 274:1237–1243.

Sánchez, R., Ballari, S. A., Bucher, E. H., and Masello, J. F. 2016. Foraging by burrowing parrots has little impact on agricultural crops in northeastern Patagonia, Argentina. *International Journal of Pest Management* 62:326–335.

Santos Neto, J. R., and Gomes, D. M. 2007. Predacão de milho por arara-azul-de-Lear, *Anodorhynchus leari* (Bonaparte, 1956) (Aves: Psittacidae) em sua área de ocorrencia no Sertão de Bahia. *Ornithologia* 2:41–46.

Schlaepfer, M. A., Sax, D. F., and Olden, J. D. 2011. The potential conservation value of non-native species. *Conservation Biology* 25:428–437.

Senar, J. C., Domènech, J., Arroyo, L., Torre, I., and Gordo, O. 2016. An evaluation of monk parakeet damage to crops in the metropolitan area of Barcelona. *Animal Biodiversity and Conservation* 39:141–145.

Seymour, M. 2013. "Support your local invasive species": Animal protection rhetoric and nonnative species. *Society & Animals* 21:54–73.

Shields, W. M., Grubb, T. C., Jr., and Telis, A. 1974. Use of native plants by monk parakeets in New Jersey. *Wilson Bulletin* 86:172–173.

Shiels, A. B., Bukoski, W. P., and Siers, S. R. 2018. Diets of Kauai's invasive rose-ringed parakeet (*Psittacula krameri*): Evidence of seed predation and dispersal in a human-altered landscape. *Biological Invasions* 20:1449–1457.

Silverman, F. 2009. In this springtime battle, the parakeets appear to be winning. *New York Times*, 19 Mar. 2009.

Simberloff, D. 2005. The politics of assessing risk for biological invasions: The USA as a case study. *Trends in Ecology & Evolution* 20:216–222.

Simberloff, D., Martin, J.-L., Genovesi, P., Maris, V., Wardle, D. A., Aronson, J., Courchamp, F., Galil, B., García-Berthou, E., Pascal, M., et al. 2013. Impacts of biological invasions: What's what and the way forward. *Trends in Ecology & Evolution* 28:58–66.

Snyder, N.F.R., Wiley, J. W., and Kepler, C. B. 1987. *The Parrots of Luquillo: Natural History and Conservation of the Puerto Rican Parrot*. Los Angeles, CA: Western Foundation of Vertebrate Zoology.

Spreyer, M. 1994. Mayor Washington's birds: The legendary monk parakeets of Chicago's Hyde Park. *Birder's World* 8:40–43.

Spreyer, M. F., and Bucher, E. H. 1998. Monk parakeet (*Myiopsitta monachus*). In *The Birds of North America*, ed. A. Poole and F. Gill, no. 322. Philadelphia, PA: Academy of Natural Sciences; Washington, DC: American Ornithologists' Union.

Strubbe, D., and Matthysen, E. 2007. Invasive ring-necked parakeets *Psittacula krameri* in Belgium: Habitat selection and impact on native birds. *Ecography* 30:578–588.

Strubbe, D., and Matthysen, E. 2009. Experimental evidence for nest-site competition between invasive ring-necked parakeets (*Psittacula krameri*) and native nuthatches (*Sitta europaea*). *Biological Conservation* 142:1588–1594.

Strubbe, D., Matthysen, E., and Graham, C. H. 2010. Assessing the potential impact of invasive ring-necked parakeets *Psittacula krameri* on native nuthatches *Sitta europeae* in Belgium. *Journal of Applied Ecology* 47:549–557.

Strubbe, D., Shwartz, A., and Chiron, F. 2011. Concerns regarding the scientific evidence informing impact risk assessment and management recommendations for invasive birds. *Biological Conservation* 144:2112–2118.

Styche, A. 2000. Distribution and behavioural ecology of the sulphur-crested cockatoo (*Cacatua galerita* L.) in New Zealand. PhD diss., Victoria Univ. of Wellington, New Zealand.

Tassin, J., Thompson, K., Carroll, S. P., and Thomas, C. D. 2017. Determining whether the impacts of introduced species are negative cannot be based solely on science: A response to Russell and Blackburn. *Trends in Ecology & Evolution* 32:230–231.

Tella, J. L., Baños, A., Hernández-Brito, D., Rojas, A., Pacífico, E., Díaz, J. A., Carrete, M., Blanco, G., and Hiraldo, F. 2015. Parrots as overlooked seed dispersers. *Frontiers in Ecology and the Environment* 13:338–339.

Thabethe, V., Wilson, A.-L., Hart, L. A., and Downs, C. T. 2015. Ingestion by an invasive parakeet species reduces germination success of invasive alien plants relative to ingestion by indigenous turaco species in South Africa. *Biological Invasions* 17:3029–3039.

Thomas, R., and Williamson, S. H. 2019. What was the U.K. GDP then? Measuring Worth. Accessed 18 Apr. 2019. http://www.measuringworth.com/ukgdp/.

Tillman, E. A., Van Doom, A., and Avery, M. L. 2000. Bird damage to tropical fruit in south Florida. In *The Ninth Wildlife Damage Management Conference Proceedings*, ed. M. C. Brittingham, J. Kays, and R. McPeake, 47–59. State College, PA.

Tompkins, D. M., Mitchell, R. A., and Bryant, D. M. 2006. Hybridization increases measures of innate and cell-mediated immunity in an endangered bird species. *Journal of Animal Ecology* 75:559–564.

Turbé, A., Strubbe, D., Mori, E., Carrete, M., Chiron, F., Clergeau, P., González-Moreno, P., Le Louarn, M., Luna, A., Menchetti, M., et al. 2017. Assessing the assessments: Evaluation of four impact assessment protocols for invasive alien species. *Diversity and Distributions* 23:297–307.

Uehling, J. J., Tallant, J., and Pruett-Jones, S. 2019. Status of naturalized parrots in the United States. *Journal of Ornithology* 160:907–921.

USFWS. 2019. Species status assessment report for red-crowned parrot (*Amazona viridigenalis*) (v.3.0.). US Fish and Wildlife Service, Albuquerque, NM.

REFERENCES

Valéry, L., Fritz, H., and Lefeuvre, J.-C. 2013. Another call for the end of invasion biology. *Oikos* 122:1143–1146.

Vall-llosera, M., Woolnough, A. P., Anderson, D., and Cassey, P. 2017. Improved surveillance for early detection of a potential invasive species: The alien rose-ringed parakeet *Psittacula krameri* in Australia. *Biological Invasions* 19:1273–1284.

van Heezik, Y., Smyth, A., and Mathieu, R. 2008. Diversity of native and exotic birds across an urban gradient in a New Zealand city. *Landscape and Urban Planning* 87:223–232.

White, R. L., Strubbe, D., Dallimer, M., Davies, Z. G., Davis, A.J.S., Edelaar, P., Groombridge, J., Jackson, H. A., Menchetti, M., Mori, E., et al. 2019. Assessing the ecological and societal impacts of alien parrots in Europe using a transparent and inclusive evidence-mapping scheme. *NeoBiota* 48:45–69.

Wilcove, D. S., Rothstein, D., Dubow, J., Phillips, A., and Losos, E. 1998. Quantifying threats to imperiled species in the United States. *Bioscience* 48:607–615.

Wiley, J. W. 1991. Status and conservation of parrots and parakeets in the Greater Antilles, Bahama Islands, and Cayman Islands. *Bird Conservation International* 1:187–214.

Williams, F., Eschen, R., Harris, A., Djeddour, D., Pratt, C., Shaw, R., Varia, S., Lamontagne-Godwin, J., Thomas, S., and Murphy, S. 2010. *The Economic Cost of Invasive Non-Native Species on Great Britain*. Wallingford, UK: CABI. Available at: https://blog.invasive-species.org/2010/12/15/the-economic-impact-of-invasive-species-on-great-britain-revealed/.

Winchell, C. S., and Doherty, P. F., Jr. 2008. Using California gnatcatcher to test underlying models in habitat conservation plans. *Journal of Wildllife Management* 72:1322–1327.

Wright, D., and Clout, M. N. 2001. The eastern rosella (*Platycercus eximius*) in New Zealand. Dept. of Conservation Science Internal Series 18. Wellington, NZ: Dept. of Conservation.

Yosef, R., Zduniak, P., and Żmihorski, M. 2016. Invasive ring-necked parakeet negatively affects indigenous Eurasian hoopoe. *Annales Zoologici Fennici* 53:281–287.

PART II
CASE STUDIES

10

GLOBAL INVASION SUCCESS OF THE ROSE-RINGED PARAKEET

Hazel A. Jackson

INTRODUCTION

Invasive species are among the most significant drivers of biodiversity loss and one of the largest global conservation challenges of today. Humans have been trading species for millennia, and contemporary global patterns of invasive species have been shaped by such human-mediated transport of plant and animal species (Hulme 2009). However, not all species introduced to regions outside their native range are considered invasive. Only species that establish, disperse, and have a detrimental impact upon native biota are considered invasive. Improvements in global transport networks over time have increased connectivity of human populations and led to increasing frequencies of biological invasions (McKinney and Lockwood 1999). Invasive alien species now represent an increasingly urgent economic, societal, and environmental problem. Their rapid spread, competitive nature, and transmission of infectious diseases pose threats to global biodiversity, and invasive species are considered one of the five main causes of global biodiversity loss, alongside climate change, pollution, overexploitation, and habitat loss (Millennium Ecosystem Assessment 2005).

The parrots (Psittaciformes) form one of the most distinctive bird groups. They are also one of the most endangered groups of birds in the world: 95 (26.8%) of the 354 known parrot species are currently threatened with extinction, and a further 36 species are classified as "near threatened" (Jetz et al. 2014). Over the past 500 years, approximately 163 avian extinctions have occurred across the globe, including some 20 parrot species (12%), half of which were island endemics (Collar 2000; Butchart et al. 2006). Major reasons for declining endemic parrot populations include invasive species, poaching, habitat loss, and the pet trade (Cheke and Hume 2009; Perrin 2012).

Interestingly, while removal from their native habitat for the pet trade has caused the decline of many endemic parrots, the popularity of parrots as pets and their global transport have contributed to some 54 species (16% of total living species) currently breeding outside their native range, with the more widely distributed species—Monk Parakeet (*Myiopsitta monachus*), Orange-winged Amazon (*Amazona amazonica*), Green Parakeet (*Psittacara holochlorus*), Rose-ringed Parakeet (*Psittacula krameri*), and Rosy-faced Lovebird (*Agapornis roseicollis*)—being the most successful at establishing populations in non-native areas (Menchetti and Mori 2014). Parrot establishment in non-native environments is a result of numerous factors; in addition to the birds' popularity as pets, and high numbers being traded and bred, they are highly synanthropic, appearing to be adapted to surviving in a wide variety of environmental conditions (Duncan et al. 2003; Cassey et al. 2004).

The world's most successful and prolific of all non-native parrots is the Rose-ringed Parakeet, a species now established in over 35 countries, across five continents (Fig. 10.1). This medium to large, emerald-green parakeet is native to Asia and sub-Saharan Africa. There are four recognized

GLOBAL INVASION SUCCESS OF THE ROSE-RINGED PARAKEET

subspecies (Forshaw 2010): two subspecies are native to Asia (*P. k. borealis*, found in eastern Pakistan, throughout northern India, Nepal, and Burma; and *P. k. manillensis*, found in southern India and Sri Lanka), and two native to Africa (*P. k. krameri*, found from Senegal to western Uganda and southern Sudan; and *P. k. parvirostris*, found in eastern Sudan and northern Ethiopia).

The Rose-ringed Parakeet is herbivorous, with a diet comprising nuts, seeds, fruits, flowers, and shoots. It is a secondary cavity nester; usually rearing just one brood per year, the female occupies a new nest prior to breeding and defends the nest aggressively. Clutch size is usually three or four eggs. Breeding season varies with latitude in the native range, occurring between November and June in Asia and August to November in Africa (del Hoyo et al. 1997). In their native range, Rose-ringed Parakeets are found in various woodland habitats, farmlands, and urban gardens and parks (Juniper and Parr 1998; Khan 2002), while in their non-native ranges, they readily inhabit forests and parks that are within or surrounded by urban habitats (Strubbe and Matthysen 2007).

Although the Rose-ringed Parakeet is firmly and widely established outside its native range, research into understanding how it thrives, and its potential impacts on native biodiversity, agriculture, and economy, is still in its infancy. Research currently underway suggests that the level of impact varies geographically and should be understood and mitigated on a case-by-case basis. This chapter discusses how the Rose-ringed Parakeet has become one of the most widely established species outside its native range, what factors have led to its ability to thrive in novel environments, what the potential impacts of this global invasion may be, and finally, what impacts this successful invader has had on native biodiversity, ecosystems, and agriculture.

NON-NATIVE DISTRIBUTION

The Rose-ringed Parakeet is one of the world's most widely introduced parrots, with breeding populations established in well over 35 countries across five continents (Menchetti et al. 2016), and is considered one of Europe's top 100 worst alien species (DAISIE 2009). Large numbers of Rose-ringed Parakeets have become established in a number of European countries since the

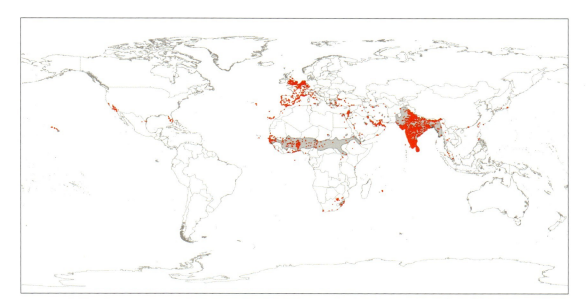

Figure 10.1. Global occurrence data of Rose-ringed Parakeets taken from the Global Biodiversity Information Facility (GBIF 2019), across native (shaded areas in southern Asia and sub-Sahel Africa; Birdlife International) and non-native (red) distributions (data courtesy of D. Strubbe).

late 1960s, including the United Kingdom (UK), Germany, the Netherlands, France, Spain, Italy, Greece, and Belgium (Lever 2005), as well as Turkey (Per 2018). There are recent reports of 90 breeding populations in 10 European countries, comprising a minimum of 85,000 birds (Pârâu et al. 2016). Outside of Europe, numerous breeding populations have been recorded in Australia (Vall-llosera et al. 2017), Mauritius, Seychelles, Hawaii (Shiels et al. 2018), and the United States (US) (Avery and Shiels 2018). There are also frequent reports of newly established populations.

POPULATION GROWTH AND SPREAD

The rapid establishment of the Rose-ringed Parakeet across Europe, combined with evidence of explosive exponential population growth (Fig. 10.2) (Butler et al. 2013; Jackson et al. 2015b), is a major cause for agro-economic and environmental concern. For example, in the Netherlands, since 1998, Rose-ringed Parakeets have increased their distribution by 239% and the number of breeding pairs by 1,582% (Pârâu et al. 2016). The species colonized the UK in the late 1960s from a few escaped pet birds; now they are the country's fastest-growing bird population, numbering >32,000 individuals in 2012 (Fig. 10.3) (Butler et al. 2013; Peck et al. 2014). An examination of the influence of climatic variables and human demographic data on rates of population growth for non-native Rose-ringed Parakeets in Europe supports the climate-matching hypothesis (Shwartz et al. 2009) by showing higher growth rates in populations in areas with climate conditions that are more similar to those of the native range (Fig. 10.4) (Jackson 2015), and, therefore, more

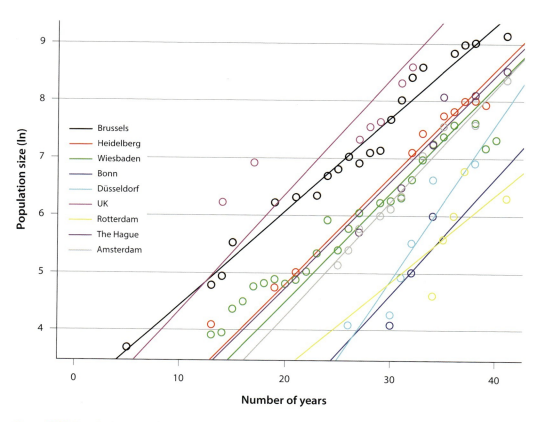

Figure 10.2. Population growth rates over time for 10 European populations of non-native Rose-ringed Parakeets.

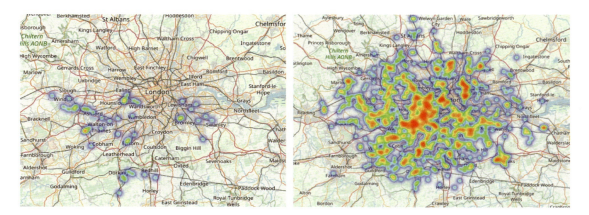

Figure 10.3. Heat maps of changes in Rose-ringed Parakeet occurrences in Greater London between 2000 (left) and 2012 (right). From ParrotNet European Monitoring Centre (ParrotNet 2018).

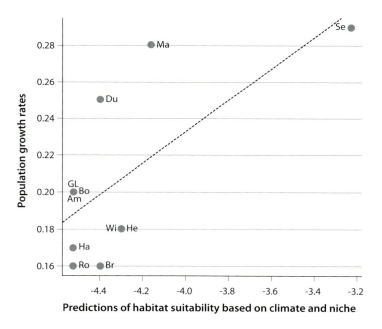

Figure 10.4. Positive relationship between Rose-ringed Parakeet population growth rates in Europe and predicted habitat suitability in the non-native range based on climate and niche structure in the native range (Jackson et al. 2015b). Am = Amsterdam, Bo = Bonn, Br = Brussels, Du = Düsseldorf, GL = Greater London, Ha = The Hague, He = Heidelberg, Ma = Marseille, Ro = Rotterdam, Se = Seville, Wi = Wiesbaden.

suitable. This result suggests that population growth in European populations relies upon the availability of climatic niches (i.e., the range of temperature and precipitation gradients) similar to those prevailing in the parakeet's native range.

Rose-ringed Parakeets in the UK were found in their largest densities around Greater London, with little evidence of spread until recently. Simultaneous roost counts conducted annually between 2010 and 2012 (Peck 2013) found that the population size appeared to be steady, at around 32,000 individuals. However, since then, new reports of parakeets are being made from all around the UK, suggesting the available nests and resources around Greater London had become saturated, pushing Rose-ringed Parakeets to spread around the country.

Predictions of Spread

To understand and mitigate the potential threats invasive species pose to native biodiversity and ecosystems, data on native distributions

Rose-ringed Parakeets (*Psittacula krameri*) returning to their roost site. These birds are part of a flock of approximately 4,000 birds that has roosted in the same tree for at least 20 years. Honolulu, Hawaii, US, October 2019. Photo by Nicholas P. Kalodimos.

and niche requirements can be examined using ecological niche models to predict where invasive species may establish in the future. Ecological niche models are often applied as a method of predicting invasion risk by using environmental variables to determine geographic areas suitable for invasive species (Jiménez-Valverde et al. 2011; Strubbe et al. 2015a, 2015b). Ecological niche models characterize a species' fundamental ecological niche (a range of environmental conditions in which a species can survive) and realized ecological niche (the range of environmental conditions in which a species is actually found) using occurrence and spatial environmental data, which are then projected onto geographical regions outside a species' native distribution (Strubbe et al. 2015a). Genetic data inferring phylogeographic structure obtained from native samples of Rose-ringed Parakeets were examined in combination with information on native distribution to predict areas across Europe that were suitable for Rose-ringed Parakeets to establish in. These models provided some insight; however, the inclusion of information on the Rose-ringed Parakeet's association with humans within its native range substantially improved the reliability and predictive power of models of invasion risk across Europe (Strubbe et al. 2015b). These predictions highlight the huge amount of available niche and therefore provide useful insight for policy makers and conservation managers on where Rose-ringed Parakeets may settle next.

Pathways of Invasion

Understanding invasion routes and ancestral origins of non-native and invasive species is essential to comprehending how they are flourishing outside their native environment. The widespread distribution of Rose-ringed Parakeets can be attributed to their popularity as companion animals (caged birds). Psittaciforms are among the most widely traded species (Cardador et al. 2017), and Rose-ringed Parakeets were exported in extensive numbers from their native range prior to the passage of regional trade bans such as the US Wild Bird Conservation Act of 1992, and the EU Wild Bird Declaration of 2005 (Jackson et al. 2015b; Cardador et al. 2017). Between 1984 and 2007 (before the EU ban on the trade of wild birds was fully implemented), a staggering 146,539 Rose-ringed Parakeets were imported into Europe. The UK alone imported over 16,000 birds, while at least 20,105 individuals were bred in captivity in the country between 1990 and 2004 (UK Parrot Society; Fletcher and Askew 2007), all to meet demands of the pet trade (Jackson et al. 2015b). Unfortunately, the implementation of regional bans did not affect the number of parrots traded internationally but instead led to the development of new trade routes (Reino et al. 2017) and changes in trade destinations (Cardador et al. 2017). In response to the risks of continual releases and escapes into the wild associated with the breeding and transport of parrots, some countries have now implemented legislation, such as Spain (Real Decreto 630/2013), that prohibits private breeding or ownership of pet parakeets.

WHAT MAKES ROSE-RINGED PARAKEETS SUCCESSFUL INVADERS?

Species that become invasive go through a multistage invasion process (Blackburn et al. 2013). Although the large majority of species fail to establish in novel environments during this process, a few succeed. To fully investigate the invasion process, it is important to examine each stage, including transportation, releases or introductions, establishment, spread, and becoming a pest, or invasive, species. Within this process, developing an understanding of genetic factors that underpin success is vitally important for the management of invasive species and prevention of future invasions. Contemporary research has focused on improving our understanding of invasive species and the evolutionary mechanisms that facilitate their successful establishment in novel environments by identifying ancestral source populations and routes of invasion and by examining levels of genetic diversity and evolutionary adaptation or plasticity (Le Roux and Wieczorek 2009; Lee 2002). Such information is important for identifying factors that play an important role in facilitating invasion success.

Genetic Diversity

As regards genetics, invasive species are intriguing because they appear able to avoid genetic difficulties associated with small population size at founding. Low genetic diversity, inbreeding, and reduced fitness are likely to occur when invasive species colonize new areas with a small number of founders (Allendorf and Lundquist 2003). The establishment of Rose-ringed Parakeets is often accompanied by anecdotal stories that describe such founding events, with just a few individuals from which large populations have established. In the UK, a number of popular stories exist: the parakeets were released from the movie set of *The African Queen* when it was being shot in London; they escaped from aviaries in the great storms of 1987; and, according to the "Hendrix hypothesis," Jimi Hendrix released a pair of Rose-ringed Parakeets on Carnaby Street to inject some psychedelic color into the streets of London (Jackson 2015, Heald et al. 2019). DNA research demonstrates that these scenarios are unlikely, as no genetic problems associated with small founder size (i.e., inbreeding depression) have been observed in any non-native populations. Rose-ringed Parakeet populations established across Europe all comprise founder sizes large enough to result in high levels of genetic diversity comparable to those observed in the native range (Jackson 2015). These high levels of genetic diversity are likely a result of multiple introductions (Jackson 2015), releases, or escapes over time, supplementing the wild gene pool and enabling long-term survival.

Ancestral Origins of Invasive Populations

Prior to the EU ban on the trade of wild birds in 2007, trade records show Rose-ringed Parakeets were imported all across Europe from both the Asian and African native ranges (Jackson et al. 2015a, 2015b). These trade records therefore suggest that populations of Rose-ringed Parakeets established across Europe would contain a high degree of mixed ancestry (comprising all four subspecies), perhaps providing some explanation for their survival and success at a genetic level. The presence of all subspecies would provide ample genetic diversity to avoid any detrimental impacts of genetic impoverishment, such as low breeding success, susceptibility to disease, and high juvenile mortality. However, a comprehensive genetic analysis of individuals spanning the entire native range in comparison to European populations discovered that the Rose-ringed Parakeets across Europe predominantly originated from the species' northern Asian range (Jackson et al. 2015a, 2015b) and largely represent just one of the four subspecies, *P. k. borealis*. This analysis suggests that African Rose-ringed Parakeets do not survive well outside their native range.

While observed levels of trade imports from the native range are likely to have influenced these signatures of ancestral origins within invasive populations (proportionally higher numbers of birds were imported from Asia), the observed patterns can also be explained by the prevalence of mitochondrial haplotypes (genotypes inherited through the maternal line) within the European invasive populations characterized by a lower cold niche limit from the native range. This association with colder parts of the native range may have preferentially assisted the successful establishment of Rose-ringed Parakeets in colder areas of their invasive range, with parakeets with warmer niche limits (i.e., those native to Africa) failing to survive in novel environments. These findings highlight the role of human-mediated transport in facilitating invasions (Meyerson and Mooney 2007; Hulme 2009) but also demonstrate how identifying ancestral origins of invasive species enables an examination of the importance of patterns of climate matching between native and invasive ranges for invasion success (Shwartz et al. 2009; Strubbe et al. 2015b). The importance of climate matching was further highlighted as population growth rates of European Rose-ringed Parakeets were driven by the availability of suitable habitat, based upon patterns of climate and niche structure from the native range (e.g., temperature and precipitation).

Phenological and Behavioral Adaptations

While patterns of climate matching favor invasion success, the successful establishment of a species outside its native range may be affected by differences in seasonality and the timing of life cycle events. Across Europe, parakeets are demonstrating phenological and behavioral plasticity that enable them to thrive in new environments. Introduced Rose-ringed Parakeets have been found to breed earlier than expected based on breeding dates from their native Asian range (Luna et al. 2017) and are experiencing high nesting success as a result. However, despite this adaptive breeding behavior, temporal mismatches were observed in more northerly European populations that negatively affect hatching success. Rose-ringed Parakeets native to Africa breed in autumn; this phenological mismatch may explain further why parakeets from sub-Saharan Africa have been less successful in establishing in comparison to their Asian counterparts (Luna et al. 2017).

Rose-ringed Parakeets are secondary cavity nesters with a strong preference for nesting and mating in trees; however, in urban spaces these intelligent parakeets have demonstrated their ability to utilize human structures for nesting opportunities (Braun and Wink 2013). In Heidelberg, Germany, where the parakeets have been firmly established since 1990, Great Spotted Woodpeckers (*Dendrocopos major*) were observed creating several small holes in the thermal Styrofoam insulation of houses. Rose-ringed Parakeets then enlarged the holes, to up to 1.5 m in length, and used them as nesting sites (Braun and Wink 2013). A nesting practice unique among all the breeding populations in Europe can be seen in Pavia, Italy, where the majority of this urban parakeet population is nesting within putlog holes of

Visconti Castle and its surrounding towers (Grandi et al. 2018). Such examples demonstrate the behavioral flexibility and adaptability of this species, characteristics that contribute to its establishment success in areas that may not provide all the resources available in the native range.

Abundance of Resources

The opportunistic nature of Rose-ringed Parakeets enables them to take advantage of available resources. The species was first observed in the wild in the UK in the late 1800s; however, these birds failed to establish and survive long term. In recent times, provisioning of bird food by humans provides a year-round, readily available energy source for non-native Rose-ringed Parakeet populations. Radio-tracking and habitat studies show Rose-ringed Parakeets prefer to forage in city parks and gardens where bird food is in ready supply (Strubbe and Matthysen 2007; Le Louarn et al. 2017). In 2008, researchers recorded 247 feeding events for an established Rose-ringed Parakeet population in Paris, France. Parakeets were observed eating seeds, buds, and fruits from trees, and spent half their feeding time at bird feeders in gardens throughout the year (Clergeau and Vergnes 2011). This year-round abundance of resources is likely to play a pivotal role in the survival and population growth of Rose-ringed Parakeet populations and may have enabled their considerable expansion into climates colder than the native range (e.g., Scotland). As tolerance of colder climates requires higher energy costs, bird feeders provide much needed energy for survival in novel climates and environments.

Lack of Native Predators?

Invasive species often thrive due to the lack of their native predators and parasites in their new environments. This is known as the "enemy release hypothesis" and is thought to play a significant role in the success of invasive species (Shwartz et al. 2009). Within the native Indian range, predation on Rose-ringed Parakeets is the main cause for a reduction in fecundity, while in non-native populations established in the UK and Israel, predation did not significantly affect fecundity (Shwartz et al. 2009), supporting the enemy release hypothesis. However, more recently in Europe, native raptor species such as Peregrine Falcons (*Falco peregrinus*) and Eurasian Sparrowhawks (*Accipiter nisus*) have begun to prey upon the abundant supply of Rose-ringed Parakeets. Regular observations in London have revealed that Rose-ringed Parakeets are preyed upon by Tawny Owls (*Strix aluco*), Eurasian Hobbies (*Falco subbuteo*), Peregrine Falcons, and Eurasian Sparrowhawks (Hancock and Martin 2015).

IMPACTS OF ROSE-RINGED PARAKEETS IN NON-NATIVE RANGES

In addition to its ranking as one of Europe's top 100 worst invasive species (DAISIE 2009), the Rose-ringed Parakeet is listed as one of the only

Rose-ringed Parakeets (*Psittacula krameri*) sharing a bird feeder with another introduced and naturalized species, the Java Sparrow (*Lonchura oryzivora*). Oahu, Hawaii, US, September 2019. Photo by Nicholas P. Kalodimos.

invasive parrot species (alongside the Monk Parakeet) that has a high impact on its invaded environments by the European Alien Species Information Network (EASIN 2020; Katsanevakis et al. 2015). Although populations of Rose-ringed Parakeets have become established outside of their native range only over the past 60 years, reports are only now beginning to document the potential impact these parakeets are having in different regions of their invaded ranges (White et al. 2019). More research is essential to enable us to truly understand how these Rose-ringed Parakeets are affecting native environments, biodiversity, and agriculture.

Competition

The Rose-ringed Parakeet's ability to exploit food sources arguably contributes to its successful survival, especially in colder regions of its invasive range. The parakeets' prevalence at garden feeders, especially in winter, when supplementary food is of great importance for survival, can lead to interspecific competition with native bird species—in particular Common Starlings (*Sturnus vulgaris*)—by displacement and aggressive behavior (Le Louarn et al. 2016), resulting in a detrimental impact upon the foraging behavior of native birds (Peck et al. 2014).

In Europe, Rose-ringed Parakeets have been shown to compete with native species such as Eurasian Nuthatches (*Sitta europaea*), starlings, Great Tits (*Parus major*), and even bats for nesting cavities (Strubbe and Matthysen 2007, 2009; Hernández-Brito et al. 2014). For example, the long-term presence of Rose-ringed Parakeets has caused the decline of a threatened noctule bat species in Seville, Spain, as a result of nest-site competition. Rose-ringed Parakeets were observed killing bats to occupy nest sites (Hernández-Brito et al. 2018). Such nest-site competition has also affected native Eurasian Hoopoes (*Upupa epops*) in Israel. The Rose-ringed Parakeets' increasing use of palm groves for nests between 2002 and 2012 led to a significant decline in hoopoe density. Parakeets breed earlier than hoopoes, which often means they use all available nest sites before the hoopoes can get to them. Higher densities of hoopoes remained stable across these years in palm groves where parakeets were not present (Yosef et al. 2016). Further afield, in Mauritius, invasive Rose-ringed Parakeets, established since 1880, compete with the native endangered Echo Parakeet (*Psittacula eques*) for nest sites and food resources (Tatayah et al. 2007; Jones et al. 2013).

Disease

Parrots are susceptible to psittacine beak and feather disease (PBFD) caused by the highly infectious beak and feather disease virus. It is likely that invasive populations of Rose-ringed Parakeets are carriers of PBFD (e.g., PBFD occurrence was reported in the UK; Sa et al. 2014), although this is problematic only in countries with endemic parrot species. Flourishing invasive Rose-ringed Parakeets in Mauritius are a suspected source of PBFD, which threatens the population of endangered endemic Echo Parakeets (Kundu et al. 2012). Rose-ringed Parakeets also became established in the Seychelles (Jones et al. 2013), presenting a serious disease threat to the endangered endemic Seychelles Black Parrot (*Coracopsis barklyi*) (Seychelles Islands Foundation 2020). In response to this threat, the Seychelles became the first country to eradicate the Rose-ringed Parakeet, in 2017.

Agriculture

Considered a severe crop pest across their native Asian range, Rose-ringed Parakeets are known to decimate maize and fruit crops in India (Ramzan and Toor 1973; Forshaw 2010; Ahmad et al. 2012). While reports of crop damage largely originate from the native range, quantitative data on the damage caused within the species' invaded ranges are scarce. Reports of damage are beginning to emerge from across regions of Europe and beyond. In Israel, the parakeets' diet includes almonds, sunflower seed, and maize, and they heavily attack plantations, causing severe crop damage (Schäckermann et al. 2014). More recently, Rose-ringed Parakeets were recorded to have damaged ~32% of an almond plantation in Rome, Italy (Mentil et al. 2018). While such instances of crop damage by Rose-ringed Parakeets have been recorded, few studies have yet to quantify the economic cost of such damage

(Menchetti and Mori 2014). In the UK, damages to vineyards in Surrey are estimated to cost ~$8,250 (£5,000) per year (Fletcher and Askew 2007).

Rose-ringed Parakeets' appetite for seeds and ability to inflict severe agricultural damage may have consequences outside the economic damage. Birds play an important role within their native ecosystems as seed dispersers, and this ecological function is vital to native flora diversity. However, in Hawaii, the diet of an invasive population of >2,000 Rose-ringed Parakeets was found to consist of 80% seed, of which 30% was invasive Common Guava (*Psidium guajava*), suggesting invasive plant species such as the guava are being dispersed by invasive parakeets (Shiels et al. 2018), further exacerbating the invasive species dilemma.

Noise Pollution

An unexplored impact is the potential noise disturbance from roost sites. Rose-ringed Parakeets can cause severe noise disturbance, as they roost at night in the thousands and have very loud vocalizations (van Kleunen et al. 2010). The largest roost recorded to date was at Hersham, UK, totaling 15,000 individual parakeets (Pârâu et al. 2016). This increase in noise level may have negative impacts on humans living close to roost sites and has the potential to disturb other wildlife, as noise pollution has been shown to have negative effects on bird communication (Nemeth et al. 2013; Arroyo-Solís et al. 2013).

A flock of Rose-ringed Parakeets (*Psittacula krameri*), one Alexandrine Parakeet (*P. eupatria*, the bird with the reddish shoulder feathers above the top of the right-hand post), and one domestic pigeon (*Columba livia*, on the top of the sacks) feeding on stored grain. Photo by Vivek rathod17–Own work, CC BY-SA 3.0 (https://commons.wikimedia.org/w/index.php?curid=33456851).

PUBLIC PERCEPTIONS OF ROSE-RINGED PARAKEETS

The impacts of invasive species are usually detrimental toward the novel habitats they have settled in; however, when it comes to charismatic, intelligent, attractive species such as Rose-ringed Parakeets, there is the potential for positive impact. While many feel that parakeets are an unwelcome addition to our parks and gardens, large numbers of people take great delight in observing these exotic birds in local areas or at their garden bird feeders. Public attitudes toward invasive Rose-ringed Parakeets are generally positive, where people enjoy the sight of a bright green parrot flying in urban areas (Berthier et al. 2017; van Kleunen et al. 2010). In South Africa, the public reported very positive reactions to the presence of this novel bird in the local area, describing the parakeets as "dynamic," "charismatic," and "beautiful," and most people took great pleasure in seeing them (Hart and Downs 2014). Research is just beginning into understanding how the public perceives the species, and the potential implications on human well-being and, conversely, on public response to attempts at conservation management action. Indeed, management of invasive species, such as Monk Parakeets within urban areas, is often challenging, as conservation actions are frequently contested or halted by local communities (Crowley et al. 2019).

CONCLUSION: A RECIPE FOR SUCCESS

While many parrot species struggle to survive within their own native ranges, Rose-ringed Parakeets are thriving not only across their extensive native ranges but also in a large number of rapidly growing non-native populations in Europe, the continental US, Hawaii, Australia, and other regions. Our long-term love of keeping parrots as pets has led to repeated and ongoing releases into the wild, creating viable populations that contain high levels of genetic diversity essential for successful reproduction and adaptation to these new ranges. Rose-ringed Parakeets are successfully exploiting the abundance of resources provided outside their native range, such as year-round available food provided by bird feeders in urban areas or large plantations in warmer climates. Combined with their ability to creatively find and utilize nesting holes in trees and man-made structures, this is a recipe for success when it comes to their establishment and long-term survival. While native predatory birds are beginning to deprecate Rose-ringed Parakeets, the parakeets' populations are so large it is unlikely this will make any impact.

Research is only just beginning into understanding the impacts of these non-native parakeets on native biodiversity, but it is clear the impacts identified to date do not occur on a broad geographical scale and appear to differ between established populations. Due to the charismatic nature of the Rose-ringed Parakeet, conservation management and mitigation can be challenging. Public perceptions of these exotic birds can lead to barriers to management action (Crowley et al. 2019; Beever et al. 2019). As people fear conservation action, they can also be reluctant to engage with researchers who are attempting to understand the impacts on native species and agriculture (Hart and Downs 2014). The spread of Rose-ringed Parakeets may be having hidden consequences on global diversity. The increased rates of extinctions and invasions due to anthropogenic impacts could lead to biotic homogenization, where a single successful species (such as the Rose-ringed Parakeet) may replace a number of unsuccessful or extinct species, which are a source of evolutionarily rich and unique phylogenetic diversity (Jackson et al. 2015b).

Despite the Rose-ringed Parakeets' abundance and rapidly growing global distribution, many people remain surprisingly unaware of the local presence of these birds. In time, the species will likely become commonplace, and while most people still consider these charismatic, colorful birds an exciting novelty, perceptions may change over time. The regular presence of Rose-ringed Parakeets in urban parks and gardens may become so normal to our children, leading to a shifting baseline syndrome, that parakeets become no more exciting than a common pigeon.

REFERENCES

Ahmad, S., Khan, H. A., and Javed, M. 2012. An estimation of rose-ringed parakeet (*Psittacula krameri*) depredations on citrus, guava and mango in orchard fruit farm. *International Journal of Agriculture and Biology* 14:149–152.

Allendorf, F. W., and Lundquist L. 2003. Introduction: Population biology, evolution and control of invasive species. *Conservation Biology* 17:24–30.

Arroyo-Solís, A., Castillo, J. M., Figueroa, E., López-Sánchez, J. L., and Slabbekoorn, H. 2013. Experimental evidence for an impact of anthropogenic noise on dawn chorus timing in urban birds. *Journal of Avian Biology* 44:288–296.

Avery, M. L., and Shiels, A. B. 2018. Monk and rose-ringed parakeets. In *Ecology and Management of Terrestrial Vertebrate Invasive Species in the United States*, ed. W. C. Pitt, J. C. Beasley, and G. W. Witmer, 333–357. Boca Raton, FL: CRC Press.

Beever, E. A., Simberloff, D., Crowley, S. L., Al-Chokhachy, R., Jackson, H. A., and Petersen, S. L. 2019. Social-ecological mismatches create conservation challenges in introduced species management. *Frontiers in Ecology and the Environment* 17:117–125.

Berthier, A., Clergeau, P., and Raymond, R. 2017. From beautiful exotic to beautiful invasive: Perceptions and appreciations of the rose-ringed parakeet *Psittacula krameri* in the metropolis of Paris. *Annales de Géographie* 4:408–434.

Blackburn, T. M., Prowse, T. A., Lockwood, J. L., and Cassey, P. 2013. Propagule pressure as a driver of establishment success in deliberately introduced exotic species: Fact or artefact? *Biological Invasions* 15:1459–1469.

Braun, M. P., and Wink, M. 2013. Nestling development of ring-necked parakeets (*Psittacula krameri*) in a nest box population. *Open Ornithology Journal* 6:9–24.

Butchart, S.H.M., Stattersfield, A. J., and Brooks, T. M. 2006. Going or gone: Defining "possibly extinct" species to give a truer picture of recent extinctions. *Bulletin of the British Ornithologist's Club* 126A:7–24.

Butler, C. J., Cresswell, W., Gosler, A., and Perrins, C. 2013. The breeding biology of rose-ringed parakeets *Psittacula krameri* in England during a period of rapid population expansion. *Bird Study* 60:527–532.

Cardador, L., Lattuada, M., Strubbe, D., Tella, J. L., Reino, L., Figueira, R., and Carrete, M. 2017. Regional bans on wild-bird trade modify invasion risks at a global scale. *Conservation Letters* 10:717–725.

Cassey, P., Blackburn, T. M., Jones, K. E., and Lockwood, J. L. 2004. Mistakes in the analysis of exotic species establishment: Source pool designation and correlates of introduction success among parrots (Aves: Psittaciformes) of the world. *Journal of Biogeography* 31:277–284.

Cheke, A., and Hume, J. P. 2009. *Lost Land of the Dodo: The Ecological History of Mauritius, Reunion and Rodrigues*. London: A & C Black.

Clergeau, P., and Vergnes, A. 2011. Bird feeders may sustain feral rose-ringed parakeets *Psittacula krameri* in temperate Europe. *Wildlife Biology* 17:248–252.

Collar, N. J. 2000. Globally threatened parrots: Criteria, characteristics and cures. *International Zoo Yearbook* 37:21–35.

Crowley, S. L., Hinchliffe, S., and McDonald, R. A. 2019. The parakeet protectors: Understanding opposition to introduced species management. *Journal of Environmental Management* 229:120–132.

DAISIE. 2009. *Psittacula krameri*. In *Handbook of Alien Species in Europe*. Dordrecht, Netherlands: Springer.

del Hoyo, J., Elliott A., and Sargatal, J., eds. 1997. *Handbook of the Birds of the World*. Vol. 4: *Sandgrouse to Cuckoos*. Barcelona, Spain: Lynx Edicions; Cambridge, UK: BirdLife International.

Duncan, R. P., Blackburn, T. M., and Sol, D. 2003. The ecology of bird introductions. *Annual Review of Ecology, Evolution, and Systematics* 34:71–98.

EASIN. 2020. European Alien Species Information Network. Accessed Oct. 2020. http://easin.jrc.ec.europa.eu/.

Fletcher, M., and Askew, N. 2007. *Review of the Status, Ecology and Likely Future Spread of Parakeets in England*. York, UK: Central Science Laboratory.

Forshaw, J. 2010. *Parrots of the World*. Princeton, NJ: Princeton Univ. Press.

GBIF. 2019. Global Biodiversity Information Facility. Accessed Mar. 2019. https://doi.org/10.15468/dl.j1fzvm.

Grandi, G., Menchetti, M., and Mori, E. 2018. Vertical segregation by breeding ring-necked parakeets *Psittacula krameri* in northern Italy. *Urban Ecosystems* 21:1011–1017.

Hancock, R., and Martin, J. 2015. Predation of rose-ringed parakeets by raptors and owls in Inner London. *British Birds* 6:349–353.

Hart, L. A., and Downs, C. T. 2014. Public surveys of rose-ringed parakeets, *Psittacula krameri*, in the Durban metropolitan area, South Africa. *African Zoology* 49:283–289.

Heald, O.J.N., Fraticelli, C., Cox, S. E., Stevens, M.C.A., Faulkner, S. C., Blackburn, T. M., and Le Comber, S. C. 2019. Understanding the origins of the ring-necked parakeet in the UK. *Journal of Zoology* 312:1–11.

Hernández-Brito, D., Carrete, M., Ibáñez, C., Juste, J., and Tella, J. L. 2018. Nest-site competition and killing by invasive parakeets cause the decline of a threatened bat population. *Royal Society Open Science* 5:172477.

Hernández-Brito, D., Carrete, M., Popa-Lisseanu, A. G. Ibáñez, C., and Tella, J. L. 2014. Crowding in the city: Losing and winning competitors of an invasive bird. *PLoS One* 9:1–11.

Hulme, P. E. 2009. Trade, transport and trouble: Managing invasive species pathways in an era of globalization. *Journal of Applied Ecology* 46:10–18.

Jackson, H. A. 2015. Evolutionary conservation genetics of invasive and endemic parrots. PhD diss., Univ. of Kent, UK.

REFERENCES

Jackson, H. A., Jones, C. G, Agapow, P.-M., Tatayah, V., and Groombridge, J. J. 2015a. Micro-evolutionary diversification among Indian Ocean parrots: Temporal and spatial changes in phylogenetic diversity as a consequence of extinction and invasion. *Ibis* 157:496–510.

Jackson, H., Strubbe, D., Tollington, S., Prys-Jones, R., Matthysen, E., and Groombridge, J. J. 2015b. Ancestral origins and invasion pathways in a globally invasive bird correlate with climate and influences from bird trade. *Molecular Ecology* 24:4269–4285.

Jetz, W., Thomas, G. H., Joy, J. B., Redding, D. W., Hartmann, K., and Mooers, A. O. 2014. Global distribution and conservation of evolutionary distinctness in birds. *Current Biology* 24:919–930.

Jiménez-Valverde, A., Peterson, A. T., Soberón, J., Overton, J. M., Aragón, P., and Lobo, J. M. 2011. Use of niche models in invasive species risk assessments. *Biological Invasions* 13:2785–2797.

Jones, C. G., Malham, J., Reuleux, A., Richards, H., Raisin, C., Tollington, S., Zuel, N., Chowrimootoo, A., and Tataya, V. 2013. Echo parakeet *Psittacula eques*. In *The Birds of Africa*, ed. F. Hawkins and R. Safford, R. Vol. 8: *Birds of the Malagasy Region*, 433–438. London: A & C Black.

Juniper, T., and Parr, M. 1998. *Parrots: A Guide to the Parrots of the World*. East Sussex, UK: Pica Press.

Katsanevakis, S., Deriu, I., D'Amico, F., Nunes, A. L., Pelaez Sanchez, S., Crocetta, F., Arianoutsou, M., Bazos, I., Christopoulou, A., Curto, G., and Delipetrou, P. 2015. European alien species information network (EASIN): Supporting European policies and scientific research. *Management of Biological Invasions* 6:147–157.

Khan, H. A. 2002 Breeding habitats of the rose-ringed parakeet (*Psittacula krameri*) in the cultivations of central Punjab. *International Journal of Agricultural Biology* 4:401–403.

Kundu, S., Faulkes, C. G., Greenwood, A. G., Jones, C. G., Kaiser, P., Lyne, O. D., Black, S. A., Chowrimootoo, A., and Groombridge, J. J. 2012. Tracking viral evolution during a disease outbreak: The rapid and complete selective sweep of a *Circovirus* in the endangered echo parakeet. *Journal of Virology* 86:5221–5229.

Lee, C. E. 2002. Evolutionary genetics of invasive species. *Trends in Ecology & Evolution* 17:386–391.

Le Louarn, M., Clergeau, P., Briche, E., and Deschamps-Cottin, M. 2017. "Kill two birds with one stone": Urban tree species classification using bi-temporal pléiades images to study nesting preferences of an invasive bird. *Remote Sensing* 9:916.

Le Louarn, M., Couillens, B., Deschamps-Cottin, M., and Clergeau, P. 2016. Interference competition between an invasive parakeet and native bird species at feeding sites. *Journal of Ethology* 34:291–298.

Le Roux, J., and Wieczorek, A. M. 2009. Molecular systematics and population genetics of biological invasions: Towards a better understanding of invasive species management. *Annals of Applied Biology* 154:1–17.

Lever, C. 2005. *Naturalised Birds of the World*. London: T. & A. D. Poyser.

Luna, A., Franz, D., Strubbe, D., Shwartz, A., Braun, M. P., Hernández-Brito, D., Malihi, Y., Kaplan, A., Mori, E., Menchetti, M., and van Turnhout, C. A. 2017. Reproductive timing as a constraint on invasion success in the ring-necked parakeet (*Psittacula krameri*). *Biological Invasions* 19:2247–2259.

McKinney, M. L., and Lockwood, J. L. 1999. Biotic homogenization: A few winners replacing many losers in the next mass extinction. *Trends in Ecology & Evolution* 14:450–453.

Menchetti, M., and Mori, E. 2014. Worldwide impact of alien parrots Aves Psittaciformes on native biodiversity and environment: A review. *Ethology Ecology & Evolution* 26:172–194.

Menchetti, M., Mori, E., and Angelici, F. M. 2016. Effects of the recent world invasion by ring-necked parakeets *Psittacula krameri*. In *Problematic Wildlife: A Cross-Disciplinary Approach*, ed. F. M. Angelici, 253–266. Switzerland and New York: Springer International Publishing.

Mentil, L., Battisti, C., and Carpaneto, G. M. 2018. The impact of *Psittacula krameri* (Scopoli, 1769) on orchards: First quantitative evidence for Southern Europe. *Belgian Journal of Zoology* 148:129–134.

Meyerson, L. A., and Mooney, H. A. 2007. Invasive alien species in an era of globalisation. *Frontiers in Ecology and the Environment* 5:199–208.

Millennium Ecosystem Assessment. 2005. Accessed Oct. 2020. https://www.millenniumassessment.org/en/index.html.

Nemeth, E., Pieretti, N., Zollinger, S. A., Geberzahn, N., Partecke, J., Catarina, A., Brumm, H., and Miranda, A. C. 2013. Bird song and anthropogenic noise: Vocal constraints may explain why birds sing higher-frequency songs in cities. *Proceedings of the Royal Society of London B: Biological Sciences* 280:1–7.

ParrotNet. 2018. European Network on Invasive Parakeets. Univ. of Kent. Accessed Dec. 2018. https://www.kent.ac.uk/parrotnet/.

Pârâu, L. G., Strubbe, D., Mori, E., Menchetti, M., Ancillotto, L., Kleunen, A. V., White, R. L., Luna, A., Hernández-Brito, D., Louarn, M. L., and Clergeau, P. 2016. Rose-ringed parakeet *Psittacula krameri* populations and numbers in Europe: A complete overview. *Open Ornithology Journal* 9:1–13.

Peck, H. L. 2013. Investigating ecological impacts of the non-native population of rose-ringed parakeets (*Psittacula krameri*) in the UK. PhD diss., Imperial College, London.

Peck, H., Pringle, H. E., Marshall, H. H., Owens, I.P.F., and Lord, A. M. 2014. Experimental evidence of impacts of an invasive parakeet on foraging behavior of native birds. *Behavioural Ecology* 25:582–590.

Per, E. 2018. The spread of the rose-ringed parakeet, *Psittacula krameri*, in Turkey between 1975 and 2015 (Aves: Psittacidae). *Zoology in the Middle East* 64:297–303.

Perrin, M. 2012. *Parrots of Africa, Madagascar and the Mascarene Islands: Biology, Ecology and Conservation*. Johannesburg, South Africa: Wits Univ. Press.

Ramzan, M., and Toor, H. S. 1973. Damage to maize crop by rose ringed parakeet, *Psittacula krameri* (Scopoli) in the Punjab. *Journal of Bombay Natural History Society* 70:201–204.

Real Decreto 630/2013. Ministerio de Agricultura, Alimentación y Medio Ambiente, Gobierno de España. Doc. BOE-A-2013-8565. https://www.boe.es/eli/es/rd/2013/08/02/630/con.

Reino, L., Figueira, R., Beja, P., Araújo, M. B., Capinha, C., and Strubbe, D. 2017. Networks of global bird invasion altered by regional trade ban. *Science Advances* 3:e1700783.

Sa, R.C.C., Cunningham, A. A., Dagleish, M. P., Wheelhouse, N., Pocknell, A., Borel, N., Peck, H., and Lawson, B. 2014. Psittacine beak and feather disease in a free-living ring-necked parakeet (*Psittacula krameri*) in Great Britain. *European Journal of Wildlife Research* 60:395–398.

Schäckermann, J., Weiss, N., von Wehrden, H., and Klein, A. M. 2014. High trees increase sunflower seed predation by birds in an agricultural landscape of Israel. *Frontiers in Ecology and Evolution* 2:35.

Seychelles Islands Foundation. 2020. Accessed Oct. 2020. https://www.sif.sc.

Shiels, A. B., Bukoski, W. P., and Siers, S. R. 2018. Diets of Kauai's invasive rose-ringed parakeet (*Psittacula krameri*): Evidence of seed predation and dispersal in a human-altered landscape. *Biological Invasions* 20:1449–1457.

Shwartz, A., Strubbe, D., Butler, C. J., Matthysen, E., and Kark, S. 2009. The effect of enemy-release and climate conditions on invasive birds: A regional test using the rose-ringed parakeet (*Psittacula krameri*) as a case study. *Diversity and Distributions* 15:310–318.

Strubbe, D., Beauchard, O., and Matthysen, E. 2015a. Niche conservatism among non-native vertebrates in Europe and North America. *Ecography* 38:321–329.

Strubbe, D., Jackson, H., Groombridge, J., and Matthysen, E. 2015b. Invasion success of a global avian invader is explained by within-taxon niche structure and association with humans in the native range. *Diversity and Distributions* 21:675–685.

Strubbe, D., and Matthysen, E. 2007. Invasive ring-necked parakeets *Psittacula krameri* in Belgium: Habitat selection and impact on native birds. *Ecography* 30:578–588.

Strubbe, D., and Matthysen, E. 2009. Establishment success of invasive rose-ringed and monk parakeets in Europe. *Journal of Biogeography* 36:2264–2278.

Tatayah. R. V., Malham, J., Haverson, P., Reuleaux, A., and Van de Wetering, J. 2007. Design and provision of nest boxes for echo parakeets *Psittacula eques* in Black River Gorges National Park, Mauritius. *Conservation Evidence* 4:6–19.

Vall-llosera, M., Woolnough, A. P., Anderson, D., and Cassey, P. 2017. Improved surveillance for early detection of a potential invasive species: The alien rose-ringed parakeet *Psittacula krameri* in Australia. *Biological Invasions* 19:1273–1284.

van Kleunen, A., van den Bremer, L., Lensink, R., and Wiersma, P. 2010. *De Halsbandparkiet, Monniksparkiet en Grote Alexanderparkiet in Nederland: Risicoanalyse en beheer*. SOVON onderzoeksrapport 2010/10. Njimegen, Netherlands: Team Invasieve Exoten van het Ministerie van Landbouw, Natuur en Voedselkwaliteit.

White, R. L., Strubbe, D., Dallimer, M., Davies, Z. G., Davis, A.J.S., Edelaar, P., Groombridge, J. J., Jackson, H. A., Menchetti, M., Mori, E., et al. 2019. Assessing the ecological and societal impacts of alien parrots in Europe using a transparent and inclusive evidence-mapping scheme. *NeoBiota* 48:45–69.

Yosef, R., Zduniak, P., and Żmihorski, M. 2016. Invasive ring-necked parakeet negatively affects indigenous Eurasian hoopoe. *Annales Zoologici Fennici* 53:281–287.

11

MONK PARAKEETS AS A GLOBALLY NATURALIZED SPECIES

Carlos E. Calzada Preston, Stephen Pruett-Jones, and Jessica R. Eberhard

INTRODUCTION

The introduction of non-native species is a global phenomenon involving thousands of species of plants and animals. For an introduced species to become invasive, it must overcome several barriers: (1) transport, (2) introduction, (3) establishment, and in the case of invasive species, (4) spread (Blackburn et al. 2011). Whether a particular species proceeds from one stage to the next depends on a variety of factors that are both intrinsic (e.g., life history characteristics, diet, physiological traits) and extrinsic (e.g., climatic conditions, interactions with native species, etc.). An introduced species may become established but fail to become invasive if local conditions inhibit survival and reproduction

Records of humans capturing, keeping, and breeding birds as pets date as far back as 4,000 years ago, but only recently has international trade allowed the annual transport of millions of birds (Carrete and Tella 2008). Most recent avian introductions tend to be birds transported as pets, followed by the birds escaping captivity and establishing local populations (Hobson et al. 2017). Some of the factors thought to influence successful introductions are propagule pressure (aka introduction effort), origin of introduced individuals as either wild-caught or captive-bred, and life history traits (Cassey et al. 2004; Carrete and Tella 2008). Establishment success of non-native species has been linked to propagule pressure in that introductions with more individuals tend to be more successful (Blackburn et al. 2009).

One of the most popular pet parrot species that has become widely established is the Monk Parakeet (*Myiopsitta monachus*). Whether this species has been so popular in the pet trade because it is actually a desirable pet parrot or because it is abundant in the wild and cheap on the market is unclear. Monk Parakeets have been some of the most frequently exported parrots globally (Russello et al. 2008) and, along with the Rose-ringed Parakeet (*Psittacula krameri*), are the most widely distributed naturalized parrots globally (Cardador et al., chap. 1 this vol.; Royle and Donner, chap. 2 this vol.).

Monk Parakeets are relatively small parrots that are native to temperate and subtropical South America, where they typically inhabit savanna woodland and open forest areas, often near human habitation (Forshaw 1989). Monk Parakeets are unique among the Psittaciformes for being the only species to construct a nest rather than nest in a cavity. Daily activity is centered at their nests, which are used year-round for roosting as well as for breeding (Bucher et al. 1991). The native range of Monk Parakeets extends from central Bolivia and southern Brazil to Uruguay and central Argentina (Forshaw 1989). During the 20th century, this range expanded into the Pampas grasslands of Argentina, as a result of the widespread practice of planting *Eucalyptus* trees near ranches and houses, which has provided the parakeets with nesting substrates in a habitat that had previously been treeless (Bucher and Aramburú 2014).

In this chapter, we illustrate the Monk Parakeet's trajectory through the invasion framework, document the species' current global distribution, and describe its natural history in both native and naturalized populations.

DATA SETS

In addition to using data available in the primary literature, in this summary we made use of data from four public databases. First, trade records for Monk Parakeets were downloaded from the Convention on International Trade in Endangered Species (CITES 2017) database for the years 1981 to 2017. The CITES Trade Database records the importing and exporting country, the source and purpose of the individuals, the year of the trade, and the number of individuals reported by each country. We refer to traded individuals as the number of live Monk Parakeets reported by the exporting country only. This ensures that counts weren't duplicated by combining the records provided by both countries involved in the trade. Regarding records for former countries (e.g., Soviet Union, Yugoslavia, Czechoslovakia), if one of the successor countries also had trade records, the records listed were simply relabeled according to the corresponding successor country (in this case: Russia, Slovenia, and Czech Republic, respectively). Lastly, only those CITES records that involved importations of live birds were included. Also, please note, by "countries," we also mean territories. Thus, Puerto Rico and the US Virgin Islands are considered separately from the United States (US).

Data for occurrence and status of the species (i.e., whether the species was established or breeding) were obtained from the Global Biodiversity Information Facility (GBIF 2019) and the Global Avian Invasion Atlas (GAVIA; Dyer et al. 2017). The GBIF database combines data from other databases (a total of 42,000) dealing with biodiversity around the world. The GAVIA database is a spatial and temporal data set referencing known avian introduction events by species and includes published mentions of a species at a specific locality, published sightings, records of museum collections, etc. Our interest was in countries in which Monk Parakeets were recorded as either breeding (observed instances of breeding but not necessarily as part of a self-sustaining population) or established (with a self-sustaining population).

Occurrences in each country were obtained from the eBird (eBird 2017) and GBIF databases. Annual country counts from eBird were estimated by summing the maximum observation counts within each country's states or provinces. This gives a conservative estimate for bird counts by assuming each state has only one population, thus avoiding duplicate counts of birds that may have moved throughout the state. We have chosen to illustrate time series of importations and introduced population size for two countries that have both imported large numbers of Monk Parakeets and have implemented restrictions on their importation for at least 10 years—the US and Spain. To avoid inflating estimates of abundance, records were rarefied to the state/province scale, where the largest observation within each locality was obtained for every year: , where $i = 1, 2, ..., k$ localities within each country. While this method for estimating abundance is certainly unreliable for obtaining precise population size estimates, the general trends reflect broad-scale patterns of population growth initiated by importation of individuals but becoming independent once populations are self-sustaining.

Lastly, we examine annual counts of Monk Parakeets by the annual National Audubon Society Christmas Bird Count (CBC) up to 2018. To account for sampling effort, we corrected raw numbers to reflect counts per party hour, and we summarize the number of different count circles (census areas) reporting Monk Parakeets during each annual census (National Audubon Society 2019).

TRANSPORT

According to the CITES database, 1,295,805 live Monk Parakeets were traded internationally from 1981 to 2017. From this total, the source of individuals was reported as 60.98% wild-caught, 30.39% unreported, and 9.63% all other categories (captive-bred, born in captivity, confiscated, pre-convention, ranched, and unknown). Similarly, individuals were reported as being imported for a particular purpose

Monk Parakeets (*Myiopsitta monachus*) at a nest entrance in a date palm tree. Melonera, Grand Canary Island, Spain. Photo by Philip McErlean (https://creativecommons.org/licenses/by-nd/2.0/).

(e.g., captive breeding, circus, commercial, law/forensic, personal, scientific, unreported, wild-release, and zoo); CITES records indicate that 84.20% of individuals were traded for commercial purposes and the import purpose was unreported for 14.62%.

A total of 66 exporting countries are listed in the CITES data set, and the five top exporters are Uruguay (1,022,616 Monk Parakeets exported), Argentina (209,391), South Africa (43,206), Singapore (18,211), and the Netherlands (7,794). Of these, only the top two are within the species' native range.

Excluding records listing countries within the native distribution and "various" or "unknown" as the importing country, 103 countries have been recorded as importing live Monk Parakeets, with anywhere between 12 and 40 countries importing parakeets in any given year (Fig. 11.1; Table 11.1). The top five countries importing Monk Parakeets were Mexico (542,751 individuals imported), US (263,654), Spain (239,522), Italy (85,665), and Taiwan (25,334). Taiwan represents an interesting example of a country with a large number of importations but in which Monk Parakeets have not become established.

The top three importing countries (Mexico, US, and Spain) have instituted different bans, which has resulted in a reduction of Monk Parakeet importations and a change in the identity of the major importing countries over time (Cardador et al. 2017). In the US, the Wild Bird Conservation Act of 1992 restricted the import of wild-caught birds in general. The act took effect in 1993, after which Monk Parakeet importations were negligible (Fig. 11.2). Importations into Spain declined after the European Union (EU) passed the Wild Bird Declaration, in 2005, prohibiting the importation of wild-caught birds (Groupo de Aves Exóticas, SEO/Birdlife 2012, Fig. 11.3).

MONK PARAKEETS AS A GLOBALLY NATURALIZED SPECIES

TABLE 11.1

Countries in which Monk Parakeets are established or breeding, along with data on importation records from CITES (2017), sightings in eBird for the years 2015–17 (eBird 2017), and records in GBIF (2019) and GAVIA (Dyer et al. 2017). Countries are listed alphabetically.

COUNTRY	CITES IMPORTS	EBIRD 2015	EBIRD 2016	EBIRD 2017	GBIF RECORD	GAVIA ESTABLISHED	GAVIA BREEDING
Australia	NA	NA	NA	NA	NA	Established	NA
Austria	313	NA	NA	NA	33	Established	NA
Belgium	977	1	15	8	729	Established	Breeding
Brazil	33	309	399	280	72	Established	NA
Canada	3,750	0	0	0	NA	Established	Breeding
Cayman Islands	45	12	16	14	1	Established	NA
Chile	15,183	224	127	272	3	Established	Breeding
Czech Republic	703	NA	NA	NA	30	Established	NA
Denmark	565	NA	NA	NA	44	Established	NA
Dominican Republic	6	NA	NA	NA	NA	Established	NA
France	10,112	0	0	0	252	Established	NA
Germany	9,580	NA	NA	NA	151	Established	NA
Guadeloupe	NA	NA	NA	NA	NA	Established	Breeding
Israel	753	54	124	61	11	Established	NA
Italy	85,665	59	124	92	185	Established	NA
Japan	6,139	NA	NA	NA	NA	Established	NA
Kenya	NA	NA	NA	NA	NA	Established	NA
Mexico	542,751	347	532	492	694	Established	Breeding
Netherlands	5,253	11	14	17	19	Established	NA
Puerto Rico	NA	699	513	526	14	Established	NA
Portugal	14,378	19	18	16	57	Established	NA
Slovakia	2	NA	NA	NA	17	Established	NA
Spain	239,522	433	602	571	4,752	Established	Breeding
United Kingdom	4,147	5	6	0	143	Established	Breeding
United States	263,654	638	506	566	1,776	Established	Breeding
Venezuela	2,046	NA	NA	NA	NA	NA	Breeding
Virgin Islands (US)	NA	4	0	0	NA	Established	NA

In Mexico, the largest importer of Monk Parakeets, demand was boosted in 2008 by regulations that banned the purchase of native parrot species (Hobson et al. 2017). In 2014–15 importations were temporarily banned due to concerns over the possible introduction of avian influenza (Hobson et al. 2017), and in 2016 the species was added to Mexico's invasive species list, permanently restricting its importation into the country (Secretaria de Medio Ambiente y Recursos Naturales 2016).

As Cardador et al. (2017) describe, bans on wild-caught birds, and Monk Parakeets in particular, have shifted exportations and

TRANSPORT

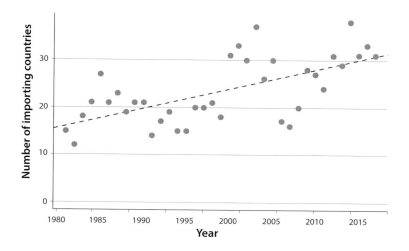

Figure 11.1. Change in number of countries importing Monk Parakeets by year.

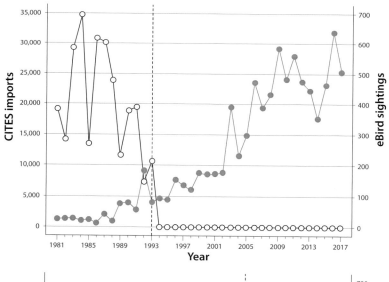

Figure 11.2. Change in importations of Monk Parakeets across years (open circles) vs. reported sightings in eBird (closed circles) for the United States. "CITES imports" refers to the number of birds reported to have been exported to the US in the CITES (2017) database. "eBird sightings" refers to a corrected measure of all the eBird reports. See "Data Sets" section of this chapter for more details. The vertical, dashed line refers to the year of the legal restriction on importation of Monk Parakeets and other parrots into the US.

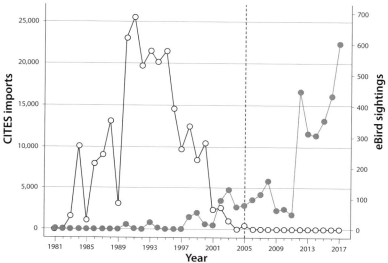

Figure 11.3. Change in importations of Monk Parakeets across years (open circles) vs. reported sightings in eBird (closed circles) for Spain. "CITES imports" refers to the number of birds reported to have been exported to that country in the CITES (2017) database. "eBird sightings" refers to a corrected measure of all the eBird reports. See "Data Sets" section of this chapter for more details. The vertical, dashed line refers to the year of the European Union's legal restriction on importation of Monk Parakeets and other parrots.

importations to distinct global locations in the span of a few decades. Nevertheless—and despite the bans by the US, the EU, and Mexico—the number of countries importing Monk Parakeets has continued to increase (Fig. 11.1).

INTRODUCTION AND ESTABLISHMENT

Having determined the countries where Monk Parakeets have been imported, we were interested in those areas where the species has been recorded breeding or is now established. The GBIF and eBird databases provide occurrence data for birds observed in the wild, and both data sets were filtered to include only those records where the country and observation data in the field were reported.

From the pool of 103 countries that reported having imported Monk Parakeets in the CITES database, 22 were included in the set of countries for which observations of breeding or establishment were reported in the GAVIA database (Table 11.1). In addition, there are five countries in which free-ranging Monk Parakeets have been observed and are likely breeding, based on the GAVIA database, but for which there are no CITES importation records. Adding these countries yields a minimum total of 27 countries (or territories) in which Monk Parakeets are or appear to be breeding or are established (Table 11.1).

In Table 11.1, we list only countries for which there is documented evidence that Monk Parakeets are now established. Nevertheless, there are many other countries in which this species is occasionally seen but for which there have not been any records of breeding or transportation. For example, in Morocco and Palestine, Monk Parakeets are recorded, but no breeding records exist. Morocco is close to Spain, and Palestine is close to Israel, and both Spain and Israel have long histories of importations of the species and large breeding populations. Thus, in the cases of Morocco and Palestine, Monk Parakeets could have been either transported there via undocumented means or dispersed from adjacent countries.

POPULATION GROWTH

Monk Parakeet populations have expanded their distribution and increased in size, in both the native range and in areas where naturalized populations have established. With respect to the species' native range, Bucher and Aramburú (2014) document that over the past 150 years, Monk Parakeets have greatly expanded their distribution in Argentina as a result of changes in land-use patterns. In particular, the increase in areas planted with *Eucalyptus* trees, the introduction of cattle, expansion of croplands, and urban development were key factors promoting rapid population growth in the parakeets (Bucher and Aramburú 2014).

With respect to naturalized populations, in each country or area in which Monk Parakeets have become established, their population has subsequently grown rapidly and often at an exponential rate (van Bael and Pruett-Jones 1996; Pruett-Jones and Tarvin 1998; Butler 2005; Pruett-Jones et al. 2012; Burgio et al. 2016; Postigo et al. 2017, 2019). Postigo et al. (2019) document the presence of more than 23,750 Monk Parakeets in 179 municipalities in eight EU countries. Furthermore, in each municipality for which there are historical records, the species' populations have grown exponentially, with Mediterranean countries seeing faster growth in Monk Parakeet populations than Atlantic countries (Postigo et al. 2019).

In the US, after the species became established in the early 1970s, the population grew at an exponential rate until the early 2000s (Fig. 11.4) (van Bael and Pruett-Jones 1996; Pruett-Jones and Tarvin 1998; Butler 2005; Pruett-Jones et al. 2012). The population decline in the early 2000s deserves considerably greater attention than it has received to date. There are two interesting aspects about this decline. First, although the total size of the population has declined (Fig. 11.4), the number of locations in the US where Monk Parakeets are recorded has not declined (Fig. 11.5). Thus, it appears that the geographic range or distribution of the species in the US has not gotten smaller, even if the numbers of birds in each area have declined.

The timing of the decline in Monk Parakeet population size in the US is the second interesting aspect of this population change. Figure 11.4

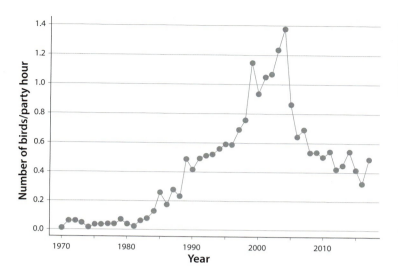

Figure 11.4. Changes in numbers of Monk Parakeets recorded by the National Audubon Society's annual Christmas Bird Count for the United States from 1970 to 2017.

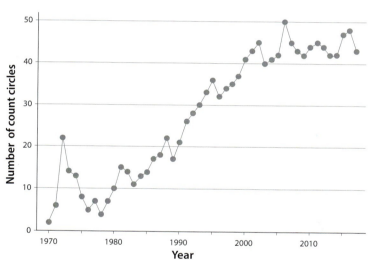

Figure 11.5. Number of count circles reporting Monk Parakeets in the National Audubon Society's annual Christmas Bird Count for the United States from 1970 to 2017.

illustrates the CBC numbers for the national population. However, when we examine the population trends in the states where Monk Parakeets are common (Calzada Preston, unpubl. data), a different trend emerges. The first state to see a decline in numbers was New York, where the population decline started in 1999. In contrast, the Florida and Illinois populations began to decline in 2004. This difference in timing of the decline in New York mirrors the timing of the spread of West Nile virus in bird populations, particularly corvids. West Nile virus was first seen in crows in New York in 1999, and the virus then began spreading south and west, reaching Illinois around 2002. There are no samples of Monk Parakeets from this time period that have been examined for West Nile virus, but it is possible that either this virus or another disease caused the widespread population decline.

NATURAL HISTORY OF MONK PARAKEETS IN THEIR NATIVE AND INTRODUCED RANGES

Myiopsitta monachus was previously divided into four subspecies: *M. m. monachus*, *M. m. calita*, *M. m. cotorra*, and *M. m. luchsi* (Forshaw 1989);

however, a study using molecular data called these subspecies' distinctions into question. Russello et al. (2008) analyzed sequence data from the mitochondrial control region of 73 museum specimens and found no evidence of differentiation among *calita*, *cotorra*, and *monachus*. On the other hand, the *luchsi* sequences were distinct and monophyletic in both network and phylogenetic analyses, and *luchsi* has since been elevated to species status (*M. luchsi*) (del Hoyo et al. 1997; Gill and Donsker 2018). *M. luchsi* differs from *M. monachus* in plumage, body size, and also in behavior. *M. luchsi* constructs its stick nests on cliffs rather than in trees (Lanning 1991) and is found at higher elevations than *M. monachus* (Forshaw 1989).

Most of our knowledge of Monk Parakeets' basic biology and behavior comes from field studies that have been conducted in Argentina. Comparable information about introduced populations is more scattered, though quantitative studies have been carried out at a few locations. Here we briefly review the natural history of the species in both its natural range and that of the introduced populations.

Timing of Breeding

Monk Parakeets are nonmigratory, and in their native range they breed during the austral summer. Mean clutch initiation dates fall in early November in both Córdoba, Argentina (Navarro et al. 1992), and Entre Ríos, Argentina (Eberhard 1998), and in mid- to late October in Buenos Aires province, Argentina (Peris and Aramburú 1995). Introduced populations in the US breed during the boreal spring and summer (Kibbe and Cutright 1973; Neidermyer and Hickey 1977; Hyman and Pruett-Jones 1995). In Florida, eggs were found in nests from late March through early July (Avery et al. 2012), and in Chicago, young fledged from early July through mid-August (Hyman and Pruett-Jones 1995). Reports of Monk Parakeets in Mexico indicate the breeding season may be even longer there, from February to September (Ramírez-Albores 2012; Tinajero and Rodríguez Estrella 2015; Romero-Figueroa et al. 2016). In Barcelona, Spain, Batllori and Nos (1985) report that eggs were found in a nest in late August; however, in a detailed study of breeding phenology, Senar et al. (2019) report a mean initiation date of 25 April (range: 5 March–8 August) in the city. Similar timing of breeding has been observed in southwestern France, where Adde (1998) documented a successful breeding in which eggs were laid in July and nestlings fledged in mid-August. Similarly, Monk Parakeets in Brussels, Belgium, fledged young in May and June, indicating that egg-laying begins in March (Weiserbs and Jacob 1999). At their main study site in Córdoba, Navarro et al. (1992) found that only 20% of pairs laid replacement clutches after an unsuccessful breeding attempt, and second clutches were rare. For naturalized Monk Parakeets, second breeding attempts are mentioned by Weiserbs and Jacob (1999) in Brussels and by Neidermyer and Hickey (1977) in Illinois, US. Senar et al. (2019) found that in Barcelona, the percentage of pairs that attempted a second brood was three times that observed in the native range.

Breeding Productivity

Sizes of first clutches were found to vary slightly across three sites in Argentina, with yearly means for first clutches ranging from 5.1 to 6.1 eggs (overall range 1–11 eggs) recorded for a study in Córdoba (Navarro et al. 1992); an average of 6.9 eggs (range 5–12) for a study in Buenos Aires province (Peris and Aramburú 1995); and a mean of 4.8 eggs (range 3–7) in Entre Ríos (Eberhard 1998). The Córdoba study, which included data from six years at a single site, found that the average breeding success (number of fledglings/number of eggs) was 42% (range 14%–65%) but varied significantly among years (Navarro and Bucher 1992). At this site, breeding pairs fledged an average of 1.3 young per breeding attempt. Breeding success was only 8.9% in the Buenos Aires study (Peris and Aramburú 1995); and 20% in Entre Ríos, where breeding pairs produced an average of 0.9 young per year (Eberhard 1998). In the three studies, breeding failure was most often caused by predation, failure to hatch, or nest abandonment (Navarro et al. 1992; Peris and Aramburú 1995; Eberhard 1998).

Until recently, information on clutch sizes and breeding success for introduced Monk Parakeets was limited. In a case of double brooding reported in Illinois, no clutch-size data were given for the first clutch, but the second clutch contained four eggs (Neidermyer and Hickey 1977). A note about the species in New York reports a single clutch size of six (Kibbe and Cutright 1973). A study of Monk Parakeets in Barcelona includes mention of a nest containing three eggs (Batllori and Nos 1985). In the southern coastal area of Santa Catarina, Brazil, where Monk Parakeets have been present since 2006 (Viana et al. 2017), clutch sizes over four breeding seasons averaged 5.5 eggs for two nests in *Eucalyptus* trees (Viana et al. 2016), and average breeding success was 84.1% (about 4.6 fledglings/pair per breeding attempt). While they do not present clutch-size data for Monk Parakeets nesting on utility distribution poles in southern Florida, Avery et al. (2012) estimated that breeding pairs produce an average of three fledglings per breeding attempt. Taken together, these limited data suggest that breeding success of naturalized Monk Parakeets may be higher than in native habitats. A long-term study of parakeets in Barcelona followed a total of 651 breeding attempts over five years and found that fledging success of first broods was double that observed in the native range (Senar et al. 2019). Senar et al. (2019) also found that over half (55%) of first-year birds in Barcelona made breeding attempts, which contrasts with observations from Argentina that indicate parakeets typically wait at least a year before breeding.

Parental Care and Dispersal

A study of marked birds in Argentina found that in breeding pairs, the male does most of the nest building and nest maintenance, and the female is responsible for incubating eggs and brooding young hatchlings (Eberhard 1998). The female is often fed by her partner prior to, and during, the incubation period. In the nestling phase, both the male and female feed the young. After fledging, most young remain with their parents until they are approximately three months old and leave by the following breeding season, waiting at least two years before starting to breed (Martín and Bucher 1993).

Based on recaptures of banded parakeets in Córdoba, Martín and Bucher (1993) recorded a median dispersal distance (from natal nest to first breeding site) of 1,230 m (range 300–2,000 m). Gonçalves da Silva et al. (2010) used microsatellite DNA analysis to estimate natal dispersal using two sets of field-collected samples—one from Entre Ríos, Argentina, and a second set from Florida. Their estimated dispersal distances range from 0.5–9.6 km (Gonçalves da Silva et al. 2010) for Entre Ríos, and 0.052–106 km for the introduced population in Florida. While the results from Florida indicate that Monk Parakeets disperse over long distances, the limited geographic sampling affecting both the Córdoba and Entre Ríos estimates makes it impossible to say whether dispersal distances are different for native vs. introduced populations (Gonçalves da Silva et al. 2010). The only other published reports of dispersal by naturalized parakeets are an observation by Senar et al. (2016) of two parakeets that were banded in Barcelona's city center and later seen approximately 10 km away, and a reference to an unpublished study that found a mean dispersal distance of 1.1 km for parakeets in Barcelona (Rodríguez-Pastor et al. 2012).

Diet and Foraging Behavior

A factor that has undoubtedly been important in Monk Parakeets' ability to successfully establish populations in a variety of habitats is their generalist and opportunistic diet. In its native habitat, the species forages in trees and shrubs as well as on the ground, consuming seeds, fruits, berries, nuts, leaf buds, blossoms, and occasional insects and insect larvae (Forshaw 1989; Spreyer and Bucher 1998). Aramburú (1997) examined the crop and gizzard contents of 166 parakeets collected from January 1987 through March 1989 in Buenos Aires province, and found that over 99% of the contents' dry weight consisted of seeds. The types of seeds consumed varied through the year, with seeds of sedges (*Cyperus* spp.), Common Thistle (*Cirsium vulgare*), Common Sunflower (*Helianthus annuus*), and maize (*Zea mays*) constituting significant proportions of the diet; cultivated species were primarily consumed during the winter months (Aramburú 1997). During two breeding seasons in Entre Ríos, parakeets were observed feeding on seeds from trees (*Prosopis affiinis*, *P. nigra*, and *Acacia caven*) as well as grasses and thistles, tender young leaves, galls (which contained cynipid wasp larvae), and opportunistically on spilled corn and bits of fat and flesh on sheep hides (Eberhard 1997). Monk Parakeets have long been maligned as crop pests in their native range, implicated in damage to corn, sunflower, sorghum, and other grains, as well as fruit crops such as peaches, pears, and grapes (Mott 1973; Bucher 1992; Spreyer and Bucher 1998). However, the magnitude of this damage appears to be overestimated, possibly because crop damage by parakeets tends to be very patchy and sporadic (Bucher 1992; Canavelli et al. 2013).

Observations of naturalized Monk Parakeets have also noted their generalist foraging habits and have documented diets very similar to those described for the parakeets in their native range (Freeland 1973; Shields and Grubb 1974; Neidermyer and Hickey 1977; De Schaetzen and Jacob 1985; Weiserbs and Jacob 1999; South and Pruett-Jones 2000; Di Santo et al. 2013, 2017a; Tinajero and Rodríguez Estrella 2015). Although introduced Monk Parakeets have the potential to cause agricultural damage (e.g., see Davis 1974; MacGregor-Fors et al. 2011; Postigo et al. 2017), in most naturalized populations they have thus far remained restricted to urban areas and pose little threat to agricultural crops (Minor et al. 2012; Pruett-Jones et al. 2012; Sevenair 2012). In areas with a cold winter, this association with urban areas has been attributed to the parakeets' reliance on seed provided at bird feeders during the winter months (Bull 1973; Shields and Grubb 1974; Weiserbs and Jacob 1999; South and Pruett-Jones 2000; Minor et al. 2012). In only a few cases have they been observed to damage crops: fruit of longan (*Euphoria longana*) in south Florida (Tillman et al. 2000); minor damage to corn crops in Mexico (Muñoz-Jiménez and Alcántara-Carabajal 2017); and variable but not insignificant damage to tomatoes, corn, and orchard fruits around Barcelona (Senar et al. 2016).

In large populations, Monk Parakeet foraging flocks can comprise several hundred individuals during the winter nonbreeding season (Bucher et al. 1991), but during the breeding season, foraging groups are smaller; a study in Entre Ríos recorded a mean group size of about three parakeets (Eberhard 1998). In their study of the species in Chicago, US, South and Pruett-Jones (2000) found that flock sizes were variable, with parakeets usually forming small groups of ≤5 individuals. They also noted that flock size varied seasonally, with the largest flocks observed during the nonbreeding season. In Rome, Italy, flock sizes of foraging parakeets followed during July–September averaged 4.3 parakeets (Di Santo et al. 2013). Observations made during April–August of a small population near Mexico City, Mexico, indicated a mean foraging group size of 2.6 parakeets (Muñoz-Jiménez and Alcántara-Carabajal 2017).

The only detailed information available on daily movements of parakeets comes from a radio-tracking study in Texas, which found that males and females have similar-size core activity areas (Reed et al. 2013). Males and females also traveled similar distances to forage. During the winter, parakeets foraged over greater distances (winter mean 579 m, summer mean 339 m) and had larger core activity areas (Reed et al. 2013). Spreyer and Bucher (1998) report that parakeets typically travel 3.2–8 km from the nest to forage, and up to 24 km during the nonbreeding season, though the methods used to obtain these estimates are not provided.

Nesting Behavior

Monk Parakeets' bulky stick nests are easy to spot, even at a distance, and as a result there is more published information on nesting than on any other aspect of their natural history. In the parakeets' native range, nests are typically built using thorny twigs (Bucher et al. 1991), which are interlaced to form a domed chamber with a single opening that is sometimes extended to form an entrance tunnel (Harrison 1973). Nest structures can also consist of multiple chambers, each with its own entrance and occupied by a different pair or group of parakeets (Bucher et al. 1991), and multiple nests are frequently placed in the same or neighboring trees, forming colonies (Forshaw 1989; Bucher et al. 1991). In their native range, parakeets sometimes build their nests as additions to stick nests constructed by other species, including Firewood-gatherers (*Anumbius annumbi*) (Humphrey and Peterson 1978),

Brown Cacholotes (*Pseudoseisura lophotes*) (Eberhard 1996), Jabiru (*Jabiru mycteria*), and possibly Grey-crested Cacholotes (*P. unirufa*) (Burger and Gochfeld 2005). This behavior is echoed in observations of naturalized parakeets in Brussels, where parakeets added a nest on to a (presumably unoccupied) Eurasian Magpie (*Pica pica*) nest (Weiserbs and Jacob 1999). Nest construction and maintenance is carried out by both breeding and nonbreeding birds (Bucher et al. 1991; Eberhard 1998), and this activity increases as pairs initiate breeding attempts, which are often associated with the construction of new nest chambers (Martella and Bucher 1993; Eberhard 1998). Within breeding pairs, males are responsible for adding most of the twigs to the nest structure, and females add twigs (often green, leafy ones) to the interior, shredding them to line the nest chamber (Eberhard 1998; Aramburú et al. 2002, 2003).

In Argentina, nest structures have been found to be relatively small, with one to four chambers in Córdoba (Navarro et al. 1992), and an average of about two chambers (range one to six) in Entre Ríos (Eberhard 1997, 1998). In the Córdoba study, nests were in native trees (*Celtis tala*, *Geoffroea decorticans*, *Aspidosperma quebracho-blanco*, and *Prosopis* spp.) as well as *Eucalyptus*. At the Entre Ríos study site, one nest was on a utility pole, but the parakeets preferentially nested in *Prosopis nigra* and *Eucalyptus* trees, which were the tallest trees available, often placing their nests on fairly small branches at the edge of the tree's crown (Eberhard 1997, 1998). In the Brazilian Pantanal, Burger and Gochfeld (2005) found that Monk Parakeets built somewhat larger nests, often attached to nests of the Jabiru. Nest structures there contained an average of about seven chambers when built separate from Jabiru nests, and 13.8 chambers when attached to Jabiru nests. At the Pantanal site, parakeets built their nests in trees that were taller and with larger trunk diameters than neighboring trees, preferring *piúva* (*Tabebuia* spp.) and *manduvi* (*Sterculia apetala*) trees, and placed the nests close to the trunk or on large branch bifurcations (Burger and Gochfeld 2005). This difference in nest placement, as compared with the observations from Argentina, likely explains the larger nest sizes observed in the Pantanal (Burger and Gochfeld 2005).

Two studies in Argentina have examined nest substrate use and nest size in urban habitats within the Monk Parakeet's native range. Volpe and Aramburú (2011) studied nest preferences at two urban parks in La Plata, whose tree cover is dominated by *Eucalyptus* spp., and found that the parakeets preferred eucalypts that were taller and had larger diameters than neighboring trees, and tended to place their nests near the trunks or near the bases of branches rather than near the tips. Nest structures at this site contained a mean of 1.8 compartments (range one to seven compartments). Romero et al. (2015) censused nests and available trees in five Buenos Aires parks and found that parakeets preferred certain species (cedars, araucaria, and palms), even when they were not the most common, and also noted that *Eucalyptus* trees were not used, even though they were available and are clearly preferred in rural areas (Bucher and Aramburú 2014). Romero et al. (2015) attribute the parakeets' nest substrate choice to a preference for evergreen foliage and particular tree architecture features (main central trunk with supporting horizontal lateral branches) and noted that nests were usually placed near the center of the tree.

To date, introduced populations of Monk Parakeets are all in urban or suburban areas (Muñoz and Real 2006; Pranty 2009; Strubbe and Matthysen 2009), which offer a mix of natural and man-made nesting substrates. A literature survey of nest substrate use, summarized in Table 11.2, shows a great deal of variation across populations and also through time within populations. At some locations (e.g., Lake Worth and Miami Springs in Florida, US, as well as all sites in Mexico and Spain), parakeets built most of their nests in trees, while at others (e.g., Dallas, Texas, US) they show a clear preference for artificial substrates (Table 11.2). Pranty (2009) notes that in Florida, nests on power-line towers are more frequent in the southern portion of the peninsula than in its center, while nests on electrical substations are more frequent in the central peninsula. In New Jersey, US, both trees and man-made structures were used, and nests on trees and utility poles were similar in size and shape, at about the same distance from the ground, and similar distances from houses or neighboring buildings (Burger and Gochfeld 2009).

MONK PARAKEETS AS A GLOBALLY NATURALIZED SPECIES

TABLE 11.2

Nest substrates used by Monk Parakeets nesting in urban areas. Studies included in the table are ones that present quantitative data on nest substrate use. Percentages were rounded to nearest whole number. With the exception of the two sites in Argentina, all are naturalized populations.

STUDY LOCATION	YEAR	SUBSTRATE TYPE %		SAMPLE SIZE	REFERENCE
		TREES	ARTIFICIAL		
Chicago, IL, US	1992–93	20	80	156	Hyman and Pruett-Jones (1995)
Chicago, IL, US	2010	43	57	249	Minor et al. (2012)
Edgewater, NJ, US	2006	55	45	51	Burger and Gochfeld (2009)
New Orleans, LA, US	2003–4	73	27	150	Sevenair (2012), pers. comm.
New Orleans, LA, US	2008–9	33	67	186	Sevenair (2012), pers. comm.
Pittsburgh, PA, US	1972	60	40	5	Freeland (1973)
FL (19 counties), US	1999–2000	49	51	1,046	Pranty (2009)
FL (Lake Worth, Miami Spr.), US	2000	100	0	28	Burger and Gochfeld (2000)
South FL, US	2001–4	20	80	156	Newman et al. (2004)
TX (Dallas, Tarrant), US	2010–12	25	75	235	Reed et al. (2014)
Chametla, Mexico	2013–14	100	0	8	Tinajero and Rodríguez (2015)
Chihuahua, Mexico	2012–13	67	33	3	Soto-Cruz et al. (2014)
Guerrero, Mexico	2015	100	0	2	Sierra-Morales and Almazán-Núñez (2017)
Guerrero Negro, Mexico	2011	100	0	15	Guerrero-Cárdenas et al. (2012)
Hidalgo, Mexico	2011–14	33	67	3	Zuria et al. (2017)
Mexico City, Mexico	2008–12	100	0	5	Ramírez-Albores (2012)
Oaxaca, Mexico	2008	100	0	4	Pablo López (2009)
Texcoco, Mexico	2010–11	100	0	8	Muñoz-Jiménez and Alcántara-Carbajal (2017)
Toluca/Metepec, Mexico	2012	100	0	3	Salgado-Miranda et al. (2016)
Santa Catarina, Brazil	2010–13	100	0	140	Viana et al. (2016)
Chile	2003–13	100	0	393	Tala et al. (2005)
Buenos Aires, Argentina	2012–13	100	0	128	Romero et al. (2015)
La Plata, Argentina	2008–9	100	0	42	Volpe and Aramburú (2011)
Tel Aviv, Israel	2004–7	100	0	511	Postigo et al. (2017)
Rome, Italy	2009	100	0	73	Di Santo et al. (2017b)
Spain (nationwide)	2015	80	1	5,321[1]	Molina et al. (2016)
Barcelona, Spain	1985	100	0	17	Batllori and Nos (1985)
Barcelona, Spain	1992–94	100	0	77	Sol et al. (1997)
Cádiz, Spain	2013	100	0	75	Barrena and Jiménez-Cintado (2014)
Catalonia (incl. Barcelona), Spain	2001	98	2	492	Domènech et al. (2003)

STUDY LOCATION	YEAR	SUBSTRATE TYPE %			REFERENCE
		TREES	ARTIFICIAL	SAMPLE SIZE	
Madrid, Spain	2005	>99	<1	287	Martín Pajares (2005)
Málaga, Spain	2013	99	1	541	Postigo (2013)
Parentis-en-Born, France	1997	0	100	1	Adde (1998)
Brussels, Belgium	1984	0	100	1	De Schaetzen and Jacob (1985)
Brussels, Belgium	1998–99	45	55	11	Weiserbs and Jacob (1999)
Brussels, Belgium	2016	98	2	139	Weiserbs and Paquet (2016)

[1] Sample size inferred from data in the text. For 17% of nests, the substrate was listed as "undefined."

Site-to-site differences in the relative usage of trees vs. man-made structures undoubtedly reflect, at least in part, the characteristics of the trees and structures that are locally available. For this reason, changes in usage over time at given sites are particularly intriguing. Such data are available for Chicago and New Orleans in the US, Barcelona, and Brussels (Table 11.2). In Chicago, data from 1992–93 and 2010 show that substrate usage is quite evenly divided between trees and man-made substrates, shifting only slightly from a 57%–43% (trees vs. man-made) to a 43%–57% split (Hyman and Pruett-Jones 1995; Minor et al. 2012). In Barcelona, surveys conducted in 1985, 1992–94, and 2001 also show that parakeets' substrate preferences held steady, but in this case nearly all nests are built in trees (Batllori and Nos 1985; Sol et al. 1997; Domènech et al. 2003). A 2003–4 nest census in New Orleans found that most nests (73%) were built in trees, but approximately five years later (following Hurricane Katrina), a majority of nests (67%) were in man-made structures (Sevenair 2012). Finally, three Monk Parakeet censuses in Brussels tracked the growth of a population from 1984, starting with a single nest structure built on a ventilation duct, growing to 11 nest structures in 1998–99 approximately evenly divided between trees and artificial structures, and eventually to 139 nest structures in 2016, nearly all of which were in trees (De Schaetzen and Jacob 1985; Weiserbs and Jacob 1999; Weiserbs and Paquet 2016). A decrease in nest size (number of compartments) accompanied this shift in substrate use (Weiserbs and Paquet 2016), possibly because trees are unable to support as much weight as man-made structures (Weiserbs and Jacob 1999).

Some authors also note temporal shifts in the usage of different tree species. For a population in Málaga, Spain, Postigo (2013) found that in 2013, 44% of nests were in palms, 33.5% in eucalypts, and 17% in pines. However, in this same population 20 years previous, in 1994, 94% of nests had been in *Eucalyptus* trees, leading the author to suggest that nest site selection could be "contagious." Similarly, Molina et al. (2016) note that eucalypts were a preferred nest substrate early in Monk Parakeets' colonization of Andalucía, Spain, but the birds later shifted to using palms. In contrast, parakeets nesting in New Jersey persistently rebuilt nests on utility poles, even when they were repeatedly torn down (Burger and Gochfeld 2009). In the Madrid area, no nests were found in Mediterranean Cypress (*Cupressus sempervivens*) trees, even though there were tall ones available and they are preferred in other parts of Spain (Martín Pajares 2005). Burger and Gochfeld (2000) noted strong local substrate preferences in southeastern Florida, where they found a parakeet colony with nearly all nests built in *Melaleuca quinquenervia* trees (even though there were palms nearby), and at a second colony <100 km away, all nests were in palms (though there were broadleaf trees available). In a study of parakeets nesting in urban parks in Argentina, preferences seemed to be somewhat idiosyncratic; for example, palms were preferred in two parks where palms were rare but not used in a third park where palms were common (Romero et al. 2015). These observations suggest that, while some aspects of nest site selection are consistent across habitats (e.g., preference for trees/structures that are relatively tall; Bucher et al. 1991; Sol et al. 1997; Eberhard 1998; Burger and Gochfeld 2000, 2009; Volpe and Aramburú 2011; Di Santo et al. 2017b), nest substrate preferences can be locally variable (Burger and Gochfeld 2005). This may reflect variation in the rates of nest removal for different substrates (Minor et al. 2012) or differences in exposure to wind (Burger and Gochfeld 2000). The observed local and temporal variation could also result from cultural transmission of substrate preference, and that culture may shift over time (possibly in response to nest removal or environmental conditions) as the parakeet population becomes established and expands.

Ectoparasites

As an introduced species, the Monk Parakeet has the potential to affect the dynamics of parasite and disease transmission in its adopted habitats, by bringing co-introduced parasites with it and/or being involved in the spread of endemic parasites and diseases (Ancillotto et al. 2018). Also, since Monk Parakeet nests are used year-round, and often for several years, nest structures themselves have the potential to build up populations of arthropod parasites and scavengers. Indeed, several authors have suggested that the parakeets' habit of lining the nest chamber with green leaf material might be an attempt to repel ectoparasites (Aramburú et al. 2002; Viana et al. 2016). A study that sampled 44 nests in Buenos Aires province found that the community of arthropods that inhabits Monk Parakeet nests is diverse and includes hematophagous species as well as predators, detritivores, and others (Aramburú et al. 2009). The samples included two species of hematophagous mites, *Ornithonyssus bursa* and an unidentified cheyletid, and two bloodsucking insects, *Psitticimex uritui* and *Lyctocoris campestris*. In a separate study, Aramburú et al. (2003) examined 52 nestlings from an unspecified number of nests and found that half of them were parasitized, usually by at least one species. Hematophagous mites (*O. bursa*) were found on nearly half of the nestlings, while cimicid bugs (*P. uritui*) and ischnoceran chewing lice (*Paragoniocotes fulvofasciatus*) were each found on about 10% of nestlings. In naturalized populations, adult Monk Parakeets appear to experience similar parasite loads. Briceño et al. (2017) surveyed parasites of 92 Monk Parakeets collected in central Chile and found that 55% carried parasitic arthropods, about half of which were lice (*P. fulvofasciatus*). Similarly, Ancillotto et al. (2018) sampled arthropod parasites from 127 individuals brought to a rescue center in Rome and found that 38.6% were parasitized, primarily by ischnoceran chewing lice (*P. fulvofasciatus*) and a hematophagous mite (*O. bursa*), both of which are from the parakeets' native range. Overall, half of the parasite species found to parasitize the parakeets in this study were native to the introduced area and included generalist bird parasites as well as parasites that specialize on mammals. Parasite prevalence was higher for parakeets from more urbanized areas and in areas with higher Monk Parakeet density (Ancillotto et al. 2018). While directly comparable prevalence data are not available from the native range (Aramburú et al. 2003 sampled nestlings, and the studies from the naturalized range sampled adults), studies from naturalized populations show that introduced parakeets do not escape their parasites upon introduction to a new area (Ancillotto et al. 2018).

Interactions with Other Species

In Argentina, a number of bird species, including ducks and birds of prey, have been recorded using Monk Parakeet nest structures as nests of their own (Aramburú 1990; Martella and Bucher 1984; Martella et al. 1985; Eberhard 1998; Port and Brewer 2004). In many of these cases, the Monk Parakeets were evicted from an occupied nest, but in others the interlopers moved into an unoccupied chamber and were tolerated by the parakeets in neighboring chambers. Wagner (2012) observed a large nest structure in Rio Grande do Sul, Brazil, where five male House Sparrows (*Passer domesticus*) evicted resident parakeets (likely incubating females) from three nest chambers. Spot-winged Falconets

The Rock Creek power transmission corridor in Miami-Dade County, Florida, illustrating the concentration of nesting structures of Monk Parakeets (*Myiopsitta monachus*) in this area. There are least 20 structures, supporting 30+ pairs of parakeets (some of the structures are hard to see in this photo). Miami-Dade County, Florida, US, 2001. Photo by James Lindsay.

(*Spiziapteryx cirumcincta*) were found to be frequent adopters of Monk Parakeet nests in Córdoba, Argentina, and even though they are known to prey on parakeets, their presence was tolerated as long as they were in a separate nest structure, even as close as 5 m from an occupied parakeet nest (Martella and Bucher 1984). A study of breeding parakeets in Argentina found that parakeets are generally tolerant of other birds (both conspecifics and other species) in the vicinity of their nests but do occasionally chase birds close to (on average 3 m from) their nests, and chases are more frequent against conspecifics than heterospecific individuals (Eberhard 1998).

Similar tolerance of other species was noted in observations of introduced Monk Parakeets in Mexico (Muñoz-Jiménez and Alcántara-Carbajal 2017), Belgium (Weiserbs and Jacob 1999), and Italy (Di Santo et al. 2016). In Florida, Tracey and Miller (2018) found two pairs of American Kestrels (*Falco sparverius*) nesting within Monk Parakeet nests, and in one of those cases, parakeets were occupying a chamber on the opposite side of the nest structure. The authors note that no instances were observed of parakeets or kestrels chasing or displacing individuals of the other species, and that they often perched within 5 m of each other. Freeland (1973) reports a territorial dispute between Monk Parakeets and introduced House Sparrows in Pennsylvania, US, after which a dead sparrow was found beneath the nest, but this type of aggression appears to be rare. Studies in Spain and Italy observed House Sparrows, Italian Sparrows (*Passer italiae*), and Rock Doves (*Columba livia*) nesting within Monk Parakeet nest structures without conflict (Batllori and Nos 1985; Di Santo et al. 2017a; Postigo 2013), and in Spain, abandoned Monk Parakeet nests are sometimes used by a variety of other bird species, including Western Jackdaw (*Coloeus monedula*), Spotless Starling (*Sturnus unicolor*), Rock Dove, House

Sparrow, and Common Wood Pigeon (*Columba palumbus*) (Martín Pajares 2005). In Brussels, feral pigeons (*C. livia domestica*) and possibly Stock Doves (*C. oenas*) use parakeet nests for breeding, and the only observed interspecific aggression by parakeets was directed toward Common Blackbirds (*Turdus merula*) (Weiserbs and Jacob 1999). The few instances of reported Monk Parakeet aggression generally involve defense against potential predators such as jackdaws (Postigo 2013) and Hooded Crows (*Corvus cornix*) (Di Santo et al. 2016). Naturalized Monk Parakeets could potentially compete with native birds for food, but data from Chicago suggest that in spite of the parakeets' heavy use of bird feeders during the winter, they do not appear to affect local species diversity (Pruett-Jones et al. 2012). In Spain, Monk Parakeets were seen chasing Common Blackbirds from feeding on palm fruits (Batllori and Nos 1985) but were also observed feeding on the ground alongside other species without any aggression (Postigo 2013). Observations of foraging parakeets in Italy also note a lack of interspecific aggression at foraging sites (Di Santo et al. 2017a).

CONCLUSION

Although the Rose-ringed Parakeet is the most globally successful naturalized parrot in the world, the Monk Parakeet is a close second. In most countries where large numbers of Monk Parakeets have been imported, the species is now established and breeding. Furthermore, in their native range, Monk Parakeets appear to be modifying their distribution and biology to human-modified habitats. The species' success in establishing populations throughout the world can be attributed to a variety of factors related to behavioral plasticity, reproductive biology, and human activity. Their ability to construct stick nests instead of relying on tree cavities is partly responsible for enabling their establishment in urban areas. Monk Parakeets seem equally likely to construct nests in trees as on man-made substrates, although individual populations may show tendencies to use particular nest substrates. Another contribution to the success of Monk Parakeets may be their broad diet and flexibility to seasonal changes in food availability (South and Pruett-Jones 2000).

International trade data from CITES validate the importance of propagule pressure for increasing a species' probability of subsequent establishment and invasion. The data for Monk Parakeets illustrate the dynamic nature of how human commercial activity responds quickly to importation bans. Even with the current trade bans, the market for the species is growing in more countries, as the focus of imports shifts to the available markets of countries with no trade restrictions. It appears that Monk Parakeets will continue to be traded internationally on a massive scale.

There are a number of important areas in which research on Monk Parakeets would be useful (as with all naturalized parrots). The interaction between nest-site preferences (trees vs. man-made structures) and habitat preferences (natural vs. human-modified) of individuals will be particularly important to quantify. Similarly, in areas where the species is naturalized, we don't know whether nest-site preferences are culturally transmitted within a family or a population. It would be very interesting to better understand how diet and nest substrate preferences may shift over time for naturalized populations, especially related to geographic expansion in those populations. This may facilitate a better understanding of how and why Monk Parakeet populations become (or don't become) invasive. The data we summarize is a useful start to this, but there is much work that still needs to be done. Virtually every aspect of the biology and life history of the Monk Parakeet provides considerable opportunities for research into basic aspects of ecology, evolution, and behavior.

REFERENCES

Adde, C. 1998. Conures veuves *Myiopsitta monachus* nidificatrices dans le sud-ouest de la France. *Alauda* 66(1):66–67

Ancillotto, L., Studer, V., Howard, T., Smith, V. S., McAlister, E., Beccaloni, J., Manzia, F., Renzopaoli, F., Bosso, L., Russo, D., and Mori, E. 2018. Environmental drivers of parasite load and species richness in introduced parakeets in an urban landscape. *Parasitology Research* 117:3591–3599.

Aramburú, R. M. 1990. Observaciones sobre posturas del pato barcino Anas flavirostris, en nidos de cotorra común *Myiopsitta monachus*. *Neotropica* 36:101–105.

Aramburú, R. M. 1997. Ecología alimentaria de la cotorra (*Myiopsitta monachus monachus*) en la provincial de Buenos Aires, Argentina (Aves: Psittacidae). *Physis, Sección C* 53:29–32.

Aramburú, R. M., Calvo, S., Alzugaray, M. E., and Cicchino, A. 2003. Ectoparasitic load of monk parakeet (*Myiopsitta monachus*) nestlings. *Ornitología Neotropical* 14:415–418.

Aramburú, R. M., Calvo, S., Carpintero, D. L., and Cicchino, A. C. 2009. Artrópodos presentes en nidos de cotorra *Myiopsitta monachus monachus* (Aves: Psittacidae). *Revista del Museo Argentino de Ciencias Naturales* 11:1–5.

Aramburú, R. M., Cicchino, A., and Bucher, E. 2002. Material vegetal fresco en cámaras de cría de la cotorra argentina *Myiopsitta monachus* (Psittacidae). *Ornitología Neotropical* 13:433–436.

Avery, M. L., Tillman, E. A., Keacher, K. L., Arnett, J. E., and Lundy, K. J. 2012. Biology of invasive monk parakeets in south Florida. *Wilson Journal of Ornithology* 124:581–588.

Barrena, P., and Jiménez-Cintado, M. 2014. Estima de la abundancia de la población reproductora de la cotorra argentina *Myiopsitta monachus* en la ciudad de Cádiz. *Revista de la Sociedad Gaditana de Historia Natural*. 8:1–4.

Batllori, X., and Nos, R. 1985. Presencia de la cotorra gris (*Myiopsitta monachus*) y de la cotorrita de collar (*Psittacula krameri*) en el área metropolitana de Barcelona. *Miscellania Zoologica* 9:407–411.

Blackburn, T. M., Lockwood, J. L., and Cassey, P. 2009. *Avian Invasions: The Ecology and Evolution of Exotic Birds*. Oxford: Oxford Univ. Press.

Blackburn, T. M., Pyšek, P., Bacher, S., Carlton, J. T., Duncan, R. P., Jarošík, V., Wilson, J.R.U., and Richardson, D. M. 2011. A proposed unified framework for biological invasions. *Trends in Ecology & Evolution* 26:7

Briceño, C., Surot, D., González-Acuña, D., Martínez, F. J., and Fredes, F. 2017. Parasitic survey on introduced monk parakeets (*Myiopsitta monachus*) in Santiago, Chile. *Brazilian Journal of Veterinary Parasitology* 26:129–135.

Bucher, E. H., 1992. Neotropical parrots as agricultural pests. In *New World Parrots in Crisis: Solutions from Conservation Biology*, ed. S. R. Beissinger and N.F.R. Snyder, 201–219. Washington, DC: Smithsonian Institution Press.

Bucher, E. H., and Aramburú, R. M. 2014. Land-use changes and monk parakeet expansion in the Pampas grasslands of Argentina. *Journal of Biogeography* 41:1160–1170.

Bucher, E. H., Martín, L. R., Martella, M. B., and Navarro, J. L. 1991. Social behavior and population dynamics of the monk parakeet. *Proceedings of the International Ornithological Congress* 20:681–689.

Bull, J. 1973. Exotic birds in the New York City area. *Wilson Bulletin* 85:501–505.

Burger, J., and Gochfeld, M. 2000. Nest site selection in monk parakeets (*Myiopsitta monachus*) in Florida. *Bird Behavior* 13:99–105.

Burger, J., and Gochfeld, M. 2005. Nesting behavior and nest site selection in monk parakeets (*Myiopsitta monachus*) in the Pantanal of Brazil. *Acta Ethologica* 15:23–34.

Burger, J., and Gochfeld, M. 2009. Exotic monk parakeets (*Myiopsitta monachus*) in New Jersey: Nest site selection, rebuilding following removal, and their urban wildlife appeal. *Urban Ecosystems* 12:185–196.

Burgio, K. R., van Rees, C. B., Block, K. E., Pyle, P., Patten, M. A., Spreyer, M. F., and Bucher, E. H. 2016. Monk parakeet (*Myiopsitta monachus*) (v.3.0). In *The Birds of North America*, ed. P. G. Rodewald. Ithaca, NY: Cornell Lab of Ornithology.

Butler, C. J. 2005. Feral parrots in the continental United States and United Kingdom: Past, present and future. *Journal of Avian Medicine and Surgery* 19:142–149.

Canavelli, S. B., Aramburú, R., and Zaccagnini, M. E. 2013. Aspectos a considerer para disminuir los conflictos originados por los daños de la cotorra (*Myiopsitta monachus*) en cultivos agrícolas. *El Hornero* 27:89–101.

Cardador, L., Lattuada, M., Strubbe, D., Tella, J. L., Reino, L., Figueira, R., and Carrete, M. 2017. Regional bans on wild-bird trade modify invasion risks at a global scale. *Conservation Letters* 10:717–725.

Carrete, M., and Tella, J. L. 2008. Wild-bird trade and exotic invasions: A new link of conservation concern? *Frontiers in Ecology and the Environment* 6:207–211.

Cassey, P., Blackburn, T. M., Russell, G. J., Jones, K. E., and Lockwood, J. L. 2004. Influences on the transport and establishment of exotic bird species: An analysis of the parrots (Psittaciformes) of the world. *Global Change Biology* 10:417–426.

CITES. 2017. CITES Trade Database. Accessed Dec. 2019. https://trade.cites.org/.

Davis, L. R. 1974. The monk parakeet: A potential threat to agriculture. *Proceedings of the 6th Vertebrate Pest Conference*, 253–256. Lincoln: Univ. of Nebraska.

del Hoyo, J., Elliott A., and Sargatal, J., eds. 1997. *Handbook of the Birds of the World*. Vol. 4: *Sandgrouse to Cuckoos*. Barcelona, Spain: Lynx Edicions; Cambridge, UK: BirdLife International.

De Schaetzen, R., and Jacob, J. P. 1985. Installation d'une colonie de perriches jeune-veuve (*Myiopsitta monachus*) à Bruxelles. *Aves* 22:127–130.

Di Santo, M., Battisti, C., and Bologna, M. A. 2017a. Interspecific interactions in nesting and feeding urban sites among introduced monk parakeet (*Myiopsitta monachus*) and syntopic bird species. *Ethology Ecology & Evolution* 29:138–158.

Di Santo, M., Bologna, M. A., and Battisti, C. 2017b. Nest tree selection in a crowded introduced population of monk parakeet (*Myiopsitta monachus*) in Rome (central Italy): Evidence for selectivity. *Zoology and Ecology*, doi:10.1080/21658005.2017.1366293.

Di Santo, M., Vignoli, L., Battisti, C., and Bologna, M. A. 2013. Feeding activity and space use of a naturalized population of monk parakeet, *Myiopsitta monachus*, in a Mediterranean urban area. *Revue D'Ecologie: La Terre et la Vie* 68:275–282.

Domènech, J., Carrillo, J., and Senar, J. C. 2003. Population size of the monk parakeet *Myiopsitta monachus* in Catalonia. *Revista Catalana d'Ornitologia* 20:1–9.

Dyer, E. E., Redding, D. W., and Blackburn, T. M. 2017. The Global Avian Invasions Atlas: A database of alien bird distributions worldwide. *Scientific Data* 4:170041.

Eberhard, J. R. 1996. Nest adoption by monk parakeets. *Wilson Bulletin* 108:374–377.

Eberhard, J. R. 1997. The evolution of nest-building and breeding behavior in parrots. PhD diss., Princeton Univ., Princeton, NJ.

Eberhard, J. R. 1998. Breeding biology of the monk parakeet. *Wilson Bulletin* 110:463–473.

eBird. 2017. eBird Basic Dataset (v.EBD_relNov-2017). Cornell Lab of Ornithology, Ithaca, NY. Accessed Nov. 2017. https://www.ebird.org.

Forshaw, J. M. 1989. *Parrots of the World*. 3rd (rev.) ed. Melbourne, Australia: Lansdowne Editions.

Freeland, D. B. 1973. Some food preferences and aggressive behavior by monk parakeets. *Wilson Bulletin* 85:332–334.

GBIF. 2019. Global Biodiversity Information Facility. Accessed Mar. 2019. https://doi.org/10.15468/dl.j1fzvm.

Gill, F., and Donsker, D., eds. 2018. IOC World Bird List (v.8.1). Accessed Dec. 2019. http://www.worldbirdnames.org/.

Gonçalves da Silva, A., Eberhard, J. R., Wright, T. F., Avery, M. L., and Russello, M. A. 2010. Genetic evidence for high propagule pressure and long-distance dispersal in monk parakeet (*Myiopsitta monachus*) invasive populations. *Molecular Ecology* 19:3336–3350.

Grupo de Aves Exóticas, SEO/Birdlife. 2012. *Legislación Sobre Aves Exóticas*, 4–7. N.p.: SEO/Birdlife. https://www.seo.org/wp-content/uploads/2012/05/docgae_legislacion2012.pdf.

Guerrero-Cárdenas, I., Galina-Tessaro, P., Caraveo-Patiño, J., Tovar-Zamora, I., Cruz Andrés, O. R., and Alvarez-Cárdenas, S. 2012. First record of the monk parakeet (*Myiopsitta monachus*) in Baja California Sur, Mexico. *Huitzil, Revista Mexicana de Ornitología* 13:156–161.

Harrison, C.J.O. 1973. Nest-building behavor of Quaker parrots *Myiopsitta monachus*. *Ibis* 115:124–128.

Hobson, E. A., Smith-Vidaurre, G., and Salinas-Melgoza, A. 2017. History of nonnative monk parakeets in Mexico. *PLoS One* 12:e0184771.

Humphrey, P. S., and Peterson, R. T. 1978. Nesting behavior and affinities of monk parakeets of southern Buenos Aires province, Argentina. *Wilson Bulletin* 90:544–552.

Hyman, J., and Pruett-Jones, S. 1995. Natural history of the monk parakeet in Hyde Park, Chicago. *Wilson Bulletin* 107:510–517

Kibbe, D. P., and Cutright, N. J. 1973. The monk parakeet in New York. *Bird Control Seminars Proceedings* 101:73–78.

Lanning, D. V. 1991. Distribution and nest sites of the monk parakeet in Bolivia. *Wilson Bulletin*. 103:366–372.

MacGregor-Fors, I., Calderón-Parra, R., Meléndez-Herrada, A., López-López, S., and Schondube, J. E. 2011. Pretty, but dangerous! Records of non-native monk parakeets (*Myiopsitta monachus*) in Mexico. *Revista Mexicana de Biodiversidad* 82:1053–1056.

Martella, M. B., and Bucher, E. H. 1984. Nesting of the spot-winged falconet in monk parakeet nests. *Auk* 101:614–615.

Martella, M. B., and Bucher, E. H. 1993. Estructura del nido y comportamiento de nidificación de la cotorra *Myiopsitta monachus*. *Boletín de la Sociedad Zoológica del Uruguay* 8:211–217.

Martella, M. B., Navarro, J. L., and Bucher, E. H. 1985. Vertebrados asociados a los nidos de la cotorra *Myiopsitta monachus* en Córdoba y La Rioja. *Physis, Sección C* 43:49–51.

Martín, L. F., and Bucher, E. H. 1993. Natal dispersal and first breeding age in monk parakeets. *Auk* 110:930–933.

Martín Pajares, M. 2005. La cotorra argentina (*Myiopsitta monachus*) en la ciudad de Madrid: Expansión y hábitos de nidificación. *Anuario Ornitológico de Madrid* 2005:76–95.

Minor, E. S., Appelt, C. W., Grabiner, S., Ward, L., Moreno, A., and Pruett-Jones, S. 2012. Distribution of exotic monk parakeets across an urban landscape. *Urban Ecosystems* 15:979–991.

Molina, B., Postigo, J. L., Román-Muñoz, A., and del Moral, J. C. 2016. *La cotorra argentina en España: Población reproductora en 2015 y método de censo*. Madrid: SEO/BirdLife.

Mott, D. F. 1973. Monk parakeet damage to crops in Uruguay and its control. *Bird Control Seminars Proceedings*. 102:79–81.

Muñoz, A. R., and Real, R. 2006. Assessing the potential range expansion of the exotic monk parakeet in Spain. *Diversity and Distributions* 12:656–665.

Muñoz-Jiménez, J. L., and Alcántara-Carbajal, J. L. 2017. La cotorra argentina (*Myiopsitta monachus*) en el Colegio de Postgraduados: ¿Una especie invasiva? *Huitzil, Revista Mexicana de Ornitología* 18:38–52.

National Audubon Society. 2019. The Christmas Bird Count historical results (Online). Accessed Aug. 2019. http://www.christmasbirdcount.org

Navarro, J. L., and Bucher, E. H. 1992. Annual variation in the timing of breeding of the monk parakeet in relation to climatic factors. *Wilson Bulletin* 104:545–549.

Navarro, J. L., Martella, M. B., and Bucher, E. H. 1992. Breeding season and productivity of monk parakeets in Córdoba, Argentina. *Wilson Bulletin* 104:413–424.

REFERENCES

Neidermyer, W. J., and Hickey, J. J. 1977. The monk parakeet in the United States, 1970–75. *American Birds* 31:273–278.

Newman, J. R., Newman, C. M., Lindsay, J. R., Merchant, B., Avery, M. L., and Pruett-Jones, S. 2008. Monk parakeets: An expanding problem on power lines and other electrical utility structures. In *8th International Symposium on Environmental Concerns in Rights-of-Way Management*, ed. J. W. Goodrich-Mahoney, L. P. Abrahamson, J. L. Ballard, and S. M. Tikalsky, 343–354. New York: Elsevier Science.

Pablo López, R. E. 2009. Primer registro del perico argentina (*Myiopsitta monachus*) en Oaxaca, México. *Huitzil, Revista Mexicana de Ornitología* 10:48–51.

Peris, S. J., and Aramburú, R. M. 1995. Reproductive phenology and breeding success of the monk parakeet (*Myiopsitta monachus monachus*) in Argentina. *Studies on Neotropical Fauna and Environment* 30:115–119.

Port, J. L., and Brewer, G. L. 2004. Use of monk parakeet (*Myiopsitta monachus*) nests by speckled teal (*Anas flavirostris*) in eastern Argentina. *Ornitología Neotropical* 15:209–218.

Postigo, J. L. 2013. Censo y análisis de invasibilidad de la cotorra argentina *Myiopsitta monachus*, en la cuenca del río Guadalhorce, Málaga. MS thesis, Univ. de Granada, Spain.

Postigo, J. L., Shwartz, A., Strubbe, D., and Muñoz, A. R. 2017. Unrelenting spread of the alien monk parakeet in Israel. Is it time to sound the alarm? *Pest Management Science* 73:349–353.

Postigo, J. L., Strubbe, D., Mori, E., Ancillotto, L., Carneiro, I., Latsoudis, P., Menchetti, M., Pârâu, L. G., Parrott, D., Rein, L., Weiserbs, A., and Senar, J. C. 2019. Mediterranean versus Atlantic monk parakeets *Myiopsitta monachus*: Towards differentiated management at the European scale. *Pest Management Science* 75:915–922.

Pranty, B. 2009. Nesting substrates of monk parakeets (*Myiopsitta monachus*) in Florida. *Florida Field Naturalist* 37:51–57.

Pruett-Jones, S., Appelt, C. W., Sarfaty, A., Van Vossen, B., Leibold, M. A., and Minor, E. S. 2012. Urban parakeets in northern Illinois: A 40-year perspective. *Urban Ecosystems* 15:709–719.

Pruett-Jones, S., and Tarvin, K. A. 1998. Monk parakeets in the United States: Population growth and regional patterns of distribution. In *Proceedings of the 18th Vertebrate Pest Conference*, ed. R. O. Baker and A. C. Crabb, 55–58. Davis: Univ. of California.

Ramírez-Albores, J. E. 2012. Registro de la cotorra argentina (*Myiopsitta monachus*) en la Ciudad de México y áreas adyacentes. *Huitzil, Revista Mexicana de Ornitología* 13:110–115.

Reed, J. E., McCleery, R. A., Silvy, N. J., Smeins, F. E., and Brightsmith, D. J. 2014. Monk parakeet nest-site selection of electric utility structures in Texas. *Landscape Urban Planning* 129:65–72.

Reed, J. E., Silvy, N. J., McCleery, R. A., Smeins, F. E., and Brightsmith, D. J. 2013. Habitat use of monk parakeets in Dallas and Tarrant counties, Texas. In *Proceedings of the 15th Wildlife Damage Management Conference*, ed. J. B. Armstrong and G. R. Gallagher, paper 157. Lincoln: Univ. of Nebraska.

Rodríguez-Pastor, R., Senar, J. C., Ortega, A., Faus, J., Uribe, F., and Montalvo, T. 2012. Distribution patterns of invasive monk parakeets (*Myiopsitta monachus*) in an urban habitat. *Animal Biodiversity and Conservation* 35:107–117.

Romero, I. P., Codesido, M., and Bilenca, D. N. 2015. Nest building by monk parakeets *Myiopsitta monachus* in urban parks in Buenos Aires, Argentina: Are tree species used randomly? *Ardeola* 62:323–333.

Romero-Figueroa, G., Ortiz-Avila, V., Lozano-Cavazos, E. A., and Heredia-Pineda, F. J. 2016. Primer registro de la cotorra argentina (*Myiopsitta monachus*) en Coahuila, México. *Huitzil, Revista Mexicana de Ornitología* 18:81–86.

Russello, M. A., Avery, M. L., and Wright, T. F. 2008. Genetic evidence links invasive monk parakeet populations in the United States to the international pet trade. *BMC Evolutionary Biology* 8:217.

Salgado-Miranda, C., Medina, J. P., Sánchez-Jasso, J. M., and Soriano-Vargas, E. 2016. Registro altitudinal más alto en México para la cotorra argentina (*Myiopsitta monachus*). *Huitzil, Revista Mexicana de Ornitología* 17:155–159.

Secretaria de Medio Ambiente y Recursos Naturales, Mexico. 2016. Acuerdo por el que se determina la Lista de las Especies Exóticas Invasoras para México. *Diario Oficial de la Federación*, 7 Dec. 2016, 3rd sect.:35.

Senar, J. C., Carrillo-Ortiz, J. G., Ortega-Segalerva, A., Dawson Pell, F.S.E., Pascual, J., Arroyo, L., Mazzoni, D., Montalvo, T., and Hatchwell, B. J. 2019. The reproductive capacity of monk parakeets *Myiopsitta monachus* is higher in their invasive range. *Bird Study* 66:136–140.

Senar, J. C., Domènech, J., Arroyo, L., Torre, I., and Gordo, O. 2016. An evaluation of monk parakeet damage to crops in the metropolitan area of Barcelona. *Animal Biodiversity and Conservation* 39:141–145.

Sevenair, J. P. 2012. Monk parakeet (*Myiopsitta monachus*) nests in metropolitan New Orleans before and after hurricane Katrina. *Journal of Louisiana Ornithology* 9:7–14.

Shields, W. M., and Grubb, T. C., Jr. 1974. Use of native plants by monk parakeets in New Jersey. *Wilson Bulletin* 86:172–173.

Sierra-Morales, P., and Almazán-Núñez, R. C. 2017. Nesting site record of the monk parakeet (*Myiopsitta monachus*) in the state of Guerrero, Mexico. *Acta Zoológica Mexicana* 33:126–129.

Sol, D., Santos, D. M., Feria, E., and Clavell, J. 1997. Habitat selection by the monk parakeet during colonization of a new area in Spain. *Condor* 99:39–46.

Soto-Cruz, R. A., Lebgue-Keleng, T., Espinoza-Prieto, J. R., Quintana-Martínez, R. M., Quintana-Martínez, G., Balderrama, S., Zamudio-Mondragón, F. R., Quintana-Chávez, M. A., and Mondaca-Fernández, F. 2014. Primer registro de la cotorra argentina (*Myiopsitta monachus*) en Chihuahua, México. *Huitzil, Revista Mexicana de Ornitología* 15:1–5.

South, J. M., and Pruett–Jones, S. 2000. Patterns of flock size, diet, and vigilance of naturalized monk parakeets in Hyde Park, Chicago. *Condor* 102:848–854.

Spreyer, M. F., and Bucher, E. H. 1998. Monk parakeet (*Myiopsitta monachus*). In *The Birds of North America*, ed. A. Poole and F. Gill, no. 322. Philadelphia, PA: Academy of Natural Sciences; Washington, DC: American Ornithologists' Union.

Strubbe, D., and Matthysen, E. 2009. Establishment success of invasive ring-necked and monk parakeets in Europe. *Journal of Biogeography* 36:2264–2278.

Tala, C., Guzmán, P., González, S. 2005. Cotorra argentina (*Myiopsitta monachus*) convidado de piedra en nuestras ciudades y un invasor potencial, aunque real, de sectores agrícolas. Boletín DIPROREN. Santiago, Chile: Servicio Agrícola y Ganadero, División de Protección de los Recursos Naturales Renovables.

Tillman, E. A., Van Doorn, A., and Avery, M. L. 2000. Bird damage to tropical fruit in south Florida. In *The Ninth Wildlife Damage Management Conference Proceedings*, ed. M. C. Brittingham, J. Kays, and R. McPeake, 47–59. State College, PA.

Tinajero, R., and Rodríguez Estrella, R. 2015. Cotorra argentina (*Myiopsitta monachus*), especie anidando con éxito en el sur de la península de Baja California. *Acta Zoológica Mexicana* 31:190–197.

Tracey, K. F., and Miller, K. E. 2018. Monk parakeets provide nesting opportunities for the threatened southeastern American Kestrel. *Journal of Raptor Research* 52:389–392.

van Bael, S., and Pruett-Jones, S. 1996. Exponential population growth of monk parakeets in the United States. *Wilson Bulletin* 108:584–588.

Viana, I. R., Prevedello, J. A., and Zocche, J. J. 2017. Effects of landscape composition on the occurrence of a widespread invasive bird species in the Brazilian Atlantic Forest. *Perspectives in Ecology and Conservation* 15:36–41.

Viana, I. R., Strubbe, D., and Zocche, J. J. 2016. Monk parakeet invasion success: A role for nest thermoregulation and bactericidal potential of plant nest material? *Biological Invasions* 18:1305–1315.

Volpe, N. L., and Aramburú, R. M. 2011. Preferencias de nidificación de la cotorra argentina (*Myiopsitta monachus*) en un área urbana de Argentina. *Ornitología Neotropical* 22:111–119.

Wagner, N. 2012. Occupation of monk parakeet (*Myiopsitta monachus*) nest cavities by house sparrows (*Passer domesticus*) in Rio Grande do Sul, Brazil. *Boletín SAO* 20:72–78.

Weiserbs, A., and Jacob, J. P. 1999. Etude de la population de perriche jeune-veuve (*Myiopsitta monachus*) à Bruxelles. *Aves* 36:207–223.

Weiserbs, A., and Paquet, A. 2016. Recensement de la conure veuve *Myiopsitta monachus* à Bruxelles en 2016. *Aves* 53:19–28.

Zuria, I., Castellanos, I., Valencia-Herverth, R., and Carbó-Ramírez, P. 2017. Primeros registros de la cotorra argentina (*Myiopsitta monachus*) en el estado de Hidalgo, México. *Huitzil, Revista Mexicana de Ornitología* 18:33–37.

12

INTRODUCED AND NATURALIZED PARROTS IN THE CONTIGUOUS UNITED STATES

Jennifer J. Uehling, Jason Tallant, and Stephen Pruett-Jones

INTRODUCTION

Exotic parrots are now common in widely varying geographical areas of the United States (US), from the Chicago shores of Lake Michigan to the palm-lined boulevards of southern California (Butler 2005; Runde et al. 2007; Uehling et al. 2019). The Monk Parakeet (*Myiopsitta monachus*) is the most abundant naturalized parrot in the US (Pruett-Jones et al. 2012; Burgio et al. 2016), but it is just one of many species of exotic parrots now established (Uehling et al. 2019). States with particularly diverse sets of naturalized parrots include Florida, California, and Texas (Garrett 1997, 1998; Pranty and Epps 2002; Butler 2005; Runde et al. 2007; Uehling et al. 2019). Additionally, parrots have been sighted regularly in many other states, including Arizona, Connecticut, Illinois, Louisiana, New Jersey, and New York (Uehling et al. 2019).

As the most common naturalized parrot species in the US, the Monk Parakeet has received considerable study (Hyman and Pruett-Jones 1995; van Bael and Pruett-Jones 1996; South and Pruett-Jones 2000; Pruett-Jones et al. 2007, 2012; Davis et al. 2014; Burgio et al. 2016). However, little research has focused beyond the Monk Parakeet on the large suite of other parrot species in the country, although there are notable exceptions (Garrett et al. 1997; Mabb 1997; Garrett 1997, 1998; Pranty and Epps 2002; Pranty and Garrett 2003; Butler 2005; Runde et al. 2007; Uehling et al. 2019). Here, we explore the status and distribution of non-native parrots in the contiguous US, review possible explanations behind their presence and establishment, and explore some geographic regions where parrots have been particularly successful at establishment. Additionally, we review summaries of naturalized parrots in Hawaii (VanderWerf and Kalodimos, chap. 13 this vol.) and Puerto Rico (Falcón and Tremblay 2018). Together with our findings, these works suggest that there are no fewer than 29 species of parrots currently breeding in the mainland US, Hawaii, and Puerto Rico combined, and it is likely that more will soon establish breeding populations. For now, parrots are certainly in the US to stay, and their presence is a source of fascination for both ecologists and the general public alike.

DIVERSITY AND DISTRIBUTION OF PARROTS IN THE CONTIGUOUS UNITED STATES

Parrots are now regularly sighted in the wild in the US. Based on data from citizen science databases, Uehling et al. (2019) report, from 2002 to 2016, 118,744 unique observations of 56 species of parrots at 19,812 unique localities in 43 of the contiguous states. Uehling et al. (2019) defined a unique observation as a sighting of a parrot species at a unique locality, time, and date. Unique observations could include any number of individuals of that species of parrot at that locality. Unique localities were defined as locations with a unique set of latitude and longitude coordinates. Based on these data, parrots are commonly sighted in the wild and have been seen outside of captivity in almost every state in the contiguous US.

Because Monk Parakeets are the most common and widely distributed parrots in the US, we compared the distribution of sightings of Monk Parakeets to that of other species. In the Uehling et al. (2019) study, more than 35% of all unique observations were of Monk Parakeets. We took the observations in Figure 1 from Uehling et al. (2019) and separated the records for Monk Parakeets from those of all other species. These data are illustrated in Figures 12.1 and 12.2. For Monk Parakeets, there were 42,981 observations at 9,199 localities, and for all other species there were 75,763 observations at 11,711 localities. As is evident, Monk Parakeet sightings are common and widespread, but sightings of species other than Monk Parakeets are also common and widespread.

Though parrots are commonly sighted across the US, parrot diversity is highest in a few select southern regions. Conversely, northern localities with parrot sightings often have just one species. Separate from the locality data in Figures 12.1 and 12.2, this point is made more striking when the distribution of parrot diversity is plotted by county (Fig. 12.3). There are 3,118 counties in the contiguous United States, and parrots have been seen in 822 (26.4%) of these. At the more northerly sites, most parrot sightings are of a single species, the Monk Parakeet, which has been particularly successful at establishing breeding populations in cold climates, such as Chicago and New York City (van Bael and Pruett-Jones 1996; Butler 2005; Burgio et al. 2016). If they are not Monk Parakeets, the parrots sighted in more northerly counties are often escaped pets, which likely do not breed in the wild.

Monk Parakeet (*Myiopsitta monachus*) at its nest site. Overpeck County Park, Bergen County, New Jersey, US, June 2019. Photo by Carole Hughes.

DIVERSITY AND DISTRIBUTION OF PARROTS IN THE CONTIGUOUS UNITED STATES

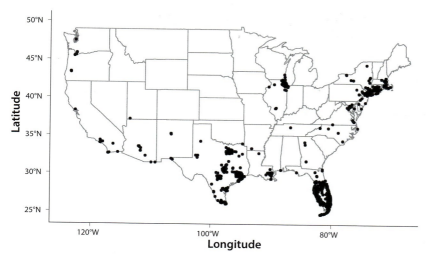

Figure 12.1. Distribution of sightings of Monk Parakeets in the contiguous United States during the 15-year period 2002–2016. See Uehling et al. (2019) for methods. For Monk Parakeets, the map illustrates 42,981 unique sightings at 9,199 unique localities.

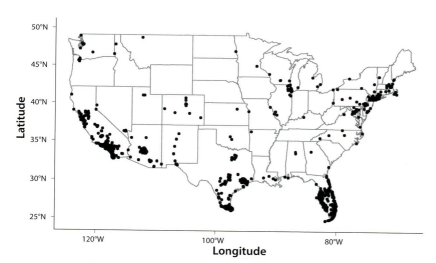

Figure 12.2. Distribution of sightings of all species of parrots excluding Monk Parakeets in the contiguous United States during the 15-year period 2002–2016. See Uehling et al. (2019) for methods. The map illustrates 75,763 unique sightings at 11,711 unique localities.

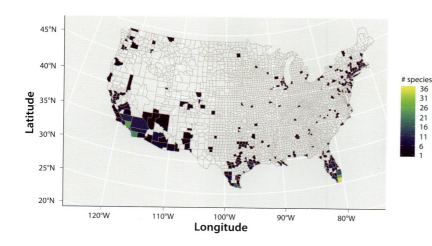

Figure 12.3. The distribution of parrot diversity across all counties in the United States during the 15-year period 2002–2016. The color of each county corresponds to the relative number of parrot species sighted in that county.

INTRODUCED AND NATURALIZED PARROTS IN THE CONTIGUOUS UNITED STATES

Although parrots may often be seen in the wild across wide geographic localities, most species do not have self-sustaining, breeding populations across the country. Of the 56 parrot species that have been sighted outside of captivity in the contiguous US, 25 species have records of breeding in the wild (Uehling et al. 2019), and at least one species, the Monk Parakeet, is or has bred in 23 different states (Burgio et al. 2016; Table 12.1). In some states (mainly California, Florida, and Texas), a number of other parrot species are also present and breeding (Table 12.1). Arizona, notably, supports a breeding population of Rosy-faced Lovebirds (*Agapornis roseicollis*), a species not confirmed to be breeding elsewhere in the US, though there is some suggestion that it may be attempting to breed in Florida. Uehling et al. (2019) state that California, Florida, and Texas collectively support breeding populations of all the parrot species established in the US, though Arizona should be added to this list, given the uncertainty of Rosy-faced Lovebirds breeding in Florida.

Numerous challenges prevent reliable population estimates of all parrot species across the US. For example, if a parrot is sighted outside of captivity, it is unclear whether that individual is a recently escaped pet or a true "wild" bird breeding outside of captivity. This can lead to inconsistencies in how non-native bird species are reported in community science databases (eBird Basic Dataset 2017; Garrett 2018). Additionally, some birders will submit records for species they see in zoological parks, or for individuals that are clearly recently escaped pets, and it can be very difficult to parse these sightings from observations of birds truly established "in the wild" (J. Uehling, pers. obs.).

Nevertheless, local or statewide surveys have suggested dramatic increases or decreases in population sizes of certain species. For example, the Budgerigar (*Melopsittacus undulatus*) was once present in the thousands in Florida; in 1978, the total population size was estimated to exceed 20,000 individuals (Pranty 2001; Runde et al. 2007). However, by the 1990s, the population numbered only 100–200 individuals (Pranty 2001; Butler 2005). Additionally, although the Monk Parakeet remains the most common species in the contiguous US, its population declined steeply between 2003 and 2006 (Pruett-Jones et al. 2012; Avery and Shiels 2018; see below).

ARRIVAL OF PARROTS IN THE US

Parrot Importation

How did so many non-native parrots originally arrive in the US? Most species were imported as part of the pet trade and were subsequently released into

Budgerigar (*Melopsittacus undulatus*) at a nest box. This individual is one of the last remnants of a once large population in Florida. Hernando Beach, Florida, US, May 2011. Photo by Roelant Jonker.

TABLE 12.1

Number of sightings of each species of naturalized parrot in the contiguous United States during the 15-year period 2002–2016, and the states in which there is any evidence of breeding. Modified from Uehling et al. (2019). See Uehling et al. (2019) for information on specific species in particular states. Species are listed in alphabetical order by common name.

SPECIES	NUMBER OF SIGHTINGS	STATES WITH EVIDENCE OF BREEDING
Blue-and-yellow Macaw (*Ara ararauna*)	476	FL
Blue-crowned Parakeet (*Thectocercus acuticaudatus*)	2,847	CA, FL
Budgerigar (*Melopsittacus undulatus*)	1,264	IL, FL, TX
Chestnut-fronted Macaw (*Ara severus*)	371	FL
Green Parakeet (*Psittacara holochlorus*)	5,645	FL, TX
Finsch's Parakeet (*Psittacara finschi*)	67	FL
Lilac-crowned Amazon (*Amazona finschi*)	1,863	CA, FL, TX
Mitred Parakeet (*Psittacara mitratus*)	3,861	CA, FL
Monk Parakeet (*Myiopsitta monachus*)	42,981	23 states (Burgio et al. 2016; Uehling et al. 2019)
Nanday Parakeet (*Aratinga nenday*)	14,137	CA, FL
Orange-winged Amazon (*Amazona amazonica*)	724	FL
Red-crowned Amazon (*Amazona viridigenalis*)	15,746	CA, FL, TX
Red-lored Amazon (*Amazona autumnalis*)	1,216	CA
Red-masked Parakeet (*Psittacara erythrogenys*)	7,311	CA, FL
Rose-ringed Parakeet (*Psittacula krameri*)	1,662	CA, FL
Rosy-faced Lovebird (*Agapornis roseicollis*)	6,375	AZ, FL
Scarlet-fronted Parakeet (*Psittacara wagleri*)	50	FL
Turquoise-fronted Amazon (*Amazona aestiva*)	104	CA, FL
White-eyed Parakeet (*Psittacara leucophthalmus*)	178	FL
White-fronted Amazon (*Amazona albifrons*)	474	FL
White-winged Parakeet (*Brotogeris versicolurus*)	765	CA (past evidence, may not continue today), FL
Yellow-chevroned Parakeet (*Brotogeris chiriri*)	8,104	CA, FL
Yellow-collared Lovebird (*Agapornis personatus*)	55	AZ
Yellow-crowned Amazon (*Amazona ochrocephala*)	52	FL
Yellow-headed Amazon (*Amazona oratrix*)	1,958	CA, FL, TX

the wild, either accidentally or on purpose. There are two possible exceptions: the Red-crowned Amazon (*Amazona viridigenalis*) and the Green Parakeet (*Psittacara holochlorus*). The Red-crowned Amazon is endemic to northeastern Mexico, and a population breeding in the southernmost part of Texas may be a combination of escapees and dispersers from the native population (Enkerlin-Hoeflich and Hogan 1997; Benson and Arnold 2001; Berg 2019). Similarly, the Green Parakeet population in southern Texas may be a result of individuals dispersing from the species' native range in northeastern Mexico, where there has been extensive habitat loss. However, escaped cage birds may have also contributed to this southern Texas population (Lockwood and Freeman 2014).

For the parrot populations present and breeding in the US without any dispersal from their native ranges, it is difficult to pinpoint exactly where and when they were released into the wild. To begin exploring this question, we delved into importation records to determine when parrots arrived in the country. We focused on publicly available parrot trade data from the Convention on International Trade in Endangered Species of Wild Fauna and Flora (CITES 2018). It is important to note that these data have their limitations. First, they do not contain parrot importation records from before 1977, and parrots were imported before that year. Additionally, they do not contain records of illegal importations. Finally, CITES relies on self-reporting from participating countries, which could result in incomplete records (Smith et al. 2009; Phelps et al. 2010; Panter et al. 2019). Nevertheless, CITES provides a good starting point for understanding importation patterns.

To explore parrot importation patterns in the US, we downloaded from the CITES Trade Database importation records of all parrot species (taxon Psittaciformes) imported into the US from 1977 to 2017 (CITES 2018). We then filtered the data set in two ways. First, the records include the importer-reported quantity of individuals imported—i.e., the number of individuals reported by the US as coming into the country. We used this importer-reported quantity unless it was unavailable, in which case we used the exporter-reported quantity. Second, the full CITES data set includes records of parrots imported for a number of purposes, including breeding in captivity, educational, zoo, scientific, and commercial. We included only importations that listed "commercial" or did not list a specific purpose in our filtered data set. All analyses and figures discussed hereafter were performed with this filtered data set.

Figure 12.4 plots these importation records. Throughout the 1980s, over 200,000 individual parrots were imported each year. These importation numbers represented a large proportion of global imports; indeed, between 1981 and 1985, US imports made up 25% of worldwide psittacine imports (Iñigo-Elias and Ramos 1991). However, there was a sharp decline in importations in the early 1990s, which can be explained by several factors. In the early 1990s, many countries created national biodiversity institutions after signing the Convention on Biological Diversity and as a result reduced or banned wild bird export (Convention on Biological Diversity 2020; E. Iñigo-Elias, pers. comm.). Additionally, several avian disease outbreaks associated with parrot importations, as well as pressure from and lobbying by researchers and nongovernmental organizations dedicated to conservation, led to decreased importations (E. Iñigo-Elias, pers. comm). Finally, in the US, the Wild Bird Conservation Act (WBCA) was passed in October 1992 (USFWS 2016). This law was enacted in response to the unsustainable harvest of birds for the pet industry, and in response

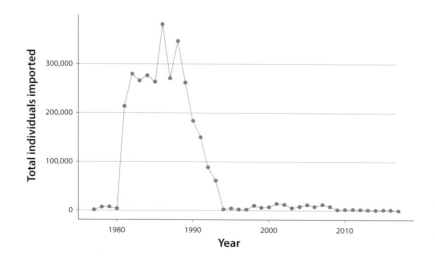

Figure 12.4. Total number of individual parrots, of all species, imported into the United States since 1977.

to the fact that the US was the largest single importer of live wild birds (Beissinger 2001). The act prohibits the importation of all species of birds listed in CITES Appendices 1 and 2. All parrot species are included in these CITES appendices and cannot be imported unless the birds come from a licensed captive breeder or sustainable harvest program (Beissinger 2001). The WBCA went into effect in October 1993 (Beissinger 2001). By 1994, parrot importation numbers in the US had declined dramatically (Fig. 12.4).

ESTABLISHMENT OF PARROTS IN THE US

After parrots were imported into the US in large numbers, some of them escaped or were purposefully released from captivity and then sighted "in the wild." Here, we explore the factors that may lead to sightings of some parrot species in the wild and the ultimate establishment of breeding and self-sustaining populations in the US. The success of any exotic species in a non-native habitat can depend on a variety of factors (Blackburn and Duncan 2001a, 2001b; Blackburn et al. 2009). We discuss here the possible roles of legal importations, illegal importations, characteristics of parrots as pets, and characteristics of the habitats into which parrots are introduced. The relative importance of these factors in promoting establishment likely differs by parrot species, and this list is certainly not exhaustive; we hope this will serve as a starting point for further lines of research on parrots in the US.

Legal Importations

Propagule pressure—the number of individuals released (propagule size) and the number of independent release events (propagule number)—plays a large part in establishment (Blackburn et al. 2009; Lockwood et al. 2013). Previous work on introduced birds in New Zealand, for example, found that propagule pressure had an effect on probability of establishment, whereas other variables such as body size, diet breadth, and range size did not (Lockwood et al. 2013). Though we cannot determine the exact propagule pressure for parrot species in the US, because we do not know exactly when and where release events occurred, we can use the number of individuals imported as a rough metric of the number of individuals that have some chance of being released into the wild. In the case of parrots, some species were imported to the US in much higher numbers than others, and so the number of individuals that ultimately escaped and became established in the wild may be related to the number of individuals originally imported into the country.

In Figure 12.5, we explore the relationship between the number of individuals of each (of 16) species imported (using the aforementioned CITES data set) and the number of times that species has been sighted in the wild. For the number of sightings in the wild, we used the data set of all sightings (or, as defined above, unique observations) of non-native parrots in the US from 2002 to 2016 recorded in eBird and by the National Audubon Society's annual Christmas Bird Count (eBird Basic Dataset 2017; National Audubon Society 2018; see Uehling et al. 2019 for full details). As is noted in Uehling et al. (2019), community science data are not perfect records of all non-native parrots seen in the US. Additionally, as is noted above, each "sighting" could include any number of individuals of that species of parrot. However, these data are a good starting point for understanding trends in naturalized parrot sightings across the US. In Figure 12.5, we include the 14 species that have more than 1,000 sightings (Table 12.1). We also added the Grey Parrot (*Psittacus erithacus*) and Fischer's Lovebird (*Agapornis fischeri*), which did not have more than 1,000 sightings outside of captivity, as representatives of species that have been extensively imported but have very few sightings in the wild.

As is shown in Figure 12.5, certain parrot species were imported in much higher numbers than others, but the number of importations is not linearly related to the number of times that species has been sighted in the wild. The Monk Parakeet was one of the most commonly imported species and was sighted the most times of any parrot species. In contrast, the Grey Parrot had a similar number of importations

INTRODUCED AND NATURALIZED PARROTS IN THE CONTIGUOUS UNITED STATES

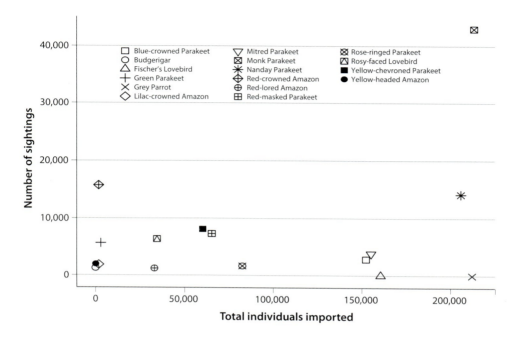

Figure 12.5. Number of sightings of 16 parrot species in the wild in the United States from 2002 to 2016 as a function of the total number of individuals imported.

but was sighted only once outside of captivity. The two lovebirds, Rosy-faced and Fischer's, illustrate unexpected differences in the success of similar species. The Fischer's Lovebird was sighted just a few times in the wild, whereas the Rosy-faced Lovebird has successfully established large breeding populations in Arizona and is now commonly seen in the wild.

In summary, for some species, such as the Monk Parakeet, high importation numbers are accompanied by numerous sightings outside of captivity; however, Figure 12.5 demonstrates that high importation numbers are not always predictive of frequent sightings in the wild.

Illegal Importations

In addition to legal importations, illegal importations may affect non-native parrot propagule pressure, and importing parrots illegally is likely easier from some localities than others. The geographic proximity of Central and South America, as well as their physical connectedness with the US, may increase propagule pressure via illegal importations for species from these regions. Conversely, illegal smuggling from Africa, Asia, and Australia may be more challenging due to their geographic distance from the US.

The Red-crowned Amazon provides a useful case study on this topic. As Figure 12.5 shows, this species has a low number of reported imported individuals but has a very high number of sightings relative to other species with similar importation numbers. The true number of individuals legally imported is likely not fully captured in the CITES data set; indeed, Iñigo-Elias and Ramos (1991) note that from 1981 to 1982, 4,372 Red-crowned Amazons were imported into the US from Mexico, whereas the CITES data note 1,802 individuals imported for the entire period from 1977 to 2017. Nevertheless, it is also possible that, in addition to dispersing individuals from Mexico, this naturalized population could be the result of legal importations that are not fully captured in any available data set. This species' native range extends very close to the edge of Texas and may even extend into Texas, so it could be much easier to illegally bring this species into the US.

Characteristics of Parrots as Pets

In addition to the number of individual parrots brought into the US, characteristics of parrots as pets could relate to their probability of being released and establishing populations outside of captivity, thereby leading to higher propagule pressure for some species over others. One specific characteristic of importance could be cost: if a parrot is a larger financial investment, its owner may be more vigilant in preventing escape and may be less likely to release the animal on purpose. Another factor could be the intelligence of the species and the degree to which it bonds with humans. More intelligent species, and/or species that are more bonded to their human caretakers, may be more likely to return home if they escape by accident. Grey Parrots, for example, can cost thousands of dollars and are renowned for their intelligence (Pepperberg 1983, 2012) and ability to mimic human speech (BirdLife International 2018). Grey Parrots are imported in high numbers, which could lead to high propagule pressure if some subset of these birds escaped or was released intentionally. Nevertheless, the species has a low number of sightings outside of captivity in the US (Fig. 12.5) and is not breeding in the wild in the US (Uehling et al. 2019).

Similar to Grey Parrots, Budgerigars are a very popular cage bird (Pranty and Epps 2002). However, unlike Grey Parrots, Budgerigars are reliably found at most pet stores across the US for incredibly low prices (less than $25). This species has very few importations to the US according to the CITES data (Fig. 12.5), but it has been raised in captivity since the early 1900s, when it was originally bred on farms in California and Florida (Grier 2006). Though the species has not been sighted in extremely high numbers (Table 12.1), it has been sighted in a wide range of geographic localities across the US, including states with colder weather, such as Minnesota, North Dakota, and Wisconsin (eBird Basic Dataset 2017; National Audubon Society 2018). Such birds are likely the result of accidental or purposeful releases and do not survive long in the wild in most localities (Pranty and Epps 2002), though they do survive long enough to be observed and recorded by avid community scientists (eBird Basic Dataset 2017; National Audubon Society 2018). Budgerigars breed in the wild in Florida and may breed in the wild in Texas and Illinois (Table 12.1; Uehling et al. 2019); however, other than these localities, the geographically widespread sightings of this species are likely due to an abundance of release events.

Characteristics of New Habitat

Once parrots are actually released into the wild, the characteristics of the environment into which they are released can play a large role in their probability of establishment. If the climate or habitat of a new region is similar to the climate or habitat of an exotic species' native region, it is easier for the exotic species to survive in this

Red-crowned (*Amazona viridigenalis*) and Lilac-crowned (*A. finschi*) Amazons feeding in a garden. Malibu, California, US, August 2009. Photo by Roelant Jonker.

new environment (Hayes and Barry 2008). Some evidence suggests that non-native birds tend to become established in a new locality if the temperature and habitat are similar to that of their native habitat (Kolar and Lodge 2001).

Parrots' thermal tolerances and specific habitat requirements may affect their chances of establishing naturalized populations in the US. Generally, parrot species are common in lowland tropical rain forest (Forshaw 2010). As might be expected of tropical species, non-native parrot populations in the US are most often found in the warmer climates of the southern states, suggesting that favorable thermal conditions may play an important part in establishment.

Thermal conditions may also affect the presence of food resources; in more temperate parts of the US, seasonal changes in food availability may prevent parrots from establishing. Indeed, the Monk Parakeet, the only species of parrot that is widespread in the northern US, is known to rely heavily on food from backyard bird feeders during the winter months (South and Pruett-Jones 2000). Its distribution in the northern US is limited to urban areas, such as Chicago and New York City, where the birds likely rely on human-provided food supplementation to survive in these harsher climates (Davis et al. 2014).

Presence or absence of suitable nesting sites may also play a role in promoting establishment. Most parrots nest in cavities (Forshaw 2010), and therefore availability of cavities may be a limiting factor in allowing some species to become established. Again, one possible reason behind the massive success of Monk Parakeets in the US is that they build their own nests, on natural or man-made structures, and therefore are not limited by cavity availability (Forshaw 2010; Bucher and Aramburú 2014; Uehling et al. 2019).

Davis et al. (2014) delved into Monk Parakeet distribution patterns in the US, and their work provides a useful case study for understanding how various habitat characteristics affect parrot establishment. Davis et al. (2014) examined the distribution of Monk Parakeets in the contiguous US as it relates to biophysical and anthropogenic variables. They report that in the southern US, the species' distribution is best explained by the environmental variables of temperature and forest cover. This contrasts with their finding in the northern US, where variables like housing density and distance to cities best explain Monk Parakeet distribution, and the parakeets are mostly restricted to urban environments.

Within these urban environments, Monk Parakeets in northern areas have specific preferences for nesting habitats. Minor et al. (2012) examined the distribution of Monk Parakeet nests in Chicago. They mapped out 249 nesting structures and examined the relationship between different environmental variables and the presence of Monk Parakeet nests. Their findings showed that nests were found most often in areas with less than 84% residential zoning. Therefore, though Monk Parakeets are associated with urban regions in the northern US, the density of humans in these areas affects the likelihood that parakeets will nest in a particular location within the city.

The case study of the Monk Parakeet highlights the large effect that human activities and human-altered landscapes can have on parrot establishment patterns. The Monk Parakeet is not unique in this regard—many exotic species become established in urban areas. Urban areas serve as centers of commerce to which exotic species are imported (McKinney 2006). Urban areas also often have a high level of human-caused disturbance, which promotes the establishment of exotic species (Lockwood et al. 2013; Sakai et al. 2001) by altering the environment such that native species' prior adaptations no longer give them competitive advantages over non-native colonizers (Byers 2002). Many parrot species, not just the Monk Parakeet, are found in urbanized regions, and specific associations with urban and suburban habitats are discussed below, in the "Vignettes" section.

Plainly, the reasons behind the successful establishment of some parrots and not others in various localities across the US are still unclear. We suggest this as a fruitful area of research with numerous unanswered questions. Parrots, unlike many other invasive species, have a unique relationship with humans because they are commonly kept as pets, and therefore we suggest that researchers think creatively about how anthropogenic factors may affect non-native parrot distribution and establishment success across the US.

VIGNETTES OF LOCALITIES WHERE PARROTS ARE COMMON

Here, we review the data from three areas with particularly diverse sets of parrot species, and from one particular location where the history of colonization by parrots is well known.

Southern California

There are 12 parrot species known to be breeding in southern California: the Monk Parakeet, Yellow-chevroned Parakeet (*Brotogeris chiriri*), Lilac-crowned Amazon (*Amazona finschi*), Red-lored Amazon (*A. autumnalis*), Red-crowned Amazon, Yellow-headed Amazon (*A. oratrix*), Turquoise-fronted Amazon (*A. aestiva*), Nanday Parakeet (*Aratinga nenday*), Blue-crowned Parakeet (*Thectocercus acuticaudatus*), Mitred Parakeet (*Psittacara mitratus*), Red-masked Parakeet (*P. erythrogenys*), and Rose-ringed Parakeet (*Psittacula krameri*) (Table 12.1; Garrett 2018; Uehling et al. 2019). White-winged Parakeets (*Brotogeris versicolurus*) formerly bred in southern California, but their numbers have declined, and it is unclear whether they have bred there recently (Garrett 2018; Uehling et al. 2019). Similarly, White-fronted Amazons (*Amazona albifrons*) have been sighted in southern California, but it is unclear whether they are breeding there (Garrett 2018; Uehling et al. 2019).

As elsewhere in the US, parrots occur in southern California because of escaped or purposefully released pets, and possibly because of illegal importations followed by purposeful releases into the wild (Horspool 2019). There are also unusual specific events that may have led to the start of the wild parrot populations in southern California. For example, in 1959, Simpson's Garden Town Nursery in Pasadena caught fire, and an employee, with the help of firefighters, released 65–70 parrots to save them from burning (Horspool 2019). Another possible release event occurred in 1979, when Busch Gardens, a theme park in the San Fernando Valley, was moving its headquarters. Employees were unable to find new homes for some of the parrots, and some accounts state that they released them into the wild (Horspool 2019). Despite such "urban legends" about release events, the source individuals for the current naturalized populations in southern California may never be completely known, and it is likely that the birds originated from several independent release events.

Dr. Kimball Garrett of the Natural History Museum of Los Angeles County pioneered much of the research on the non-native parrots in southern California (Garrett et al. 1997; Garrett 1997, 1998, 2018) and established the California Parrot Project (http://www.californiaparrotproject.org). Garrett has found that most parrots in southern California occur in highly modified urban and suburban habitats (Garrett et al. 1997), with the exception of a population of Nanday Parakeets found in more natural habitat in the Santa Monica Mountains (Horspool 2019). Additionally, Garrett has observed that most non-native parrots in southern California feed on non-native plants (K. Garrett, quoted in Horspool 2019). This suggests that these introduced parrots do not threaten native species (Good 2015), or at least that they are not likely to spread beyond human-dominated landscapes.

The Red-crowned Amazon, one of the breeding species in southern California, is of particular interest to conservationists because it is endangered in its native range in Mexico, where its decline is due to a mixture of poaching for the pet trade and habitat loss (Heise 2018). The Red-crowned Amazon is now common in southern California, in contrast to its status in its native range, and the species has been on the state's official bird checklist since 2001 (California Bird Records Committee 2019; Heise 2018). Indeed, Garrett (2018) identifies the Red-crowned Amazon as the dominant parrot in urban areas of the coastal slope of southern California. There are currently about 2,000 to 3,000 individuals in the Los Angeles area, which may equal or exceed the number of individuals left in the species' native range (Heise 2018). The Red-crowned Amazon's population in southern California may help to conserve the species if the population in Mexico continues to decline.

The public response to naturalized parrots in southern California varies widely, from very negative reactions to equally strong positive ones (Paskin 2018). One example of a positive reaction meant to protect these parrots is SoCal Parrot (https://www.socalparrot.org), a non-native parrot rescue and rehab facility east of San Diego (Good 2015). Run by Brooke Durham and Josh Bridwell, SoCal Parrot rehabilitates injured naturalized parrots from southern California, with the goal of ultimately releasing them back into the wild. In a strict reading of the law, releasing parrots into the wild is not legal; however, California Fish and Wildlife has agreed to not enforce the law for SoCal Parrot, given the lack of evidence that naturalized parrots pose a threat to native species (Good 2015).

Texas

There are three parrot species known to be breeding in Texas: the Monk Parakeet, Red-crowned Amazon, and Green Parakeet. As previously discussed, the populations of Red-crowned Amazons and Green Parakeets in Texas may actually contain some dispersers from native populations in Mexico. There are also other species that may be establishing breeding populations in Texas, including the Lilac-crowned Amazon, Yellow-headed Amazon, and Budgerigar (Table 12.1; Uehling et al. 2019). Though parrot sightings are common in Texas (Figs. 12.1, 12.2; Uehling et al. 2019), the taxonomic diversity of breeding parrots in this state is much lower than in California and Florida (discussed next).

The population of Red-crowned Amazons in Texas is of particular interest to conservationists. As of 2018, the population in Texas was estimated to number around 680 individuals (Bjork 2018), and it could be an important source of genetic diversity for the population in Mexico (Berg 2015). Like many of the parrots found in California, the Red-crowned Amazons in southern Texas are frequently sighted in suburban habitat, often eating from backyard bird feeders (Bjork 2018).

Studies of parrots in Texas focus mainly on Red-crowned Amazons. Donald Brightsmith and Simon Kiacz, both of Texas A&M University, are pioneering the Tejano Parrot Project, with an accompanying social media site (https://www.facebook.com/TejanoParrots/), to gain a better understanding of the movements and roosting behaviors of Red-crowned Amazons in southern Texas (Bjork 2018). Additionally, Dr. Karl Berg of the University of Texas Rio Grande Valley has pioneered a project to put nesting tubes up around campus to encourage Red-crowned Amazon breeding in the region (KVEO-TV 2019).

The Red-crowned Amazon is likely native to Texas (unlike California), and Texas Parks and Wildlife lists the species as native. In addition, the US Fish and Wildlife Service considers the Red-crowned Amazon a native species, but because parrots are not mentioned in the Migratory Bird Treaty Act, it receives no federal protection, and any protection measures must rely on state and local laws. Regardless of whether the species is eventually federally listed as an endangered species, the state (or federal) protection of a species that is now dependent on human-modified habitats is complicated and very unusual. Red-crowned Amazons rely on neighborhood trees and gardens in suburban habitats in Texas (Bjork 2018). It remains unresolved how the management of threatened or endangered species will be handled when those species are dependent on neighborhood trees or the backyards of private homes.

Florida

Of the 25 species identified as breeding in the US by Uehling et al. (2019), there is evidence or suggestion of breeding for 23 of these species in Florida (Table 12.1). In other words, Florida has the highest diversity of non-native parrots in the US.

Many of Florida's non-native parrots occur in suburban or urban habitats that have been highly modified by humans (Pranty and Epps 2002; Minta 2018; Uehling et al. 2019), and non-native parrots seem to rely on non-native vegetation for food and habitat (Pranty and Epps 2002; Minta 2018). Work in southeastern Florida suggests that cavity-nesting parrots do not have negative impacts on native species via competition for nesting cavities, primarily because they nest in urban areas (Diamond and Ross 2019).

Florida is one of the only sites in the US where there is evidence of economic impacts by introduced parrots, specifically the Monk Parakeet. Naturalized parrots can have both

Lilac-crowned Amazons (*Amazona finschi*) on the grounds of the Biltmore Hotel. Coral Gables, Florida, US, June 2011. Photo by Roelant Jonker.

ecological and economic impacts in their new habitats (Mori and Menchetti, chap. 6 this vol.; Brightsmith and Kiacz, chap. 9 this vol.). Monk Parakeets in Florida are known to cause damage both to agriculture and to electrical utility structures. As regards agriculture, Monk Parakeets have damaged longan fruit orchards (Tillman et al. 2000). In terms of damage to electrical utility structures, Monk Parakeets often build their large stick nests on human-made structures (Minor et al. 2012), including bridge infrastructure, supports for lighting arrays, church steeples, cell-phone towers, telephone poles, power substations, transmission lines, and transformers (Avery et al. 2002; Burger and Gochfeld 2009; Minor et al. 2012; Burgio et al. 2014, 2016; Reed et al. 2014). Nests on electrical structures specifically can cause electrical shortages, fires, power outages, and very costly corrections (Avery et al. 2002; Pruett-Jones et al. 2012; Reed et al. 2014; Burgio et al. 2016). In Florida, the utility company Florida Power and Light spends between $340,000 and $940,000 each year to remove nests of the Monk Parakeet (Avery et al. 2008). Another study (Newman et al. 2008) documented that during a five-month period in 2001, nests of Monk Parakeets caused 198 power outages, affecting more than 10,000 customers.

Floridians respond to non-native parrots in a variety of ways. Florida's state laws do not prohibit the poaching of non-native parrots. Some residents view this as a business opportunity and will poach parrots from nests in order to sell them in the pet trade, whereas others decry this practice and are working to prevent it (Minta 2018). Other Floridians view the wild parrots as a good tourist attraction, and there is a field guide to the parrots of south Florida (Epps 2007), helping to facilitate the public's acceptance of these birds.

Chicago: The Tale of Monk Parakeets in One City

The following account of Monk Parakeets in Chicago is summarized from Pruett-Jones et al. (2012). In the late 1960s, Monk Parakeets were simultaneously observed in the wild in a number of cities in the eastern and southeastern US (Spreyer and Bucher 1998). Chicago was one of these cities. In terms of our knowledge of the establishment of naturalized parrots in the US, our data set for Monk Parakeets in Chicago is the best available for any species of parrot for any region in the US.

Monk Parakeets were first seen in the wild in Chicago in 1968, and the first nests were seen in the early 1970s. During the early 1970s, there were repeated reports of nesting birds, but none of these nests persisted from one year to the next (Ingersoll 1973; Larson 1973; Neidermyer and Hickey 1977). Everything changed in Chicago in 1979, when the parakeets established a nesting colony along the lakefront of Lake Michigan in the community of Hyde Park, where the University of Chicago is located. This group of nesting birds persisted, and it grew in size and became the center of a very large population of parakeets in Chicago. Thus, from the first sighting of a Monk Parakeet in the wild in Chicago, it was only three to four years before birds were observed breeding and another six to seven years before there were sufficient birds to establish a stable colony.

In the mid-1980s, there were at least 17 Monk Parakeets in Hyde Park (Walsten 1988), and by the early 1990s this population had grown to 60+ birds (Hyman and Pruett-Jones 1995). The population of parakeets in Hyde Park grew rapidly after this, following an exponential growth rate throughout the 1990s and into the mid-2000s. The peak count of Monk Parakeets in Hyde Park was more than 300 birds in 2006 (Pruett-Jones et al. 2012).

After 2006, the population of Monk Parakeets in Hyde Park, and in Chicago generally, began to decline. This decline was also seen in most populations around the country (Pruett-Jones et al. 2012). In their review of the possible reasons for this decline, Pruett-Jones et al. (2012) suggested that it was the result of birds moving into more suburban and rural areas, rather than an actual reduction in numbers. This explanation may have seemed reasonable at the time but is almost certainly incorrect. Yes, the population in Hyde Park declined, and this decline continues today, but the explanation for the national decline in numbers of Monk Parakeets cannot easily be explained by local circumstances. It is much more likely that the decline in Monk Parakeets in Chicago and nationally was due to a disease, and the most likely candidate is West Nile virus. No samples of Monk Parakeets from this time period have been obtained or analyzed, but it is our suggestion that the nationwide decline of Monk Parakeets between 2005 and 2015 was due to a disease.

After 2015, the population of Monk Parakeets in Chicago and in the nation stabilized, but it has not yet begun to increase to numbers seen before the decline. In Chicago, Monk Parakeets still persist and survive but in smaller numbers. The population in Hyde Park is more or less extinct, though there are hundreds of birds still nesting in the greater Chicago region. But why the birds left Hyde Park remains a mystery.

ELSEWHERE IN THE US

Our concern in this review has been the contiguous US. Nevertheless, this focus ignores two very important areas for parrots: Hawaii and Puerto Rico. Vanderwerf and Kalodimos (chap. 13 this vol.) summarize the status of parrots in Hawaii, and Falcón and Tremblay (2018) have summarized the status of introduced parrots in Puerto Rico. We believe it is relevant to briefly encapsulate the results of these two studies here.

As VanderWerf and Kalodimos (chap. 13 this vol.) document, there were no native parrot species on the Hawaiian Islands, but currently at least 39 different species of parrots have been seen in the wild, on various islands, and four of these (Mitred Parakeet, Red-crowned Amazon, Red-masked Parakeet, and Rose-ringed Parakeet) are fully naturalized. A fifth species (Rosy-faced Lovebird) is likely to be confirmed as naturalized once adequate surveys are completed, and an additional five species have a status of "incipient," meaning that the species is at least locally common and breeding or thought to be breeding in small numbers. Two of these incipient species, the Salmon-crested Cockatoo (*Cacatua moluccensis*) and the White Cockatoo (*C. alba*), are interesting because they are not breeding on the mainland (see Table 13.1, chap. 13).

Falcón and Tremblay (2018) summarize the current status of introduced parrots in Puerto Rico. They document that of 46 different species imported to the island, 29 different species have been reported in the wild, and of these, 12 species are breeding in self-sustaining populations. Eight of the 12 species are also established in the mainland US (and mentioned in Table 12.1), whereas four are not. Those four unique species are: Hispaniolan Amazon (*Amazona ventralis*), Scarlet Macaw (*Ara macao*), Orange-fronted Parakeet (*Eupsittula canicularis*), and Hispaniolan Parakeet (*Psittacara chloropterus*). Separate from the 12 species that are naturalized in Puerto Rico, Falcón and Tremblay (2018) documented an additional 21 species that are breeding in localized areas but not yet in self-sustaining numbers. Nevertheless, it is entirely possible these species will become fully naturalized in the future. Among the 12 naturalized species, the clearly invasive species (according to Falcón and Tremblay's criteria) were the Monk Parakeet, White-winged Parakeet, Red-masked Parakeet, and Orange-fronted Parakeet. Niche modeling for these species suggested that future range expansions were highly likely (Falcón and Tremblay 2018).

Combining the data from Hawaii and Puerto Rico with our other surveys in the US, we reach the rather startling conclusion that at present, there are 29 species of naturalized parrots breeding in the US as a whole. If current trends in Hawaii and Puerto Rico continue, this number will almost certainly increase.

CONCLUSIONS

Naturalized parrots are widespread and, in some regions of the US, becoming quite common. In most cases, these parrots arrived via the pet trade, and their accidental or purposeful release has created high enough propagule pressure to allow some species to become established.

In California, Texas, and Florida, the three mainland states where naturalized parrots are most common, the parrots tend to occur primarily in highly modified urban and suburban habitats with non-native plants and plentiful bird feeders. Therefore, there is minimal concern about non-native parrot species harming native species or causing economic damage at these sites. There are a few exceptions, including the Nanday Parakeet, which has moved into more "natural" habitat in the Santa Monica Mountains, California, and the Monk Parakeet, which has caused economic damage and has the potential to cause future problems.

We highly encourage future researchers to study the ecology and behavior of these parrot species rather than simply noting their presence. Parrots in the US highlight the intersection of bird ecology, human-wildlife interactions, and management of introduced species. However, little work published in academic journals has gone beyond simply noting the presence of individual species. Indeed, some of the most up-to-date information about non-native parrots in the US comes from local news outlets. The strong association of parrots with human landscapes in the US suggests that the birds may be of interest to the general public and ecologists alike as their populations continue to grow.

REFERENCES

Avery, M. L., Greiner, E. C., Lindsay, J. R., Newman, J. R., and Pruett-Jones, S. 2002. Monk parakeet management at electric utility facilities in south Florida. In Proceedings of the 20th Vertebrate Pest Conference, ed. R. M. Timm and R. H. Schmidt, 140–145. Davis: Univ. California.

Avery, M. L., and Shiels, A. B. 2018. Monk and rose-ringed parakeets. In *Ecology and Management of Terrestrial Vertebrate Invasive Species in the United States*, ed. W. C. Pitt, J. C. Beasley, and G. W. Witmer, 333–357. Boca Raton, FL: CRC Press.

Avery, M. L., Yoder, C. A., and Tillman, E. A. 2008. Diazacon inhibits reproduction in invasive monk parakeet populations. *Journal of Wildlife Management* 72:1449–1452.

Beissinger, S. 2001. Trade of live wild birds: Potentials, principles, and practices of sustainable use. In *Conservation of Exploited Species*, ed. J. D. Reynolds, G. M. Mace, K. H. Redford, J. G. Robinson, 182–202. Cambridge: Cambridge Univ. Press.

Benson, K. L. P., and Arnold, K. A. 2001. *The Texas Breeding Bird Atlas*. Texas A&M Univ., College Station/Corpus Christi. Accessed 18 Oct. 2019. http://txtbba.tamu.edu.

Berg, K. S. 2015. Wild thing: Wild parrots. *Texas Parks and Wildlife Magazine*, Dec.

Berg, K. S. 2019. Red-crowned parrot. In *Texans on the Brink: Threatened and Endangered Animals*, ed. B. R. Chapman and W. I. Lutterschmidt, 150–152. College Station: Texas A&M Univ. Press.

BirdLife International. 2018. *Psittacus erithacus*. IUCN Red List of Threatened Species 2018: e.T22724813A129879439. Accessed 29 Oct. 2019. https://dx.doi.org/10.2305/IUCN.UK.2018-2.RLTS.T22724813A129879439.en.

Bjork, R. 2018. Nature notes: Red-crowned parrots wander into Texas. *Victoria (TX) Advocate*, 30 Aug. 2018.

Blackburn, T. M., and Duncan, R. P. 2001a. Determinants of establishment success in introduced birds. *Nature* 414:195–197.

Blackburn, T. M., and Duncan, R. P. 2001b. Establishment patterns of exotic birds are constrained by non-random patterns in introduction. *Journal of Biogeography* 28:927–939.

Blackburn, T. M., Lockwood, J. L., and Cassey, P. 2009. *Avian Invasions: The Ecology and Evolution of Exotic Birds*. Oxford: Oxford Univ. Press.

Bucher, E. H., and Aramburú, R. M. 2014. Land-use changes and monk parakeet expansion in the Pampas grasslands of Argentina. *Journal of Biogeography* 41:1160–1170.

Burger, J., and Gochfeld, M. 2009. Exotic monk parakeets (*Myiopsitta monachus*) in New Jersey: Nest site selection, rebuilding following removal, and their urban wildlife appeal. *Urban Ecosystems* 12:185–196.

Burgio, K. R., Rubega, M. A., and Sustaita, D. 2014. Nest-building behavior of monk parakeets and insights into potential mechanisms for reducing damage to utility poles. *PeerJ* 2:e601.

Burgio, K. R., van Rees, C. B., Block, K. E., Pyle, P., Patten, M. A., Spreyer, M. F., and Bucher, E. H. 2016. Monk parakeet (*Myiopsitta monachus*) (v.3.0). In *The Birds of North America*, ed. P. G. Rodewald. Ithaca, NY: Cornell Lab of Ornithology.

Butler, C. J. 2005. Feral parrots in the continental United States and United Kingdom: Past, present, and future. *Journal Avian Medicine and Surgery* 19:142–149.

Byers, J. E. 2002. Impact of non-indigenous species on natives enhanced by anthropogenic alteration of selection regimes. *Oikos* 97:449-458.

California Bird Records Committee. 2019. Official California checklist. Accessed Dec. 2019. http://californiabirds.org/ca_list.asp.

CITES. 2018. CITES Trade Database. Accessed June 2018. http://trade.cites.org.

Convention on Biological Diversity. 2020. History of the convention. Accessed Aug. 2020. https://www.cbd.int/history/.

Davis, A. Y., Malas, N., and Minor, E. S. 2013. Substitutable habitats? The biophysical and anthropogenic drivers of an exotic bird's distribution. *Biological Invasions* 16:415–427.

Diamond, J. M., and Ross, M. S. 2019. Exotic parrots breeding in urban tree cavities: Nesting requirements, geographic distribution, and potential impacts on cavity nesting birds in southeast Florida. *Avian Research* 10:39.

eBird. 2017. eBird Basic Dataset (v.EBD_relNov-2017). Cornell Lab of Ornithology, Ithaca, NY. Accessed Nov. 2017. https://www.ebird.org.

Enkerlin-Hoeflich, E. C., and Hogan, K. M. 1997. Red-crowned parrot (*Amazona viridigenalis*) (v.2.0). In *Birds of North America*, ed. A. F. Poole and F. B. Gill. Ithaca, NY: Cornell Lab of Ornithology.

Epps, S. A. 2007. *Parrots of South Florida*. Sarasota, FL: Pineapple Press.

Falcón, W., and Tremblay, R. L. 2018. From the cage to the wild: Introductions of Psittaciformes to Puerto Rico. *PeerJ* 6:e5669.

Forshaw, J. M. 2010. *Parrots of the World*. Princeton, NJ: Princeton Univ. Press.

Garrett, K. L. 1997. Population status and distribution of naturalized parrots in southern California. *Western Birds* 28:181–195.

Garrett, K. L. 1998. Population trends and ecological attributes of introduced parrots, doves, and finches in California. In *Proceedings of the 18th Vertebrate Pest Conference*, ed. R. O. Baker and A. C. Crabb, 46–54. Davis: Univ. of California.

Garrett, K. L. 2018. Introducing change: A current look at naturalized bird species in western North America. In *Trends and Traditions: Avifaunal Change in Western North America*, ed. W. D. Shuford, R. E. Gill Jr., and C. M. Handel, 116–130. Studies of Western Birds no. 3. Camarillo, CA: Western Field Ornithologists.

Garrett, K. L., Mabb, K. T., Collings, C. T., and Kares, L. M. 1997. Food items of naturalized parrots in southern California. *Western Birds* 28:196–201.

REFERENCES

Good, D. 2015. Jailbird parrots return to the wild ... as fugitives. *Aubudon*, July–Aug. iss.

Grier, K. C. 2006. *Pets in America: A History*. Chapel Hill: Univ. of North Carolina Press.

Hayes, K. R., and Barry, S. C. 2008. Are there any consistent predictors of invasion success? *Biological Invasions* 10:483–506.

Heise, U. K. 2018. Creating an "urban ark" for endangered species in Los Angeles. KCET: Earth Focus, 7 June 2018.

Horspool, S. 2019. Wild parrots multiplying in southern California. PetHelpful. Accessed Dec. 2019. https://pethelpful.com/wildlife/Wild-Parrots-Multiplying-in-Southern-California.

Hyman, J., and Pruett-Jones, S. 1995. Natural history of the monk parakeet in Hyde Park, Chicago. *Wilson Bulletin* 107:510–517.

Ingersoll, B. 1973. Pretty parakeet a future pest? *Chicago Sun-Times*, 26 Feb. 1973.

Iñigo-Elias, E. E., and Ramos, M. A. 1991. The psittacine trade in Mexico. In *Neotropical Wildlife Use and Conservation*, ed. J. G. Robinson and K. H. Redford, 380–392. Chicago: Univ. of Chicago Press.

Kolar, C. S., and Lodge, D. M. 2001. Progress in invasion biology: Predicting invaders. *Trends in Ecology & Evolution* 16:199–204.

KVEO-TV. 2019. UTRGV's effort to protect the red-crowned parrot. 29 Mar. 2019.

Larson, G. E. 1973. The monk parakeet in Illinois: New views of alarm. *Illinois Audubon Bulletin* 166:29–30.

Lockwood, M. W., and Freeman, B. 2014. *The TOS Handbook of Texas Birds*. 2nd ed. Texas College Station: Texas A&M Univ. Press.

Lockwood, J. L., Hoopes, M. F., and Marchetti, M. P. 2013. *Invasion Ecology*. 2nd ed. West Sussex, UK: Wiley and Sons.

Mabb, K. T. 1997. Roosting behavior of naturalized parrots in the San Gabriel Valley, California. *Western Birds* 28:202–208.

McKinney, M. L. 2006. Urbanization as a major cause of biotic homogenization. *Biological Conservation* 127:247–260.

Minor, E. S., Appelt, C. W., Grabiner, S., Ward, L., Moreno, A., and Pruett-Jones, S. 2012. Distribution of exotic monk parakeets across an urban landscape. *Urban Ecosystems* 15:979–991.

Minta, M. 2018. Miami's wild parrots are being poached, and there's no law to protect them. *Miami New Times*, 5 June 2018.

National Audubon Society. 2018. The Christmas bird count historical results (online). Accessed Mar. 2019. http://www.christmasbirdcount.org.

Neidermyer, W. J., and Hickey, J. J. 1977. The monk parakeet in the United States, 1970–1975. *American Birds* 31:273–278.

Newman, J. R., Newman, C. M., Lindsay, J. R., Merchant, B., Avery, M. L., and Pruett-Jones, S. 2008. Monk parakeets: An expanding problem on power lines and other electrical utility structures. In *8th International Symposium on Environmental Concerns in Rights-of-Way Management*, ed. J. W. Goodrich-Mahoney, L. P. Abrahamson, J. L. Ballard, and S. M. Tikalsky, 343–354. New York: Elsevier Science.

Panter, C. T., Atkinson, E. D., and White, R. L. 2019. Quantifying the global legal trade in live CITES-listed raptors and owls for commercial purposes over a 40-year period. *Avocetta* 43:23–36.

Paskin, J. 2018. Pasadena's screaming parrots are super annoying but may save their species from extinction. *LAist*, 10 July 2018.

Pepperberg, I. M. 1983. Cognition in the African grey parrot: Preliminary evidence for auditory/vocal comprehension of the class concept. *Animal Learning and Behavior* 11:179–185.

Pepperberg, I. M. 2012. Further evidence for addition and numerical competence by a grey parrot (*Psittacus erithacus*). *Animal Cognition* 15:711–717.

Phelps, J., Webb, E. L., Bickford, D., Nijman, V., and Sodhi, N. S. 2010. Boosting CITES. *Science* 330(6012):1752–1753.

Pranty, B. 2001. The budgerigar in Florida: Rise and fall of an exotic psittacid. *North American Birds* 55:389–397.

Pranty, B., and Epps, S. 2002. Distribution, population status, and documentation of exotic parrots in Broward County, Florida. *Florida Field Naturalist* 30:111–150.

Pranty, B., and Garrett, K. L. 2003. The parrot fauna of the ABA area: A current look. *Birding* 35:248–261.

Pruett-Jones, S., Appelt, C. W., Sarfaty, A., Van Vossen, B., Leibold, M. A., and Minor, E. S. 2012. Urban parakeets in northern Illinois: A 40-year perspective. *Urban Ecosystems* 15:709–719.

Pruett-Jones, S., Newman, J. R., Newman, C. M., Avery, M. L., and Lindsay, J. R. 2007. Population viability analysis of monk parakeets in the United States and examination of alternative management strategies. *Human-Wildlife Conflicts* 1:35–44.

Reed, J. E., McCleery, R. A., Silvy, N. J., Smeins, F. E., and Brightsmith, D. J. 2014. Monk parakeet nest-site selection of electric utility structures in Texas. *Landscaping and Urban Planning* 129:65–72.

Runde, D. E., Pitt, W. C., and Foster, J. 2007. Population ecology and some potential impacts of emerging populations of exotic parrots. In *Managing Vertebrate Invasive Species: Proceedings of an International Symposium*, ed. G. W. Witmer, W. C. Pitt, and K. A. Fagerstone, 338–360. Fort Collins, CO: USDA/APHIS Wildlife Services, National Wildlife Research Center.

Sakai, A. K., Allendorf, F. W., Holt, J. S., Lodge, D. M., Molofsky, J., With, K. A., Baughman, S., Cabin, R. J., Cohen, J. E., Ellstrand, N. C., et al. 2001. The population biology of invasive species. *Annual Review of Ecology and Systematics* 32:305–332.

Smith, K. F., Behrens, M., Schloegel, L. M., Marano, N., Burgiel, S., and Daszak, P. 2009. Reducing the risks of the wildlife trade. *Science* 324:594–595.

South, J. M., and Pruett-Jones, S. 2000. Patterns of flock size, diet, and vigilance in naturalized monk parakeets in Hyde Park, Chicago. *Condor* 102:848–854.

Spreyer, M. F., and Bucher, E. H. 1998. Monk parakeet (*Myiopsitta monachus*). In *The Birds of North America*, ed. A. Poole and F. Gill, no. 322. Philadelphia, PA: Academy of Natural Sciences; Washington, DC: American Ornithologists' Union.

Tillman, E. A., Van Doorn, A., and Avery, M. L. 2000. Bird damage to tropical fruit in south Florida. In *The Ninth Wildlife Damage Management Conference Proceedings*, ed. M. C. Brittingham, J. Kays, and R. McPeake, 47–59. State College, PA.

Uehling, J. J., Tallant, J., and Pruett-Jones, S. 2019. Status of naturalized parrots in the United States. *Journal of Ornithology* 160:907–921.

USFWS 2016. *Wild Bird Conservation Act: Summary of Regulations and Effects*. Falls Church, VA: US Fish and Wildlife Service.

van Bael, S., and Pruett-Jones, S. 1996. Exponential population growth of monk parakeets in the United States. *Wilson Bulletin* 108:584–588.

Walsten, D. M. 1988. Hyde Park's parakeets: These green-winged arrivals from Argentina appear to be settling in. *Field Museum of Natural History Bulletin*, Apr.:23–29.

13
STATUS OF NATURALIZED PARROTS IN THE HAWAIIAN ISLANDS

Eric A. VanderWerf and Nicholas P. Kalodimos

INTRODUCTION

The avifauna of the Hawaiian Islands is one of the most invaded in the world. At least 211 species of non-native birds are known to have been released or have been identified in the wild in Hawaii (Pyle and Pyle 2017). Approximately 54 of these are regarded as naturalized, having established wild, self-sustaining breeding populations (Foster 2009; Pyle and Pyle 2017; VanderWerf et al. 2017). Most of the remaining endemic Hawaiian bird species are restricted to higher-elevation montane areas because of habitat loss and their vulnerability to mosquito-borne avian diseases (Scott et al. 2001). In most low-elevation areas and urban and suburban environments in Hawaii, the only land birds present are naturalized non-native species.

Several previous authors have described the history and status of non-native birds in Hawaii, including parrots. Pyle and Pyle (2017) summarized the status of all non-native bird species known to have been released or observed in Hawaii, and Foster (2009) details the effects of introduced birds on the environment and native bird species. There are no native parrots in Hawaii, but at least 39 parrot species have been observed in the wild in the Hawaiian Islands (Table 13.1) (Pyle and Pyle 2017). Determining which species have established wild breeding populations is not entirely clear or easy. The Hawaii Bird Records Committee (HBRC) regarded non-native species to be established if they have maintained self-sustaining populations in the wild for at least 15 years (VanderWerf et al. 2018). However, parrots in particular can live much longer than 15 years, so populations may appear to remain stable for many years despite a lack of reproduction. In such cases, the HBRC requires evidence of reproduction or an increasing population trend indicative that reproduction must be occurring.

Four parrot species are regarded as naturalized in Hawaii following the criteria described above: Mitred Parakeet (*Psittacara mitratus*), Red-masked Parakeet (*P. erythrogenys*), Red-crowned Amazon (*Amazona viridigenalis*), and Rose-ringed Parakeet (*Psittacula krameri*) (VanderWerf et al. 2017). Of these species, the Rose-ringed Parakeet is the most common, and it is growing in population size and impacting the environment (Shiels et al. 2018). The Rosy-faced Lovebird (*Agapornis roseicollis*) has increased in number and spread dramatically on Maui and Hawaii Island in the past five years and appears to be established; its official status warrants reevaluation. Two other species, Blue-crowned Parakeet (*Thectocercus acuticaudatus*) and Burrowing Parrot (*Cyanoliseus patagonus*), have smaller populations and narrower ranges but are observed regularly and may be increasing in number, indicating they may become established in the near future, if they are not already. One additional species, the Pale-headed Rosella (*Platycercus adscitus*), was established on Maui for over 50 years but was last observed in 1928 and is now extirpated (Pyle and Pyle 2017). This species is not included in the American Ornithological Society (formerly the American Ornithologists' Union) checklist of North American birds (AOU 1998; Chesser et al. 2018), but it still is regarded as having been established in the past (VanderWerf et al. 2017). In this chapter, we review the distribution and

STATUS OF NATURALIZED PARROTS IN THE HAWAIIAN ISLANDS

TABLE 13.1

Summary of parrots in Hawaii by island. Islands are listed from west to east. Blank = species not known to occur; Observed = occasional or historical sightings, but the species is not known to be present regularly; Present = the species is regularly present at least locally but is not known to be breeding; Incipient = the species is at least locally common and breeding or thought to be breeding in small numbers; Established = the species has been common for years and is known to be breeding; ? = status uncertain; Hist. Est. = Historically established. Species in red type are covered in the species accounts. Species are listed in alphabetical order by common name.

SPECIES	KAUAI	OAHU	MOLOKAI	MAUI	HAWAII
Blue-and-yellow Macaw (*Ara ararauna*)	Observed	Observed			Observed
Blue-crowned Parakeet (*Thectocercus acuticaudatus*)		Incipient			
Budgerigar (*Melopsittacus undulatus*)		Observed		Observed	
Burrowing Parrot (*Cyanoliseus patagonus*)					Incipient
Cockatiel (*Nymphicus hollandicus*)		Observed	Observed		
Eclectus Parrot (*Eclectus roratus*)	Observed	Observed			
Galah (*Eolophus roseicapilla*)		Observed			
Grey Parrot (*Psittacus erithacus*)					Observed
Jandaya Parakeet (*Aratinga jandaya*)		Observed			
Meyer's Parrot (*Poicephalus meyeri*)		Observed			
Military Macaw (*Ara militaris*)				Observed	
Mitred Parakeet (*Psittacara mitratus*)		Observed		Established[2]	Observed
Monk Parakeet (*Myiopsitta monachus*)		Observed			
Nanday Parakeet (*Aratinga nenday*)	Observed	Observed			
Orange-chinned Parakeet (*Brotogeris jugularis*)		Observed			
Orange-fronted Parakeet (*Eupsittula canicularis*)		Observed			
Orange-winged Amazon (*Amazona amazonica*)		Observed			
Pale-headed Rosella (*Platycercus adscitus*)				Hist. Est.	
Palm Cockatoo (*Probosciger aterrimus*)		Observed			
Plum-headed Parakeet (*Psittacula cyanocephala*)		Observed			
Red-crowned Amazon (*Amazona viridigenalis*)	Observed	Established		Observed	
Red-crowned Parakeet (*Cyanoramphus novaezelandiae*)		Observed			
Red-lored Amazon (*Amazona autumnalis*)		Observed			
Red-masked Parakeet (*Psittacara erythrogenys*)		Established			Established
Rose-ringed Parakeet (*Psittacula krameri*)	Established	Established		Observed	Observed
Rosy-faced Lovebird (*Agapornis roseicollis*)	Observed	Observed		Established	Incipient

SPECIES ACCOUNTS

SPECIES	KAUAI	OAHU	MOLOKAI	MAUI	HAWAII
Salmon-crested Cockatoo (*Cacatua moluccensis*)		Incipient[1]		Observed	
Scarlet Macaw (*Ara macao*)		Observed			Observed
Scarlet-fronted Parakeet (*Psittacara wagleri*)					?
Senegal Parrot (*Poicephalus senegalus*)	Observed	Observed			Observed
Slender-billed Parakeet (*Enicognathus leptorhynchus*)					Observed
Sulphur-crested Cockatoo (*Cacatua galerita*)		Observed			
Tanimbar Corella (*Cactua goffiniana*)		Observed			Incipient
Turquoise-fronted Amazon (*Amazona aestiva*)		Observed			
White Cockatoo (*Cacatua alba*)		Incipient[1]			
White-fronted Amazon (*Amazona albifrons*)		Observed			
Yellow-collared Lovebird (*Agapornis personatus*)			Observed		
Yellow-crested Cockatoo (*Cacatua sulphurea*)			Observed		
Yellow-headed Amazon (*Amazona oratrix*)			Observed		

[1] The Salmon-crested and White Cockatoos at Lyon Arboretum on Oahu have interbred, and most of the individuals currently present are hybrids.
[2] The Mitred Parakeet became established on Maui in about 1995, but a control program begun in 2006 has reduced the population to about 15 birds as of 2018.

occurrence of parrots in Hawaii, focusing on established and incipient species (species that are locally common and breeding or thought to be breeding in small numbers). After reviewing the status of each species, we discuss the environmental effects that parrots are having in Hawaii. Please note that in the accounts below, any unreferenced observations are personal observations by the authors or communicated to the authors by S. Kondo, H. D. Pratt, A. Radford, or A. Shiels.

SPECIES ACCOUNTS

Tanimbar Corella
(*Cacatua goffiniana*)

Endemic to the Tanimbar Islands of eastern Indonesia, this species has an incipient population on Hawaii Island, and a small number are present on Oahu. A flock of about a dozen birds was first noted in the 1980s in the Kaloko Mauka area above Kailua-Kona in western Hawaii Island. This population has persisted in the same area and in 2019 numbered about 15 birds (Pyle and Pyle 2017; eBird 2019). On Oahu, a single bird was first reported at Lyon Arboretum in upper Manoa Valley in spring 1987, presumably having escaped from a nearby theme park along with other cockatoos, and up to five or six individuals were reported at that location from 1991 to 2012 (Pyle and Pyle 2017). Two individuals have persisted at Lyon Arboretum as of 2019. There have been several reports of one to three birds in other areas of Oahu, including Haleiwa in August 1987, Makiki in April 1988 and March 2000, Kahuku in August 1996, Punaluu in 2003, and Kaneohe in 2003 (Pyle and Pyle 2017). Four birds were present in Waiahole Valley in March 2004, but area residents indicated they damaged papaya crops, and by 2019 they were gone. Two of the Tanimbar Corellas observed at Lyon Arboretum from 2003 to 2010 had a USDA stainless-steel quarantine-station leg band, indicating they likely were wild-caught individuals from Indonesia. The Tanimbar Corellas at Lyon Arboretum socialize with the larger Salmon-crested Cockatoo × White Cockatoo hybrids and have learned to mimic the calls of the larger hybrids when interacting with them, as well as using their own species' vocalizations. This and other species of cockatoos at Lyon Arboretum frequently feed on the seeds of Moluccan Albizia (*Falcataria moluccensis*), killing the seeds in the process; on mature stands of *Macaranga* spp.; on the ripe fruit of *Archontophoenix alexandrae* palms, scraping off the pulp and then discarding the seed unharmed; and on the fruit of Blue Marble (*Elaeocarpus angustifolius*).

Salmon-crested Cockatoo (*Cacatua moluccensis*)

Native to Seram and Ambon Islands of eastern Indonesia, the Salmon-crested Cockatoo is an incipient species on Oahu and was observed for several years on Maui but was extirpated in 2002. On Oahu, up to 13 individuals have been reported regularly at Lyon Arboretum in upper Manoa Valley and in adjacent Makiki Valley since at least 1972. The presence of cockatoos in this area is attributable to the escape of two Salmon-crested Cockatoos and two White Cockatoos from a nearby theme park (Pyle 1984; Pyle and Pyle 2017). The exact number of individuals present over the years has been confusing, because of both identification difficulties and hybridization. As of December 2019, there was only a single Salmon-crested Cockatoo present, and the remaining 12 birds were Salmon-crested × White Cockatoo hybrids that have characteristics of both species. Hybrid pairs have been observed feeding single chicks over the past 15 years. All cockatoos at Lyon Arboretum use the same call types and have similar vocal repertoires, which most closely resemble those of the Salmon-crested Cockatoo, based on comparison with recordings of wild birds in their native range.

The cockatoos present in the Manoa–Makiki area are strongly associated with lowland tropical rainforest habitat that is composed of non-native

A hybrid Salmon-crested Cockatoo (*Cacatua moluccensis*) × White Cockatoo (*C. alba*) feeding on coconut as a source of food for its nestling. It appears that cockatoos in Hawaii feed on coconuts only when they have nestlings. Honolulu, Hawaii, US, August 2019. Photo by Nicholas P. Kalodimos.

tree species, especially the very tall, spreading canopy of Moluccan Albizia, two species of *Macaranga*, banyans (*Ficus* spp.), *Veitchia* spp., Umbrella Tree (*Schefflera actinophylla*), and *Livistona*, and *Archontophoenix* palms. These cockatoos have shown very limited dispersal behavior and have not been observed to move more than about 1 km within the forest habitat and never outside of it. The cockatoos probably exhibit limited movements because Lyon Arboretum contains a higher abundance than the surrounding areas of the tree species preferred for foraging.

Two Salmon-crested Cockatoos also have been reported from Hauula, on the northeastern side of Oahu, approximately 32 km from the Manoa Valley population. These two birds are almost certainly from an adjacent private aviary that housed a dozen or more of this species in 2006. On Maui, four birds were collected in Waikapu Valley in 2002 and were deposited as specimens in the Bernice Pauahi Bishop Museum in Honolulu (Pyle and Pyle 2017).

White Cockatoo
(*Cacatua alba*)

Endemic to Halmahera Island of eastern Indonesia, the White Cockatoo may be an incipient species on Oahu. Although single birds have been reported elsewhere on Oahu, most sightings have been at Lyon Arboretum (Pyle and Pyle 2017), and the species has been present there since 1991. However, as of 2018, it appears that all individuals present are White Cockatoo × Salmon-crested Cockatoo hybrids.

Over the nearly 50 years that several cockatoo species have been observed on Oahu and Hawaii Island, they have persisted in only two locations (Lyon Arboretum on Oahu and Kaloko Mauka on Hawaii), and they have shown only marginal recruitment and no range expansion. Unlike Australian cockatoo species, Moluccan forest cockatoos are quite sedentary. Lack of suitable large nest cavities, slow time to maturation (five years for the larger species), low fertility, poaching, and possibly lethal control to prevent crop depredation are all plausible factors for the low degree of persistence and population growth of these cockatoo species compared with some of the species to be discussed next.

Red-crowned Amazon
(*Amazona viridigenalis*)

The Red-crowned Amazon is native to the Atlantic slope of northeastern Mexico from eastern Nuevo Leon and Tamaulipas south to San Luis Potosí and northern Veracruz (Howell and Webb 1995). The species is listed as endangered by BirdLife International, based on the decline of its small population size, caused by trapping for the pet trade and habitat loss (BirdLife International 2018). Naturalized populations have become established in Puerto Rico, southern Florida, southern Texas, southern California, other areas of Mexico, and Oahu in the Hawaiian Islands (Garrett 1997; Lever 2005; Uehling et al. 2019). In Hawaii, separate from an established population on Oahu, individuals have been observed on Kauai and Maui (Table 13.1).

Two separate populations of Red-crowned Amazons were present on Oahu during the 1970s and 1980s: one in the Pearl City area of Central Oahu (where they still occur) and a smaller population in southeastern Honolulu, centered around Kapiolani Park. The small Kapiolani Park population was present from 1969 until 1987, after which it permanently joined the Pearl City population (Pyle and Pyle 2017). Details about the establishment and status of the two populations are chronicled below.

Three birds reported at the Honolulu Zoo and adjacent areas of Kapiolani Park in 1969 were the first free-flying Red-crowned Amazons observed on Oahu. The number of birds had increased to 14 by 1972 (Pyle and Pyle 2017). The flock continued to gradually increase in size, reaching a peak of at least 33 birds in 1986, and spread to nearby areas, observed east to Diamond Head and north to Makiki and Manoa Valleys (Pyle and Pyle 2017). A single Blue-crowned Parakeet and one or two Red-lored Amazons (*A. autumnalis*) were occasionally reported with this flock (Pyle and Pyle 2017).

A separate population of Red-crowned Amazons has been present in the Pearl City area of Oahu since at least 1982 (Pyle and Pyle 2017) and probably as early as 1972. Additional birds were introduced to the Pearl City population when Hurricane Iwa struck Oahu on 23–24 November 1982, and three or four pairs escaped from a mall aviary in Aiea. This Pearl City population continued to increase, with counts of 26–40 birds from 1987 to 1993 and 80–110 birds from 2001 to 2006 (Pyle and Pyle 2017).

The Kapiolani Park flock was gone in November 1986, and by May 1987 the same 33 birds, presumably, had relocated to the Pearl City area, still accompanied by a single Blue-crowned Parakeet (Kalodimos 2013b; Pyle and Pyle 2017). Separate from the Pearl City population, Red-crowned Amazons have occasionally been seen at other localities on Oahu over the past 30 years (Pyle and Pyle 2017).

The Red-crowned Amazon population in Central Oahu has been monitored and studied intensively by N. Kalodimos since 2004, including regular evening roost counts, visual tracking of birds to and from roost sites, and observation of foraging and nesting behavior (Kalodimos 2013b). This effort showed that the population has continued to increase, from 120–200 birds in 2003, to 550–600 in 2008–9. Since 2010, however, evening counts have dropped substantially, to 200–300 birds. There have been occasional reports that farmers were killing some birds, which may partially explain the lower numbers in recent years. The communal evening roost of this species is in tall stands of non-native *Eucalyptus* trees on the slopes of Manana and Waimano ridges above Pacific Palisades and Pearl City, at 200–300 m elevation (Kalodimos 2013b). During the day, the birds forage in disturbed and secondary forest in valleys and parkland with large trees in the island's central plateau. They only briefly visit residential areas, preferring to spend the majority of their time in forested gulches and valleys. Estimated maximum daily movement distances from the evening roost is up to 15.3 km, but typically is 5–6.5 km. Movements from ridgetop roosts are always downslope out into the central plateau or across the forested foothills and valleys to adjacent suburban areas. Movement patterns vary seasonally, depending on which plant species are fruiting, and dispersal distances tend to be shorter and flock sizes smaller during the nesting season, from April to July, when females stay closer to the nest (Kalodimos 2013b). Throughout this area the parrots commonly foraged on fruits of Java Plum (*Syzygium cumini*) and Mango (*Mangifera indica*) and the seeds of passion fruit (*Passiflora* spp.),

Moluccan Albizia, Formosan Koa (*Acacia confusa*), and Cook Pine (*Araucaria columnaris*).

A nest of this species in Aiea was located about 8 m above the ground, in a cavity where a limb had broken off a Moluccan Albizia tree (Kalodimos 2013b). Observations at the nest site showed that both parents entered and inspected the nest cavity, but only the female spent the night inside; that the male visited the nest site twice a day, an hour after sunrise and an hour before sunset, to provision the female outside the nest; and that both parents fed a nestling at the cavity entrance (Kalodimos 2013b). The male was larger than the female and had roughly twice the amount of red plumage on the crown.

Burrowing Parrot (*Cyanoliseus patagonus*)

Native to Chile and Argentina, where it occurs in arid grassy woodland and shrubland, watercourses, and towns (del Hoyo et al. 1997; Tella et al. 2014), the Burrowing Parrot has an incipient population on the island of Hawaii. Free-flying, banded Burrowing Parrots were first reported from the Kona Airport on the western side of Hawaii Island in 1990 and were thought to be of the Chilean subspecies, *bloxami* (Pyle and Pyle 2017). Flocks of up to 15 birds have been reported from 2015 to 2018, with one report of 20 individuals. Flocks have been reported south to Honaunau Bay, north to Hapuna Beach State Park, and inland to Puu Waa Waa. The habitat used by this species in Hawaii is generally similar to that in the native range, consisting of dry forest and shrubland and suburban areas, with non-native *Prosopis pallida* one of the most common trees. In its native range, the Burrowing Parrot nests primarily in burrows or crevices in rocky or earthen cliffs but also is known to use anthropogenic nesting sites (del Hoyo et al. 1997; Tella et al. 2014). In Hawaii, most or all of the birds roost, and presumably nest, in cavities on sea cliffs just south of the Kona Surf Hotel, and most reports on eBird have been centered around the hotel. If urban nesting behavior is adopted by birds in Hawaii, it could allow considerable range expansion. Of 15 birds seen in September 2017, three had dull plumage color with no red on the belly, characteristic of juveniles, and photographs did not show any birds with bands, demonstrating the species is reproducing. All photos of adults showed them to have the large red belly patch and more extensive white pectoral markings typical of the Chilean subspecies, *bloxami*.

Blue-crowned Parakeet
(*Thectocercus acuticaudatus*)

Widely distributed in South America, the Blue-crowned Parakeet has an incipient population on Oahu. During the 1980s, a single Blue-crowned Parakeet was observed regularly with a flock of Red-crowned Amazons around Kapiolani Park and Diamond Head. This flock relocated to the Pearl City area of Central Oahu in May 1987, and the single Blue-crowned Parakeet was used to help identify the flock and monitor its movements (Pyle and Pyle 2017). A single Blue-crowned Parakeet was again sighted in downtown Honolulu in March 1999, and one was reported paired and nesting with a Red-masked Parakeet and producing hybrids in Honolulu from 2004 to 2006 (Pyle and Pyle 2017).

More recently, up to 26 Blue-crowned Parakeets have been seen and photographed regularly around the Turtle Bay Resort on the northern coast of Oahu, starting in February 2012 and continuing through 2019 (eBird 2019). Three individuals were observed in the Pupukea-Paumalu Forest Reserve, about 8 km from Turtle Bay and 5 km from the coast, in May 2015, indicating the species may be spreading and moving into montane forested areas. One of the birds observed at Pupukea in 2015 had only a small amount of dull blue coloration on the head, which is characteristic of juvenile plumage of birds less than one year old. In October 2016, 26 Blue-crowned Parakeets were seen foraging on ripe Cook Pine seeds at the Turtle Bay Resort. Photos, video, and detailed observations of most of the birds revealed that they were the nominate subspecies, *T. a. acuticaudatus*, and that all were adults. In September 2019, a juvenile was observed foraging with adults on Cook Pine seeds, confirming successful wild reproduction. It is possible that Blue-crowned Parakeets exhibit the same daily movement patterns as Red-crowned Amazons on Oahu, roosting and nesting in upland areas and dispersing to forage during the day at lower elevations.

Mitred Parakeet
(*Psittacara mitratus*)

The Mitred Parakeet is native to the eastern Andes Mountains from central Peru south through central Bolivia and northern Argentina (Forshaw 1989; del Hoyo et al. 1997). In its native range, it uses a variety of tropical and subtropical habitats, including primary broadleaf forest, secondary forest, cloud forest, and high-elevation arid shrubland and grassland (Forshaw 1989; del Hoyo et al. 1997). Escaped birds and/or naturalized populations have been reported since at least the 1980s in Puerto Rico, Florida, several areas of California, and in Barcelona, Spain (Garrett 1997; Butler 2005; Lever 2005; Uehling et al. 2019).

In the Hawaiian Islands, Mitred Parakeets occur on northeastern Maui, a small area of eastern Oahu, and possibly the western side of Hawaii Island. The Maui population has been the subject of control efforts and currently is small. The status on Hawaii Island is not certain, because of difficulty in distinguishing this species from the similar Red-masked Parakeet, especially juveniles, and possible hybridization with Red-masked Parakeets.

The Mitred Parakeet population on Maui originated in 1986, when at least one pair escaped or was liberated from a private aviary in Huelo on the northern coast. Unless otherwise noted, the following observations and information on this species on Maui come from Waring (1998); Loope et al. (2001); Gassmann-Duvall (2002); Runde and Pitt (2008a, 2008b); and Radford and Penniman (2014). The population grew rapidly from about 30 birds by 1995, to 150–200 by 2002. The birds remained close (within 0.5 km) to the original release point, and roosted and nested in rock crevices and cavities about 30 m high on sea cliffs 50–60 m tall around Huelo Point. In 2002, two pairs split off and were observed nesting in a cave near Waipio Bay, 1.1 km west of the Huelo Point colony; this group had grown to 40 birds by 2006 and also roosted and nested on cliffs and in steep forested areas. Birds from the two groups often merged at dawn to forage in fruit and nut trees, primarily between Huelo and Hoolawa Valley, up to about 1,300 m elevation, and sporadically as far west as Peahi Gulch and east to the Keanae Peninsula.

Concern began to grow about possible negative impacts of Mitred Parakeets on Maui, including disturbance and competition with seabirds for nesting sites on the cliffs, seed predation and fruit and seed loss of rare native plants, damage to agricultural crops, and spread of seeds of noxious invasive plants, particularly of *Miconia calvescens* (Runde and Pitt 2008a, 2008b). Various methods of controlling the birds were investigated, including capture of live birds in mist nets or traps; using food, audio playback, and live parakeets as lures; and shooting. Control efforts for this species began in 2006, and at least 199 parrots were removed from 2006 to 2014, leaving approximately 20 birds in the wild in 2014 and about 15 birds at the end of 2018. Some of the collected individuals were sent as specimens to the Bishop Museum in Honolulu, where their plumage was determined to be consistent with the widespread subspecies *P. m. mitratus* (Pyle and Pyle 2017).

On Oahu, a small population of approximately 21 Mitred Parakeets was observed in the area around Kaaawa and Kahana Bay on the eastern side of the island until 2015–16. This population has not previously been reported, but no birds have been observed since 2016. Interviews with area residents indicated that the birds had been present in the wild since the late 1980s and that their numbers increased over time. In other areas of Oahu, sightings of single individuals or small flocks occurred in the late 1980s.

On Hawaii Island, Mitred Parakeets were first reported in July 1988 near Keaau on the eastern side of the island and in November 1989 above Kailua-Kona on the western side of the island (Pyle and Pyle 2017). From 1999 to 2006, a flock of 30–40 free-flying *Psitticara* parakeets was reported regularly in Kona and along the coast. These birds were identified by different observers as being either Mitred or Red-masked Parakeets or a mix of both species (Pyle and Pyle 2017). The status of this species on Hawaii Island remains uncertain.

Red-masked Parakeet
(*Psittacara erythrogenys*)

Red-masked Parakeets are native to the dry coastal zone of southwestern Ecuador and northwestern Peru (Forshaw 1989). The species is listed as "near threatened" by BirdLife International (2018), based on its small population size and declining population trend, which is being caused by trapping for the pet trade and habitat loss. Naturalized populations, or at least wild individuals, occur in Spain, California, southern Florida, and the Hawaiian Islands (Garrett 1997; Lever 2005). This species is considered to be established in the Hawaiian Islands by Pyle and Pyle (2017) and VanderWerf et al. (2018), but it is not yet included as an established species by the American Ornithological Society (AOU 1998; Chesser et al. 2018). Two naturalized populations of Red-masked Parakeets occur in Hawaii, one in southeastern Oahu and one on the western side of Hawaii Island, both of which were first identified in the late 1980s and have been increasing in size over time, indicating they are reproducing.

On Oahu, the first reports of wild Red-masked Parakeets were of a single bird that escaped from a theme park in upper Manoa Valley in 1987 (Pyle and Pyle 2017), along with another single bird at Pearl Harbor in January 1988 (Bremer 1988). Five individuals were observed in Kapiolani Park in December 1989, and the number of birds at that site had increased to 18 by December 1994 (Ord 1995). After this, the population continued to increase. Maximum flock sizes increased from 21 in 2001 to more than 55 by 2004 (Kalodimos 2008). In December 2019, 71 individuals were present at the evening roost, with four young of the year visible. During 2004–6, a female Red-masked Parakeet nested with a male Blue-crowned Parakeet and produced seven wild hybrids, and up to four hybrids were noted with the flock of Red-masked Parakeets in February 2010 (Kalodimos 2013a).

On Hawaii Island, a flock of 30–40 *Psitticara* parakeets observed in 1988–2003 around Kailua-Kona was identified by different observers as either Mitred or Red-masked or both species (see Mitred Parakeet). It is possible that most, if not all, reports of Mitred Parakeets in Kona involved juvenile Red-masked Parakeets. The number and range of Red-masked Parakeets in the Kona area of Hawaii has continued to increase since then, with numerous reports in eBird of up to 40 birds

A flock of Red-masked Parakeets (*Psittacara erythrogenys*) descends toward a food tree near the Diamond Head area. Honolulu, Hawaii, US, January 2020. Photo by Nicholas P. Kalodimos.

ranging from Makalawena Beach in the north to Puʻuhonua O Hōnaunau National Historical Park in the south. Occasional reports of Red-masked Parakeets from Maui in the 1990s–2000s were likely misidentified Mitred Parakeets (Pyle and Pyle 2017).

Rose-ringed Parakeet
(*Psittacula krameri*)

The Rose-ringed Parakeet is native to two geographically disjunct regions, sub-Saharan Africa north of the equator, and southern Asia from Pakistan east to India, Myanmar, and southeastern China (Forshaw 1989; del Hoyo et al. 1997). Four subspecies of Rose-ringed Parakeets are recognized, two of which occur in Africa and two in Asia (Forshaw 1989; del Hoyo et al. 1997). The Rose-ringed Parakeet is the most widely naturalized parrot species, with established populations in at least 35 countries throughout the world (Long 1981; Garrett 1997; Lever 2005; Avery and Shiels 2018).

In the Hawaiian Islands, Rose-ringed Parakeets are now established on Oahu and Kauai and have been observed on Maui and Hawaii Island. In Honolulu and elsewhere, the species has been seen since the 1930s and probably was present even earlier (Caum 1933; Munro 1944). Relatively large populations have become established on Kauai and Oahu, each numbering several thousand birds. No self-sustaining wild populations are present on Maui or Hawaii Island, but wild individuals have been observed on those islands for several decades, and breeding attempts have been documented (Paton et al. 1982). All birds observed at roosts on Oahu were of the Asian subspecies, *P. k. manillensis* or *P. k. borealis*, with none of the birds showing characteristics of the African subspecies.

On Kauai, Rose-ringed Parakeets reportedly were released in the 1960s near Lawai by workers at a bed-and-breakfast inn. By 1981 they were being recorded regularly foraging in nearby Hanapepe Valley and roosting nightly at Kukuiolono Park in Kalaheo (Pyle and Pyle 2017). By November 1982, counts reached 30–40 birds at Kukuiolono Park, and the population in Kalaheo and nearby areas was estimated to be 50 or more (Paton et al. 1982). In 1987, damage caused by the birds to corn, mango, and lychee crops in

Hanapepe Valley drew complaints from farmers, who obtained state permits to destroy the parakeets (Pyle and Pyle 2017). In early 1990, the flock abandoned the Kukuiolono Park roost and was reported more frequently a few kilometers to the east in the National Tropical Botanical Garden near Lawai and around Waita Reservoir near Koloa. In April 1994, a flock of 150–200 birds was reported, and in March 1997 the parakeets were well established from Hanapepe to Mahaulepu east of Koloa, with roosts in Omao, Kalaheo, and Lihue (Denny 1999). This population has continued to increase, despite control efforts that have removed 100–200 birds per year around agricultural fields (Avery and Shiels 2018). In the late 2000s, the population was estimated to be 500–1,000 birds (Gaudioso et al. 2012), and two roosts near Lihue and at Prince Kuhio Park near Poipu were estimated to contain approximately 1,000 birds each (Avery and Shiels 2018; Shiels et al. 2018). The total population in January 2018 was estimated to be >6,000 birds (A. Shiels, pers. comm.).

On Oahu, small groups of Rose-ringed Parakeets were first reported in several areas during the 1970s, including Waimanalo, Kapiolani Park, and elsewhere in Honolulu (Paton et al. 1982; Pyle and Pyle 2017), though they probably were present before then. The number of birds and their distribution increased steadily during the 1980s through the early 2000s (Fig. 13.1 and Fig. 13.3). The first known large roost site was located on the grounds of the Central Union Church in Honolulu, where 75 birds were observed roosting in 2000. The number of birds increased steadily through 2018, by which time there were an estimated 1,000–1,500 birds using this roost. Nesting on Oahu has been documented in cavities of large trees, often Monkeypod (*Samanea saman*), Australian Pine (*Casuarina equisetifolia*), and African Tulip (*Spathodea campanulata*), with concentrations in

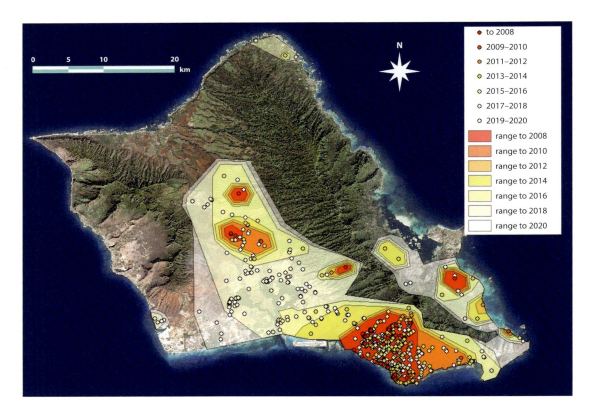

Figure 13.1. Distribution and range expansion of Rose-ringed Parakeets on Oahu, Hawaii, as indicated by reports in eBird (2019).

downtown Honolulu, Kapiolani Park, and Manoa Valley. The total population size on Oahu in 2018 was estimated to be approximately 3,000–4,000 birds, and the range area in 2020 was estimated to be 675 km^2.

On Maui, presumably escaped Rose-ringed Parakeets have been reported three separate times between 1988 and 2014. On Hawaii Island, Rose-ringed Parakeets were first noted at a nest near Keaau (Paton et al. 1982). Beginning in the early 1980s, single free-flying individuals appeared around Hilo and the Kona area, the latter increasing to a flock of 25–30 birds in 1988–90 and smaller numbers through 2016. Small populations also were noted around Kuamoo, Kalapana, and Kona south to Kealakekua, but they have been reported only occasionally near Kona (Pyle and Pyle 2017).

Rosy-faced Lovebird
(*Agapornis roseicollis*)

This species is native to semiarid woodland and watercourses in southwestern Africa from Angola to South Africa (del Hoyo et al. 1997). Naturalized populations occur in the continental US (Uehling et al. 2019) and in South Africa (Symes et al., chap. 17 this vol.). In Hawaii, lovebirds recently become numerous and appear to be established on Maui and incipient on Hawaii Island (Table 13.1). The status of this species in the Hawaiian Islands has not been formally evaluated, and currently it is included on the nonestablished list by Pyle and Pyle (2017) and VanderWerf et al. (2017), although this needs reevaluation given its recent and rapid population growth. Escaped birds are occasionally reported on Kauai and Oahu too, and given the birds' potential for rapid increase, the status on those islands should be monitored. This species is widely kept in captivity and is easily obtained in pet stores in Hawaii, and it is likely that additional birds escape but are not reported.

On Maui, Rosy-faced Lovebirds were first reported in 2005 in the Kihei area, where they increased rapidly, with 50+ documented by December 2012 (Pyle and Pyle 2017). Data from eBird indicate that the range on Maui is expanding rapidly, first along the Kihei coast, then up the western slopes of Haleakala, to western Maui, and the Kahului area (Fig. 13.2). Use of eBird has increased in recent years, and it is likely that eBird data underestimate the actual range at various times, but they are useful for illustrating the expansion (Fig. 13.3).

On Hawaii Island, a small population was first reported in the Waikoloa area in January 2015. Some of the earliest reports already involved up to 31 birds, and several nest sites were documented, so the population must have been present earlier than 2015. Data from eBird indicate the population is spreading rapidly, with reports in Kawaihae and the Kona area beginning in 2016, in Waikoloa Beach, Mauna Lani, and Saddle Road in 2018, and Kamuela in 2019. There are large expanses of dry forest seeing little or

A Rosy-faced Lovebird (*Agapornis roseicollis*) peering out from a nest under the metal roof of a building. Various color variations of this species occur in Hawaii. Waikoloa Village, Hawaii Island, Hawaii, US, September 2016. Photo by Eric A. VanderWerf.

STATUS OF NATURALIZED PARROTS IN THE HAWAIIAN ISLANDS

ABOVE: Figure 13.2. Distribution and range expansion of Rosy-faced Lovebirds on Maui, Hawaii, as indicated by reports in eBird (2019).

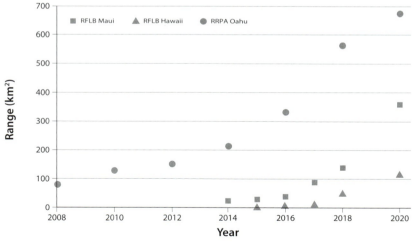

Figure 13.3. Range expansion of Rose-ringed Parakeet (RRPA) on Oahu and Rosy-faced Lovebird (RFLB) on Maui and Hawaii Island, as measured by range maps created with eBird (2019) data.

no little survey effort between areas where the species has been reported, and it is likely that the actual range is larger than eBird data indicate.

The areas on Maui and Hawaii Island where the species was first observed consist of dry, grassy woodland and shrubland, similar to the habitat in its native range in Africa. Nevertheless, the birds are increasingly using urban and suburban habitat, especially on Hawaii Island. On Maui they are also spreading to a wider variety of habitats, including mesic and wet forest at higher elevations. The range on each island is increasing at an accelerating rate, suggesting the ranges are likely to continue increasing until much of each island is occupied.

Elsewhere, on Oahu, there have been scattered reports of Rose-ringed Parakeets beginning in 1998 and continuing to 2016 (Pyle and Pyle 2017;

eBird 2019). On Kauai, there have been two reports of birds, one in 2010 and the second in 2016 (Pyle and Pyle 2017; eBird 2019).

ENVIRONMENTAL EFFECTS OF PARROTS IN HAWAII

Non-native parrots have the potential to cause serious environmental damage because of their relatively large body sizes, powerful bills, often destructive foraging behaviors, dispersal ability, and tendency to form large flocks (Runde et al. 2007; Menchetti and Mori 2014; Avery and Shiels 2018). In addition, parrots are more likely than most other groups of birds to be released into the wild and become naturalized because of their popularity in the pet industry and large international market (Menchetti and Mori 2014). These environmental impacts can be grouped into several categories: damage to human agricultural crops; damage to native plant species and ecosystems through predation on native plant seeds and dispersal of invasive alien plant seeds; competition with native birds and other animals for food or nest sites; and threats to human health and safety, including the spread of contagious diseases and collisions with aircraft. The environmental impacts of non-native parrot populations vary considerably, however, depending on the species involved, their diet, distribution, and abundance, and the nature of the environment where they occur. The impacts of non-native parrots can be particularly severe on oceanic islands that lack native parrots (Foster 2009).

In Hawaii, there has been concern about the environmental impacts of naturalized populations of two parrot species, the Mitred Parakeet on Maui and the Rose-ringed Parakeet on Kauai. Both species have been the target of control programs aimed at reducing or eliminating their populations (Radford and Penniman 2014; Avery and Shiels 2018; Shiels et al. 2018). Although Rose-ringed Parakeets also are common and widespread on Oahu, no control efforts have been undertaken on that island. The primary concerns in both cases have been damage to agricultural crops and dispersal of invasive alien plant seeds, but these and other parrot species in Hawaii have the potential to cause other types of damage.

Mitred Parakeets on Maui have been regarded as an environmental problem for several reasons, including predation on fruit and seeds of rare native plants, spread of noxious invasive plant seeds, damage to agricultural crops, competition with native seabirds for nest sites on cliffs, and noise pollution (Loope et al. 2001; Runde and Pitt 2008a, 2008b; Radford and Penniman 2014). In particular, a population of *Miconia calvescens*, a highly invasive plant in the melostome family (Medeiros et al. 1997), was found near a Mitred Parakeet roost, and another population was found far from any other known populations of the plant but along the flight path of the Mitred Parakeets. It was suspected that the parakeets may have dispersed seeds to that location and could introduce them to other new areas (Runde and Pitt 2008a, 2008b; Radford and Penniman 2014). Various methods of controlling the birds were investigated by the US Department of Agriculture's National Wildlife Research Center, including capture of live birds in mist nets or traps; using food, audio playback, and live parakeets as lures; and shooting (Runde and Pitt 2008a, 2008b; Radford and Penniman 2014). Efforts by the Maui Invasive Species Committee and the Hawaii Department of Land and Natural Resources to completely remove the population began in 2006 (Radford and Penniman 2014), with shooting as the only removal method, because capturing live birds was unsuccessful and impractical (Runde and Pitt 2008b). A total of 187 parrots were removed during 90 days of work from 2006 to 2014, though there was a gap in control efforts in 2010–11, and 12 additional birds were removed by local residents who had obtained a wildlife damage control permit (Radford and Penniman 2014). Control efforts greatly reduced the population, but complete removal was not achieved, and approximately 20 birds remained in 2014 (Radford and Penniman 2014) and about 15 birds at the end of 2018.

In their native range, Rose-ringed Parakeets inhabit a variety of habitats including forest, savanna, shrubland, grassland, parks, gardens, orchards, and other agricultural areas, where they feed on seeds, fruit, flowers, and nectar of a wide range of plant species (Forshaw 1989; del Hoyo et al. 1997). Because of their broad diet, adaptability, and potential to form large flocks, Rose-ringed Parakeets are considered a serious agricultural pest in many areas, including countries in their native range and areas where they have become naturalized (Paton et al. 1982; Forshaw 1989; Menchetti et al. 2016; Shiels et al. 2018). Because of their cavity-nesting behavior, they also pose a threat to native birds and bats in some regions (Strubbe and Matthysen 2007). The state of Hawaii officially has declared the Rose-ringed Parakeet to be a pest species and an injurious wildlife species and has allocated funding to study and control its population growth and range expansion on Kauai (Shiels et al. 2018). Concern about the Rose-ringed Parakeet in Hawaii has been related to damage of agricultural crops, primarily corn (*Zea mays*); the spread of invasive alien plant seeds; and collisions with aircraft (Avery and Shiels 2018; Shiels et al. 2018). In a diet study based on gut contents of collected specimens, Shiels et al. (2018) found that Rose-ringed Parakeets on Kauai had a diverse diet that varied among locations, with corn present in 67% of birds and representing 31% by mass of the diet, demonstrating the potential for damage to crops. In addition, Common Guava (*Psidium guajava*) was eaten by 97% of parakeets and composed 30% of their diet, demonstrating the birds' potential to disperse seeds of this highly invasive tree. Rose-ringed Parakeets feed on corn kernels just before the harvest stage and may also sample corn when it is ripening by clipping the tassel (Gaudioso et al. 2012). Rose-ringed Parakeets on Kauai had much larger home ranges than those naturalized in Belgium (1,771 ha vs. 75–86 ha, on average; Avery and Shiels 2018), with dispersal distances more similar to those of birds in their native range in India, suggesting they could disperse invasive plant seeds over long distances. One of the large Rose-ringed Parakeet roosts on Kauai is located near the Lihue Airport, where the large flocks pose a collision hazard to aircraft. There have not been any documented strikes with Rose-ringed Parakeets at the Lihue Airport, but the birds are hazed from the area and occasionally removed (Avery and Shiels 2018).

Naturalized parrots are not known to be impacting any native birds or other animal or plant species in the Hawaiian Islands, but they have the potential to affect endemic and endangered native species in several ways. The Mitred Parakeet and Burrowing Parakeet, both of which nest and roost in rock cavities and crevices on coastal cliffs, potentially could compete for nest sites with native seabirds that also nest on rocky cliffs, including the Hawaiian Black Noddy (*Anous minutus melanogenys*) and the Red-tailed Tropicbird (*Phaethon rubricauda*) (VanderWerf and Young 2014), but no such competition has been documented. Competition with native land bird species for nest sites in tree cavities has not been an issue in Hawaii, because all native tree-cavity-nesting bird species in Hawaii are either extinct or restricted to high-elevation forests where invasive parrots have not yet spread (Scott et al. 2001). Foraging by parrots potentially could damage important native trees and reduce food availability for native land birds, though this has not been observed yet. All naturalized parrots in Hawaii depredate the seedpods of non-native legume trees (e.g., *Prosopis pallida*, *Pithecellobium dulce*, *Caesalpinia pulcherrima*, *Acacia formosa*, *Falcataria moluccana*); if parrots spread to areas of native forest, they would be likely to consume the legumes of the endemic Koa (*Acacia koa*) and Mamane (*Sophora chrysophylla*) trees, the latter of which is the primary food of the endangered Palila (*Loxioides bailleui*; Scott et al. 2001). Red-masked Parakeets in western Hawaii Island forage on Ohia (*Metrosideros polymorpha*) flowers and seed capsules during their daily movements between montane roosts and daytime foraging locations at sea level. Ohia is a keystone endemic tree species and the primary food source for several nectarivorous Hawaiian honeycreepers, including the Iiwi (*Drepanis coccinea*), which was recently listed under the US Endangered Species Act.

PRIORITIES FOR FUTURE RESEARCH AND MONITORING

There is awareness and concern about the potential environmental impacts of non-native parrots on native species and habitats in Hawaii, but more information is needed to adequately assess the status and impacts of several parrot species. Research about parrots in Hawaii is needed in the following areas: (1) the fate of native and invasive plant seeds eaten by different parrot species and the role of parrots as seed predators vs. dispersers; (2) movement patterns and the degree to which different parrot species could disperse viable seeds; (3) the potential for destructive foraging by parrots to result in dispersal of and infection by the fungi that cause rapid Ohia death, an emerging infectious disease that is threatening this keystone tree species on several islands (Fortini et al. 2019); and (4) the degree to which parrots serve as disease reservoirs, especially of mosquito-transmitted avian malaria (*Plasmodium relictum*) and avian poxvirus (*Poxvirus avium*), which have decimated the endemic Hawaiian avifauna (Scott et al. 2001; VanderWerf et al. 2006; Atkinson and LaPointe 2009).

Rose-ringed Parakeets are by far the most numerous parrot species in Hawaii, and they have become an agricultural pest on Kauai; control efforts have reduced their numbers locally, but the birds have responded by shifting roost sites, and their overall numbers have not declined. Research is needed into control methods effective at larger scale. On Oahu, Rose-ringed Parakeets occur primarily in the southern part of the island, away from most agricultural areas, but they appear to be spreading northward. More information is needed about their movements and potential for damaging agricultural crops on Oahu. Rosy-faced Lovebirds have spread rapidly on Maui and are incipient on Hawaii Island. There is little information about the diet and potential environmental impacts of lovebirds, in part because they rarely have established non-native populations. This species has spread to areas of native forest that are important to native forest birds on Maui; more information is needed about the birds' potential to compete with native forest birds for food and to disperse invasive plant seeds. The Blue-crowned Parakeet appears to be increasing on Oahu and spreading into montane areas, but almost all observations are from a single location, the Turtle Bay Resort. Greater monitoring effort is needed to assess the status of this species.

REFERENCES

AOU. 1998. *Check-list of North American Birds*. 7th ed. Washington, DC: American Ornithologists' Union.

Atkinson, C. T., and LaPointe, D. A. 2009. Ecology and pathogenicity of avian malaria pox. In *Conservation Biology of Hawaiian Forest Birds: Implications for Island Avifauna*, ed. T. K. Pratt, C. T. Atkinson, P. C. Banko, J. D. Jacobi, and B. L. Woodworth, 234–252. London and New Haven, CT: Yale Univ. Pres.

Avery, M. L., and Shiels, A. B. 2018. Monk and Rose-ringed parakeets. In *Ecology and Management of Terrestrial Vertebrate Invasive Species in the United States*, ed. W. C. Pitt, J. C. Beasley, and G. W. Witmer, 333–357. Boca Raton, FL: CRC Press.

BirdLife International. 2018. *Psittacara erythrogenys*. IUCN Red List of Threatened Species 2018:e.T22685672A132059066. Accessed 03 Jan. 2019. http://dx.doi.org/10.2305/IUCN.UK.2018-2.RLTS.T22685672A132059066.en.

Bremer, D. 1988. Waipio, Oahu, Christmas Bird Count, 1987. '*Elepaio* 48:39–40.

Butler, C. J. 2005. Feral parrots in the continental United States and United Kingdom: Past, present, and future. *Journal of Avian Medicine and Surgery* 19:142–149.

Caum, E. L. 1933. The exotic birds of Hawaii. *Occasional Papers of the B. P. Bishop Museum* 10:1–55.

Chesser, R. T., Burns, K. J., Cicero, C., Dunn, J. L., Kratter, A. W., Lovette, I. J., Rasmussen, P. C., Remsen, J. V., Jr., Rising, J. D., Stotz, D. F., Winger, B. M., and Winker, K. 2018. Fifty-ninth supplement to the American Ornithologist Society's *Check-list of North American Birds*. *Auk: Ornithological Advances* 135:798–813.

del Hoyo, J., Elliott A., and Sargatal, J., eds. 1997. *Handbook of the Birds of the World*. Vol. 4: *Sandgrouse to Cuckoos*. Barcelona, Spain: Lynx Edicions; Cambridge, UK: BirdLife International.

Denny, J. 1999. *The Birds of Kauai*. Honolulu: Univ. of Hawaii Press.

eBird. 2019. eBird Basic Dataset (v.EBD_relNov-2017). Cornell Lab of Ornithology, Ithaca, NY. Accessed June 2019. https://www.ebird.org.

Forshaw, J. M. 1989. *Parrots of the World*. 3rd (rev.) ed. Melbourne, Australia: Lansdowne Editions.

Fortini, L. B., Kaiser, L. R., Keith, L. M., Price, J., Hughes, R. F., Jacobi, J. D., and Friday, J. B. 2019. The evolving threat of rapid 'Ōhi'a death (ROD) to Hawai'i's native ecosystems and rare plant species. *Forest Ecology and Management* 448:376–385.

Foster, J. T. 2009. The history and impact of introduced birds. In *Conservation Biology of Hawaiian Forest Birds: Implications for Island Avifauna*, ed. T. K. Pratt, C. T Atkinson, P. C. Banko, J. D. Jacobi, and B. L. Woodworth, 312–330. New Haven, CT: Yale Univ. Press.

Garrett, K. L. 1997. Population status and distribution of naturalized parrots in southern California. *Western Birds* 28:181–195.

Gassmann-Duvall, R. 2002. Naturalized mitred conures on Maui and other parrots: Pest evaluation and management proposal for non-native, invasive bird species. Unpubl. report. Makawao, HI: Maui Invasive Species Committee.

Gaudioso, J. M., Shiels, A. B., Pitt, W. C., and Bukowski, W. P. 2012. Rose-ringed parakeet impacts on Hawaii's seed crops on the island of Kauai: Population estimate and monitoring of movements using radio telemetry. Unpubl. report, QA 1874. Hilo, HI: USDA National Wildlife Research Center.

Howell, S.N.G., and Webb, S. 1995. *A Guide to the Birds of Mexico and Northern Central America*. Oxford: Oxford Univ. Press.

Kalodimos N. 2008. Determination of Movement and Foraging Patterns of Aratinga erythrogenys (Aves: Psittacidae) on O'ahu, Hawai'i, Using Mist-Net Live Capture and Radio Telemetry. Technical report. Honolulu: Hawaii Division of Forestry and Wildlife, Dept. of Land and Natural Resources.

Kalodimos, N. P. 2013a. Red-masked parakeet (*Psittacara erythrogenys*) (v.1.0.) In *Neotropical Birds Online*, ed. T. S. Schulenberg. Ithaca, NY: Cornell Lab of Ornithology. https://birdsoftheworld.org/bow/historic/nb/rempar/1.0/overview.

Kalodimos, N. P. 2013b. The status and comparative nesting phenology of the red-crowned amazon on O'ahu, Hawai'i. *'Elepaio* 73:1–3.

Lever, C. 2005. *Naturalised Birds of the World*. London: T. & A. D. Poyser.

Long, J. L. 1981. *Introduced Birds of the World*. New York: Universe Books.

Loope, L. L., Howarth, F. G., Kraus, F., and Pratt, T. K. 2001. Newly emergent and future threats of alien species to Pacific birds and ecosystems. *Studies in Avian Biology* 22:291–304.

Medeiros, A. C., Loope, L. L., Conant, P., and McElvaney, S. 1997. Status, ecology, and management of the invasive plant Miconia calvescens DC (Melastomataceae) in the Hawaiin Islands. Records of the Hawaii Biological Survey for 1996. *Bishop Museum Occasional Papers* 48:23-36.

Menchetti, M., and Mori, E. 2014. Worldwide impact of alien parrots (Aves Psittaciformes) on native biodiversity and environment: A review. *Ethology Ecology & Evolution* 26:172–194.

Menchetti, M., Mori, E., and Angelici, F. M. 2016. Effects of the recent world invasion by ring-necked parakeets *Psittacula krameri*. In *Problematic Wildlife: A Cross-Disciplinary Approach*, ed. F. M. Angelici, 253–266. Switzerland and New York: Springer International Publishing.

Munro, G. C. 1944. *The Birds of Hawaii*. Honolulu: Tongg Publishing.

Ord, M. 1995. Red-masked conure (*Aratinga erythrogenys*). *'Elepaio* 55:13.

Paton, P.W.C., Griffin, C. R., and MacIvor, L. H. 1982. Rose-ringed parakeets nesting in Hawai'i: A potential agricultural threat. *'Elepaio* 43:37–39.

Pyle, L. 1984. Lyon Arboretum field trip report, June 1983. *'Elepaio* 44:103.

Pyle, R. L., and Pyle, P. 2017. *The Birds of the Hawaiian Islands: Occurrence, History, Distribution, and Status* (v.2). Honolulu: B. P. Bishop Museum. Accessed Dec. 2019. http://hbs.bishopmuseum.org/birds/rlp-monograph.

Radford, A., and Penniman, T. 2014. Mitred Conure Control on Maui. In *Proceedings of the 26th Vertebrate Pest Conference*, ed. R. M. Timm and J. M. O'Brien, 61–66. Davis: Univ. of California.

Runde, D. E., and Pitt, W. C. 2008a. Maui's mitred parakeets (*Aratinga mitrata*), pt. 1. *'Elepaio* 68:1–4.

Runde, D. E., and Pitt, W. C. 2008b. Maui's mitred parakeets (*Aratinga mitrata*), pt. 2. *'Elepaio* 68:16–17.

Runde, D. E., Pitt, W. C., and Foster, J. 2007. Population ecology and some potential impacts of emerging populations of exotic parrots. In *Managing Vertebrate Invasive Species: Proceedings of an International Symposium*, ed. G. W. Witmer, W. C. Pitt, and K. A. Fagerstone, 338–360. Fort Collins, CO: USDA/APHIS Wildlife Services, National Wildlife Research Center.

Scott, J. M., Conant, S., and van Riper, C. eds. 2001. Evolution, ecology, and management of Hawaiian birds: A vanishing avifauna. *Studies in Avian Biology* 22:1–428.

Shiels, A. B., Bukoski, W. P., and Siers, S. R. 2018. Diets of Kauai's invasive rose-ringed parakeet (*Psittacula krameri*): Evidence of seed predation and dispersal in a human-altered landscape. *Biological Invasions* 20:1449–1457.

Strubbe, D., and Matthysen, E. 2007. Invasive ring-necked parakeets *Psittacula krameri* in Belgium: Habitat selection and impact on native birds. *Ecography* 30:578–588.

Tella, J. L., Canale, A., Carrete, M., Petracci, P., and S.M. Zalba, P. 2014. Anthropogenic nesting sites allow urban breeding in burrowing parrots (*Cyanoliseus patagonus*). *Ardeola* 61.

Uehling, J. J., Tallant, J., and Pruett-Jones, S. 2019. Status of naturalized parrots in the United States. *Journal of Ornithology* 160:907–921.

VanderWerf, E. A., Burt, M. D., Rohrer, J. L., and Mosher, S. M. 2006. Distribution and prevalence of mosquito-borne diseases in O'ahu *'Elepaio*. *Condor* 108:770–777.

VanderWerf, E. A., David, R. E., Donaldson, P., May, R., Pratt, H. D., Pyle, P., and Tanino, L. 2017. Hawaiian Islands bird checklist, 2017. *'Elepaio* 77:33–42.

VanderWerf, E. A., David, R. E., Donaldson, P., May, R., Pratt, H. D., Pyle, P., and Tanino, L. 2018. First report of the Hawaii birds records committee. *Western Birds* 49:2–23.

VanderWerf, E. A., and Young, L. C. 2014. Breeding biology of red-tailed tropicbirds *Phaethon rubricauda* and response to predator control on O'ahu, Hawai'i. *Marine Ornithology* 42:73–76.

Waring, G. H. 1998. Free-ranging parrot population of Haiku District, Maui, Hawaii. USGS/BRDPIERC. Unpubl. report. Maui, HI: Haleakala Field Station.

14

INTRODUCED AND NATURALIZED PARROTS IN EUROPE

Michael P. Braun

INTRODUCTION

Able to mimic the human voice and form social, long-lived bonds with humans, and relatively easy to keep in aviculture, parrots are among the most widespread pets in the world. The first parrots to arrive in Europe were presumably Rose-ringed Parakeets (*Psittacula krameri*) from the Indian subcontinent, more than 2,300 years ago, during the times of Alexander the Great. Until modern history, however, Rose-ringed Parakeets did not establish breeding populations in Europe. With the advent of acclimatization societies, which encouraged the introduction of non-native species in different places in the world, this situation changed in the 19th and 20th centuries. There have been several attempts to introduce parrots to European wild fauna by aviculturists since the mid-19th century, especially in the United Kingdom (UK), Germany, and Austria. One of the most intensive attempts was undertaken by the Duke of Bedford in England (especially at his Endsleigh and Woburn estates), from the 1910s to the 1950s (Bedford 1954). The Duke of Bedford kept and bred several parrot species at liberty and tried to acclimatize the Budgerigar (*Melopsittacus undulatus*) in England in the first half of the 20th century. From the 19th century, several attempts were made to introduce Monk Parakeets (*Myiopsitta monachus*) to urban zoological and botanical gardens in many European countries (Niethammer 1963). None of those intentional releases resulted in a permanent success, presumably due to the common use of shotguns in the 19th and early 20th centuries, when unknown exotic birds were frequently killed by hunters (Bedford 1954; Niethammer 1963). The most remarkable introduction was of the now-extinct Carolina Parakeet (*Conuropsis carolinensis*) by the founder of "applied bird conservation," Hans Freiherr von Berlepsch, in Seebach, Thuringia, Germany (Kolar 1979).

After the colonialist period and two world wars, attitudes toward the general use of weapons changed, at least in Western Europe, ushering in a more stable and peaceful era. The import of wild birds still continued, but hunting of exotic birds in the wild slowed, and conservation laws became more prominent. Predators like Peregrine Falcons (*Falco peregrinus*) and Northern Goshawks (*Accipiter gentilis*) had become rare, victims to the effects of the insecticide DDT. This situation favored the establishment of several parrot species in Europe, and since the 1960s a number of species have established new populations.

Since the European Union (EU) instituted a ban of the wild bird trade in 2005, the number of imported wild birds has dropped by 90%. This has reduced the likelihood that newly escaped species will establish breeding populations but nevertheless has not slowed the population expansion of already established species (Reino et al. 2017). Indeed, today, introduced parrots are breeding in at least 11 European countries (Portugal, Spain, France, UK, Belgium, the Netherlands, Germany, Italy, Romania, Greece, Turkey). Among these are upwards of 90,000 Rose-ringed Parakeets and more than 20,000 Monk Parakeets (Pârâu et al. 2016; Braun et al. 2017; Braun et al. 2019b; Postigo et al. 2019).

In this chapter I review the introduction and establishment of naturalized parrots in Europe and discuss factors related to their continued success.

SPECIES ACCOUNTS

No fewer than 48 species of parrots have been introduced or made at least one breeding attempt in Europe. Of those, 23 species (presented here in taxonomic order) are known to have bred in Europe in recent years or currently have established populations. Please note that in the species accounts below, any unreferenced observations refer to personal observations by the author or communicated to him by O. Arnoult, N. Braun, L. Evans, D. Franz, M. Grimminger, M. Hows, R. Jonker, G. Innemee, T. Pittman, T. Krause, J. Martens, M. Schmolz, M. Teissier, or R. Zamora.

Cockatiel
(*Nymphicus hollandicus*)

The Duke of Bedford bred this small Australian cockatoo species at liberty in England after World War II, but the population did not persist (Bedford 1954). In recent years, breeding pairs have been observed in France (Dubois 2007; Dubois et al. 2016). In 2006, a small breeding population of about 10 birds was known in Puerto de la Cruz, Tenerife, Canary Islands, Spain. The Cockatiel is listed as an occasional breeding bird in Germany (Bauer and Woog 2008).

Sulphur-crested Cockatoo
(*Cacatua galerita*)

Prior to 1953, this Australian cockatoo was kept at liberty by the Duke of Bedford in the UK (Bedford 1954). In Vienna, Austria, escaped Sulphur-crested Cockatoos from the Tiergarten bred successfully and raised two chicks on an island in the River Danube. Since 1989, free-flying cockatoos, including the Australian Sulphur-crested Cockatoo, have been observed in St.-Jean-Cap-Ferrat in southern France. These birds probably originated from escapees from a former local zoo. Since 2008, a Sulphur-crested and Yellow-crested Cockatoo pair has been regularly observed in a private garden nearby. This population consisted of five to seven birds in 2010.

Yellow-crested Cockatoo
(*Cacatua sulphurea*)

Since 1989, free-flying cockatoos, including the Indonesian Yellow-crested Cockatoo, have been observed in St.-Jean-Cap-Ferrat, southern France. In 2015, two to six birds were still observed, and the last known brood was in 2012 (Dubois et al. 2016).

Senegal Parrot
(*Poicephalus senegalus*)

This African parrot species has been imported in large numbers into Europe. In 1982, a pair raised three chicks in a tree at a height of 15 m, next to a Rose-ringed Parakeet nest cavity, in Wiesbaden-Biebrich, Germany. Another bird was seen at this site in 1989–90 (Zingel 1993). In 2006, a pair bred in Ilion, Attica, Greece (Latsoudis 2007). In 1991, 2003–4, and 2005 breeding was observed on Tenerife, Canary Islands. Further probable breeding occurred on Grand Canary Island, and in Valencia on the Spanish mainland in 2006 (Santos 2008d).

Monk Parakeet
(*Myiopsitta monachus*)

The South American Monk Parakeet, or Quaker Parrot, is the only parrot species that builds large stick nests. In Europe, the first introductions of Monk Parakeets occurred as early as the late 19th century, and birds were successfully breeding in the wild shortly after release (Niethammer 1963). Throughout the 20th century, Monk Parakeets were, at various times, released in Australia, Switzerland, Germany, Italy, and Spain (Niethammer 1963). While many of these intentionally released populations did not persist, many did, especially those in southern Europe.

The most successful establishment of Monk Parakeets has been in Spain, as discussed in detail by Carrete et al. (chap. 15 this vol.). In 1975, the first birds were found in Barcelona; the population had increased to 44 birds by 1983. There are also important populations in Madrid and the Canary Islands (Batllori and Nos 1985; Santos 2008c). Since 1979, there has been a breeding population of between 200 and 250 Monk Parakeets in Brussels, Belgium (Weiserbs and Paquet 2016). This population is one of the most northerly of this species in the world. The fact that numbers remain low in Brussels suggests there is some limiting factor, probably linked to climatic conditions (Weiserbs and Paquet 2016). A recent study estimated there are over 23,000 Monk Parakeets in the wild in Europe. They breed in eight European countries—Spain, Portugal, France, UK, Belgium, the Netherlands, Italy, and Greece—exhibiting exponential growth in Mediterranean countries but not in Atlantic countries (Postigo et al. 2019).

Yellow-headed Amazon
(*Amazona oratrix*)

The only established breeding population of this Mexican species outside the Americas is in Germany. Individuals were seen in 1984 and 1985, and the first brood, with three fledglings, occurred in 1986 (Hoppe 1997). By 2003 this population had increased to at least 46 birds (Mahler 2001; Braun 2004). Since 2004, mixed broods of Yellow-headed and Turquoise-fronted Amazons with fertile offspring have occurred in Stuttgart. The population, which breeds mainly in cavities of exotic trees like London Plane Tree (*Platanus × hispanica*), did not significantly increase during the next few years (Martens and Woog 2017). By 2019, the population was 50–60 birds.

Turquoise-fronted Amazon
(*Amazona aestiva*)

This South American species is often kept as a companion bird. During the 1890s, there were occasional records and a successful breeding of this species in Switzerland (Niethammer 1963). From 1991 to 1999, broods of a Turquoise-fronted Amazon and Orange-winged Amazon (*A. amazonica*) hybrid pair occurred in Hürth, Germany (Braun 2004), and a similar hybrid pair produced offspring in Wiesbaden, Germany, in 1998–2003 (Zingel 2000; Braun 2004). Similarly, since 2004, mixed broods of Turquoise-fronted and Yellow-headed Amazons with fertile offspring have occurred in Stuttgart, Germany. In 1999, the Turquoise-fronted Amazon bred in Valencia, Spain, and it is frequently observed in the wild in that country (Sánchez 2005). In 1999, a presumably breeding population of 20 birds was found in Genoa, Italy (Borgo 1999).

Burrowing Parrot
(*Cyanoliseus patagonus*)

The Burrowing Parrot has been a commonly traded South American species, known from extremely large colonies in cliffs of El Cóndor, Patagonia, Argentina (Masello and Quillfeldt 2002, 2004). From 1990 to 2007, this species was observed in many locations in the Canary Islands, Mallorca, Portugal, and mainland Spain (Costa and Schäffer 2000; Santos 2008b).

Nanday Parakeet
(*Aratinga nenday*)

The Nanday Parakeet is a South American species that was first recorded breeding in Europe in 1960s, although that initial population did not persist (Niethammer 1963). Beginning in 1989 and continuing to the present, this species is occasionally observed in the wild in Spain and has been recorded as breeding in numerous locations, including Andalusia, Catalonia, and the Canary Islands (García del Rey 2007; Carrete et al., chap. 15 this vol.). Elsewhere in Spain, the species is occasionally seen, but those individuals appear to be new escapes or releases (Santos 2006c).

Blue-winged Macaw
(*Primolius maracana*)

The Blue-winged Macaw is a South American species, whose introduction into Europe is uncertain with respect to the exact dates. This species has been observed breeding in public parks in 2003 and 2004 on Tenerife, Canary Islands.

Scarlet Macaw
(*Ara macao cyanopterus*)

This well-known species has bred several times in the Netherlands since the late 1980s and early 2000s. The maximum number of birds counted was seven in Haarlem near Amsterdam. One pair was still present into the 2000s and bred successfully from 2005 to 2007 (Jonker and Tamis 2012). In 2010, the last female died during a poaching event, and only one male was left. In southern England, a bird park kept free-flying parrots for nearly 30 years, from 1970 until 2001, including two to three pairs of Scarlet Macaw, which returned every night into their aviaries.

Blue-crowned Parakeet
(*Thectocercus acuticaudatus*)

This South American species was first observed in Europe in the 1990s. In 1990, nine birds were observed in Barcelona, Spain, and this population increased to 100 by 2006 (Santos 2006d). The species was first observed in the wild in the UK in 1997, in Lewisham, Kent, and by 1995 at least 15 birds were recorded there. This population then began to decline, and by 2001, no successful reproduction was recorded (Butler et al. 2002). Although occasional birds were seen up to 2008, by 2009 the species was regarded as extinct in the UK. In 2005, there was a successful nesting of this species in Gibraltar (Cortes 2006) and observations from the Canary Islands and Portugal (Costa and Schäffer 2000).

Mitred Parakeet
(*Psittacara mitratus*)

This South American species has been observed in Europe since the 1980s. Five birds were detected in Barcelona in 1991, and the species has been confirmed to be breeding there since 1996. Birds were seen in Palma de Mallorca, Spain, in 1992, and the first brood in Mallorca was found in 1993. By 2006, the population in Catalonia had increased to more than 100 birds. Records are also known from other locations in Spain (Santos 2006b).

Red-masked Parakeet
(*Psittacara erythrogenys*)

This South American species has been observed in Europe since the 1980s. By 1989, the first birds had been detected in Valencia, and the species has bred in Barcelona since 1993. By 2006, the population had increased to more than 30 birds in Valencia and 10 birds in Barcelona (Santos 2006a).

Alexandrine Parakeet
(*Psittacula eupatria*)

This South Asian species has been observed in Europe since the late 1980s, and it is often found in association with the Rose-ringed Parakeet. It was first recorded in 1987 in Wiesbaden-Biebrich, Germany, and in the following year, 1988, two pairs successfully produced offspring (Zingel 2000). This population increased to nine pairs by 1993 and to 23 pairs by 2000 (Zingel 2000). Separately, a different population established itself in Cologne, Germany, in 1993–94 (Ernst

Alexandrine Parakeets (*Psittacula eupatria*) nesting in a London Plane Tree in Wiesbaden State Park, Germany, May 2008. Photo by Roelant Jonker.

1995). Collectively, the population in Germany was at least 500 birds by 2016 and at least 1,100 birds by 2019 (Braun et al. 2018, 2019b).

The species has been recorded in Brussels since 1998, and in 2000, nine nesting pairs were recorded. In Brussels, the individuals are integrated into the larger population of Rose-ringed Parakeets, sharing nesting sites, feeding areas, and roosting places (Weiserbs 2000).

Elsewhere, successful breeding has been observed in Rome, Italy (Angelici and Fiorillo 2016), and in Istanbul, Turkey, where the population appears to be increasing (Şahin and Arslangündoğdu 2019). In the UK, Alexandrine Parakeets have bred since the late 1990s, both in Greater London and elsewhere (Butler 2002; Ogilvie 2003; Lever 2005).

Rose-ringed Parakeet
(*Psittacula krameri*)

This common South Asian parrot species is now the most widespread, northernmost-breeding, and successful parrot species in Europe and the world (Jackson, chap. 10 this vol.). Rose-ringed Parakeets in Europe were first recorded in 1855 in Norfolk, England (Lever 2005; also see Butler, chap. 16 this vol.). Since 1966, this species has become established throughout Europe (years in parentheses represent first broods per country): Belgium (1966), Netherlands (1968), UK (1969), Germany (1969), France (1974), Italy (1970s), Spain (1982), Portugal (1986), Greece (1998), and Turkey (1997) (Braun 2009). In 2019, the total European population was estimated at >90,000 individuals (Pârâu et al. 2016; Braun et al. 2019b). It should be mentioned that all breeding populations in Europe are of Asian origin, a result, in part, of the fact that the African Rose-ringed Parakeet is a distinct species. A genetic revision of *Psittacula* revealed that both the species *P. krameri* and the genus *Psittacula* are paraphyletic (Braun et al. 2016; Braun et al. 2019a).

Red-rumped Parrot
(*Psephotus haematonotus*)

The Duke of Bedford bred this Australian species at complete liberty in England prior to World War I (Bedford 1954). Butler (chap. 16 this vol.) refers to a pair with chicks in the UK in 1998. In Germany, a pair was recorded nesting in a nest box in 2000.

Budgerigar
(*Melopsittacus undulatus*)

The Australian Budgerigar is one of the most abundant birds in aviculture. In Europe, there have been repeated intentional releases of Budgerigars since the late 1800s, and escaped pet birds are regularly seen (Russ 1926; Niethammer 1963). Some of these released birds have bred in the wild, but most perish during the winter months. In the 1970s, Budgerigars were introduced to the Isles of Scilly, UK, and a breeding population persisted and increased in size, reaching more than 100 birds by 1975. Despite this early success, however, the entire population appears to have died out during a particularly cold winter in 1976–77. In 1996, a pair bred in Würzburg, Bavaria, Germany (Fünfstück et al. 2003), which appears to represent one of the last breeding records in Europe.

Grey-headed Lovebird
(*Agapornis canus*)

This Madagascan species is quite rare in captivity. In the early 1900s, it was kept at liberty on the Woburn estate, Bedfordshire, UK, and the birds stayed on the grounds and raised young in the wild (Bedford 1954). In 2006, on Tenerife, there was a pair at liberty, but breeding was not confirmed.

Rosy-faced Lovebird
(*Agapornis roseicollis*)

This African species is frequently kept in captivity. The Duke of Bedford bred this species at liberty in England after World War I (Bedford 1954). In 2002, the first successful breeding in the wild in the UK was registered in Dunbar, Scotland, where approximately five birds were recorded (Lever 2009). In the same year, a pair of birds bred in Bavaria, Germany (Fünfstück et al. 2003). By 2014–15, several groups had also been found in France, in Alpes-Maritimes, Gironde, Seine-et-Marn, and Bouches-du-Rhône, but breeding was not confirmed (Dubois et al. 2016).

Fischer's Lovebird
(*Agapornis fischeri*)

This African species is frequently kept in captivity, and in Europe birds are often allowed to fly free, regularly returning to their aviaries for roosting. Around 1960, two pairs with young were put at liberty in Vienna but failed to establish (Niethammer 1963). Birds released in St.-Jean-Cap-Ferrat, France, in 1992, began breeding there, increasing to about 100–300 birds (Dubois 2007; Jiguet 2007). In 2010, 32 nests were counted under roofs of houses in St.-Jean-Cap-Ferrat. In 1994, the first breeding recorded in Spain occurred on Gran Canaria (Barranco de Palmito), and by 2006, a small breeding colony was present at Mas Palomas, Gran Canaria (Santos 2008a). On the Spanish mainland, a breeding attempt was recorded for Catalonia, and a maximum of 10 birds was observed in Málaga, Andalusia (Santos 2008a).

Yellow-collared Lovebird
(*Agapornis personatus*)

This African species is frequently kept in captivity and has had breeding populations in Europe since the mid-1990s. In 2004, the birds started to hybridize with Fischer's Lovebird in different locations in France (Dubois 2007; Jiguet 2007), and hybrid pairs were seen up to 2015, but reproduction declined (Dubois et al. 2016). In 1994, the first breeding occurred in Spain on Gran Canaria, and this population increased to least 15 birds in 2008.

Additional Records

In addition to the species described above, for which breeding has been observed, there are eight parrot species for which there is historical evidence of breeding but no recent records or just casual observations of escaped pets. These species are the White Cockatoo (*Cacatua alba*; Bedford 1954), Grey Parrot (*Psittacus erithacus*; Bedford 1954; Bauer and Woog 2008), Plain Parakeet (*Brotogeris tirica*; Niethammer 1963), Maroon-bellied Parakeet (*Pyrrhura frontalis*; Bauer and Woog 2008), Blue-and-yellow Macaw (*Ara ararauna*; Niethammer 1963), Red-and-green Macaw (*A. chloropterus*; Niethammer 1963), Australian King Parrot (*Alisterus scapularis*; Bedford 1954), and Black-winged Lovebird (*Agapornis taranta*; Walder 1926; Niethammer 1963).

There are, additionally, another 16 species that the Duke of Bedford bred or released freely in England in the first half of the 20th century. These records are detailed in Bedford (1954). For some of these species, there was early evidence of breeding, but there are no recent records. For other species, free-flying birds were occasionally seen, but there was no evidence of breeding (Bedford 1954).

Lastly, although this species is now extinct, it is perhaps important to note again that the Carolina Parakeet was bred in captivity in Europe in the late 1800s and was later introduced into the wild in Seebach, Germany. The entire wild flock of approximately 20 birds was unfortunately shot, in two days, 50 km away from its home (Russ 1887; Berlepsch 1929).

Rose-ringed Parakeets (*Psittacula krameri*) eating brick mortar to obtain minerals. This is a common behavior in this species before roosting. Amsterdam, Netherlands, September 2009. Photo by Roelant Jonker.

ESTABLISHMENT OF NATURALIZED SPECIES

Although parrots are now established breeding birds in several European countries, their ability to spread is restricted. Colonies in new cities are usually not started by established populations but by descendants of locally released or escaped birds (see also Carrete et al., chap. 15 this vol.). In 1969, Rose-ringed Parakeets bred for the first time in Germany, but their population appears to be geographically limited by climate, particularly winter temperatures (Fig. 14.1). Since the first establishment of Rose-ringed Parakeets, city populations have spread little. The parakeets fly every night to their roost sites in the cities where they escaped or were released. Real dispersal, with the founding of new roost sites, is a rare event (Braun 2015).

While European Rose-ringed Parakeet populations stayed quite small until around 2000, since then they have grown very rapidly (Fig. 14.2). In contrast, the Yellow-headed Amazons of Stuttgart show a more or less stable population (Fig. 14.3) (Martens and Woog 2017).

The Alexandrine Parakeet in Germany has recently begun to increase in population size (Fig. 14.4) (Braun et al. 2018, 2019b). In this species, roost counts are more complex, because a part of

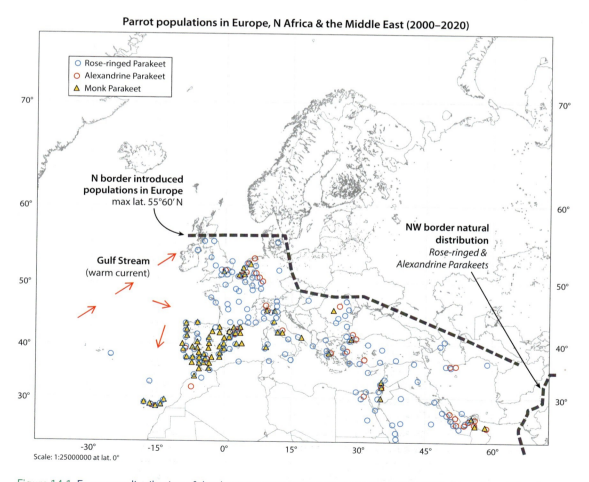

Figure 14.1. European distribution of the three most common parrot species, 2000–2020: Rose-ringed Parakeet, Alexandrine Parakeet, and Monk Parakeet. The northernmost boundary for these species worldwide is at about 55°60′ N, where the warm Gulf Stream contributes to a mild climate; the boundary decreases steeply in latitude toward continental Eurasia. (Data from Observation.org and GBIF 2019, as well as pers. obs.).

INTRODUCED AND NATURALIZED PARROTS IN EUROPE

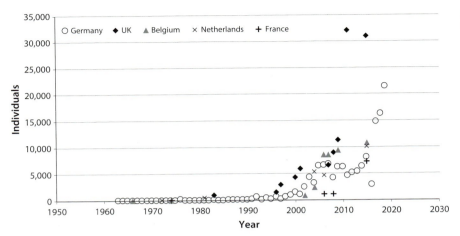

Figure 14.2. Explosive population growth of populations of Rose-ringed Parakeets in Europe (Pârâu et al. 2016, Braun et al. 2017, 2019b).

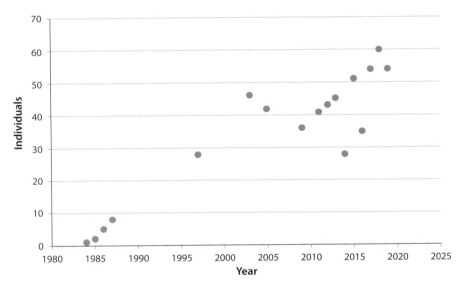

Figure 14.3. Population growth of Yellow-headed Amazons in Stuttgart, Germany (Martens and Woog 2017).

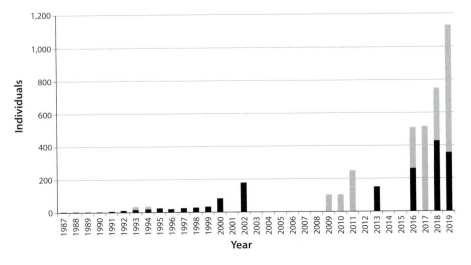

Figure 14.4. Population development of Alexandrine Parakeets in Wiesbaden (black shading) and Cologne (gray shading), Germany (Braun et al. 2018, 2019b).

the population is known to roost in the breeding cavities. The roost site is not as stable or reliable as that of the Rose-ringed Parakeet, and accurate counts are more difficult.

IMPACTS OF NATURALIZED PARROTS

Ecological competition between Rose-ringed Parakeets and other cavity-nesting species has been investigated and is reported on at length in this volume by Mori and Menchetti (chap. 6) and Brightsmith and Kiacz (chap. 9). Czajka et al. (2011) reported a study of interactions between Rose-ringed Parakeets and native Common Starlings (*Sturnus vulgaris*) but concluded that little competition appeared to occur. The two species had different preferences for nesting trees and local habitats, and direct competition was minor. In an area with an abundance of nesting cavities, just two parakeet nests were taken over by starlings, and only one starling nest was taken over by the parakeets.

The Eurasian Nuthatch (*Sitta europaea*) is also dependent on nest cavities and, like the Rose-ringed Parakeet, readily uses nest boxes. There is disagreement as to whether the Eurasian Nuthatch has been negatively impacted by the increasing Rose-ringed Parakeet population (Strubbe and Matthysen 2007, 2009; Newson et al. 2011). Nuthatches are known to interact with Rose-ringed Parakeets at the same nest boxes, and they can change the size of the opening to exclude the parakeets (as shown in photo).

As parrots are secondary cavity nesters, their ability to enlarge tree cavities introduces a new role in the European avian cavity-nesting community. Great Spotted Woodpecker (*Dendrocopos major*) often makes initial holes in trees, and Rose-ringed Parakeets are able to use those cavities for breeding. Rose-ringed Parakeets are known to breed in the same cavity for at least 18 years (Zingel 1993). For Alexandrine Parakeets, woodpecker cavities are far too small, but the parakeets are able to enlarge the openings. Cavity numbers also increase with time: in a case study, the number of medium-size and large cavities in a landscape park in Wiesbaden-Biebrich, Germany, increased from 80–100 in 1990 (Zingel 1990) to 264 in 2008, during the presence of both Rose-ringed and Alexandrine Parakeet colonies (Czajka et al. 2011). The reason for such behavior lies in the colonial breeding system. In a tree used by Rose-ringed Parakeets over a period of 35 years, up to nine pairs of birds used the tree at the same time (Czajka et al. 2011). Alexandrine Parakeets themselves are known to breed and roost in the same tree holes year-round (Zingel 2000), so cavity size and numbers will increase over the years.

Alexandrine Parakeets are larger than Rose-ringed Parakeets and appear to be able to outcompete the smaller species for nest sites (Braun et al. 2018; Bresser 2018). In parks in

A nest box set up for Rose-ringed Parakeets (*Psittacula krameri*) in an effort to prevent them from nesting in facades with thermal insulation. In this case, after the parakeets bred in the box, Eurasian Nuthatches (*Sitta europaea*) placed mud around the entrance of the nest, making it too small for the parakeets. Heidelberg, Germany, May 2009. Photo by Michael P. Braun.

Cologne, Germany, where both species have bred since the 1990s, Alexandrine Parakeets occupied more than 90% of all parakeet nests, most of which were found in London Plane Trees. The size of each species' population also reflects competition. At the Cologne Zoo, the population of Rose-ringed Parakeets has not exceeded 10 pairs over the years, whereas the population of Alexandrine Parakeets has steadily increased, numbering 45 pairs in 2018. The local Rose-ringed Parakeet population decreased to two pairs in 2018 (Braun et al. 2018; Bresser 2018).

Despite the situation at the Cologne Zoo, in the greater Cologne area the population of Rose-ringed Parakeets is still much larger than that of Alexandrine Parakeets. This can be explained by two factors: (1) replacement at breeding colonies takes place only in certain localities but not in the whole range; and (2) the Rose-ringed Parakeet has shifted its breeding sites from the old landscape parks in the city center of Cologne to the outskirts of the city, mainly to single breeding pairs in small parks and house facades.

But what is the situation with other cavity breeders? The Stock Dove (*Columba oenas*), which is found in deciduous forests with old trees, often breeds in woodpecker holes in European Beech (*Fagus sylvatica*). It depends on large cavities, mainly those created by the Black Woodpecker (*Dryocopus martius*). In cities, Black Woodpeckers are absent, but Stock Doves are attracted by the abundance of large cavities in London Planes, especially where parakeet colonies are present. In the grounds of Cologne Zoo, three pairs of Stock Doves regularly nest within the Alexandrine Parakeet colony, from which they are not chased away (Braun et al. 2018; Bresser 2018). Not only do the Stock Doves seem to increase in those parks (Krause 2001), but a flock of 15 Stock Doves was observed to share a roost site with 60 Alexandrine Parakeets and a single Green Parakeet (*Psittacara holochlorus*) in Cologne (Braun et al. 2018). The doves benefit from the active and passive defense of parakeet colonies against predators like the Northern Goshawk.

Zoonoses on Rose-ringed Parakeets were studied in Germany in 2007–10. We tested for avian flu (H5N1), *Salmonella*, *Chlamydia/Chlamydophila*; all results were negative (M. Lange, J. Tycka, M. Braun, and M. Wink, unpubl. data). *Haemoproteus minutus*, a blood parasite highly virulent for Australasian and South American parrots, could not be detected in wild Rose-ringed Parakeets in Germany (Ortiz-Catedral et al. 2019). Tests within the population of Alexandrine Parakeets and Rose-ringed Parakeets in Cologne did not show infectious diseases or parasites in the years 2017–19 (S. Marcordes, and B. Marcordes, unpubl. data). Generally, there is a low prevalence of hemoparasites in wild parrots, which is attributed in part to antiparasitic secondary metabolites in the diet (Masello et al. 2018).

ECOLOGICAL NICHE EXPANSION BY THE ROSE-RINGED PARAKEET

Rose-ringed Parakeets are one of the common species breeding in house facades with thermal insulation in Europe (Braun 2004, 2007). Prior to legislation by the EU that buildings had to be thermally insulated, the parakeets bred in natural cavities. The EU Energy Savings Ordinance, which came into force in 2002, led to a great increase in facade insulation. Woodpeckers, including the Great Spotted Woodpecker, have long been known to damage such facades; they readily create nest cavities in the insulation, and the parakeets are attracted by the woodpecker holes in such facades.

In Germany, when Rose-ringed Parakeets began to establish themselves in the late 1960s, they typically nested in tree cavities and only rarely in nest boxes. By the early 2000s, the breeding population was about 50–70 pairs in Heidelberg, Germany, and for the first time, pairs began to breed in the thermal insulation in a building facade. This change in behavior led to an expansion of the species' ecological niche, achieved by exploring and using new breeding sites. This shift occurred quickly: by 2003, more than 50% of the known breeding population was breeding in thermal insulation instead of tree cavities (Braun 2004, 2007). This behavior may have been in response to the shortage of old, hollow trees and large parks in Germany, as well as the favorable microclimate of these facades (temperatures are at least 2–3°C higher in facades than in natural cavities).

Once building facades included thermal insulation, damage to the facades was usually caused by woodpeckers. The Rose-ringed Parakeets did not excavate the cavities themselves and had to rely on cavities created by woodpeckers. But the parakeets could enlarge the cavities and use them after they were enlarged. Once the size preferences of Rose-ringed Parakeets were determined, it was discovered that the birds could be relocated to nest boxes (Braun 2004, 2007).

CONCLUSIONS

At least 48 parrot species have been intentionally released in Europe since the 19th century in many locations, and at least 40 parrot species have bred there at least once. Only since the 1960s, however, have parrots become established as new breeding birds. In the 19th and early 20th centuries, all the introduced populations in Europe—including that of the Carolina Parakeet—were shot or died out due to weather or other causes, but that situation has drastically changed in recent years. There are now many species breeding in many localities in Europe, with Monk Parakeets and Rose-ringed Parakeets the most abundant species. The impact of naturalized parrots on native birds is not necessarily negative, and their presence can even support native cavity breeders like the Stock Dove, due to enlargement and increase of tree cavities. The Eurasian Nuthatch is capable of defending its nest against parakeets by using mud—the same behavior it uses against woodpeckers and other predators. Starlings and parakeets use similar-size cavities for breeding but differ in preferred tree species and tree sizes. Nevertheless, there is evidence that Rose-ringed Parakeets can impact native bats (in this vol., Mori and Menchetti, chap. 5, and Brightsmith and Kiacz, chap. 9). The occurrence of parrots in Europe is strictly related to metropolitan areas, which underlines their dependence on food, climate, and breeding sites in human-modified areas. Therefore, their impact on native European fauna and flora should be locally restricted (White et al. 2019).

ACKNOWLEDGMENTS

I thank Nicole Braun and S. Pruett-Jones for proofreading and making valuable comments.

REFERENCES

Angelici, F., and Fiorillo, A. 2016. Repeated sightings of Alexandrine parakeet *Psittacula eupatria* in Rome (Central Italy) and its likely acclimatization. *Rivista Italiana di Ornitologia: Research in Ornithology* 85:33.

Batllori, X., and Nos, R. 1985. Presence of monk parakeet *Myiopsitta monachus* and rose ringed parakeet *Psittacula krameri* in the metropolitan area of Barcelona, Spain. *Miscellania Zoologica* 9:407–411.

Bauer, H. G., and Woog, F. 2008. Nichtheimische Vogelarten (Neozoen) in Deutschland, Teil I: Auftreten, Bestände und Status. *Vogelwarte* 46:157–194.

Bedford, Duke of. 1954. *Parrots and Parrot-like Birds*. Fond du Lac, WI: All-Pets Books.

Berlepsch, H. F. von. 1929. *Der gesamte Vogelschutz*. Staufenberg, Gerrmany: J. Newmann.

Borgo, E. 1999. Speciale specie alloctone: Pappagalli a spasso per la città. *Quaderni di Birdwatching* 1(1).

Braun, M. 2004. Neozoen in urbanen Habitaten: Ökologie und Nischenexpansion des Halsbandsittichs (*Psittacula krameri*; Scopoli, 1769) in Heidelberg. Diploma thesis, Dept. of Biology, Philipps-Univ. Marburg, Germany.

Braun, M. 2007. Welchen Einfluss hat die Gebäudedämmung im Rahmen des EU-Klimaschutzes auf die Brutbiologie tropischer Halsbandsittiche (*Psittacula krameri*) im gemäßigten Mitteleuropa? Ornithologische Jahreshefte Baden-Württemberg 23:39–56.

Braun, M. 2009. Die Bestandssituation des Halsbandsittichs (*Psittacula krameri*) in Europa, Deutschland und der Rhein-Neckar-Region (Baden-Württemberg, Rheinland-Pfalz, Hessen), 1962–2008. *Vogelwelt* 130:77–89.

Braun, M. 2015. Neue Halsbandsittich-Schlafplätze in der Rhein-Neckar-Region entdeckt. *Gefiederte Welt* 10:19–21.

Braun, M. P., Bahr, N., and Wink, M. 2016. Phylogenie und Taxonomie der Edelsittiche (Psittaciformes: Psittaculidae: *Psittacula*), mit Beschreibung von drei neuen Gattungen. *Vogelwarte* 54:322–324.

Braun, M. P., Bruslund, N., Bruslund, S., Sauer-Gürth, H., Dreyer, W., Laucht, S., Kragten, S., Pârâu, L. G., Gross, B., Franz, D., et al. 2017. Ökologie und Bestandsentwicklung des Asiatischen Halsbandsittichs (*Alexandrinus manillensis*) in Deutschland und Europa mit aktuellen Bestandszahlen. *Vogelwarte* 55:307–309.

Braun, M.P., Datzmann, T., Arndt, T., Reinschmidt, M., Schnitker, H., Bahr, N., Sauer-Gürth, H., and Wink, M. 2019a. A molecular phylogeny of the genus *Psittacula* sensu lato (Aves: Psittaciformes: Psittacidae: *Psittacula, Psittinus, Tanygnathus, Mascarinus*) with taxonomic implications. *Zootaxa* 4563:547–562.

Braun, M. P., Franz, D., Braun, N., Koch, E., Walter, C., Bresser, A., Ziegler, T., and Marcordes, B. 2018. Aktuelle Bestandserfassung des Großen Alexandersittichs (*Palaeornis eupatria*) in Deutschland und Europa. *Vogelwarte* 56:383–385.

Braun, M. P., Franz, D., Braun, N., Walter, C., Romero, J., Herder, B., Baranowski, A., Thissen, A., Kemper, A., Hillebrand, J., et al. 2019b. Vogelneozoen und ihre Populationen in Deutschland, Stand 2019. *Vogelwarte* 57(4).

Bresser, A. 2018. Bestandserfassung, Revierkartierung und angepasste Autökologie des Großen Alexandersittichs (*Psittacula eupatria*), eines Neozoen, in der Stadt Köln. Master's thesis, Faculty of Mathematics and Natural Sciences, Univ. of Cologne, Germany.

Butler, C. 2002. Breeding parrots in Britain. *British Birds* 95:345–348.

Butler, C., Hazlehurst, G., and Butler, K. 2002. First nesting by blue-crowned parakeet in Britain. *British Birds* 95:17–20.

Cortes, J. 2006. Probable nesting of the blue-crowned conure *Aratinga acuticauda* in Gibraltar. Bird Report 2005, Society GOIaNG, Gibraltar: 59–61.

Costa, H., and Schäffer, N. 2000. Portugal—"Einfallstor" exotischer Vogelarten nach Europa? *Der Falke* 47:360–377.

Czajka, C., Braun, M., and Wink, M. 2011. Resource use of non-native ring-necked parakeets (*Psittacula krameri*) and native starlings (*Sturnus vulgaris*) in Central Europe. *Open Ornithology Journal* 4:17–22.

Dubois, P. J. 2007. Les oiseaux allochtones en France: Statut et interactions avec les espèces indigènes. *Ornithos* 14:329–364.

Dubois, P. J., Maillard, J. F., and Cugnasse, J. M. 2016. Les populations d'oiseaux allochtones en France en 2015 (4e enquête nationale). *Ornithos* 23:129–141.

Ernst, U. 1995. Afro-asiatische Sittiche in einer mitteleuropäischen Großstadt: Einnischung und Auswirkung auf die Vogelfauna. *Jahrbuch für Papageienkunde* 1:23–114.

Fünfstück, H., von Lossow, G., and Schöpf, H. 2003. Rote Liste gefährdeter Brutvögel (Aves) Bayerns. Rote Liste gefährdeter Tiere Bayerns. *Schriftenreihe* 166:39–44.

García del Rey, E. 2007. Exotic, introduced and invasive avifauna on Tenerife: Are these species a serious threat? *Vieraea* 35:131–138.

GBIF. 2019. Global Biodiversity Information Facility. Accessed Mar. 2019. https://doi.org/10.15468/dl.j1fzvm.

Hoppe, D. 1997. Exoten im Park: Amazonenpapageien über Stuttgart. *Wellensittich & Papageien Magazin* 4 (Jul./Aug.). https://www.vogelbrunnen.de/de/WP-Magazin-4-2019.html.

Jiguet, F. 2007. Lovebirds in south-eastern France. *Ornithos* 14:376–381.

Jonker, R., and Tamis, W. 2012. Introduction, breeding and poaching of scarlet macaws (*Ara macao* L.) in a temperate country: A case study from the Netherlands. *Open Ornithology Journal* 5:1–4.

Kolar, K. 1979. Die Papageien. In *Grzimeks Tierleben*, ed. B. Grzimek, 280–288. Augsburg, Germany: Weltbild (reprt. dtv 1979–80).

Krause, T. 2001. Zur Verbreitung des Halsbandsittichs (*Psittacula krameri*) im Rheinland im Kontext der gesamten westeuropäischen Verbreitung. Diploma thesis, Institute of Geography, Rheinische Friedrich-Wilhelms Univ., Bonn, Germany.

Latsoudis, P. 2007. Preliminary distribution study of the ring-necked parakeet *Psittacula krameri* (Scopoli 1769) in Greece. Hellenic Ornithological Society. Accessed Dec. 2019. http://www.ornithologiki.gr/page_cn.php?aID=929.

Lever, C. 2005. *Naturalised Birds of the World*. London: T. & A. D. Poyser.

Lever, C. 2009. *The Naturalized Animals of Britain and Ireland*. London: New Holland.

Mahler, U. 2001: *Amazona oratrix*: Gelbkopfamazone. In *Die Vögel Baden-Württembergs*, ed. J. Hölzinger. Stuttgart, Germany: Ulmer.

Martens, J. M., and Woog, F. 2017. Nest cavity characteristics, reproductive output and population trend of naturalised Amazon parrots in Germany. *Journal of Ornithology* 158:823–832.

Masello, J. F., Martínez, J., Calderón, L., Wink, M., Quillfeldt, P., Sanz, V., Theuerkauf, J., Ortiz-Catedral, L., Berkunsky, I., and Brunton, D. 2018. Can the intake of antiparasitic secondary metabolites explain the low prevalence of hemoparasites among wild Psittaciformes? *Parasites and Vectors* 11:357.

Masello, J. F., and Quillfeldt, P. 2002. Chick growth and breeding success of the burrowing parrot. *Condor* 104:574–586.

Masello, J. F., and Quillfeldt, P. 2004. Consequences of La Niña phase of ENSO for the survival and growth of nestling burrowing parrots on the Atlantic coast of South America. *Emu* 104:337–346.

Newson, S. E., Johnston, A., Parrott, D., and Leech, D. I. 2011. Evaluating the population-level impact of an invasive species, ring-necked parakeet *Psittacula krameri*, on native avifauna. *Ibis* 153:509–516.

Niethammer, G. 1963. *Die Einbürgerung von Säugetieren und Vögeln in Europa*. Hamburg and Berlin, Germany: Parey.

Ogilvie, M. 2003. Non-native birds breeding in the United Kingdom in 2001. *British Birds* 96:620–625.

REFERENCES

Ortiz-Catedral, L., Brunton, D., Stidworthy, M. F., Elsheikha, H. M., Pennycott, T., Schulze, C., Braun, M., Wink, M., Gerlach, H., Pendl, H., et al. 2019. *Haemoproteus minutus* is highly virulent for Australasian and South American parrots. *Parasites and Vectors* 12:1–10.

Pârâu, L. G., Strubbe, D., Mori, E., Menchetti, M., Ancillotto, L., Kleunen, A. V., White, R. L., Luna, A., Hernández-Brito, D., Louarn, M. L., and Clergeau, P. 2016. Rose-ringed parakeet *Psittacula krameri* populations and numbers in Europe: A complete overview. *Open Ornithology Journal* 9:1–13.

Postigo, J. L., Strubbe, D., Mori, E., Ancillotto, L., Carneiro, I., Latsoudis, P., Menchetti, M., Pârâu, L. G., Parrott, D., Reino, L., Weiserbs, A., and Senar, J. C. 2019. Mediterranean versus Atlantic monk parakeets *Myiopsitta monachus*: Towards differentiated management at the European scale. *Pest Management Science* 75:915–922.

Reino, L., Figueira, R., Beja, P., Araújo, M. B., Capinha, C., and Strubbe, D. 2017. Networks of global bird invasion altered by regional trade ban. *Science Advances* 3:e1700783.

Russ, K. 1926. *The Budgerigar: Its Natural History, Breeding, and Management.* London: Cage Birds and Bird Fancy.

Şahin, D., and Arslangündoğdu, Z. 2019. Breeding status and nest characteristics of rose-ringed (*Psittacula krameri*) and Alexandrine parakeets (*Psittacula eupatria*) in Istanbul's city parks. *Applied Ecology and Environmental Research* 17:2461–2471.

Sánchez, J. 2005. *Amazona aestiva*. In *Fichas de aves introducidas en España*. Grupo de Aves Exóticas (SEO/BirdLife). Accessed Dec. 2019. https://www.seo.org/grupo-de-aves-exoticas/.

Santos, D. M. 2006a. *Aratinga erythrogenys*. In *Fichas de aves introducidas en España*. Grupo de Aves Exóticas (SEO/BirdLife). Accessed Dec. 2019. https://www.seo.org/grupo-de-aves-exoticas/.

Santos, D. M. 2006b. *Aratinga mitrata*. In *Fichas de aves introducidas en España*. Grupo de Aves Exóticas (SEO/BirdLife). Accessed Dec. 2019. https://www.seo.org/grupo-de-aves-exoticas/.

Santos, D. M. 2006c. *Nandayus nenday*. In *Fichas de aves introducidas en España*. Grupo de Aves Exóticas (SEO/BirdLife). Accessed Dec. 2019. https://www.seo.org/grupo-de-aves-exoticas/.

Santos, D. M. 2006d. *Aratinga acuticaudata*. In *Fichas de aves introducidas en España*. Grupo de Aves Exóticas (SEO/BirdLife). Accessed Dec. 2019. https://www.seo.org/grupo-de-aves-exoticas/.

Santos, D. M. 2008a. *Agapornis fischeri*. In *Fichas de aves introducidas en España*. Grupo de Aves Exóticas (SEO/BirdLife). Accessed Dec. 2019. https://www.seo.org/grupo-de-aves-exoticas/.

Santos, D. M. 2008b. *Cyanoliseus patagonus*. In *Fichas de aves introducidas en España*. Grupo de Aves Exóticas (SEO/BirdLife). Accessed Dec. 2019. https://www.seo.org/grupo-de-aves-exoticas/.

Santos, D. M. 2008c. *Myiopsitta monachus*. In *Fichas de aves introducidas en España*. Grupo de Aves Exóticas (SEO/BirdLife). Accessed Dec. 2019. https://www.seo.org/grupo-de-aves-exoticas/.

Santos, D. M. 2008d. *Poicephalus senegalus*. In *Fichas de aves introducidas en España*. Grupo de Aves Exóticas (SEO/BirdLife). Accessed Dec. 2019. https://www.seo.org/grupo-de-aves-exoticas/.

Strubbe, D., and Matthysen, E. 2007. Invasive ring-necked parakeets *Psittacula krameri* in Belgium: Habitat selection and impact on native birds. *Ecography* 30:578–588.

Strubbe, D., and Matthysen, E. 2009. Experimental evidence for nest-site competition between invasive ring-necked parakeets (*Psittacula krameri*) and native nuthatches (*Sitta europaea*). *Biological Conservation* 142:1588–1594.

Walder, H. R. 1926. *Agapornis taranta* und Klima. *Gefiederte Welt* 55:97–100.

Weiserbs, A. 2000. Une troisieme perruche nicheuse en Region bruxelloise: La perruche alexandre *Psittacula eupatria*. *Aves* 37:115–120.

Weiserbs, A., and Paquet, A. 2016. Recensement de la conure veuve *Myiopsitta monachus* à Bruxelles en 2016. *Aves* 53:19–28.

White, R. L., Strubbe, D., Dallimer, M., Davies, Z. G., Davis, A.J.S., Edelaar, P., Groombridge, J. J., Jackson, H. A., Menchetti, M., Mori, E., et al. 2019. Assessing the ecological and societal impacts of alien parrots in Europe using a transparent and inclusive evidence-mapping scheme. *NeoBiota* 48:45–69.

Zingel, D. 1990. Zum Vorkommen des Halsbandsittichs (*Psittacula krameri*) im Schloßpark von Wiesbaden-Biebrich. *Jahrbücher des Nassauischen Vereins für Naturkunde* 112:7–23.

Zingel, D. 1993. Zum Vorkommen des Halsbandsittichs im Schloßpark von Wiesbaden-Biebrich. *Gefiederte Welt* 117:64–66, 96–98.

Zingel, D. 2000. 25 Jahre frei lebende Papageien in Wiesbaden. *Jahrbücher des Nassauischen Vereins für Naturkunde* 121:129–141.

15

THE FATE OF MULTISTAGE PARROT INVASIONS IN SPAIN AND PORTUGAL

Martina Carrete, Pedro Abellán, Laura Cardador, José D. Anadón, and José L. Tella

INTRODUCTION

Invasive species are widely accepted as one of the most important causes of biodiversity loss (Clavero and García-Berthou 2005; Clavero et al. 2009) and a major driver of human-induced global change (Olden 2004). The large number of species transported and the range of pathways that move species among areas have greatly increased the number and geographic range extent of exotic species globally (Hulme 2009). For vertebrates, international trade has been identified as the most important and increasing source of invasive species worldwide (Cardador et al. 2019). Parrots (Psittaciformes) are among the most traded bird taxa (Beissinger 2001). At least 259 species of psittaciforms have been trafficked worldwide, involving millions of individuals in recent decades (Cardador et al. 2017). This trade has strongly contributed to the decline of many parrot species in their native ranges (Collar and Juniper 1992; Tella and Hiraldo 2014), while it has also made parrots among the most widespread invasive birds in the world (Cassey et al. 2004). Their colorful plumage and their ability to imitate human speech have made parrots heavily popular as pets and explain why certain species are preferred over others (Tella and Hiraldo 2014), but there is scarce information on the factors facilitating invasion success.

Invasion is often described as a process, in which an exotic species progresses through a series of stages—transport, introduction, establishment, and spread—each characterized by its own dynamics and dependent on different factors (Lockwood et al. 2007). Typically, only a subset of the species moved beyond their native range and gained access to the wild, and only some of those birds become established and subsequently spread to become invasive (Blackburn et al. 2011). Williamson (1996) proposed the "10s rule," which posits that approximately 10% of species are successful at each stage in the invasion process. Although the general tendency to report species that have become invasive relative to those that have not has resulted in an overestimation of establishment success (Jeschke et al. 2012; Rodriguez-Cabal et al. 2013), data sets including transported and introduced, but not necessarily established, species support the idea that most species fail to transit the different stages of the invasion process (Abellán et al. 2016). Furthermore, avian species passing through each of the invasion stages are not a random sample of all birds with regard to their taxonomic affiliation (mainly family) (Lockwood 1999; Blackburn and Duncan 2001; Blackburn and Cassey 2007; Abellán et al. 2016), which may affect the general conclusions about the characteristics of species associated with successful establishment and spread.

In this chapter, we document the fate of parrots transported and introduced in Spain and Portugal during the past century. We rely on one of the largest and most complete data sets on exotic birds at a regional level, which compiles information on species observed in the wild and in captivity from 1912 to 2015 in the Iberian Peninsula and archipelagos (the Canary, Balearic, Madeira, and Azores islands). The data set was built through a systematic review of scientific and so-called gray literature and citizen science platforms, and completed with our own data and unpublished observations from other researchers (Abellán et al. 2016). Transported species were also assessed by compiling information on exotic species kept in captivity in Spain and Portugal from several information sources, including the Convention on International Trade in Endangered Species of Wild Fauna and Flora (CITES 2019). The whole data set, in summary, comprises a total of 1,026 exotic bird species (29 orders and 76 families) transported to Spain and Portugal between 1912 and 2015. Additionally, there are 13,222 observations of exotic species observed in the wild (11,878 for Spain and 1,342 for Portugal; ~75,000 individuals), of 377 species, from 52 families and 23 orders. These data were examined in a geographic grid of 1,911 different cells, each 5 × 5 km.

TRANSPORT

Long-distance and international trade of exotic birds into Europe, including parrots, began centuries ago (Blackburn et al. 2009; Tella 2011). For example, Alexander the Great took Rose-ringed Parakeets (*Psittacula krameri*) back home to Greece after the Indian campaign, and the Romans taught parrots (apparently *Psittacula* parakeets from India) to speak Latin following instruction published by Pliny the Elder (77 CE). Archaeological findings of Roman mosaics of parrots suggest that the birds were present in Spain at least from the second half of the second century. In 1493, Christopher Columbus brought Queen Isabella a pair of Cuban Amazons (*Amazona leucocephala*) in gratitude for paying for his travels, while the Portuguese frequently brought back Grey Parrots (*Psittacus erithacus*) from their trips in the mid-1400s. During the 19th century, species such as the Budgerigar (*Melopsittacus undulatus*) were brought from Australia to England, where successful breeding of the Budgerigar in captivity accelerated its conquest of the continent's birdcages, including in Spain and Portugal.

However, the bulk of parrot transportation to Spain and Portugal took place during recent decades, paralleling the expansion of the international flight network (Hulme 2009) and the growing demand for parrots as pets (Tella and Hiraldo 2014). According to CITES, Spain and Portugal have legally imported more than 2 million parrots from 1975 to 2017, with peaks of almost 150,000 individuals from 120 species in a single year (Fig. 15.1).

Monk Parakeets (*Myiopsitta monachus*) have become so common in urban parks in Spain that local governments are beginning culling programs. Santa Ponsa, Mallorca, Spain, February 2010. Photo by Tamara k, CC BY-SA 3.0 (https://creativecommons.org/licenses/by-sa/3.0/legalcode).

THE FATE OF MULTISTAGE PARROT INVASIONS IN SPAIN AND PORTUGAL

Our survey of parrot species kept in captivity, which complements data provided by CITES, shows that 248 species out of the ~400 extant parrot species in the world have been transported to Spain and Portugal during the last century. Most of them are from the Neotropics (116

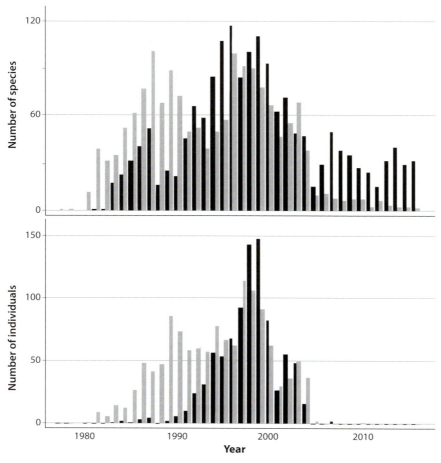

Figure 15.1. Number of parrot species (top) and individuals (bottom; in thousands) imported into Spain and Portugal until 2017, according to CITES (2017). Captive-bred (black bars) and wild-caught (gray bars) imported parrots are separately shown.

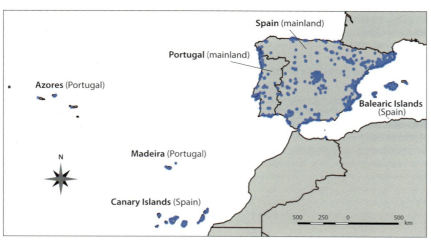

Figure 15.2. Distribution of sightings of introduced parrots in Spain and Portugal.

species), while most of the rest originate in the Australasian (78 species), Indomalayan (32 species), and Afrotropical (19 species) realms. Public installations such as zoos house 247 of these species, while 219 are accommodated in private facilities, and 218 are in both public and private accommodations.

PARROT INTRODUCTIONS

We compiled observations in Spain and Portugal of approximately 26,000 individual parrots belonging to 70 species (Fig. 15.2). Observations were more common on the continent, although there are many records of escaped parrots in all the islands. The first observation of a parrot seen in the wild in Spain and Portugal dates to 1970. Figure 15.3 illustrates the patterns of importations of parrots compared to other bird orders since the early 1900s. The most frequent biogeographic origin of parrots introduced in the wild has been the Neotropics (38 species), followed by Australasia (12 species), the Afrotropics (10 species), and the Indomalayan region (10 species).

Most records of parrots introduced in the wild corresponded to accidental escapes from public installations (five species) or private facilities (44 species); only a few species were deliberately released by people (six species). It is worth noting that birds from all six deliberately released species also escaped accidentally from captivity.

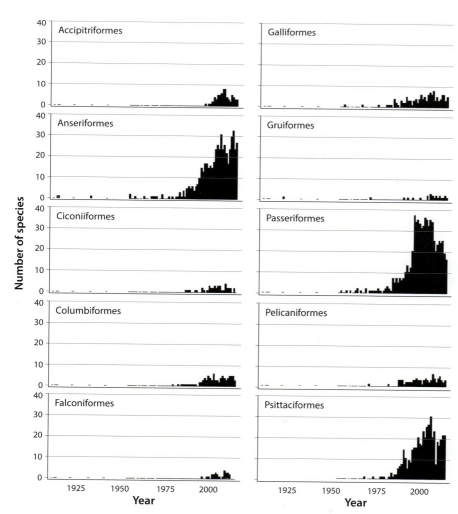

Figure 15.3. Temporal changes in the number of bird species introduced in Spain and Portugal grouped by avian orders. Modified from Abellán et al. (2016).

THE FATE OF MULTISTAGE PARROT INVASIONS IN SPAIN AND PORTUGAL

ESTABLISHMENT

Up until 2015, of all parrot species introduced into the wild, only 16 have been recorded breeding (8.5% of species) (Abellán et al. 2016). In 2016, there were new records of the Scaly-headed Parrot (*Pionus maximiliani*) breeding in an urban park in Málaga, Spain, and four individuals and three hybrids of Rose-ringed Parakeets × Alexandrine Parakeets (*Psittacula eupatria*) in the same area (Postigo 2016).

Rose-ringed and Monk Parakeets (*Myiopsitta monachus*) are the species with the largest breeding populations in Spain, mainly in central, southern, and eastern regions (Fig. 15.4). The reproduction of other parrot species is anecdotal. Species such as Fischer's Lovebird (*Agapornis fischeri*), Yellow-collared Lovebird (*A. personatus*), Cockatiel (*Nymphicus hollandicus*), and Budgerigar have been recorded as breeding but only after supplementary feeding and nest provisioning by people (Mori et al. 2019). Others, such as Senegal Parrot (*Poicephalus senegalus*) and Blue-crowned Parakeet (*Thectocercus acuticaudatus*), are breeding at very low numbers, possibly due to the low number of individuals introduced and the typical lag phases associated with invasive species (Abellán et al. 2017).

SPREAD

Most parrot species present in Spain and Portugal initially established populations in cities, possibly due to the higher abundance of cage birds and, thus, the higher rate of accidental

Rose-ringed Parakeets (*Psittacula krameri*) are the most widely distributed naturalized parrot in the world, although in Spain and Portugal, Monk Parakeets (*Myiopsitta monachus*) are more abundant. Maharashtra, India, March 2020. Photo by Phadke09, CC BY-SA 3.0 (https://creativecommons.org/licenses/by-sa/3.0/legalcode).

escapes. This fact, combined with other local factors such as the association with humans and the climatic match between native and introduced areas, has determined the rate of spatial spread and invaded-range size of parrots (Fig. 15.4; Abellán et al. 2017). However, parrot populations established in the different Spanish and Portuguese cities are not connected through dispersal processes (Ascensão et al. 2020). Typically, these populations increase in size within the limits of the cities, and the individuals breed in urban areas but move to rural areas to forage. This is the case of Monk and Rose-ringed Parakeets in Seville and Barcelona, where crop damage has been assessed (Senar et al. 2016; authors' unpubl. data). Although they rarely do, urban parrot populations can expand into neighboring rural areas. For instance, Rose-ringed and Monk Parakeets have occupied rural areas around Madrid (Hernández-Brito et al. 2020). Interestingly, the Monk Parakeet case shows that predation pressure is the main determinant of the urban confinement of most parrot populations, and that an association with an aggressive native species such as the White Stork (*Ciconia ciconia*) can allow the birds to colonize rural areas (Hernández-Brito et al. 2020). Although rare, these expansions of urban parrot populations into rural neighbourhoods alert us to the potential for these species to spread and occupy natural areas once they overcome the typical lag phase characterizing invasive species.

DETERMINANTS OF INVASION SUCCESS

Understanding factors driving successful establishment and spread is one of the cornerstones of invasion biology. For parrots, establishment success is determined largely by propagule pressure and bird origin. Increasing propagule pressure enhances establishment probability, because small populations are more likely to suffer from effects of demographic and genetic stochasticity, to be extirpated by environmental stochasticity, or to suffer from the Allee effect (the correlation between population density and mean individual fitness). Bird origin was the best predictor of establishment success among parrots, in accordance with Carrete and Tella (2008). Wild-caught birds were more likely to establish viable populations than captive-bred ones, even after controlling for other factors such as the number of imported birds or their availability in the pet market. This difference in the invasive potential of captive-bred and wild-caught individuals is explained by differences in their ability to cope with new environments. For instance,

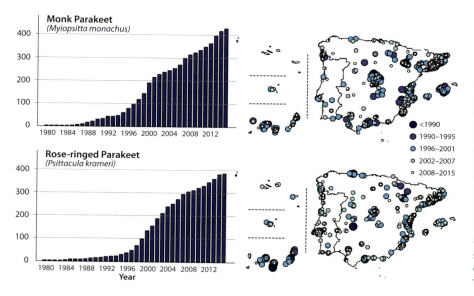

Figure 15.4. Left: Numbers of naturalized Monk and Rose-ringed Parakeets observed in the wild (y-axis = number of individuals) in Spain and Portugal since 1982. Right: Geographical spread of these observations over time.

THE FATE OF MULTISTAGE PARROT INVASIONS IN SPAIN AND PORTUGAL

Yellow-collared Lovebirds (*Agapornis personatus*) have been recorded as breeding in Spain, but the species has not established a naturalized population there. Serengeti, Tanzania, May 2009. Photo by Demetrius John Kessy, CC BY-SA 3.0 (https://creativecommons.org/licenses/by-sa/3.0/legalcode).

individuals bred in captivity have lost antipredatory and escape behaviors (Carrete and Tella 2015), while wild-caught birds show longer responses to acute stress (Cabezas et al. 2013). These differences explain why the latter are more prone to escape from cages and survive better when facing challenges in the new wild environments than the former, thus be more invasive.

Regarding spread, the longer the time since transport, the greater the likelihood of new accidentally released individuals at different locations or new colonizations of dispersers from previously occupied sites, which would imply a greater number of cumulative occupied areas (cells) (Fig. 15.4). However, the main determinant of spread was climate matching, so that once initial colonization and establishment have occurred, the degree to which the invaded region resembles the species' native range is a critical factor regulating the spread of parrot species. Cardador et al. (2016) have shown that the European distribution of Rose-ringed Parakeets can be explained by the environmental similarity between the native and the invasive range. Indeed, although Asian and African Rose-ringed Parakeets were transported to Europe in similar numbers (Cardador et al. 2016), genetic markers of established invaders mostly match those of Asian populations (Jackson et al. 2015), whose niche is more similar to Europe. African Rose-ringed Parakeets, conversely, are extremely rare and found only in some Spanish locations (authors' pers. obs.) where habitat is suitable for them (Cardador et al. 2016).

Will New Species of Parrots Become Naturalized?

Based on our previous knowledge, there are no serious risks of introducing new parrot species in the Iberian Peninsula in the near future. In 2005, a European wild bird trade ban was implemented to counter the spread of avian flu. The beginning of the ban is largely associated with a drastic decline in the number of individuals of wild origin imported into Spain and Portugal and an increment in the availability of captive-bred individuals at pet shops (Cardador et al. 2019). Importantly, records of new exotic species introduced into the wild dropped immediately after the implementation of the ban. Contrary to some predictions (Rivalan et al. 2007), this ban did not promote an upsurge of illegal parrot trade in the years after its institution. Since 2005, there have been only a few cases of individuals or eggs of very valuable species, such as Lear's Macaw (*Anodorhynchus leari*) and rare *Amazon* species, illegally transported from Brazil to Portugal, and several dozen Red-headed Lovebirds (*Agapornis pullarius*) and Senegal Parrots shipped from Senegal to Spain. Although illegal trade is extremely difficult to track due to its nature, it can never equal the thousands of individuals annually imported legally (Fig. 15.1). Furthermore, the European prohibition of importing wild-caught birds, implemented for health reasons, was incorporated in 2011 into Spanish law to reduce invasion risk, which ensures that this scenario will be stable in the long term.

Regarding parrots actually kept in captivity, it is worth mentioning that the drastic reduction of imports from 2005 onward has meant that currently most cage birds are of captive origin. Species such as Burrowing Parrot (*Cyanoliseus patagonus*), Blue-crowned Parakeet, and Nanday Parakeet (*Aratinga nenday*), which were imported by the thousands in the past, are rarely bred in captivity. Other species, such as Senegal Parrot or Grey Parrot, are largely bred in captivity, so most individuals kept as pets are of captive origin. In all cases, the low availability of wild individuals kept in captivity ensures low propagule pressures and reduced risk to reinforcement of current wild invasive cores.

REFERENCES

Abellán, P., Carrete, M., Anadón, J. D., Cardador, L., and Tella, J. L. 2016. Non-random patterns and temporal trends (1912–2012) in the transport, introduction and establishment of exotic birds in Spain and Portugal. *Diversity and Distributions* 22:263–273.

Abellán, P., Tella, J. L., Carrete, M., Cardador, L., and Anadón, J. D. 2017. Climate matching drives spread rate but not establishment success in recent unintentional bird introductions. *Proceedings of the National Academy of Sciences (USA)* 114:9385–9390.

Ascensão, F., Latombe, G., Anadón, J. D., Abellán, P., Cardador, L., Carrete, M., Tella, J. L., and Capinha, C. 2020. Drivers of compositional dissimilarity for native and alien birds: The relative roles of human activity and environmental suitability. *Biological Invasions* 22:1447–1460.

Beissinger, S. 2001. Trade of live wild birds: Potentials, principles, and practices of sustainable use. In *Conservation of Exploited Species*, ed. J. D. Reynolds, G. M. Mace, K. H. Redford, J. G. Robinson, 182–202. Cambridge: Cambridge Univ. Press.

Blackburn, T. M., and Cassey, P. 2007. Patterns of nonrandomness in the exotic avifauna of Florida. *Diversity and Distributions* 13:519–526.

Blackburn, T. M., and Duncan, R. P. 2001. Establishment patterns of exotic birds are constrained by non-random patterns in introduction. *Journal of Biogeography* 28:927–939.

Blackburn, T. M., Lockwood, J. L., and Cassey, P. 2009. *Avian Invasions. The Ecology and Evolution of Exotic Birds*. Oxford: Oxford Univ. Press.

Blackburn, T. M., Pyšek, P., Bacher, S., Carlton, J. T., Duncan, R. P., Jarošík, V., Wilson, J.R.U., and Richardson, D. M. 2011. A proposed unified framework for biological invasions. *Trends in Ecology & Evolution* 26:333–339.

Cabezas, S., Carrete, M., Tella, J. L., Marchant, T. A., and Bortolotti, G. R. 2013. Differences in acute stress responses between wild-caught and captive-bred birds: A physiological mechanism contributing to current avian invasions? *Biological Invasions* 15:521–527.

Cardador, L., Carrete, M., Gallardo, B., and Tella, J. L. 2016. Combining trade data and niche modelling improves predictions of the origin and distribution of non-native European populations of a globally invasive species. *Journal of Biogeography* 43:967–978.

Cardador, L., Lattuada, M., Strubbe, D., Tella, J. L., Reino, L., Figueira, R., and Carrete, M. 2017. Regional bans on wild-bird trade modify invasion risks at a global scale. *Conservation Letters* 10:717–725.

Cardador, L., Tella, J. L., Anadón, J., Abellán, P., and Carrete, M. 2019. The European trade ban on wild birds reduced invasion risks. *Conservation Letters* 12:e12631.

Carrete, M., and Tella, J. L. 2008. Wild-bird trade and exotic invasions: A new link of conservation concern? *Frontiers in Ecology and the Environment* 6:207–211.

Carrete, M., and Tella, J. L. 2015. Rapid loss of antipredatory behaviour in captive-bred birds is linked to current avian invasions. *Scientific Reports* 5:18274.

Cassey, P., Blackburn, T. M., Russell, G. J., Jones, K. E., and Lockwood, J. L. 2004. Influences on the transport and establishment of exotic bird species: An analysis of the parrots (Psittaciformes) of the world. *Global Change Biology* 10:417–426.

CITES. 2017. CITES Trade Database. Accessed June 2017. http://trade.cites.org.

CITES. 2019. CITES Trade Database. Accessed Dec. 2019. http://trade.cites.org.

Clavero, M., Brotons, L., Pons, P., and Sol, D. 2009. Prominent role of invasive species in avian biodiversity loss. *Biological Conservation* 142:2043–2049.

Clavero, M., and García-Bethou, E. 2005. Invasive species are a leading cause of animal extinctions. *Trends in Ecology & Evolution* 20:110.

Collar, N. J., and Juniper, A. T. 1992. Dimensions and causes of the parrot conservation crisis. In *New World Parrots in Crisis: Solutions from Conservation Biology*, ed. S. R. Beissinger and N.F.R. Snyder, 1–24. Washington, DC: Smithsonian Institution Press.

Hernández-Brito, D., Blanco, G., Tella, J. L., and Carrete, M. 2020. A protective nesting association with native species counteracts biotic resistance for the spread of an invasive parakeet from urban into rural habitats. *Journal of Zoology* 17:13.

Hulme, P. E. 2009. Trade, transport and trouble: Managing invasive species pathways in an era of globalization. *Journal of Applied Ecology* 46:10–18.

Jackson, H., Strubbe, D., Tollington, S., Prys-Jones, R., Matthysen, E., and Groombridge, J. J. 2015. Ancestral origins and invasion pathways in a globally invasive bird correlate with climate and influences from bird trade. *Molecular Ecology* 24:4269–4285.

Jeschke, J., Aparicio, L. G., Haider, S., Heger, T., Lortie, C., Pyšek, P., and Strayer, D. 2012. Support for major hypotheses in invasion biology is uneven and declining. *NeoBiota* 14:1–20.

Lockwood, J. L. 1999. Using taxonomy to predict success among introduced avifauna. *Conservation Biology* 13:560–567.

Lockwood, J. L., Hoopes, M. F., and Marchetti, M. P. 2007. *Invasion Ecology*. Oxford, UK: Blackwell Publishing.

Mori, E., Cardador, L., Reino, L., White, R. L., Hernández-Brito, D., Le Louarn, M., Mentil, L., Edelaar, P., Pârâu, L. G., Nikolov, B. P., and Menchetti, M. 2019. Lovebirds in the air: Trade patterns, establishment success and niche shifts of *Agapornis* parrots within their non-native range. *Biological Invasions* 22:421–435.

Olden, J. D. 2004. Biotic homogenization: A new research agenda for conservation biogeography. *Journal of Biogeography* 33:2027–2039.

Postigo, J. L. 2016. New records of invasive parakeet hybrids in Spain: A great opportunity to apply the rapid response mechanism. *European Journal of Ecology* 2:19–22.

Rivalan, P., Delmas, V., Angulo, E., Bull, L. S., Hall, R. J., Courchamp, F., Rosser, A. M., and Leader-Williams, N. 2007. Can bans stimulate wildlife trade? *Nature* 447:529–530.

Rodriguez-Cabal, M. A., Williamson, M., and Simberloff, D. 2013. Overestimation of establishment success of non-native birds in Hawaii and Britain. *Biological Invasions*, 15:249–252.

Senar, J. C., Domènech, J., Arroyo, L., Torre, I., and Gordo, O. 2016. An evaluation of monk parakeet damage to crops in the metropolitan area of Barcelona. *Animal Biodiversity and Conservation* 39:141–145.

Tella, J. L. 2011. The unknown extent of ancient bird introductions. *Ardeola* 58:399–404.

Tella, J. L., and Hiraldo, F. 2014. Illegal and legal parrot trade shows a long-term, cross-cultural preference for the most attractive species increasing their risk of extinction. *PLoS One* 9:e107546.

Williamson, M. 1996. *Biological Invasions*. London: Chapman and Hall.

16

NATURALIZED PARROTS IN THE UNITED KINGDOM

Christopher J. Butler

INTRODUCTION

The movement of species from their original range to new locations has occurred for millennia, with evidence for these translocations in the United Kingdom (UK) dating back to at least 6,000 years ago (Webb 1985). However, most introductions into the UK have occurred after maritime transport between continents became widespread during the 16th century (di Castri et al. 1990). More than 55,000 species are estimated to have been introduced into the British Isles, the vast majority of which are plants (Manchester and Bullock 2000).

Psittacines have long been brought into the UK, where breeding feral parrots has been noted as far back as 1855 (Lever 1977). Several species have been recorded breeding in the UK, but only the Rose-ringed Parakeet (*Psittacula krameri*) has become firmly established. As recently as 1999, it was estimated that there were only ~1,500 Rose-ringed Parakeets in the UK (Pithon and Dytham 1999). During the 21st century, however, the population has increased exponentially, and approximately 31,000 Rose-ringed Parakeets were present by 2014 (Pârâu et al. 2016). This chapter will discuss the historical changes in species richness of parrots in the UK, along with aspects of their breeding biology, population trajectory, ecology, and impacts.

HISTORICAL CHANGES IN NUMBER AND COMPOSITION

Six parrot species are known to or may have recently bred in the UK: Rose-ringed Parakeet, Alexandrine Parakeet (*Psittacula eupatria*), Monk Parakeet (*Myiopsitta monachus*), Blue-crowned Parakeet (*Thectocercus acuticaudatus*), Rosy-faced Lovebird (*Agapornis roseicollis*), and Budgerigar (*Melopsittacus undulatus*) (Holling and the Rare Breeding Birds Panel 2011). A presumed family party of Red-rumped Parrots (*Psephotus haematonotus*) was observed in Northamptonshire in 1998, but it is more likely that the birds escaped together than that they have bred in the wild (Holling and the Rare Breeding Birds Panel 2011). Braun (chap. 14 this vol.) also reviews historical records from early in the 20th century.

Rose-ringed Parakeet (*Psittacula krameri*)

The Rose-ringed Parakeet is the most widely introduced parrot in the world, with populations apparently established in 35 countries on five continents (only Australia and Antarctica remain uncolonized) (Butler 2003). Establishment success of Rose-ringed Parakeets, as well as Monk Parakeets, in Europe is positively correlated with human population density and negatively correlated with the number of frost days (Strubbe and Matthysen 2009). Populations in many western and southern European countries are growing at a rate of 19% per year (Pârâu et al. 2016).

Rose-ringed Parakeets were first reported breeding in the UK in 1855 in Norfolk, but this colony soon disappeared (Lever 1977). During the 1930s, parakeets were present in Epping Forest in Essex, but this colony did not persist (Morgan 1993). It was not until 1969 that a family group of parakeets was noted, this time in Southfleet,

Kent (Hudson 1974a). By 1973, however, the species was widespread, and its range expansion continued throughout the 1970s (Hudson 1974b; Lever 1977; Morrison 1997).

The origins of Rose-ringed Parakeets in the UK have generally been attributed to escaped or released birds (Butler 2003). However, there has been speculation about the exact circumstances. Heald et al. (2019) summarize some of the more popular suggestions, including the claims that the birds were released during the filming of *The African Queen* (despite no parrots appearing in the movie), that a pair was released by Jimi Hendrix, and that the birds escaped from pet shops or an aviary in Syon Park. However, Heald et al. (2019) used a Dirichlet process mixture model of geographic profiling to demonstrate a lack of support for any of these possibilities and instead suggest that the establishment of Rose-ringed Parakeets in the UK is most likely due to repeated escapes and introductions.

From 1984 through 2007, a total of 16,500 Rose-ringed Parakeets were legally imported into the UK; 4,627 (approximately 28%) came from Asia (4,607 of these from India), and 10,396 (approximately 72%) came from Senegal, as reported by the Convention on International Trade in Endangered Species of Wild Fauna and Flora (CITES 2019). The UK birds appear to be a mixture of two of the four subspecies of Rose-ringed Parakeets (*P. k. borealis* and *P. k. manillensis*, Morgan 1993). In a genetic survey, 60% of the observed Rose-ringed Parakeet haplotypes in the UK came from the Asian population, a disproportionately large percentage (Jackson et al. 2015). That haplotypes from the northern areas of the Rose-ringed Parakeet native range are disproportionately prevalent in Western Europe suggests that birds from colder climes may be better adapted to climatic conditions in this region (Jackson et al. 2015).

In 1983, the British Ornithologists' Union (BOU) accepted the Rose-ringed Parakeet as a category C (established exotic) species and estimated that the population consisted of 500 birds (BOU 1983). By 1986, the population was estimated to consist of 500–1,000 birds (Lack 1986). Although imports reported by CITES peaked during the late 1980s, the number of reported birds remained relatively low during this period (Fig. 16.1). A simultaneous count of the known roosts in 1996 revealed that the population had increased to 1,508 individuals (Pithon and Dytham 1999). Since that time, however, there has been a dramatic increase in the number of Rose-ringed Parakeets. By 1999, there were approximately 2,500 parakeets at a single roost (Butler 2002).

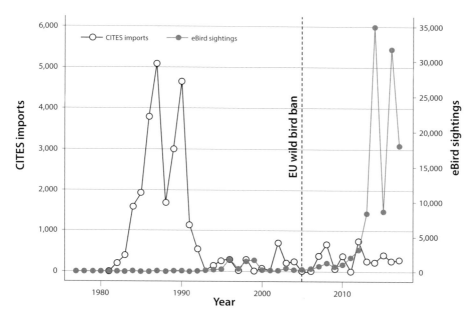

Figure 16.1. The importation of Rose-ringed Parakeets into the United Kingdom peaked between 1985 and 1990 and declined thereafter, according to CITES (2017). However, the number of Rose-ringed Parakeets reported annually into eBird did not begin to increase exponentially until 2010.

By 2005, large numbers of Rose-ringed Parakeets were seen at roosts in Kent and Surrey, and smaller numbers were noted in Buckinghamshire, Berkshire, Essex, Greater London, Hertfordshire, and Sussex (Holling and the Rare Breeding Birds Panel 2007). Since 2004, Rose-ringed Parakeets have been deemed sufficiently numerous to be monitored by the Breeding Bird Survey (BBS) spearheaded by the British Trust for Ornithology, Joint Nature Conservation Committee, and Royal Society for the Protection of Birds, and the parakeets were found in 3% (n = 110) of the monitored BBS squares in 2009 (Risely et al. 2010). Risely et al. (2010) noted a 696% increase in Rose-ringed Parakeet numbers in the BBS between 1995 and 2008. Musgrove et al. (2013) put the population of Rose-ringed Parakeets at approximately 8,600 pairs. By 2014, Rose-ringed Parakeets were considered to be "abundant" in parts of Greater London and Surrey and appeared to be establishing populations as far north as Newcastle (Holling and the Rare Breeding Birds Panel 2017). They currently occupy approximately 5,000 km² of the Greater London area and surrounding environs (Arnold et al. 2017). Pârâu et al. (2016) estimated that there are currently 31,000 Rose-ringed Parakeets in the UK. Figure 16.2 illustrates the spread of Rose-ringed Parakeets throughout the UK since they were first introduced.

Alexandrine Parakeet (*Psittacula eupatria*)

A pair of Alexandrine Parakeets bred in Lancashire and North Merseyside in 1997–99, and a second pair was noted in 1998 (Holling and the Rare Breeding Birds Panel 2011). These breeding attempts were successful, but most of the birds were shot in 1998 (Butler 2005), and the species

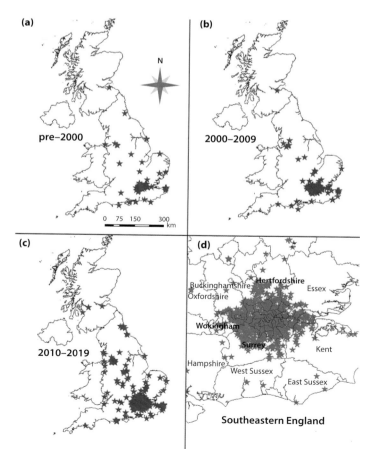

Figure 16.2. This series of maps, based on observations submitted to eBird, shows how the distribution of Rose-ringed Parakeets in the United Kingdom has continued to expand during the 21st century. By 2000 (a), Rose-ringed Parakeets were found primarily in the Greater London area and in northeastern Kent. After expanding over the next decade (b), their range began its greatest expansion after 2010 (c). By 2019 (d), the parakeets occupied a more or less continuous area from Hertfordshire south to West Sussex, east to Kent, and west to Oxfordshire.

has not been observed in this area since 1999 (Holling and the Rare Breeding Birds Panel 2011). In 2001, an Alexandrine Parakeet paired with a Rose-ringed Parakeet in Kent and attempted to breed (Butler 2002), but no further records have been submitted to the Rare Breeding Birds Panel. Occasional Alexandrine Parakeets were being reported in Northumberland and the West Midlands through 2019 (Fig. 16.3a).

Monk Parakeet
(*Myiopsitta monachus*)

The first breeding record for Monk Parakeets in the UK was during the 1930s at Whipsnade, where 31 birds were released into the wild (Marchant 2016). A small population was present near Tiverton, Devon, from 1987 to 1998 (Glaves and Darlaston 2000), but Monk Parakeets are currently restricted to the Greater London area, where populations have been identified in four general areas during the 2000s: Borehamwood, Barnes, Southall, and Mudchute/Isle of Dogs (Fig. 16.3b) (Arnold et al. 2017; Dyer et al. 2017).

In Borehamwood (Hertfordshire), Monk Parakeets were first reported in 1993, and the population increased to 51 birds by 2006 (Arnold et al. 2017). Nevertheless, despite this increase, only a single nest of Monk Parakeets was reported in the literature during 2003–5 (Holling and the Rare Breeding Birds Panel 2007). The population of Monk Parakeets persists in Borehamwood, although the number had declined to just 13 birds in 2015 (Arnold et al. 2017).

Both the Barnes and Southall Monk Parakeet populations were small at their outset and have not persisted. The Barnes population disappeared by 2002, and the Southall

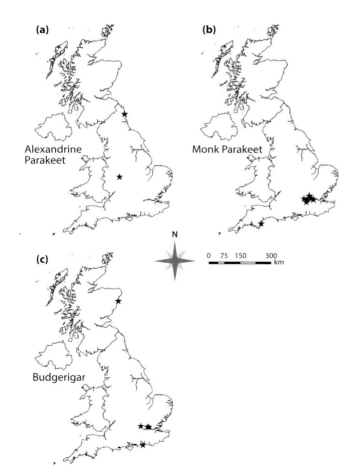

Figure 16.3. (a) Alexandrine Parakeet sightings reported in eBird through early 2019. Birds were reported only from Northumberland and the West Midlands. (b) Monk Parakeet sightings reported in eBird through 2018. Most observations were clustered in the Greater London area. (c) Budgerigar sightings reported in eBird through 2018. Most observations were from the Greater London area.

population by 2009 (Arnold et al. 2017; Holling and the Rare Breeding Birds Panel 2017). The Mudchute/Isle of Dogs Monk Parakeet population was first described in 2003; 25 birds were reported in 2006, 36 birds in 2010, and 20 birds in 2015 (Arnold et al. 2017).

As of 2014, breeding Monk Parakeets were restricted to the Mudchute/Isle of Dogs and Borehamwood populations (Holling and the Rare Breeding Birds Panel 2017). Monk Parakeets were still being reported on eBird at Mudchute Park and Farms in the Greater London area through at least 2016.

Blue-crowned Parakeet
(*Thectocercus acuticaudatus*)

Blue-crowned Parakeets were first reported in 1997, and juveniles were reported in 1999 (Arnold et al. 2017). One pair nested in Kent in 2001, although the eggs were depredated (Butler et al. 2002). No further records have been reported to the Rare Breeding Birds Panel (Holling and the Rare Breeding Birds Panel 2011), although Butler (2005) reported that the population was believed to be increasing through at least 2004. The highest count reported was 15 individuals in 1999, but by 2014 only five individuals were documented (Arnold et al. 2017).

Rosy-faced Lovebird
(*Agapornis roseicollis*)

Rosy-faced Lovebirds were present in Dunbar, Lothian, from 2002 through 2006 and apparently bred during two of those years (Holling and the Rare Breeding Birds Panel 2011). Although occasionally reported in London, these birds are no longer thought to breed in the UK (Arnold et al. 2017).

Budgerigar
(*Melopsittacus undulatus*)

A breeding population of Budgerigars was present at Tresco in the Isles of Scilly (off

Free-ranging macaws at a residence in Kirkby Stephen, Cumbria, UK. Shown are the Blue-and-yellow Macaw (*Ara ararauna*, lower center and lower right) and the Scarlet Macaw (*A. macao*, left and upper center) Some macaws are reported to have nested in chimneys in the town. April 2008. Photo © Roelant Jonker.

England's west coast) from 1969 through 1976 (Butler 2005). Four pairs were released in 1969, and six additional pairs were released in 1970 (King 1978). Food and nest boxes were provided for these individuals, and within five years the population increased to approximately 100 birds, with 35 breeding pairs (Hunt and Robinson 1976). However, the resident who provided food and nesting sites moved in 1975, and the population declined to just one bird by 1976 and disappeared entirely by 1977 (King 1978; Butler 2005). Escaped and released Budgerigars are still occasionally reported in the UK (Fig. 16.3c), but they are not currently believed to be reproducing in the wild (Dyer et al. 2017).

BIOLOGY OF THE ROSE-RINGED PARAKEET IN THE UK

Rose-ringed Parakeets are obligate cavity nesters, breeding primarily in tree cavities (Khan et al. 2004), but have also been reported to occasionally use crevices in rocks or walls (Lamba 1966, Mori et al. 2017). As of the early 2000s, all UK nests reported in the literature have been restricted to tree cavities or nest boxes (Butler 2003).

Parakeets in the UK nested in trees of 12 different genera, with *Fraxinus* spp. accounting for 33% of the trees used and *Quercus* spp. accounting for 21.7% (Butler et al. 2013). Butler et al. (2013) examined characteristics of 108 nests from the UK, including 46 nests from the Isle of Thanet (Kent), 34 from southeast London, 26 in southwest London, and two nests at Studland (Dorset). Nests were located an average of 8.1 ± 3.8 m off the ground. This is similar to heights reported in Marseilles, France, where nest cavities also were typically found between 5 m and 10 m high (Le Louarn 2017), and in California, where Rose-ringed Parakeets bred in a cavity approximately 10 m above the ground (Mabb 1997). The nests were in trees surrounded by substantial shrub and tree cover (13.1 ± 1.6 trees or shrubs ≤10 m from the nest) (Butler et al. 2013).

Butler et al. (2013) found that the Rose-ringed Parakeets bred primarily in old woodpecker nests (n = 58) but would also use natural cavities (n = 31) and nest boxes (n = 19). Parakeets were not observed excavating their own cavities, and only one instance of enlarging an existing cavity—the hole of a Great Spotted Woodpecker (*Dendrocopos major*)—was noted. Generally, Rose-ringed Parakeet nests were found in apparent European Green Woodpecker (*Picus viridis*) cavities. Nest-site reuse varied across the study sites, ranging from 26.1%–47.6% in the Greater London area, to 64.3%–80% of nests on the Isle of Thanet (Butler et al. 2013).

In the UK, eggs may be laid as early as 27 February, and a few individuals may lay eggs through mid-May; the mean date of first egg laid is 26 March ± 1.3 days (Butler et al. 2013). The 26 March date was similar to that reported in northern India (11 March; Simwat and Sidhu 1973) and Bangladesh (10 March; Hossain et al. 1993). However, some authors have suggested that Rose-ringed Parakeets breeding in Europe breed sooner than individuals at similar latitudes in their native range (e.g., Luna et al. 2017). Clutch sizes in the UK ranged from one to seven eggs, with a mean of 3.7 ± 1.2 eggs/clutch (Butler et al. 2013). In a captive outdoor colony in northeastern England, Lambert et al. (2009) found that the average first clutch size was 3.6 eggs, with an overall productivity of 2.5 fertile eggs. Although second clutches have been reported in other populations (e.g., Hossain et al. 1993), no second clutches have been reported in the wild in the UK. However, in a captive outdoor colony, 11 of 19 pairs renested, producing a similar average clutch size but an average of only 1.8 fertile eggs (Lambert et al. 2009).

Butler et al. (2013) found that an average of 72% of UK nests were successful in fledging at least one young, with an average of 1.4 ± 0.2 chicks fledging from each nest, which is approximately double the fledging rate reported by Pithon and Dytham (1999). The reason for the apparent increase is unclear, but it has been noted that reproductive output may be density-dependent in some introduced species, with low reproductive output at low densities due to the Allee effect (Lewis and Dareiva 1993). It is conceivable that the increasing density of Rose-ringed Parakeets in England since the late 1990s may have had a positive effect on the number of young fledged. The breeding success reported for the UK population (Butler et al. 2013) is comparable to that reported in India (e.g., Shivanarayan et al. 1981; Hossain et al. 1993; but see Lamba 1966). Shwartz et al. (2009) suggested that, while predation limited reproductive success in India, the colder climate of the UK was responsible for lower fledging success. Interestingly, in the UK, subadult male Rose-ringed Parakeets have been observed breeding (Butler et al. 2013), a result not reported elsewhere.

Rose-ringed Parakeets in the UK appear to molt after breeding (Butler 2003). Butler and Gosler (2004) examined the molt strategies of specimens in the Natural History Museum at Tring and found that *P. k. borealis* tended to molt primary feathers from 1 August through 24 December, and a single bird still showed primary molt on 28 December. *P. k. manillensis* specimens, in contrast, were recorded molting from 3 May through 7 August. In the UK, Rose-ringed Parakeets were recorded exhibiting primary molt between 8 May and 17 July, which closely mirrors the molting strategy of *P. k. manillensis*. Primary molt begins in feathers five or six and then spreads in both directions. Although Roselaar (1985) suggested that Rose-ringed Parakeets have a suspended molt, Butler and Gosler (2004) found no evidence of suspended molt in parakeets banded in the UK.

POPULATION TRAJECTORY

As noted earlier in the chapter, Rose-ringed Parakeet populations in the UK have rapidly increased since the end of the 20th century. Counts of parakeets at communal roosts showed a rapid increase, from 1,508 individuals in 1996 to approximately 31,000 individuals by 2014 (Pithon and Dytham 1999; Pârâu et al. 2016). Butler (2003) created a simple exponential model that predicted that there should be 28,057 birds by 2012, and a simultaneous count of all known roosts during that year found a total of 29,133 parakeets (Peck 2013). Peck (2013) suggests that the Greater London area and the Isle of Thanet population may have now reached the local carrying capacity. However, as Pârâu et al. (2016) note, it becomes difficult to monitor multiple roost sites simultaneously, and they suggest caution when extrapolating trends from counts of multiple large roosts.

Although individual birds may travel up to 7 km between nest cavities and roost sites, the rate of spread of established populations is relatively slow. Butler (2003) modeled that Rose-ringed Parakeets are spreading at a rate of approximately 0.4 km/year. Coupled with the previously reported reproductive rate, this has led to Rose-ringed Parakeets becoming locally abundant in areas where they have been established for decades. During timed counts for the London Bird Atlas, Rose-ringed Parakeets were the most commonly recorded species (per hour) for nine tetrads (Arnold et al. 2017).

It seems plausible that the spread of Rose-ringed Parakeets across the UK is due to a combination of released or escaped birds coupled with a gradual expansion away from core breeding areas in the Greater London area. At the moment, this species' northern limit in the UK is unclear. Rose-ringed Parakeets have expanded into regions considerably colder than their native range (Strubbe et al. 2015). Rose-ringed Parakeets do not appear to show evidence of hypothermia when temperatures drop to 5°C (Thabethe et al. 2013). However, subfreezing temperatures can cause physical damage or even induce mortality, which may eventually limit their northern distribution. Rose-ringed Parakeets introduced into New York City suffered from frostbite (Roscoe et al. 1976), and winter-weather-related mortality has been documented in Belgium (Temara and Arnhem 1996). In Germany, 22 of 32 adult individuals (69%) examined had toe or claw damage that was attributed to frostbite (M. Braun, pers. comm.). Çalışkan (2018) suggested that the urban heat-island effect may allow Rose-ringed Parakeets to persist in areas that are otherwise too cold for them. Luna et al. (2017) also suggested that a phenological mismatch, caused by the parakeets initiating breeding too early, may limit the spread of the species.

IMPACTS

Introduced species are widely considered to be a serious ecological problem (Bury and Luckenbach 1976; Baker 1990; Temple 1992; Manchester and Bullock 2000). They are often cited as causing serious damage to landscapes and ecosystems, particularly on islands (Savidge 1987; Williamson and Fitter 1996; Daehler and Gordon 1997), but introduced species are a global phenomenon. Indeed, introduced species have even been recorded in Antarctica (Clayton et al. 1997). The impacts of introduced species may be grouped into economic impacts, disease impacts, ecological impacts, and environmental damage.

Monk Parakeets frequently nest on electrical utility facilities such as power poles and substations, and their bulky nests can interfere with electrical service (Avery and Shiels 2018). In the UK, the Department for Environment, Food and Rural Affairs (DEFRA) conducted a trial program in 2008 to control Monk Parkaeets. Beginning in 2011, this program, coordinated by the Animal and Plant Health Agency, became a widespread effort to eliminate the species from the entire UK (Arnold et al. 2017; Holling and the Rare Breeding Birds Panel 2017). Between 2009 and 2014, the UK government spent ~$340,000 (£259,000) to kill 62 birds, destroy 212 eggs, and eliminate 21 nests (DEFRA 2014). The decline in the numbers of Monk Parakeets at Borehamwood, Mudchute, and Southall is attributed to this program (Arnold et al. 2017).

Rose-ringed Parakeets have been documented causing damage to vineyards in southeastern England (Hamilton 2004; Fletcher and Askew 2007). Saines (2002) reported that a vineyard that normally produces 3,000 bottles of wine was so impacted by Rose-ringed Parakeets feeding on the grapes that only 500 bottles could be produced. In addition, the Food and Environment Research Agency (FERA) reported that approximately 200–300 Rose-ringed Parakeets regularly visited a 1 ha market garden in Newdigate, Surrey, and damaged nearly all crops grown there (FERA 2009). The cost of damage to vineyards in Surrey (UK) is estimated to be approximately ~$8,250 (£5,000) per year (Fletcher and Askew 2007).

In general, the crops at risk of psittacine predation appear to be primarily those grown at the rural-urban interface (FERA 2009). The crops believed to be most at risk of parakeet predation include grapes, orchard fruits, and corn. However, the extent of the damage remains localized and rather limited, as licenses to kill or take Rose-ringed Parakeets have been issued to only two growers (FERA 2009). Although Rose-ringed Parakeets were initially protected along with native bird species (Butler 2003), DEFRA (2019) implemented changes to the licensing situation in 2010, and Monk Parakeets and Rose-ringed Parakeets can now be legally killed to prevent serious damage or disease, to preserve public health or public safety, to preserve air safety, and to preserve flora and fauna (Tayleur 2010).

In addition to agricultural damage, Rose-ringed Parakeets have become common in the area around Heathrow Airport, and there is the potential for bird-aircraft collisions. Fletcher and Askew (2007) report that one of 54 bird strikes in 2005 and two of 44 bird strikes in 2006 involved Rose-ringed Parakeets.

Rose-ringed Parakeets have the potential to compete with native species for nest cavities and are on the DAISIE (Delivering Alien Invasive Species Inventories Europe) list for Europe (FERA 2009; Hernández-Brito et al. 2014). However, the evidence for competition with native species is mixed. Newson et al. (2011) failed to find any evidence that Rose-ringed Parakeets negatively affected any cavity-nesting species in the UK, although they acknowledged that this could potentially be a cause for concern as populations continue to increase.

As noted earlier, Rose-ringed Parakeets do not appear to be damaging the environment in the UK at this point (Tayleur 2010). However, the potential exists for localized damage around roost sites. For example, parakeets frequently vocalize at roost sites (Butler, pers. obs.), and the resulting noise pollution could potentially affect avian or human communities around the roost. In addition, the potential accumulation of droppings under a roost could eventually affect the vegetation growing underneath the trees, through increased nitrogen enrichment, altered pH, and a potential change in vegetation biomass and composition.

FUTURE DIRECTIONS FOR RESEARCH

Rose-ringed Parakeets are able to survive in climates that are colder than their native range, but the exact tolerance limits for this species are unclear. Future research into cold tolerances, as well as the role of urban heat islands and microclimates on the persistence of Rose-ringed Parakeets may allow for a more fine-scale model of the potential future distribution of this species. In addition, research into the effects of anthropogenic climate change on the potential distribution of this species in the UK is warranted.

Furthermore, the incorporation of remote sensing to further refine species distribution models may provide additional clarity on locations where Rose-ringed Parakeets may spread in the future. The use of remote sensing with sufficient detail to resolve individual tree species has become increasingly frequent. One study demonstrated that it was possible to use remote sensing to identify 98.5% of the tree species that Rose-ringed Parakeets might use for breeding in Marseille, France (Le Louarn et al. 2017).

Another future research topic could be to explore whether interference competition with other bird species at feeders and natural foraging sites might affect populations of native bird species. For example, Fuller et al. (2008) suggested that feeding birds in an urban setting can impact avian assemblages. In contrast, Brittingham and Temple (1992) found that winter bird feeding did not affect avian survival rates. Consequently, further work needs to be done to determine whether the competitive dominance of Rose-ringed Parakeets at bird feeders affects local bird populations, as well as the effect of supplemental feeding on the parakeets' survival and reproductive success.

Several studies have examined the potential for Rose-ringed Parakeets to affect native bird species, but studies examining the effects of these birds on other native species in the UK are lacking. For example, some sources suggest that exotic parrots could potentially affect bats and dormice (for an overview, see Menchetti and Mori 2014, chap. 6 this vol.; Brightsmith and Kiacz, chap. 9 this vol.). However, to date this hypothesis has not been explicitly tested. The effects of Rose-ringed Parakeets on native and exotic flora in Europe have also not been explored in detail. For example, Rose-ringed Parakeets gather in large communal roosts (with as many as 15,000 birds occupying one roost), and to date it is unclear what effect this continuous rain of droppings may have on vegetation underneath the roost.

Finally, the frequency and prevalence of parasites and diseases in feral parrots in the UK have not been explored in depth. As Torchin et al. (2003) note, exotic species frequently have fewer parasites in areas they have been introduced to than in regions where they are native. This may allow them to experience higher survivorship and reproductive success. However, studies on parasite prevalence and associated effects on psittacines in the UK are lacking. In addition, despite the potential for several diseases to spread from feral parrots, few studies have found Rose-ringed Parakeets to have any diseases that might impact native species or humans. Further studies to explore the prevalence of diseases in Rose-ringed Parakeets in the UK will be necessary in order to gauge the potential risk.

REFERENCES

Arnold, R., Woodward, I., and N. Smith. 2017. *Parrots in the London Area: A London Bird Atlas Supplement*. London: London Natural History Society.

Avery, M. L., and Shiels, A. B. 2018. Monk and rose-ringed parakeets. In *Ecology and Management of Terrestrial Vertebrate Invasive Species in the United States*, ed. W. C. Pitt, J. C. Beasley, and G. W. Witmer, 333–357. Boca Raton, FL: CRC Press.

Baker, S. J. 1990. Escaped exotic mammals in Britain. *Mammal Review* 20:75–96.

BOU. 1983. British Ornithologists' Union Records Committee: 11th report. *Ibis* 126: 440–445.

Brittingham, M. C., and Temple, S. A. 1992. Does winter bird feeding promote dependency? *Journal of Field Ornithology* 63: 190-194.

Bury, R. B., and Luckenbach, R. A. 1976. Introduced amphibians and reptiles in California. *Biological Conservation* 10:1–14.

Butler, C. 2002. Breeding parrots in Britain. *British Birds* 95:345–348.

Butler, C. J. 2003. Population biology of the introduced rose-ringed parakeet *Psittacula krameri* in the UK. PhD diss., Oxford Univ., UK.

Butler, C. J. 2005. Feral parrots in the continental United States and United Kingdom: Past, present, and future. *Journal of Avian Medicine and Surgery* 19:142–149.

Butler, C. J., Cresswell, W., Gosler, A., and Perrins, C. 2013. The breeding biology of rose-ringed parakeets *Psittacula krameri* in England during a period of rapid population expansion. *Bird Study* 60:527–532.

Butler, C., and Gosler, A. 2004. Sexing and ageing rose-ringed parakeets *Psittacula krameri* in Britain. *Ringing & Migration* 22:7–12.

Butler, C., Hazlehurst, G., and Butler, K. 2002. First nesting by blue-crowned parakeet in Britain. *British Birds* 95:17–20.

Çalışkan, O. 2018. Rose-ringed parakeets (*Psittacula krameri*) and geographical evaluation of habitats in Turkey. *International Journal of Geography and Geography Education* 38:279–294.

CITES. 2017. CITES Trade Database. Accessed June 2017. http://trade.cites.org.

CITES. 2019. CITES Trade Database. Accessed Dec. 2019. http://trade.cites.org.

Clayton, M. N., Wiencke, C., and Kloser, H. 1997. New records of temperate and sub-Antarctic marine benthic macroalgae from Antarctica. *Polar Biology* 17:141–149.

Daehler, C. C., and Gordon, D. R. 1997. To introduce or not to introduce: Trade-offs of non-indigenous organisms. *Trends in Ecology & Evolution* 12:424–425.

DEFRA. 2014. Dept. of the Environment, Food and Rural Affairs. Request for information: Monk parakeets. Ref. RFI 6871. Accessed Dec. 2019. https://assets.publishing.service.gov.uk/government/uploads/system/uploads/attachment_data/file/355052/RFI_6871__2__amended.pdf.

DEFRA. 2019. Wild bird general licences: Defra and Natural England's approach. Dept. of the Environment, Food and Rural Affairs. Accessed Dec. 2019. https://www.gov.uk/government/publications/wild-bird-general-licences-defra-and-natural-englands-approach.

di Castri, F., Hansen, A. J., and Debussche, M. 1990. *Biological Invasions in Europe and the Mediterranean Basin*. Dordrecht, Netherlands: Kluwer Academic.

Dyer, E. E., Cassey, P., Redding, D. W., Collen, B., Franks, V., Gaston, K., Jones, K. E., Kark, S., Orme, C.D.L., and Blackburn, T. M. 2017. The global distribution and drivers of introduced bird species richness. *PLoS Biology* 15:e2000942.

FERA. 2009. *Rose-ringed Parakeets in England: A Scoping Study of Potential Damage to Agricultural Interests and Management Measures*. York: UK: Food and Environment Research Agency.

Fletcher, M., and Askew, N. 2007. *Review of the Status, Ecology and Likely Future Spread of Parakeets in England*. York, UK: Central Science Laboratory.

Fuller, R. A., Warren, P. H., Armsworth, P. R., Barbosa, O., and Gaston, K. J. 2008. Garden bird feeding predicts the structure of urban avian assemblages. *Diversity and Distributions* 14:131–137.

Glaves, D. J., and Darlaston, M. 2000. *Devon Bird Report* 71:12–144.

Hamilton, J. 2004. Pesky Polly. *Guardian* online, 02 Mar. http://society.guardian.co.uk/environment/story/0,,1160291,00.html.

Heald, O.J.N., Fraticelli, C., Cox, S. E., Stevens, M.C.A., Faulkner, S. C., Blackburn, T. M., and Le Comber, S. C. 2019. Understanding the origins of the ring-necked parakeet in the UK. *Journal of Zoology* 312:1–11.

Hernández-Brito, D., Carrete, M., Popa-Lisseanu, A. G., Ibáñez, C., and Tella, J. L. 2014. Crowding in the city: Losing and winning competitors of an invasive bird. *PLoS One* 9(6):e100593.

Holling, M., and the Rare Breeding Birds Panel. 2007. Non-native breeding birds in the United Kingdom in 2003, 2004, and 2005. *British Birds* 100:638–649.

Holling, M., and the Rare Breeding Birds Panel. 2011. Non-native breeding birds in the United Kingdom in 2006, 2007, and 2008. *British Birds* 104:114–138.

Holling, M., and the Rare Breeding Birds Panel. 2017. Non-native breeding birds in the UK, 2012–14. *British Birds* 110:92–108.

Hossain, M. T., Husain, K. Z., and Rahman, M. K. 1993. Some aspects of the breeding biology of the rose-ringed parakeet, *Psittacula krameri borealis* (Neumann). *Bangladesh Journal of Zoology* 21:77–85.

Hudson, R. 1974a. News and comment. *British Birds* 67:32–34.

Hudson, R. 1974b. News and comment. *British Birds* 67:173–175.

Hunt, D. B., and Robinson, H. P. K. 1976. Systematic list. *Isles Scilly Bird Report* 1975:7–28.

Jackson, H., Strubbe, D., Tollington, S., Prys-Jones, R., Matthysen, E., and Groombridge, J. J. 2015. Ancestral origins and invasion pathways in a globally invasive bird correlate with climate and influences from bird trade. *Molecular Ecology* 24:4269–4285.

Khan, H. A., Beg, M. A., and Khan, A. A. 2004. Breeding habitats of the ring-necked parakeet (*Psittacula krameri*) in the cultivations of central Punjab. *Pakistan Journal of Zoology* 36:133–138.

King, B. 1978. Free-winged budgerigars in the Isles of Scilly. *British Birds*. 71:82–83.

Lack, P. C. 1986. *The Atlas of Wintering Birds in Britain and Ireland*. Calton, UK: T. & A. D. Poyser.

Lamba, B.S. 1966. Nidification of some common Indian birds. 10. The rose-ringed parakeet, *Psittacula krameri* Scopoli. *Proceedings of the Zoological Society (Calcutta)* 19:77–85.

Lambert, M. S., Massei, G., Bell, J., Berry, L., Haigh, C., and Cowan, D. P. 2009. Reproductive success of rose-ringed parakeets *Psittacula krameri* in a captive UK population. *Pest Management Science* 65:1215–1218.

Le Louarn, M. 2017. Sélection de l'habitat d'une espèce exotique en milieu urbain: Le cas de la Perruche à collier *Psittacula krameri*. PhD diss., Univ. of Aix-Marseille, France.

REFERENCES

Le Louarn, M., Clergeau, P., Briche, E., and Deschamps-Cottin, M. 2017. "Kill two birds with one stone": Urban tree species classification using bi-temporal pléiades images to study nesting preferences of an invasive bird. *Remote Sensing* 9:916.

Lever, C. 1977. *The Naturalized Animals of the British Isles*. London: Hutchinson.

Lewis, M. A., and Dareiva, P. 1993. Allee dynamics and the spread of invading organisms. *Theoretical Population Biology* 43:141–158.

Luna, A., Franz, D., Strubbe, D., Shwartz, A., Braun, M. P., Hernández-Brito, D., Malihi, Y., Kaplan, A., Mori, E., Menchetti, M., and van Turnhout, C. A. 2017. Reproductive timing as a constraint on invasion success in the ring-necked parakeet (*Psittacula krameri*). *Biological Invasions* 19:2247–2259.

Mabb, K. T. 1997. Nesting behaviour of *Amazona* parrots and rose-ringed parakeets in the San Gabriel Valley, California. *Western Birds* 28:209–217.

Manchester, S. J., and Bullock, J. M. 2000. The impacts of non-native species on UK biodiversity and the effectiveness of control. *Journal of Applied Ecology* 37:845–864.

Marchant, J. 2016. Monk parakeet, *Myiopsitta monachus*. GB Non-native Species Secretariat. http://www.nonnativespecies.org/factsheet/factsheet.cfm?speciesId=2281.

Menchetti, M., and Mori, E. 2014. Worldwide impact of alien parrots (Aves: Psittaciformes) on native biodiversity and environment: A review. *Ethology Ecology & Evolution* 26:172–194.

Morgan, D. H. 1993. Feral rose-ringed parakeets in Britain. *British Birds* 86:561–564.

Mori, E., Ancillotto, L, Menchetti, M, and Strubbe, D. 2017. "The early bird catches the nest": Possible competition between scops owls and ring-necked parakeets. *Animal Conservation* 20: 463-470.

Morrison, S. 1997. *Rare Birds in Dorset*. Poole, UK: Dorset Bird Club.

Musgrove, A., Aebischer, N., Eaton, M., Hearn, R., Newson, S., Noble, D., Parsons, M., Riseley, K, and Stroud, D. 2013. Population estimates of birds in Great Britain and the United Kingdom. *British Birds* 106:64–100.

Newson, S. E., Johnston, A., Parrott, D., and Leech, D. I. 2011. Evaluating the population-level impact of an invasive species, ring-necked parakeet *Psittacula krameri*, on native avifauna. *Ibis* 153:509–516.

Pârâu, L. G., Strubbe, D., Mori, E., Menchetti, M., Ancillotto, L., Kleunen, A. V., White, R. L., Luna, A., Hernández-Brito, D., Louarn, M. L., and Clergeau, P. 2016. Rose-ringed parakeet *Psittacula krameri* populations and numbers in Europe: A complete overview. *Open Ornithology Journal* 9:1–13.

Peck, H. L. 2013. Investigating ecological impacts of the non-native population of rose-ringed parakeets (*Psittacula krameri*) in the UK. PhD diss., Imperial College, London.

Pithon, J. A., and Dytham, C. 1999. Census of the British ring-necked parakeet *Psittacula krameri* population by simultaneous count of roosts. *Bird Study* 46:112–115.

Risely, K., Baillie, S. R., Eaton, M. A., Joys, A. C., Musgrove, A. J., Noble, D. G., Renwick, A. R., and Wright, L. J. 2010. *The Breeding Bird Survey 2009*. BTO Research Report 559. Thetford, UK: British Trust for Ornithology.

Roscoe, D. E., Stone, W. B., Petrie, L., and Renkavinsky, J. L. 1976. Exotic psittacines in New York State. *New York Fish and Game Journal* 23:99–100.

Roselaar, C. S. 1985. *Psittacula krameri* ring-necked parakeet (rose-ringed parakeet). In *Handbook of the Birds of Europe, the Middle East and North Africa: The Birds of the Western Palearctic*, ed. C. S. Roselaar et al. Vol. 4: *Terns to Woodpeckers*. Oxford: Oxford Univ. Press.

Saines, K. 2002. Parakeets are a pest at Painshill Park. *Kingston Guardian*, 31 Oct.

Savidge, J. A. 1987. Extinction of an island forest avifauna by an introduced snake. *Ecology* 68:660–668.

Shivanarayan, N., Babu, K. S., and Ali, M. H. 1981. Breeding biology of roseringed parakeet *Psittacula krameri* at Maruteru. *Pavo* 19:92–96.

Shwartz, A., Strubbe, D., Butler, C. J., Matthysen, E., and Kark, S. 2009. The effect of enemy-release and climate conditions on invasive birds: A regional test using the rose-ringed parakeet (*Psittacula krameri*) as a case study. *Diversity and Distributions* 15:310–318.

Simwat, G. S., and Sidhu, A. S. 1973. Nidification of rose-ringed parakeet, *Psittacula krameri* (Scopoli), in Punjab. *Indian Journal of Agricultural Science* 43:648–652.

Strubbe, D., Jackson, H., Groombridge, J., and Matthysen, E. 2015. Invasion success of a global avian invader is explained by within-taxon niche structure and association with humans in the native range. *Diversity and Distributions* 21:675–685.

Strubbe, D., and Matthysen, E. 2009. Establishment success of invasive ring-necked and monk parakeets in Europe. *Journal of Biogeography* 36:2264–2278.

Tayleur, J. R. 2010. A comparison of the establishment, expansion and potential impacts of two introduced parakeets in the United Kingdom. *BOU Proceedings: The Impacts of Non-native Species*: 1–12.

Temara, K., and Arnhem, R. 1996. Perruches a collier (*Psittacula krameri*) victimes des conditions climatiques en region Bruxelloise. *Aves* 33:128–129.

Temple, S. A. 1992. Exotic birds: A growing problem with no easy solution. *Auk* 109:395–397

Thabethe, V., Brown, M., Downs, C. T., Hart, L. A., and Thompson, L. I. 2013. Seasonal effects on the thermoregulation of invasive rose-ringed parakeets (*Psittacula krameri*). *Journal of Thermal Biology* 38:553–559.

Torchin, M. E., Lafferty, K. D., Dobson, A. P., McKenzie, V. J., and Kuris, A. M. 2003. Introduced species and their missing parasites. *Nature* 421:628–630.

Webb, D. A. 1985. What are the criteria for presuming native status? *Watsonia* 15:231–236.

Williamson, M., and Fitter, A. 1996. The varying success of invaders. *Ecology* 77:1661–1666.

17

INTRODUCED AND NATURALIZED PARROTS OF SOUTH AFRICA: COLONIZATION AND THE WILDLIFE TRADE

Craig T. Symes, Ielyzaveta M. Ivanova, Caroline G. Howes, and Rowan O. Martin

INTRODUCTION

By virtue of their charismatic appeal, parrots are highly popular in aviculture and the cage-bird industry (Case 1996; Cassey et al. 2004; Pain et al. 2006; Blackburn et al. 2009). They are traded in vast numbers across the globe, and many species have been decimated in their natural ranges (Pires 2012; Blackburn et al. 2014; Berkunsky et al. 2017). The Grey Parrot (*Psittacus erithacus*), once widespread across the African tropics, has seen dramatic declines in its natural range, with complete extirpation in some areas (Pain et al. 2006; Martin et al. 2014; Annorbah et al. 2016; Hart et al. 2016; Lopes et al. 2019). On the other hand, some species have benefited significantly from changes to the environment by humans (Blair 1996; Chace and Walsh 2006; Blackburn et al. 2010; Bonter et al. 2010), expanding into new territories and forming new ecosystem interactions, with uncertain outcomes (Chapman 2005; Didham et al. 2005; Kumschick and Nentwig 2010; Grarock et al. 2013, 2014; Charter et al. 2016). The Rose-ringed Parakeet (*Psittacula krameri*), for example, is now recorded in at least 35 countries outside its native range and, in global terms, is viewed as a species of invasive concern (Long 1981; Lever 2005; Runde et al. 2007; Clavero et al. 2009; Strubbe and Matthysen 2009; Blackburn et al. 2014; Charter et al. 2016; Menchetti et al. 2016).

Across the globe, plants and animals moved by people are important sources for founder populations (Blackburn et al. 2013; Goss and Cumming 2013; Cassey et al. 2015; Martin-Albarracin et al. 2015; Cardador et al. 2016). Indeed, in birds, founder population size is identified as an important criterion in the successful establishment of any species outside its native range (Lockwood et al. 2005; Cardador et al. 2016). These movements are important in the context of globalization, with consequences on, for example, the spread of diseases, impacts on local biota, economies, and social well-being (Clavero and García-Berthou 2005; Blackburn et al. 2009; Clavero et al. 2009; Baker et al. 2013; Martin-Albarracin et al. 2015).

In South Africa a number of parrot species have been recorded incidentally as escapees, while some have become well established or show signs of becoming established (Symes 2014; Faulkner et al. 2017). This reporting is relatively well documented, with the majority of sightings reported by citizen scientists. In particular, the two phases of the Southern African Bird Atlas Project (SABAP1 and SABAP2) have provided valuable repositories for bird diversity and relative abundance (as indicated by reporting rate) across the subregion (Lee et al. 2017). However,

INTRODUCTION

a handful of species (usually exotic) associated with aviculture and the wild bird trade tend to be ignored and appear more likely to be reported elsewhere (newspapers, web sites, etc.) if escaped (Vall-llosera and Cassey 2017a). Also, there has been limited use of these data in understanding the role of these exotic species as passengers and drivers of natural ecosystems (but see Hugo and van Rensburg 2008, 2009; van Rensburg et al. 2009; Ivanova and Symes 2019). This is critical if informed management decisions by the relevant authorities are to be made (Andersen et al. 2004).

To understand the link between invasion processes and the large-scale trade, both legal and illegal, in this chapter we attempt to summarize the status of exotic parrot species currently reported in South Africa and parrot species regularly reported in international trade in large numbers. We highlight those species observed in the wild that have become naturalized and also show potential of naturalization/invasion. Firstly, we present a case study of the Rose-ringed Parakeet, introduced into the country at least 40 years ago and a suitable model for other introduced parrot species (Brooke 1997; Perrin and Cowgill 2005; Ivanova and Symes 2019). In the past decade, this species has increased significantly in range and abundance across the South African province of Gauteng (incorporating the greater Johannesburg and Pretoria areas (Ivanova and Symes 2019). Using data from SABAP2 we review how the species' range and abundance have changed, (1) in South Africa at large, and (2) specifically in Gauteng (where there has been a recognized establishment of the species), in the past decade. Exotic species not listed in the SABAP2 database are rarely reported, so a complete understanding of introduced species in the region cannot be obtained from this source alone. However, we also used the Southern African Rare Bird News, a biweekly Google Groups e-mail Listserv that informs birdwatchers about rare bird species in the southern African subregion (SARBN 2017), together with additional reporting platforms (e.g., ParrotAlert.com) and published accounts, to gain an understanding of escaped exotic parrot species in the country. Finally, we analyze data reported to the Convention on International Trade in Endangered Species of Wild Fauna and Flora (CITES 2019) on imports and exports of parrots to and from South Africa to gain insight into the role of recent trade, the origin of escaped species, and the establishment of exotic species. Together these analyses provide an important platform for scientists and managers to understand the rapid changes that are occurring in urban and anthropogenically transformed landscapes in South Africa (Brooke et al. 1986).

Rosy-faced Lovebirds (*Agapornis roseicollis*). This species is now naturalized in a number of countries, including South Africa. Phoenix, Arizona, May 2018. Photo by Christopher J. Butler.

INTRODUCED PARROTS OF SOUTH AFRICA

We defined status of non-native parrots in South Africa along a gradient of presence or establishment, from (1) those that have become naturalized (with established breeding populations across a wide area, e.g., Rose-ringed Parakeet) to (2) incidental reports of recently escaped individuals that are unlikely to breed or indeed survive unless recaptured to (3) those in which the majority of the population is confined to captivity (either being traded or confined to breeding and/or display facilities).

The Rose-ringed Parakeet in South Africa

The reporting data we analyzed were downloaded from SABAP2 (Harrison et al. 2008; Robertson et al. 2010; Underhill 2016) for the 10-year period July 2007–June 2017. Data were divided into annual periods (July–June), each of which spanned an entire summer period. The geographical units covered were pentads (5′ latitude × 5′ longitutde = ~9 km × 9 km). The presence of the parakeets in each pentad was used as an index of the degree of establishment (and ultimately population density) in a given area. The annual increase in area occupied by parakeets was determined by comparing the number of pentads occupied in each period with the number occupied in the previous period.

We then isolated data for Gauteng province only, for provincial level analyses. Again, parakeet distribution change over time was assessed by a count of the parakeet-reporting pentads for each data period and the increment examined. Additionally, the mean, minimum, and maximum numbers of annual cards (observations) filed for each sampled pentad were also determined. Parakeet reporting rates were compared through time using a single-factor analysis of variance. Reporting rates were used as a proxy for the species' abundance.

The Rose-ringed Parakeet's occurrence for July 2007–June 2017 has been concentrated around Johannesburg and Pretoria in Gauteng province and in the Durban city region of the KwaZulu-Natal province, according to records from the SABAP2 data. Several additional sightings of the species have been made in other provinces (in and out of other major cities), but usually in no more than two data periods, which could reflect a recent establishment or simply ad hoc presence of individual escaped birds (Fig. 17.1). The parakeet appears to be increasing in distribution across the country, with a growing number of

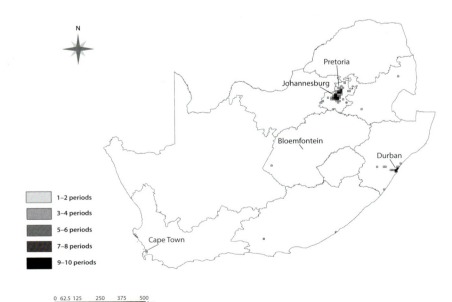

Figure 17.1. Distribution of Rose-ringed Parakeets in South Africa as indicated by SABAP2 data, July 2007–June 2017. Darker-colored pentads indicate areas where the species was recorded over a higher number of data periods (a reflection on consistency of presence). Major cities are labeled for orientation.

pentads reporting the species' presence, as well as slight but consistent increases in the proportional representation of parakeet-reporting pentads across all pentads sampled in the country (Table 17.1).

Reporting rates across parakeet-reporting pentads differed significantly across years ($F_{(9,520)}$ = 2.19; p < 0.05), but appeared to show consistent gains across the country (Table 17.1). This is interpreted as an increase in actual parakeet abundance within its growing South African range.

A highly significant correlation was found between the number of parakeet-reporting pentads and pentads sampled overall (r = 0.78; p < 0.005). However, a growth in sampling effort can account for only a fraction of the observed increase in parakeet presence, since the consistent growth in the proportion of all pentads reporting the parakeet (Table 17.1) indicates an exponential outpacing of the sampling growth. Indeed, a best-fit trend line relating numbers of sampled pentads with those of parakeet presence is an exponential curve of $y = 3.5571e^{0.0004x}$, with an r^2 = 0.82 conformity.

Cumulatively, the parakeet has been reported in 48 pentads across Gauteng (~16% of the province) over the 10-year period of sampling. The number of parakeet-reporting pentads increased from four (during July 2007–June 2008) to 35 (July 2016–June 2017) (Table 17.1; Fig. 17.2). The number of new pentads reporting the parakeet showed a steady increment across the entire period, with some data periods reporting as many as 10 new pentads for the parakeet's Gauteng range (Table 17.1).

In addition to showing a range expansion across the province, the mean reporting rates of the parakeet increased in pentads over the period, suggesting an increase in abundance within occupied areas (Table 17.1). Despite the very high standard deviation values (indicating high variability in abundance across the range), there was a significant increase in reporting rates across parakeet-reporting pentads over time ($F_{9,470}$ = 2.16; p < 0.05). No significant correlation was found between the number of sampled pentads and the number of parakeet-reporting pentads (r = 0.596, p > 0.05). Therefore, it appears that the range

TABLE 17.1

Changes in number of pentads (5′ lat. × 5′ long.) reporting Rose-ringed Parakeet, new pentads with parakeets, and changes in reporting rate across pentads with parakeets, across all of South Africa, and across Gauteng province, during annual periods from July 2007 to June 2017.

ANNUAL PERIOD (JULY–JUNE)	PENTADS SAMPLED IN GAUTENG (CARDS SAMPLED: MIN.; MEAN; MAX.)	PENTADS WITH PARAKEETS (% OF SAMPLED PENTADS)		NEW PENTADS WITH PARAKEETS		REPORTING RATE IN PARAKEET-REPORTING PENTADS (MEAN % ± SD)	
		SOUTH AFRICA	GAUTENG	SOUTH AFRICA	GAUTENG	SOUTH AFRICA	GAUTENG
2007–8	190 (1; 3; 35)	8 (0.41)	4 (2.1)	8	4	5.6 ± 17.7	4.2 ± 15.9
2008–9	268 (1; 9; 144)	19 (0.44)	14 (5.2)	11	10	7.0 ± 17.2	4.0 ± 11.2
2009–10	277 (1; 9; 156)	17 (0.33)	10 (3.6)	5	2	7.2 ± 17.6	5.2 ± 17.2
2010–11	279 (1; 9; 157)	22 (0.40)	14 (5.0)	6	4	10.9 ± 21.8	7.2 ± 18.5
2011–12	273 (1; 10; 166)	24 (0.44)	19 (7.0)	4	4	11.4 ± 21.8	8.2 ± 18.9
2012–13	258 (1; 10; 148)	23 (0.45)	16 (6.2)	4	2	13.2 ± 25.4	9.4 ± 21.0
2013–14	278 (1; 11; 192)	28 (0.54)	21 (7.6)	6	4	13.6 ± 23.8	9.3 ± 18.9
2014–15	282 (1; 14; 228)	35 (0.64)	26 (9.2)	6	4	15.0 ± 25.3	13.4 ± 23.9
2015–16	279 (1; 20; 365)	38 (0.63)	29 (10.4)	7	4	16.4 ± 26.5	13.6 ± 22.1
2016–17	273 (1; 19; 358)	46 (0.76)	35 (12.8)	15	10	20.1 ± 27.8	15.9 ± 24.1

gains for the parakeet in Gauteng (the most comprehensively sampled province in the SABAP2 project) reflect actual range increases, rather than overall SABAP2 sampling efforts. Indeed, the reason for focusing on the changes in abundance and occurrence in this province was because it is both the region where parakeets have been most successful and an intensively mapped region. During the study period, the 301 pentads in Gauteng had been sampled an average of 103 ± 218 times, with a range of 10 to 1,614 reporting cards per pentad.

Naturalized and Escaped Parrots in South Africa

We searched a total of 2,320 e-mails from July 2008 to June 2017 in the SARBN Google Group (SARBN 2017). All records of parrots were recorded, along with location, date, and observer. Photographs were downloaded when available for clarification.

A total of 10 records of parrots, mostly African species out of range, were reported in the SARBN: Meyer's Parrot (*Poicephalus meyeri*, one record), Brown-headed Parrot (*P. cryptoxanthus*, one), Rosy-faced Lovebird (*Agapornis roseicollis*, two), Yellow-collared Lovebird (*A. personatus*, one), *Agapornis* hybrid (one), Rose-ringed Parakeet (three), and Plum-headed Parakeet (*Psittacula cyanocephala*, one).

Based on current knowledge, we summarize those species that are confirmed as breeding outside of their natural ranges and have established or are likely to establish naturalized populations (Table 17.2). Records of escaped exotic parrots in the wild in South Africa are generally not reported to any central database;

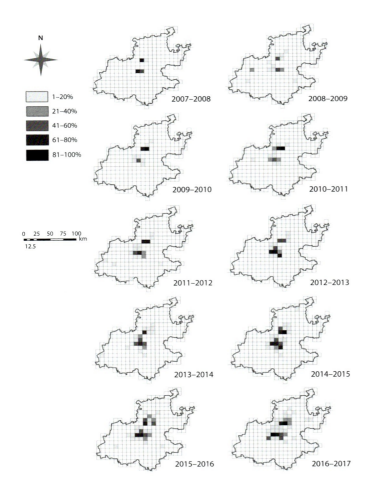

Figure 17.2. Occurrence of Rose-ringed Parakeets with reporting rates across Gauteng province, South Africa, presented in 10 annual data periods (July 2007–June 2017). Darker colors represent pentads in which the proportion of record cards containing parakeets was highest, as described by the legend.

however, because pet parrots command attention and represent some value to their owners, they are likely to be reported in some media. We therefore report those species we identified as escapees by scanning various South African–linked Facebook pages, using the keywords "parrot," "macaw," and "parakeet." Finally, and more specifically, we analyzed an online platform and screened all data available (up to 8 June 2018) on ParrotAlert.com, the "Lost and Found Parrot" website, for South Africa, using "all types" and "all dates" for species and date searches respectively in the available search options.

In addition to the confirmed establishment of the Rose-ringed Parakeet, there are a number of other species in South Africa, well outside their natural ranges (Table 17.2). Breeding has been recorded in only a few of these species, and indeed, there is some controversy about recognizing any of these additional species as becoming established or naturalized.

Numerous species are recorded as escapees in South Africa, and as previously reported, nearly every parrot species in captivity has at one time or another probably escaped the confines of captivity (W. Horsfield, pers. comm.). By scanning Facebook pages we identified records of escaped, found, or wild-sighted birds of the following species: Cockatiel (*Nymphicus hollandicus*, two records), Galah (*Eolophus roseicapilla*, one),

TABLE 17.2

Parrot species with documented or suspected breeding records outside their natural ranges in South Africa and suspected to have become established through human intervention rather than natural dispersal. South African population estimates are derived from limited publications and reports on each species in the region, to provide some working understanding of the level of invasion for the respective species (this study).

SPECIES	STATUS	ESTIMATED SOUTH AFRICAN POPULATION	NATURAL DISTRIBUTION	REFERENCES RELATING TO SOUTH AFRICAN OCCURRENCE
Rose-ringed Parakeet (*Psittacula krameri*)	Widespread and established in Gauteng province and greater Durban area (KwaZulu-Natal province)	10,000s	Africa, Asia; west to east Africa, Pakistan to Myanmar	Symes 2014; Hart and Downs 2014; Whittington-Jones 2017
Agapornis species, including Yellow-collared Lovebird (*A. personatus*), Rosy-faced Lovebird (*A. roseicollis*), and various color morphs and hybrids	Widespread and established across Gauteng province, sporadic sightings nationally	1,000s	Africa	Roche and Bedford-Shaw 2008; Symes 2014; Whittington-Jones 2016
Monk parakeet (*Myiopsitta monachus*)	Breeding records for Gauteng for late 1970s–early 1980s; likely extirpated	<10s	Argentina, Bolivia, Brazil, Paraguay, Uruguay	Symes 2014; BirdLife International 2018; G. Lockwood, pers. comm.
Amazona spp., Amazons	Historical breeding records for Pinetown (KwaZulu-Natal), likely extirpated	<10s	Neotropics	Symes 2014
Other species (see text)	Incidental escapees	<10 spp.	Various	This chapter

Grey Parrot (six), Red-fronted Parrot (*Poicephalus gulielmi*, two), Senegal Parrot (*P. senegalus*, two), Monk Parakeet (*Myiopsitta monachus*, one), Hyacinth Macaw (*Anodorhynchus hyacinthinus*, one), Sun Parakeet (*Aratinga solstitialis*, one), Blue-winged Macaw (*Primolius maracana*, one), Red-fronted Macaw (*Ara rubrogenys*, one), Rose-ringed Parakeet (two), Red-rumped Parrot (*Psephotus haematonotus*, one), and Budgerigar (*Melopsittacus undulatus*, two).

In a recent publication, Faulkner et al. (2017) report 15 parrot species introduced to South Africa and/or other parts of Africa (see references therein: Long 1981; Dean 2000; Lever 2005; Peacock et al. 2007; Picker and Griffiths 2011; van Rensburg et al. 2011). Although we do not elaborate on these further (except as reported in Table 17.2), we report them in an effort to highlight the extent to which parrot species may be recognized as an additional invasive "threat," or simply a group with potential to become naturalized outside native ranges; Rüppell's Parrot (*Poicephalus rueppellii*), Red-bellied parrot (*P. rufiventris*), Monk Parakeet, Turquoise-fronted Amazon (*Amazona aestiva*), Green-rumped Parrotlet (*Forpus passerinus*), Black-capped Parakeet (*Pyrrhura rupicola*), Burrowing Parrot (*Cyanoliseus patagonus*), Brown-throated Parakeet (*Eupsittula pertinax*), Dusky-headed Parakeet (*Aratinga weddellii*), Nanday Parakeet (*A. nenday*), Jandaya Parakeet (*A. jandaya*), Plum-headed Parakeet, Rose-ringed Parakeet, Budgerigar, and Grey-headed Lovebird (*Agapornis canus*) (Faulkner et al. 2017).

A total of 361 parrot records were extracted from ParrotAlert.com (Fig. 17.3). These records were categorized as found (67), lost (280), sightings (3), and stolen (11), in the following provinces, Western Cape (174), Gauteng (96), KwaZulu-Natal (66), Eastern Cape (21), Free State (3), and North West (1). The Grey Parrot was by far the most commonly reported parrot, making up 45.4% of all records. Rose-ringed Parakeets were also commonly recorded (16.3%).

INTERNATIONAL PARROT TRADE IN SOUTH AFRICA BETWEEN 2007 AND 2016

Data on the international trade of live parrots into and out of South Africa were downloaded directly from the CITES (2019) database. This database includes records of import, export, and reexport of CITES-listed species, as reported by parties to the convention. The majority of parrot species are listed in the CITES appendices, with the exception of Budgerigar, Rosy-faced Lovebird, Rose-ringed Parakeet, and Cockatiel. However, many parties, including South Africa, submit trade reports for Rosy-faced Lovebirds and Rose-ringed Parakeets, which have at times been listed in the appendices. Two data sets were downloaded from the database for the 10-year period between 2007 and 2016 (Table 17.3): (1) imports into South Africa, and (2)

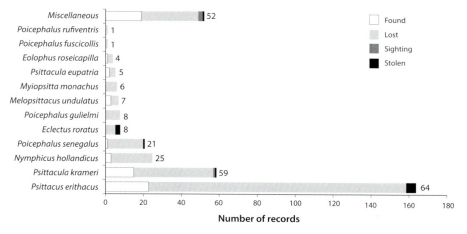

Figure 17.3. All records (total number per species) of parrots on ParrotAlert.com identified to species-level, and also including reports of: conure (10), *Pionus* (2), lovebird (9), parakeet (9), macaw (8), cockatoo (5), bird/parrot (3), *Amazona* spp. (2), parrot (2), rosella (2).

exports out of South Africa. These data included trade in parrots from all sources (including wild-sourced and captive-bred) and for all purposes (including both commercial and noncommercial). Where discrepancies existed between numbers reported by the importing vs. exporting country, the highest value was used. There are multiple reasons for such discrepancies, including possible underreporting by importing countries and reports of permits issued rather than permits used. As such, reported statistics should be considered only indicative of the scale of trade (Robinson and Sinovas 2018).

To explore the relationship between the international parrot trade and escaped and established parrots in South Africa, data on imports and exports were compared with information on parrots reported to ParrotAlert.com and populations of naturalized parrots. Data were aggregated by genus, as often reports to ParrotAlert.com did not provide descriptions to species level (e.g., "lovebirds"). Cockatiels and Budgerigars were excluded from this analysis due to the lack of reliable data on trade volumes.

In the 10-year period between 2007 and 2016, South Africa reported issuing CITES import permits for 38,647 parrots, while countries exporting to South Africa reported 82,854. The vast majority of all importer-reported imports were wild-sourced *Psittacus* spp. (63.3%), but significant numbers of other species endemic to Africa (but not South Africa) were reported, including those in the genera *Agapornis* (11.8%), *Poicephalus* (3.2%), and *Psittacula* (2.1%). Fourteen non-native genera were reported in quantities of more than 1,000 specimens (Table 17.3).

Since 2008, South Africa has consistently been the largest global exporter of parrots, and over this period its global dominance has increased. Exports have risen from just over 40,000 parrots in 2007 to almost 357,000 in 2016—almost nine times the volume of exports of the second-highest exporter, the Philippines, in 2016. Volumes of parrots exported over this period vastly exceeded imports—South Africa reported issuing CITES export permits for 1,343,625 specimens, while reported imports to South Africa were 304,164. South Africa reported exports of 198 species. Of these, 73 species were exported in numbers of more than 1,000 specimens.

The Grey Parrot (*Psittacus erithacus*), native to equatorial Africa, is one of the most heavily trafficked parrots in the world. Nevertheless, although escaped Grey Parrots are occasionally seen in the wild, the species is not established or naturalized in any part of the world. Malombe, Angola, 19 September 2019. Photo by Rowan O. Martin/World Parrot Trust.

The numbers of reported escaped parrots aggregated by genera was positively correlated with numbers reported exported by South Africa (Fig. 17.4; $F = 36.6$, $R^2 = 0.38$, $p < 0.001$, $n = 60$ genera). Not surprisingly, the genera of parrots that have established breeding populations (Table 17.2) are also among the birds most frequently reported in trade (Table 17.3; Fig. 17.4). However, it is notable that some of the species exported from South Africa in the highest volumes, including those in the genera *Aratinga*, *Ara*, *Pyrrhura*, and *Cacatua* have only seldom (or

INTRODUCED AND NATURALIZED PARROTS OF SOUTH AFRICA: COLONIZATION AND THE WILDLIFE TRADE

TABLE 17.3

Parrot genera imported to and/or exported from South Africa during 2007–2016, recorded in CITES (2019). Genera are ranked by gross imports/exports, which, in instances where there are discrepancies in the annual totals reported by importers and exporters, takes the larger of the two. * indicates escaped birds reported in South Africa. † indicates exotic populations established in South Africa (see Table 17.2).

GENUS	SOUTH AFRICAN IMPORTS			SOUTH AFRICAN EXPORTS		
	IMPORTER-REPORTED QUANTITY	EXPORTER-REPORTED QUANTITY	GROSS IMPORTS	IMPORTER-REPORTED QUANTITY	EXPORTER-REPORTED QUANTITY	GROSS EXPORTS
*Psittacus** (Parrots)	24,465	16,931	24,465	114,619	341,010	341,020
Trichoglossus (Lorikeets)	273	13,735	13,736	–	–	–
*Pyrrhura** (Parrots)	143	8,960	8,960	10,560	54,994	54,998
*Psittacula**† (Parakeets)	800	5,892	5,903	4,174	38,991	38,991
*Agapornis**† (Lovebirds)	4,575	4,648	5,619	46,736	367,771	367,771
*Platycercus** (Rosellas)	903	5,325	5,363	3,619	21,322	21,589
*Aratinga** (Parakeets)	50	5,163	5,163	32,064	132,690	132,690
*Myiopsitta**† (Parakeets)	185	3,556	3,556	2,521	17,176	17,176
Glossopsitta (Lorikeets)	10	2,576	2,576	–	–	–
*Eolophus** (Cockatoo)	47	2,079	2,079	5,578	28,376	28,376
Neophema (Parrots)	374	1,438	1,502	–	–	–
*Poicephalus** (Parrots)	1,236	500	1,311	6,535	29,979	30,031
*Cacatua** (Cockatoos)	393	1,096	1,268	14,684	39,168	39,273
Forpus (Parrotlets)	391	1,205	1,205	–	–	–
Barnardius (Ringneck)	50	1,092	1,092	–	–	–
*Eclectus** (Parrots)	–	–	–	4,184	25,262	25,262
*Pionus** (Parrots)	–	–	–	2,846	15,344	15,344
Psephotus (Parrots)	–	–	–	1,665	11,968	11,968
*Ara** (Macaws)	–	–	–	16,845	57,702	57,707
*Amazona**† (Amazons)	–	–	–	23,428	83,967	84,038

never) been reported as escaped parrots. This is despite several of these, such as members of the *Ara* and *Cacatua*, being readily recognizable as exotic birds. Of particular note is the relatively low numbers of *Agapornis* reported escaped despite the very high volumes produced for export.

DISCUSSION

The Rose-ringed Parakeet in South Africa

The Rose-ringed Parakeet has established itself as an integral species in the urban avifauna in two large metropolitan areas in South Africa: namely Durban, in KwaZulu-Natal province, and the greater Johannesburg-Pretoria region of Gauteng province (Brooke 1997; Perrin and Cowgill 2005; Hart and Downs 2014; Symes 2014; Symes et al. 2017). The Gauteng population has experienced an eight to nine times increase in range (as determined by pentad presence), and an approximately four times increase in abundance (as indicated by reporting rate), in the 10-year period July 2007–June 2017. These two core regions, separated by a distance of ~550 km, likely represent separate invasion events, and through molecular

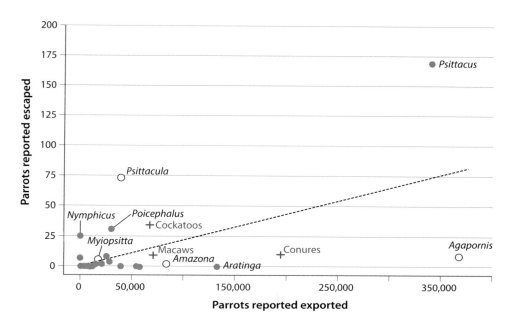

Figure 17.4. Relationship between numbers of parrots exported from South Africa (exporter-reported data) and the number of escaped birds reported to ParrotAlert.com between 2007 and 2016. Open circles represent genera that have established breeding populations in South Africa, and solid circles those that have not. + symbols represent records to ParrotAlert.com reported using alternative descriptions of taxonomic classification that could not be identified to the genus level. These include conures, macaws, and cockatoos (numerous genera), none of which have an established breeding population. The dashed line indicates the predictions of a fitted linear model based on data to the genus level.

techniques, their relatedness and origin may be easily established. The population that has spread across much of the urban Durban area likely originated in Durban North (Mt. Edgecombe area) sometime in the 1970s (Perrin 2012; Hart and Downs 2014; Symes 2014). In Gauteng, the period of establishment appears to be similar, with the first published breeding record occurring in Johannesburg in the late 1970s (Weissenbacher and Allan 1985). However, a localized population present in the Centurion area of southern Pretoria in the 1970s may have bred earlier (C. Symes, various pers. comm.) and from this, together with additional escaped birds, established the current and broader distribution evident today (Symes 2014; Symes et al. 2017; Ivanova and Symes 2019).

While the abundance and distribution of the Rose-ringed Parakeet has increased significantly in the past decade, in regions where it occurs, there appears to be little to no evidence of it outcompeting native species (Ivanova and Symes 2019). Indeed, the avifauna co-occurring with the Rose-ringed Parakeet is already one derived in a significantly transformed urban landscape, and co-occurring species coexist in response to filling vacant niches in a fragmented natural habitat within urban landscape (Symes et al. 2017; Ivanova and Symes 2019). The urban landscape in which the Rose-ringed Parakeet has become established is unique in many respects. The high-elevation city of Johannesburg (1,500–1,600 m) is situated in the grassland biome, little of which now remains where urbanization has occurred, while extensive afforestation by both exotic and savanna biome native trees has taken place in suburbia—areas where the parakeet is most common. The slightly lower-elevation city of Pretoria (1,300–1,400 m) to the north borders marginally the savanna biome (see Fig. 16.1 in Symes et al. 2017). In the urban landscape of Pretoria, the Rose-ringed Parakeet is currently limited to the most transformed suburbs. Its presence and establishment in a niche are thus achieved alongside a cohort of numerous other species that are similarly defining a niche in this urban metropolis.

This turnover of species in the face of ongoing habitat transformation (Beissinger and Osborne 1982) results in winners and losers (Carrete et al. 2010; Hernández-Brito et al. 2014), whether exotic or native to this region. Currently, the majority of observations of Rose-ringed Parakeet–involved aggressive interactions are with other exotic bird species (D. Hernández-Brito, unpubl. data). This, with the limited evidence in SABAP data of parakeets affecting potential competitor bird species (Ivanova and Symes 2019), lead us to suggest that at this time, the presence of the Rose-ringed Parakeet is defined by the filling of a vacant niche in a modified urban landscape. Indeed, in this landscape it co-occurs with a number of other exotic species already well established—e.g., Mallard (*Anas platyrhynchos*), Rock Dove (*Columba livia*), Common Starling (*Sturnus vulgaris*), Common Myna (*Acridotheres tristis*), and House Sparrow (*Passer domesticus*) (Symes et al. 2017; Hobbs 2018).

The Rose-ringed Parakeet is a vocal and easy-to-identify species, and as a charismatic parrot, endears itself to the humans with which it co-occurs (CAB International 2019). In this it is unlike the Common Myna, which has in the past century spread across human-dense and anthropogenically transformed areas of South Africa (Dean 2000; Peacock et al. 2007) and is regarded as a nuisance species. Such perception bias is a common phenomenon, in which prevailing positive feelings among the public toward charismatic exotic species result in lack of interest in management (Sharp et al. 2011). The Rose-ringed Parakeet is known to feed on economically important crops in its natural range (Dhindsa and Saina 1994; Ahmad et al. 2011), but evidence of the impact by invasive populations is scarce (White et al. 2019). For the past ~40 years, it has been present in two densely populated urban areas of South Africa; however, over that time it has not been reported feeding on economically important crops, besides pecans (*Carya illinoinensis*) in northern Johannesburg (M. Rudman, pers. comm.). Nevertheless, the species may still be in the early stages of an "invasion," and the currently urban-confined population may eventually extend into agricultural regions (Ivanova and Symes 2019).

Anthropogenically transformed environments have created situations that may force species that do not naturally coexist into competitive interactions (e.g., Carrete et al. 2010), but the precise effects of numerous introduced species on native and co-occurring fauna have produced conflicting outcomes and interpretations. This situation may be due to the nature of competitive interactions between alien and native species and the anthropogenic modifications made to habitats (Byers 2002), which would likely differ on a case-by-case basis. These studies therefore emphasize how complex ecological understanding can be and, similarly, how uninformed management responses still are. Indeed, in South Africa, so-called control programs of these avian species are based on very little (to nonexistent) scientific evidence (Ivanova and Symes 2019).

In Germany, introduced Rose-ringed Parakeets and native Common Starlings were able to coexist without direct competition through niche partitioning (Czajka et al. 2011), while in Italy, nesting differences between the two species were argued to be the starling's adaptations to cope with parakeet competition (Dodaro and Battisti 2014). In Belgium, of two native hole nesters considered to be the most vulnerable to competition with the parakeet—Eurasian Nuthatch (*Sitta europaea*) and Common Starling—the nuthatch was likely the more vulnerable (Strubbe and Matthysen 2007), showing a negative relationship with the parakeet in abundance (Strubbe and Matthysen 2009). These findings differ from those of Newson et al. (2011), who suggest that there is no evidence for parakeet nest-site competition with nuthatch populations or any other cavity-nesting species with which the parakeet co-occurs in the United Kingdom. Additionally, across its range in Belgium, the impact of the parakeet is predicted to be limited (Strubbe et al. 2010). Currently, it is difficult to conclude whether the parakeets are having a negative impact; however, as their numbers continue to increase, so too might their effects on native avifauna (Newson et al. 2011). In addition, interactions among species in an avian community are complex, making predictions very tenuous. These complexities are demonstrated in a study by Orchan et al. (2013), which showed that invasive Rose-ringed Parakeets in Tel Aviv,

Israel, could potentially enhance breeding of the invasive Common Myna by enlarging previously unavailable holes. The myna in turn excludes the invasive Vinous-breasted Starling (*Acridotheres burmannicus*), which is also a direct competitor of the native Syrian Woodpecker (*Dendrocopos syriacus*). Therefore, control efforts toward the Common Myna alone may simply reduce its competition with the Vinous-breasted Starling, thereby increasing the impact of the Vinous-breasted Starling on other native species with which it competes (Orchan et al. 2013). As such, it is particularly important to understand interaction networks, because control efforts targeted at single species may have indirect effects on other, nontarget species in the ecosystem.

Diversity of Naturalized and Escaped Parrots in South Africa

Psittaciformes is one of the most threatened orders of birds worldwide (Juniper and Parr 1998). For centuries parrots have appealed to humans and as a result have been wild-caught and traded across great distances (Olah et al. 2016). With this parrot-human association, it is inevitable that birds will escape. Less inevitable is that these birds will breed and establish themselves in populations well beyond their natural ranges. Our attempt to present those parrot species occurring outside their natural ranges in South Africa has been limited and in no way represents a comprehensive review of the topic. Rather, by scanning the literature and paying attention to social networks and the broader avicultural and birding community, we highlight the extent to which a diversity of parrot species gets reported in the wild. In addition, as a professional ornithologist, the senior author, in an academic position at a university close to the Johannesburg Central Business District, is often approached regarding bird identifications, especially of "unusual or obscure" birds. For example, a report of a Blue-and-yellow Macaw (*Ara ararauna*) on 20 August 2015, on the southern border of the Kruger National Park (C. Prat, pers. obs.), highlights the point that the most unexpected species can appear in the most unlikely places. Far from any human-modified habitat (or aviary), the image of this Neotropical macaw set against the dry late-winter savanna, highlighted the effect that the wild-bird trade has in how far and wide birds are moved.

The paucity of parrot records reported to the SARBN demonstrates how inattentive a particular focus group (primarily interested in wild birds that in nearly all cases are native) is to that of escapees. Thus, an individual birder observing an escaped parrot (or any other exotic for that matter) is unlikely to report such a sighting to a database, and the record will pass unnoticed unless it is a target species for a focused study or it is present on an atlas list of recordable species (e.g., SABAP). If the ratio of reports to actual escaped birds is low, there may indeed be a greater number of escaped parrots unreported, thereby increasing the potential for numerous parrot species to become established outside their native ranges (Vall-llosera and Cassey 2017a).

However, only a small proportion of potential founder populations are successful, and for those that are established, it is often unclear what criteria are most important in driving their success. Therefore, as long as a feeder source of propagules exists, with the persistent influx of avicultural species into the wild, so too will the potential exist for additional parrot species to become naturalized beyond their natural ranges. In the next section, considering the degree to which parrots are traded across the globe, both legally and illegally, we attempt to address the role of the parrot trade as a source of naturalized parrot populations.

The Role of Trade in the Introduction and Establishment of Parrots in South Africa

The pet industry, including the practice of keeping caged parrots, is recognized as a significant source of new invasive species around the world (Cassey et al. 2004; Vall-llosera and Cassey 2017a, 2017b). Although no current statistics on levels of parrot ownership in South Africa exist, the keeping and breeding of parrots, the vast majority of which are non-native to South Africa, is a popular pastime. In addition, the breeding of parrots for commercial export has seen a rapid expansion over the past decade, with South Africa emerging as a central hub in the global parrot trade (Martin 2018a, 2018b). In recent years, there has been considerable investment in the development of avicultural facilities (Kriel 2018), including the development of a number of large-scale, export-oriented mega-facilities containing over 1,000 breeding pairs (Parrot Breeders' Association of South Africa, pers. comm.). As well as producing large volumes of captive-bred parrots for export, the practice also involves the importation of large numbers of wild-sourced and captive-bred parrots as breeding stock for export-oriented production as well as for the local pet trade.

We found statistical support for a positive correlation between numbers of parrots reported exported at the genera level and numbers of parrots reported escaped, suggesting a possible link between the avicultural industry and numbers of escaped parrots. This lends support to the assumption that has been made previously in some studies (e.g., Cardador et al. 2016, 2017; Mori et al. 2017) that trade volumes of parrots reported in CITES statistics can be a useful proxy measure of propagule pressure. However, our simple linear model explained only a small proportion of variance in numbers of escaped parrots (0.38), and there were some notable outliers, suggesting that the relationship may be complex and influenced by a number of additional factors. Firstly, it should be noted that the data used on parrots reported escaped may be biased in significant but difficult to quantify ways, by the financial worth and ease of recognition of different species. For instance, there appear to be relatively few records of escaped lovebirds (*Agapornis*), despite these parrots being very widely kept and bred in captivity. Members of the genus *Agapornis* are relatively small and may not be recognized as exotic by the casual observer. They are also relatively cheap and readily available and so might be underreported as missing by owners. The same might also apply to members of the genus *Aratinga*, which are among the most imported and exported, yet are absent from reported escapees.

It is also notable that some highly traded species, such as members of the genus *Psittacus*, have not established naturalized populations—despite *Psittacus* parrots being imported into and exported from South Africa in large volumes and also being the group by far the most frequently reported as escaped in South Africa.

The reasons for differences between taxa are unclear, but it would appear likely that niche suitability and physiological plasticity also play a role in determining which species become established (Thabethe et al. 2013; Cardador et al. 2016; Mori et al. 2017). In Europe, the role of propagule pressure in increasing the probability of naturalized populations becoming established was detectable only once niche suitability was taken into account (Cardador et al. 2016). Niche modeling by Cardador et al. (2017) found few potentially suitable areas for *Psittacus* parrots in South Africa, and despite the popularity of *Psittacus* parrots in the pet trade worldwide, there are few naturalized populations (Cardador et al. 2017). The best documented case, in Kampala, Uganda, occurs within the species' native range (Irumba et al. 2016; Chamberlain et al. 2018), where some have shown signs of adapting to the urban environment (Twanza and Pomeroy 2011). In contrast, for the Rose-ringed Parakeet, which has established the largest naturalized populations of any parrot in South Africa, climatic suitability was considered high (Cardador et al. 2017).

Implications for the Spread of Infectious Diseases

There is increasing evidence that the global trade in parrots is linked to the spread of infectious diseases, such as psittacine beak and feather disease (PBFD) (Fogell et al. 2016, 2018). Such diseases pose a potential threat to many wild parrot populations (Raidal and Peters 2018), including the Cape Parrot (*Poicephalus robustus*), a South African endemic classified as globally vulnerable (Regnard et al. 2015). Naturalized populations of Rose-ringed Parakeets in South Africa do not currently occur within the range of the Cape Parrot, but further population expansions should be closely monitored for potential disease risks for local bird populations.

CONCLUSION

A globally successful invader, the Rose-ringed Parakeet is showing a slow but consistent expansion through South Africa, due to the growth of successfully established wild populations. These populations appear to be currently restricted to highly populated urban areas (which would have been sites of individual escapees from the pet trade), concentrating in the Johannesburg-Pretoria region as well as in the city of Durban. The effects of the growing range and abundance within that range on other avifauna (and on agriculture) are yet to be determined, given the uncertainty surrounding the species' competitive outcomes in its introduced ranges (Berruti 1991; Ivanova and Symes 2019). These questions are particularly timely given the anthropogenically driven habitat change that has happened within these urban areas and the likely changes in the array of niches this has brought about.

Avian introductions are currently derived mostly from the international transport and accidental escape of birds (Carrete and Tella 2008; Blackburn et al. 2010; Cardador et al. 2016). It is reasonable to assume that the growth of this trade has increased exponentially with the growth of human populations across the globe, and if that is so, it is likely to be a major contributor to the passive establishment of bird populations outside their natural ranges. Parrots are unlikely to lose appeal as a traded species and are therefore a taxon most likely to see establishment in areas of high human density. However, a number of criteria, as with any invasive species, need to be met before successful establishment is attained. Therefore, we predict further additions to the list of bird species already naturalizing in significantly human-transformed landscapes, with the list of species presented in this study as a clue to which those species may be.

ACKNOWLEDGMENTS

We acknowledge the use of data from the Southern African Bird Atlas Project, and the contributions of numerous observers to this database. Conversations with Geoff Lockwood provided valuable perspective on the changing avifauna of Gauteng. The Southern African Rare Bird News Listserv is coordinated by Trevor Hardaker and is similarly acknowledged. Dailos Hernández-Brito is thanked for providing preliminary data from intensive observations conducted at Delta Park in Johannesburg. Caryl Prat kindly reported, with photographs, the *Ara ararauna* sighting.

REFERENCES

Ahmad, S., Khan, H. A., Javed, M., and Ur-Rehman, K. 2011. Roost composition and damage assessment of rose-ringed parakeet (*Psittacula krameri*) on maize and sunflower in agro-ecosystem of Central Punjab, Pakistan. *International Journal of Agriculture and Biology* 13:731–736.

Andersen, M. C., Adams, H., Hope, B., and Powell, M. 2004. Risk analysis for invasive species: General framework and research needs. *Risk Analysis* 24:893–900.

Annorbah, N. D., Collar, N. J., and Marsden, S. J. 2016. Trade and habitat change virtually eliminate the grey parrot *Psittacus erithacus* from Ghana. *Ibis* 158:82–91.

Baker, J., Harvey, K. J., and French, K. 2013. Threats from introduced birds to native birds. *Emu* 114:1–12.

Beissinger, S. R., and Osborne, D. R. 1982. Effects of urbanization on avian community organization. *Condor* 84:75–83.

Berkunsky, I., Quillfeldt, P., Brightsmith, D. J. et al. 2017. Current threats faced by Neotropical parrot populations. *Biological Conservation* 214:278–287.

Berruti, A. 1991. Alien birds. SAOS policy statement. *Birding in South Africa* 44:7.

BirdLife International. 2018. *Myiopsitta monachus*. Accessed June 2019. https://www.birdlife.org.

Blackburn, T. M., Essl, F., Evans, T., Hulme, P. E., Jeschke, J. M., Kühn, I., Kumschick, S., Markova, Z., Mrugała, A., Nentwig, W., et al. 2014. A unified classification of alien species based on the magnitude of their environmental impacts. *PLoS Biology* 12:e1001850.

Blackburn, T. M., Gaston, K. J., and Parnell, M. 2010. Changes in non-randomness in the expanding introduced avifauna of the world. *Ecography* 33:168–174.

Blackburn, T. M., Lockwood, J. L., and Cassey, P. 2009. *Avian Invasions: The Ecology and Evolution of Exotic Birds*. Oxford: Oxford Univ. Press.

Blackburn, T. M., Prowse, T. A., Lockwood, J. L., and Cassey, P. 2013. Propagule pressure as a driver of establishment success in deliberately introduced exotic species: Fact or artefact? *Biological Invasions* 15:1459–1469.

Blair, R. B. 1996. Land use and avian species diversity along an urban gradient. *Ecological Applications* 6:506–519.

Bonter, D. N., Zuckerberg, B., and Dickinson, J. L. 2010. Invasive birds in a novel landscape: Habitat associations and effects on established species. *Ecography* 33:494–502.

Brooke, R. K. 1997. Rose-ringed parakeet. In *The Atlas of Southern African Birds*, ed. J. A. Harrison, D. G. Allan, L. G. Underhill, M. Herremans, A. J. Tree, V. Parker, and C. J. Brown, vol. 1, 536. Johannesburg: BirdLife South Africa.

Brooke, R. K., Lloyd, P. H., and de Villiers, A. L. 1986. Alien and translocated terrestrial vertebrates in South Africa. In *The Ecology and Management of Biological Invasions in Southern Africa*, ed. I.A.W. MacDonald, F. J. Kruger, and A. A. Ferrar, 63–74. Cape Town, South Africa: Oxford Univ. Press.

Byers, J. E. 2002. Impact of non-indigenous species on natives enhanced by anthropogenic alteration of selection regimes. *Oikos* 97:449–458.

CAB International. 2019. *Psittacula krameri* (rose-ringed parakeet). Invasive Species Compendium. CABI. Accessed 22 Jan. 2019. www.cabi.org/ISC/datasheet/45158.

Cardador, L., Carrete, M., Gallardo, B., and Tella, J.L. 2016. Combining trade data and niche modelling improves predictions of the origin and distribution of non-native European populations of a globally invasive species. *Journal of Biogeography* 43:967–978.

Cardador, L., Lattuada, M., Strubbe, D., Tella, J. L., Reino, L., Figueira, R., and Carrete, M. 2017. Regional bans on wild-bird trade modify invasion risks at a global scale. *Conservation Letters* 10:717–725.

Carrete, M., Lambertucci, S. A., Speziale, K., Ceballos, O., Travaini, A., Delibes, M., Hiraldo, F., and Donázar, J. A. 2010. Winners and losers in human-made habitats: Interspecific competition outcomes in two Neotropical vultures. *Animal Conservation* 13:390–398.

Carrete, M., and Tella, J. L. 2008. Wild-bird trade and exotic invasions: A new link of conservation concern? *Frontiers in Ecology and the Environment* 6:207–211.

Case, T. J. 1996. Global patterns in the establishment and distribution of exotic birds. *Biological Conservation* 78:69–96.

Cassey, P., Blackburn, T. M., Russell, G. J., Jones, K. E., and Lockwood, J. L. 2004. Influences on the transport and establishment of exotic bird species: An analysis of the parrots (Psittaciformes) of the world. *Global Change Biology* 10:417–426.

Cassey, P., Vall-llosera, M., Dyer, E., and Blackburn, T. M. 2015. The biogeography of avian invasions: History, accident and market trade. In *Biological Invasions in Changing Ecosystems: Vectors, Ecological Impacts, Management and Predictions*, ed. J. Canning-Clode, 37–54. Warsaw and Berlin: De Gruyter Open.

Chace, J. F., and Walsh, J. J. 2006. Urban effects on native avifauna: A review. *Landscape and Urban Planning* 74:46–69.

Chamberlain, D., Kibuule, M., Skeen, R. Q., and Pomeroy, D. 2018. Urban bird trends in a rapidly growing tropical city. *Ostrich* 89:275–280.

Chapman, T. 2005. *The Status and Impact of the Rainbow Lorikeet (Trichoglossus haematodus moluccanus) in South-west Western Australia*. Perth, Western Australia: Dept. of Agriculture and Food.

Charter, M., Izhaki, I., Mocha, Y. B., and Kark, S. 2016. Nest-site competition between invasive and native cavity nesting birds and its implication for conservation. *Journal of Environmental Management* 181:129–134.

CITES. 2019. CITES Trade Database. Accessed Jan. 2019. http://trade.cites.org.

Clavero, M., Brotons, L., Pons, P., and Sol, D. 2009. Prominent role of invasive species in avian biodiversity loss. *Biological Conservation* 142:2043–2049.

Clavero, M., and García-Berthou, E. 2005. Invasive species are a leading cause of animal extinctions. *Trends in Ecology & Evolution* 20:110.

Czajka, C., Braun, M. P., and Wink, M. 2011. Resource use by non-native rose-ringed parakeets (*Psittacula krameri*) and native starlings (*Sturnus vulgaris*) in central Europe. *Open Ornithology Journal* 4:17–22.

Dean, W. R. J. 2000. Alien birds in southern Africa: What factors determine success? *South African Journal of Science* 96:9–14.

Dhindsa, M. S., and Saina, H. K. 1994. Agricultural ornithology: An Indian perspective. *Journal of Biosciences* 19(4):391–402.

Didham, R. K., Tylianakis, J. M., Hutchison, M. A., Ewers, R. M., and Gemmell, N. J. 2005. Are invasive species the drivers of ecological change? *Trends in Ecology & Evolution* 20:470–474.

Dodaro, G., and Battisti, C. 2014. Rose-ringed parakeet (*Psittacula krameri*) and starling (*Sturnus vulgaris*) syntopics in a Mediterranean urban park: Evidence for competition in nest-site selection? *Belgian Journal of Zoology* 144:5–14.

Faulkner, K. T., Hurley, B. P., Robertson, M. P., Rouget, M., and Wilson, J.R.U. 2017. The balance of trade in alien species between South Africa and the rest of Africa. *Bothalia* 47:a2157.

Fogell, D. J., Martin, R. O., Bunbury, N., Lawson, B., Sells, J., and McKeand, A. M. 2018. Trade and conservation implications of new beak and feather disease virus detection in native and introduced parrots. *Conservation Biology* 32:1325–1335.

Fogell, D. J., Martin, R. O., and Groombridge, J. J. 2016. Beak and feather disease virus in wild and captive parrots: An analysis of geographic and taxonomic distribution and methodological trends. *Archives of Virology* 161: 2059–2074

Goss, J. R., and Cumming, G. S. 2013. Networks of wildlife translocations in developing countries: An emerging conservation issue? *Frontiers in Ecology and the Environment* 11:243–250.

Grarock, K., Tidemann, C. R., Wood, J. T., and Lindenmayer, D. B. 2014. Are invasive species drivers of native species decline or passengers of habitat modification? A case study of the impact of the common myna (*Acridotheres tristis*) on Australian bird species. *Austral Ecology* 39:106–114.

Harrison, J. A., Underhill, L. G., and Barnard, P. 2008. The seminal legacy of the Southern African Bird Atlas Project. *South African Journal of Science* 104:82–84.

Hart, L. A., and Downs, C. T. 2014. Public surveys of rose-ringed parakeets, *Psittacula krameri*, in the Durban Metropolitan Area, South Africa. *African Zoology* 49:283–289.

REFERENCES

Hart, J., Hart, T., Salumu, L., Bernard, A., Abani, R., and Martin, R. O. 2016. Increasing exploitation of grey parrots in eastern DRC drives population declines. *Oryx* 50:16–17.

Hernández-Brito, D., Carrete, M., Popa-Lisseanu, A. G., Ibáñez, C., and Tella, J. L. 2014. Crowding in the city: Losing and winning competitors of an invasive bird. *PLoS One* 9:e100593.

Hobbs, R. J. 2018. Novel ecosystems: Can't we just pretend they're not there? In *Effective Conservation Science: Data Not Dogma*, ed. P. Kareiva, M. Marvier, and B. Silliman, 45–50. Oxford: Oxford Univ. Press.

Hugo, S., and van Rensburg, B. J. 2008. The maintenance of a positive spatial correlation between South African bird species richness and human population density. *Global Ecology and Biogeography* 17:611–621.

Hugo, S., and van Rensburg, B. J. 2009. Alien and native birds in South Africa: Patterns, processes and conservation. *Biological Invasions* 11:2291–2302.

Irumba, I-O., Pomeroy, D., and Perrin, M. 2016. Grey parrots *Psittacus erithacus* in Kampala, Uganda: Are they becoming suburbanised? *Ostrich* 87:193–195.

Ivanova, I. M., and Symes, C. T. 2019. Invasion of *Psittacula krameri* in Gauteng, South Africa: Are other birds impacted? *Biodiversity and Conservation* 28:3633.

Juniper, T., and Parr, M. 1998. *Parrots: A Guide to the Parrots of the World*. East Sussex, UK: Pica Press.

Kriel, G. 2018. Exotic birds: Getting into business. *Farmer's Weekly* 18019:52–54.

Kumschick, S., and Nentwig, W. 2010. Some alien birds have as severe an impact as the most effectual alien mammals in Europe. *Biological Conservation* 143:2757–2762.

Lee, A.T.K., Altwegg, R., and Barnard, P. 2017. Estimating conservation metrics from atlas data: The case of southern African endemic birds. *Bird Conservation International* 27:323–336.

Lever, C. 2005. *Naturalised Birds of the World*. London: T. & A. D. Poyser.

Lockwood, J. L., Cassey, P., and Blackburn, T. M. 2005. The role of propagule pressure in explaining species invasions. *Trends in Ecology & Evolution* 20:223–228.

Long, L. J. 1981. *Introduced Birds of the World*. London: David and Charles.

Lopes, D., Martin, R., Henriques, M., Monteiro, H., Cardoso, P., Tchantchalam, Q., Pires, A. J., Regalla, A., and Catry, P. 2019. Combining local knowledge and field surveys to determine status and threats to Timneh parrots *Psittacus timneh* in Guinea-Bissau. *Bird Conservation International* 29:400–412.

Martin, R. O. 2018a. Grey areas: Temporal and geographical dynamics of international trade of grey and Timneh parrots (*Psittacus erithacus* and *P. timneh*) under CITES. *Emu* 118:113–125.

Martin, R. O. 2018b. The wild bird trade and African parrots: Past, present and future challenges. *Ostrich* 89:139–143.

Martin, R. O., Perrin, M. R., Boyes, R. S., Abebe, Y. D., Annorbah, N. D., Asamoah, A., Bizimana, D., Bobo, K. S., Bunbury, N., Brouwer, J., et al. 2014. Research and conservation of the larger parrots of Africa and Madagascar: A review of knowledge gaps and opportunities. *Ostrich* 85:205–233.

Martin-Albarracin, V. L., Amico, G. C., Simberloff, D., and Nuñez, M. A. 2015. Impact of non-native birds on native ecosystems: A global analysis. *PLoS One* 10:e0143070.

Menchetti, M., Mori, E., and Angelici, F. M. 2016. Effects of the recent world invasion by ring-necked parakeets *Psittacula krameri*. In *Problematic Wildlife: A Cross-Disciplinary Approach*, ed. F. M. Angelici, 253–266. Switzerland and New York: Springer International Publishing.

Mori, E., Grandi, G., Menchetti, M., Tella, J. L., Jackson, H. A., Reino, L., van Kleunen, A., Figueira, R., and Ancillotto, L. 2017. Worldwide distribution of non-native Amazon parrots and temporal trends of their global trade. *Animal Biodiversity and Conservation* 40:49–63.

Newson, S., Johnston, A., Parrott, D., and Leech, D. I. 2011. Evaluating the population-level impact of an invasive species, ring-necked parakeet, *Psittacula krameri*, on a native avifauna. *Ibis* 153:509–516.

Olah, G., Butchart, S.H.M., Symes, A., Guzmán, I. M., Cunningham, R., Brightsmith, D. J., and Heinsohn, R. 2016. Ecological and socioeconomic factors affecting extinction risk in parrots. *Biodiversity and Conservation* 25:205–223.

Orchan, Y., Chiron, F., Shwartz, A., and Kark, S. 2013. The complex interaction network among multiple invasive bird species in a cavity-nesting community. *Biological Invasions* 15:429–445.

Pain, D. J., Martins, T. L. F., Boussekey, M., Diaz, S. H., et al. 2006. Impact of protection on nest take and nesting success of parrots in Africa, Asia and Australasia. *Animal Conservation* 9:322–330.

Peacock, D. S., van Rensburg, B. J., and Robertson, M. P. 2007. The distribution and spread of the invasive alien common myna, *Acridotheres tristis* L. (Aves: Sturnidae), in southern Africa. *South African Journal of Science* 103:465–473.

Perrin, M. R. 2012. *Parrots of Africa, Madagascar and the Mascarene Islands: Biology, Ecology and Conservation*, 612. Johannesburg, South Africa: Wits Univ. Press.

Perrin, M. R., and Cowgill, R. 2005. Rose-ringed parakeet *Psittacula krameri*. In *Roberts' Birds of Southern Africa*, 7th ed., ed. P. A. R. Hockey, W. R. J. Dean, and P. G. Ryan, 229–230. Cape Town, South Africa: Trustees of the John Voelcker Bird Book Fund.

Picker, M., and Griffiths, C. L. 2011. *Alien and Invasive Animals: A South African Perspective*. Cape Town, South Africa: Struik Nature.

Pires, S. F. 2012. The illegal parrot trade: A literature review. *Global Crime* 13:176–190.

Raidal, S. R., and Peters, A. 2018. Psittacine beak and feather disease: Ecology and implications for conservation. *Emu* 118:80–93.

Regnard, G. L., Boyes, R. S., Martin, R. O., Hitzeroth, I. I., and Rybicki, E. P. 2015. Beak and feather disease viruses circulating in Cape parrots (*Poicephalus robustus*) in South Africa. *Archives of Virology* 160:47–54.

Robertson, M. P., Cumming, G. S., and Erasmus, B.F.N. 2010. Getting the most out of atlas data. *Diversity and Distributions* 16:363–375.

Robinson, J. E., and Sinovas, P. 2018. Challenges of analyzing the global trade in CITES-listed wildlife. *Conservation Biology* 32:1203–1206.

Roche, C., and Bedford-Shaw, A. 2008. Escapee cage birds in suburbs of Johannesburg. *Bird Numbers* 14:2–3.

Runde, D. E., Pitt, W. C., and Foster, J. 2007. Population ecology and some potential impacts of emerging populations of exotic parrots. In *Managing Vertebrate Invasive Species: Proceedings of an International Symposium*, ed. G. W. Witmer, W. C. Pitt, and K. A. Fagerstone, 338–360. Fort Collins, CO: USDA/APHIS Wildlife Services, National Wildlife Research Center.

Sharp, R. L., Larson, L. R., and Green, G. T. 2011. Factors influencing public preferences for invasive alien species management. *Biological Conservation* 144:2097–2104.

SABAP. 2019. South African Bird Atlas Project. Accessed June 2019. http://sabap2.birdmap.africa.

SARBN. 2017. Southern African Rare Bird News, Google Groups. Accessed June 2017. https://groups.google.com/g/sa-rarebirdnews.

Strubbe, D., and Matthysen, E. 2007. Invasive ring-necked parakeets *Psittacula krameri* in Belgium: Habitat selection and impact on native birds. *Ecography* 30:578–588.

Strubbe, D., and Matthysen, E. 2009. Establishment success of invasive ring-necked and monk parakeets in Europe. *Journal of Biogeography* 36:2264–2278.

Strubbe, D., Matthysen, E., and Graham, C. H. 2010. Assessing the potential impact of invasive ring-necked parakeets *Psittacula krameri* on native nuthatches *Sitta europaea* in Belgium. *Journal of Applied Ecology* 47:549–557.

Symes, C. T. 2014. Founder populations and the current status of exotic parrots in South Africa. *Ostrich* 85:235–244.

Symes, C. T., Roller, K., Howes, C., Lockwood, G., and van Rensburg, B. J. 2017. Grassland to urban forest in 150 years: Avifaunal response to an African metropolis. In *Ecology and Conservation of Birds in Urban Environments*, ed. E. Murgui, and M. Hedblom, 309–343. Switzerland: Springer International Publishing.

Thabethe, V., Thompson, L. J., Hart, L. A., Brown, M., and Downs, C. T. 2013. Seasonal effects on the thermoregulation of invasive rose-ringed parakeets (*Psittacula krameri*). *Journal of Thermal Biology* 38:553–559.

Twanza, L., and Pomeroy, D. 2011. Grey Parrots *Psittacus erithacus* successfully nesting in a suburban house in Kampala, Uganda. *Bulletin of the African Bird Club* 18:81–82.

Underhill, L. G. 2016. The fundamentals of the SABAP2 protocol. *Biodiversity Observations* 7:1–12.

Vall-llosera, M., and Cassey, P. 2017a. Leaky doors: Private captivity as a prominent source of bird introductions in Australia. *PLoS One* 12:e0172851.

Vall-llosera, M., and Cassey, P. 2017b. "Do you come from a land down under?" Characteristics of the international trade in Australian endemic parrots. *Biological Conservation* 207:38–46.

Van Rensburg, B. J., Peacock, D. S., and Robertson, M. P. 2009. Biotic homogenization and alien bird species along an urban gradient in South Africa. *Landscape and Urban Planning* 92:233–241.

Van Rensburg, B. J., Weyl, O.L.F., Davies, S. J., van Wilgen, L. J., Peacock, D. S., Spear, D., and Chimimba, C. T. 2011. Invasive vertebrates of South Africa. In *Biological Invasions: Economic and Environmental Costs of Alien Plant, Animal, and Microbe Species*, 2nd ed., ed. D. Pimentel, 325–378. Boca Raton, FL: CRC Press.

Weissenbacher, B.K.H., and Allan, D. 1985. Rose-ringed parakeet breeding attempts in the Transvaal. *Ostrich* 56:169.

White, R. L., Strubbe, D., Dallimer, M., Davies, Z. G., Davis, A.J.S., Edelaar, P., Groombridge, J. J., Jackson, H. A., Menchetti, M., Mori, E., et al. 2019. Assessing the ecological and societal impacts of alien parrots in Europe using a transparent and inclusive evidence-mapping scheme. *NeoBiota* 48:45–69.

Whittington-Jones, C. 2016. *The Status of the Yellow-collared Lovebird (Agapornis personatus) in Gauteng*. Annual report. Scientific Services Gauteng, Dept. of Agriculture and Rural Development.

Whittington-Jones, C. 2017. *The Status of Rose-ringed Parakeets in Gauteng*. Second annual report. Scientific Services Gauteng, Dept. of Agriculture and Rural Development.

18

AUSTRALIA'S URBAN CAVITY NESTERS AND INTRODUCED PARROTS: PATTERNS, PROCESSES, AND IMPACTS

Andrew M. Rogers and Salit Kark

INTRODUCTION

Australia's native parrot communities are exceptionally diverse and occur in all habitats across the continent. The number of species and their relative abundances in most habitats have undergone significant changes as a result of habitat change, species introductions, and the movement of species beyond their historic ranges via natural range expansions and by trade (Keast 1995; Gibbons and Lindenmayer 2002). Despite the recognition of these changes, with few exceptions (e.g., Koch et al. 2008), there has been little work examining how parrot introductions have changed local communities across the continent. At a continental scale, there appear to be three trends for parrot communities: (1) a few widespread species have become highly urbanized; (2) native species of parrots are being moved around through the pet trade and escaping captivity in new locations; and (3) non-native parrots from other countries are escaping from the pet trade. Here, we review the current patterns of non-native parrots across Australia, with a particular focus on urban areas, as these are hot spots for species introductions and establishment.

Introduced non-native parrots in this chapter include both Australian species that have been introduced to parts of the continent in which they are non-native and non-Australian species that originated from other parts of the world, mostly via the pet trade. While most introduced parrots on the continent are native to some region of Australia (Gibbons and Lindenmayer 2002), there is growing acknowledgment that several non-Australian species have potential to establish breeding populations in Australia in the near future (Vall-llosera and Cassey 2017a; Vall-llosera et al. 2017). However, most of these non-Australian parrots remain understudied, and the barriers to establishment, potential ranges within Australia, and impact they might have on native species remain mostly unknown.

Australia has a wide range of established alien (also referred to as "non-native" or "introduced") birds (Duncan et al. 2001; Kark and Sol 2005; Garnett et al. 2015; Dyer et al. 2017b; McKinney and Kark 2017). However, surprisingly, none are parrots. The exact reasons for the lack of non-Australian parrots established in the country are not well understood, especially given the success of introduced parrots on other continents. This may be due, at least partly, to the fact that the native Australian bird communities include a large number of parrots. Many of these native parrots are well established in urban centers, and high levels of competition for food, breeding cavities, and other resources may create barriers to establishment for introduced non-Australian species (Bacher et al. 2011; Rogers 2019). In Australia, lessons from other invasive birds may inform the potential for species establishment as well as how widespread novel invasive species will become (McKinney and Kark 2017) and their potential impacts (Linz et al. 2007; Evans et al. 2014).

During European settlement of Australia, in the late 18th to early 19th century, a number of acclimatization societies were formed in the new territory, whose goals included introducing species from Europe to the island continent. Such historical species introductions in Australia have been fairly well documented (Blackburn and Duncan 2001a, 2001b; Duncan et al. 2001, 2014; Blackburn et al. 2015a), and the societies responsible for the introduction attempts often recorded the number of individuals, the number of introduction attempts, and whether or not a species became successfully established (Case 1996). Recent species introductions differ in that they are largely the result of birds escaping from the pet trade (Vall-llosera and Cassey 2017a, 2017b). For these species, there is growing mismatch between the number of non-native birds escaping into the wild and how often they are detected after they escape (Vall-llosera and Cassey 2017a). The differences in introduction effort between historical and recent introductions may hint at the recent barriers to establishment. The historical introductions were intentional, and in some cases, many attempts were made at establishing self-sustaining populations. Recent introductions differ in that they comprise small numbers of birds escaping at any one time. Given the importance of propagule pressure in successful introductions (Chiron et al. 2009; Duncan et al. 2014), the lack of established parrots may be the result of too few birds surviving the initial introduction and thus failing to establish in the wild (Cassey et al. 2005; Bacher et al. 2011; Dyer et al. 2017a; Vall-llosera and Cassey 2017a).

Other limitations to species establishment in Australia are local environmental factors, such as local species richness (McKinney and Kark 2017). This is in contrast to places like Europe, where the key factors were found to be human-related (Chiron et al. 2009). Non-Australian parrots introduced into the wild in Australia, especially those that escape into urban environments, must contend with a large number of native parrots and other aggressive species, for both breeding sites and foraging space (Parsons et al. 2006; Griffin et al. 2012). Escaping pet birds must also contend with a large number of native cavity-using mammals, which both compete with and prey on birds in hollows (MacDonald and Kirkpatrick 2003; Clergeau and Vergnes 2011; Hernández-Brito et al. 2014). It may be that chances of establishment are reduced by the predation pressure and a lack of available breeding niche space, but this has not been explicitly studied in Australia.

Understanding the barriers to establishment and the potential impacts of new species introductions is especially important in Australia, as the continent is home to many cavity-nesting species likely to be impacted by the introduction of non-native parrots. Around 15% of all terrestrial vertebrate species and 15% of all native Australian birds use cavities for nesting or shelter (Gibbons and Lindenmayer 2002). Of the native, cavity-nesting birds, 40 species are true parrots (Psittacoidea) and 14 species are cockatoos (Cacatuidae). The high rates of cavity use by Australia's birds occur in the absence of vertebrate cavity excavators, such as woodpeckers (Gibbons and Lindenmayer 2002); most tree hollows are formed by fungus or termites and decay over a long time. Therefore, older trees more often contain cavities suitable for use by larger vertebrates such as birds (Gibbons et al. 2000; Lindenmayer et al. 2000; Koch and Woehler 2007; Stojanovic et al. 2014a; Koch et al. 2018). The loss of such large old trees from human-dominated habitats is of particular conservation concern for parrots and other hollow-nesting species (Tidemann et al. 1992; Webb et al. 2012; Manning et al. 2013). Of Australia's native parrots, 12 species are endangered (Dooley and Vine 2013), among them the world's only migratory parrots, the Orange-bellied Parrot (*Neophema chrysogaster*) and the Swift Parrot (*Lathamus discolor*). While a variety of threats affect each of these 12 species, the impacts of habitat loss (and associated loss of breeding sites) and invasive species are significant threatening processes for many of these birds (Cameron 2006; Holdsworth 2006; Baker et al. 2013; Stojanovic et al. 2014b, 2016; Le Roux et al. 2016a; Lindenmayer et al. 2017).

AUSTRALIA'S CHANGING CAVITY-NESTING COMMUNITIES

The conversion of native habitats to agricultural and urban land uses is currently impacting much of Australia's native biodiversity and changing the distribution and abundance of birds at landscape scales (Clavero and García -Berthou 2005; Tulloch et al. 2016). While habitat change is associated with the decline of many species (Keast 1995; Ford et al. 2009), habitat conversion has also contributed to the range expansion of a number of the larger-bodied native parrots, such as the Galah (*Eolophus roseicapilla*), Long-billed Corella (*Cacatua tenuirostris*), Little Corella (*C. sanguinea*), and Rainbow Lorikeet (*Trichoglossus moluccanus*) (Gibbons and Lindenmayer 2002). Species such as the Long-billed Corella and Little Corella, whose range used to be confined largely to the more arid interior of Australia, now extend eastward to the coast (including many large urban centers) and south to the island of Tasmania (Gibbons and Lindenmayer 2002; Koch et al. 2008). The impact of breeding populations of these large parrots to environments where they did not previously occur across Australia has largely been unstudied but has important implications for understanding the potential for non-Australian parrot establishment as well as the impact of future invasive species.

Further changes in urban cavity-breeding communities, more generally, have come from the addition of non-Australian species. Some of the most widespread non-Australian birds include the Common Myna (*Acridotheres tristis*), Common Starling (*Sturnus vulgaris*), and House Sparrow

Both native and introduced birds have become urbanized and in some cases invasive in Australia. Examples of urbanized native species are (a) Rainbow Lorikeet (*Trichoglossus moluccanus*); (b) Little Corella (*Cacatua sanguinea*); and (c) Sulphur-crested Cockatoo (*C. galerita*). An example of an introduced urbanized species is (d) Common Myna (*Acridotheres tristis*). Photos by Andrew M. Rogers.

(*Passer domesticus*). Importantly, the invasion dynamics for these species are continuously changing; the House Sparrow has recently been declining in urban areas in Australia (Woodall 2002), while the Common Myna is continuing to expand its range (Martin 1996). Both the Common Starling and Common Myna have had reported impacts on native parrots, with the most direct impact being increased competition for nesting sites (Pell and Tidemann 1997a, 1997b; Holdsworth 2006; Haythorpe 2013; Grarock et al. 2014). However, the impacts of such species are not consistent across habitats, and most studies have been restricted to the southeast of Australia in a small part of the invasive range of these species (Sol et al. 2011; Grarock et al. 2012, 2013). Generally, competition between invasive and native cavity-nesting species is likely to be highest for species that are close in size and share preferences for nest sites (Diamond and Ross 2020; Rogers et al. 2020), providing insight into the potential impacts for future invasive species.

NATIVE PARROTS IN AUSTRALIAN CITIES

Australia is a diverse continent with large coastal cities across a range of ecoregions. Nearly 90% of Australians live in urban areas, and around 60% of the country's 24.6 million people live in capital cities (Australian Bureau of Statistics 2016), making Australia one of the most urbanized countries. These cities are also home to a large number of native cavity-nesting species. Between the years 2000 and 2018, 26 species of native parrots were recorded in at least two of Australia's major cities (eBird 2018), representing nearly half (46.4%) of all native Australian parrots (Table 18.1). While it is unclear how many of these species breed in cities, their presence means that cities should be considered an important place for native species conservation (Dearborn and Kark 2009; Miller and Hobbs 2013).

One of the paradoxes in Australia's urban ecology is that parrots are some of the most abundant urban species, despite the apparent lack of suitable nesting locations within urban environments (Shukuroglou and McCarthy 2006).

The decline of large trees in human-dominated environments is of major conservation concern, as it reduces the number of natural tree hollows suitable for nesting (Gibbons and Lindenmayer 2002; Lindenmayer et al. 2012; Le Roux et al. 2016a, 2016b; Stojanovic et al. 2016). For some species, nest boxes can supplement the lack of natural tree hollows (Tidemann et al. 1992; Heinsohn et al. 2003; Brazill-Boast et al. 2010; Webb et al. 2012; Goldingay et al. 2015). However, boxes that are accessible to invasive species or native predators may not be suitable for many native species (Lindenmayer et al. 2017), creating challenges for native cavity-nesting species conservation in urban areas.

Despite the limitations to breeding, Rainbow Lorikeets, Sulphur-crested Cockatoos (*Cacatua galerita*), and Galah were among the top 10 most frequently observed species nationally, in a citizen science program aimed at allowing people to count and report birds around cities (Birdlife Australia 2018). In the same project, various parrot species were in the top three most observed species in all states (Birdlife Australia 2018). The success of select native parrots may be due to important traits that are common in other species that are able to exploit urbanized environments, including large relative brain size and behavioral flexibility (Bennett and Harvey 1985; Sol et al. 2002; Lefebvre et al. 2013). Both of these traits allow for high levels of feeding innovation, which is the ability to adapt to novel food resources, and such resources are common in urban environments (White et al. 2005; Anderies et al. 2007; Sol et al. 2011; Griffin et al. 2012).

The abundance of some parrot species in urban areas is also the result of complex interactions between species and habitat at multiple spatial scales. At local scales, many parrots have benefited from people's choice of plants in urban landscapes, many of which provide nectar, fruits, and seeds (Daniels and Kirkpatrick 2006; Shukuroglou and McCarthy 2006; Ikin et al. 2013; Davis et al. 2014, 2015). At larger regional spatial scales, species have been shown to use cities as refuges during extreme weather events and environmental disturbance such as wildfire and drought (Keast 1995; Mawson and Long 1995; Parsons et al. 2003; Davis et al. 2012). Specifically, in a review of parrot abundance in Sydney, Davis

TABLE 18.1

The 26 Australian parrot species reported within city limits of major urban centers across Australia since 2000 (eBird 2018). These 26 species represent nearly half (26/56 = 46.4%) of all Australian native parrot species. X marks presences that may be due to escaped pets or natural range expansions (Gibbons and Lindenmayer 2002; Dyer et al. 2017b). Species are listed in alphabetical order by common name. Urban centers are arranged by total number of species present there.

SPECIES	MAJOR URBAN CENTERS								
	BRISBANE	CAIRNS	MELBOURNE	ADELAIDE	ALICE SPRINGS	DARWIN	HOBART	SYDNEY	PERTH
Australian King Parrot (*Alisterus scapularis*)	178	1						1	
Australian Ringneck (*Barnardius zonarius*)					54				24
Bourke's Parrot (*Neopsephotus bourkii*)					1				
Budgerigar (*Melopsittacus undulatus*)					10				
Cockatiel (*Nymphicus hollandicus*)	5	3			2	1			
Crimson Rosella (*Platycercus elegans*)	11	1	3	10				4	
Double-eyed Fig Parrot (*Cyclopsitta diophthalma*)		113							
Eastern Rosella (*Platycercus eximius*)	4		9	1X				20	
Galah (*Eolophus roseicapilla*)	77X	4	12	9	60	5	X	X	2X
Glossy Black Cockatoo (*Calyptorhynchus lathami*)	1								
Little Corella (*Cacatua sanguinea*)	2X	3		2	10	13		X	1X
Little Lorikeet (*Parvipsitta pusilla*)	179	2							
Long-billed Corella (*Cacatua tenuirostris*)	2X	2	X	X	4		X		X
Mulga Parrot (*Psephotellus varius*)					4				
Musk Lorikeet (*Glossopsitta concinna*)	17		6	10			6	1	
Northern Rosella (*Platycercus venustus*)						1			
Pale-headed Rosella (*Platycercus adscitus*)	652	1							
Rainbow Lorikeet (*Trichoglossus moluccanus*)	2290	274	69	24		74	X	104	5X
Red-rumped Parrot (*Psephotus haematonotus*)	2		19	3					
Red-tailed Black Cockatoo (*Calyptorhynchus banksia*)		13			7	25			3
Red-winged Parrot (*Aprosmictus erythropterus*)		1				23			
Scaly-breasted Lorikeet (*Trichoglossus chlorolepidotus*)	1071	39							
Sulphur-crested Cockatoo (*Cacatua galerita*)	927	156	11	5		24	15	94	
Swift Parrot (*Lathamus discolor*)			1				1		
Varied Lorikeet (*Psitteuteles versicolor*)						6			
Yellow-tailed Black Cockatoo (*Calyptorhynchus funereus*)	23		3				6		
TOTAL SPECIES	16	14	10	9	9	9	8	7	6

et al. (2011) found that six parrots increased in abundance in urban areas during periods of low rainfall and wildfire in the surrounding landscape. Interestingly, the higher abundance in urban areas persisted even after the recovery of the disturbed area (Davis et al. 2011). How native species can maintain such high abundances in the apparent absence of suitable breeding areas, and with the high levels of competition for remaining breeding sites, is not clear.

Across Australia, cavity-nesting species as a group are declining in abundance, following the general pattern for birds (Birdlife Australia 2015), but different species show high variation in their ability to persist in urban areas (Joyce et al. 2018). It is clear that some Australian parrots do well in cities, as evidenced by the number of parrots found in multiple Australian cities across the continent (Table 18.1). However, if they follow the patterns shown in other urban assemblages, it is likely that cities will become dominated by a few highly abundant species (McKinney and Lockwood 1999; Anderies et al. 2007; van Rensburg et al. 2009).

In Australia, both habitat change and interspecific competition are important drivers of species persistence in select habitats (Kath et al. 2009; Maron et al. 2013; Mac Nally et al. 2014). In particular, competition between cavity-nesting species for critical resources, such as nesting sites in urban areas, is already high (Davis 2013; Davis et al. 2013) and is likely to increase with the ongoing additions of non-Australian parrots. Cities are hot spots for non-native species because the introduction rate is higher where there are more people (Gaertner et al. 2017; Vall-llosera and Cassey 2017a). As Australia's population grows (the projected population is expected to reach between 30 and 40 million by 2056; Buckmaster and Simon-Davies 2018), it is likely that the number of pets escaping into the wild will also grow. However, as the vast majority of escaped species are not recorded by citizen science surveys (Vall-llosera and Cassey 2017a), there is a significant risk that species may establish before monitoring programs detect significant numbers of the birds. This is especially true for long-lived, highly adaptable and behaviorally flexible species such as parrots, which are likely to establish in cities first (Blackburn and Duncan 2001a; Cassey et al. 2004).

NON-AUSTRALIAN PARROTS: PATTERNS AND POTENTIAL IMPACTS

Additions to the Australian cavity-breeding community increasingly result from escaped pet birds. In a recent review, Vall-llosera and Cassey (2017a) documented, between 1993 and 2013, over 5,000 reports of escaped birds (all birds, including parrots), with 49 native and 42 non-native species escaping in similar proportions (45.7% and 53.3% of total pet escapes, respectively). This list included 38 native parrot

Little Corellas (*Cacatua sanguinea*) form large flocks in the middle of the day in outback Queensland. They tend to roost around towns where there are ample supplies of water and lawns to graze on. Longreach, Queensland, Australia, July 2013. Photo by Gemma Deavin.

species (Table 18.2) and 27 non-Australian parrot species (Table 18.3). The five most commonly reported escaped parrots were the non-native Rose-ringed Parakeet (*Psittacula krameri*), Alexandrine Parakeet (*P. eupatria*), Green-cheeked Parakeet (*Pyrrhura molinae*), Monk Parakeet (*Myiopsitta monachus*), and Sun Parakeet (*Aratinga solstitialis*) (Vall-Ilosera and Cassey 2017a). Importantly, the number of escaped non-native parrot species (27) and the sheer number of records reported in that study is much higher than the number of records in other available databases. For instance, in eBird (2018) records for mainland Australia, there are only 114 records, of 10 species, of non-native parrots (Table 18.4). Dyer et al. (2017b), in their Global Avian Invasions Atlas (GAVIA), reported four species of non-Australian parrots and 21 species of Australian parrots introduced around Australia. Within the GAVIA database, only the Rose-ringed Parakeet is listed as breeding in Western Australia, but studies of breeding individuals are sparse, and

TABLE 18.2

Escaped pet birds that are native to Australia reported in Vall-Ilosera and Cassey (2017a). Species are ranked by how commonly they were reported as escaped.

#	COMMON NAME	SCIENTIFIC NAME	#	COMMON NAME	SCIENTIFIC NAME
1	Cockatiel	*Nymphicus hollandicus*	21	Turquoise Parrot	*Neophema pulchella*
2	Galah	*Eolophus roseicapilla*	22	Western Corella	*Cacatua pastinator*
3	Budgerigar	*Melopsittacus undulatus*	23	Elegant Parrot	*Neophema elegans*
			24	Musk Lorikeet	*Glossopsitta concinna*
4	Eclectus Parrot	*Eclectus roratus*	25	Pale-headed Rosella	*Platycercus adscitus*
5	Rainbow Lorikeet	*Trichoglossus moluccanus*	26	Red-winged Parrot	*Aprosmictus erythropterus*
6	Sulphur-crested Cockatoo	*Cacatua galerita*	27	Western Rosella	*Platycercus icterotis*
			28	Crimson Rosella	*Platycercus elegans*
7	Princess Parrot	*Polytelis alexandrae*	29	Yellow-tailed Black Cockatoo	*Calyptorhynchus funereus*
8	Little Corella	*Cacatua sanguinea*			
9	Major Mitchell's Cockatoo	*Lophochroa leadbeateri*	30	Eastern Bluebonnet	*Northiella haematogaster*
10	Australian King Parrot	*Alisterus scapularis*	31	Blue-winged Parrot	*Neophema chrysostoma*
11	Long-billed Corella	*Cacatua tenuirostris*			
12	Scaly-breasted Lorikeet	*Trichoglossus chlorolepidotus*	32	Gang-gang Cockatoo	*Callocephalon fimbriatum*
13	Scarlet-chested Parrot	*Neophema splendida*	33	Golden-shouldered Parrot	*Psephotellus chrysopterygius*
14	Red-tailed Black Cockatoo	*Calyptorhynchus banksii*	34	Hooded Parrot	*Psephotus dissimilis*
15	Superb Parrot	*Polytelis swainsonii*	35	Mulga Parrot	*Psephotellus varius*
16	Bourke's Parrot	*Neopsephotus bourkii*	36	Red-capped Parrot	*Purpureicephalus spurius*
17	Red-rumped Parrot	*Psephotus haematonotus*	37	Carnaby's Black Cockatoo	*Calyptorhynchus latirostris*
18	Eastern Rosella	*Platycercus eximius*	38	Baudin's Black Cockatoo	*Calyptorhynchus baudinii*
19	Australian Ringneck	*Barnardius zonarius*			
20	Regent Parrot	*Polytelis anthopeplus*			

TABLE 18.3

Escaped pet parrots that are non-native to Australia reported in Vall-llosera and Cassey (2017a). Species are ranked by how commonly they were reported as escaped.

#	COMMON NAME	SCIENTIFIC NAME	#	COMMON NAME	SCIENTIFIC NAME
1	Rose-ringed Parakeet	*Psittacula krameri*	16	Red-shouldered Macaw	*Diopsittaca nobilis*
2	Alexandrine Parakeet	*Psittacula eupatria*	17	Plum-headed Parakeet	*Psittacula cyanocephala*
3	Green-cheeked Parakeet	*Pyrrhura molinae*	18	Ochre-marked Parakeet	*Pyrrhura cruentata*
4	Monk Parakeet	*Myiopsitta monachus*	19	Pearly Parakeet	*Pyrrhura lepida*
5	Sun Parakeet	*Aratinga solstitialis*	20	Red-and-green Macaw	*Ara chloropterus*
6	Rosy-faced Lovebird	*Agapornis roseicollis*	21	Black-capped Lory	*Lorius lory*
7	Blue-and-yellow Macaw	*Ara ararauna*	22	Crimson-bellied Parakeet	*Pyrrhura perlata*
8	Grey Parrot	*Psittacus erithacus*	23	Golden-capped Parakeet	*Aratinga auricapillus*
9	Jandaya Parakeet	*Aratinga jandaya*	24	Blue-winged Parakeet	*Psittacula columboides*
10	Red-breasted Parakeet	*Psittacula alexandri*	25	Scarlet Macaw	*Ara macao*
11	Red-crowned Parakeet	*Cyanoramphus novaezelandiae*	26	Senegal Parrot	*Poicephalus senegalus*
12	Fischer's Lovebird	*Agapornis fischeri*	27	Yellow-headed Amazon	*Amazona oratrix*
13	Nanday Parakeet	*Aratinga nenday*			
14	Yellow-collared Lovebird	*Agapornis personatus*			
15	Turquoise-fronted Amazon	*Amazona aestiva*			

it is unlikely that this represents an established breeding population (Table 18.5). The Atlas of Living Australia (2018) lists no non-Australian parrots in the updated version of the data set.

Given the increasing frequency of reported pet escapees and the importance of propagule pressure in establishing alien populations (Blackburn and Duncan 2001a; Duncan et al. 2014; Blackburn et al. 2015b), it is likely just a matter of time before a number of introduced parrots establish breeding populations in Australia's cities. However, the lack of information on these species in national and global data sets limits our understanding of where and when these birds might establish. While some established invasive birds, such as the Common Myna, have attracted a lot of research attention and control efforts by community groups (Grarock et al. 2012), other potentially invasive species, such as the Rose-ringed Parakeet, have received less attention in Australia to date (but see Vall-llosera et al. 2017). However, as this species and others that have become established in other parts of the world, including many cities, Australia needs better approaches to monitor emerging invasive birds.

Determining the ecological impacts of invasive birds—while difficult, due to complex interactions between species and their environments (Grarock et al. 2014)—could help to prioritize the management of emerging and likely invasive parrot species. If we assume that future invasive parrots are likely to have impacts similar to those they do abroad, and to those of other invasive birds in Australia, it is likely that non-native parrots that establish will further increase competition for cavity nests (Ingold 1998; Orchan et al. 2013; Peck 2013; Menchetti and Mori 2014).

TABLE 18.4

eBird records of non-Australian parrots recorded from 2000 to 2013. The world eBird database was downloaded in November 2018 and filtered by country, and then native species listed in Garnett et al. (2015) were filtered out. Species are listed in alphabetical order by common name.

COMMON NAME	SCIENTIFIC NAME	STATE	CITIES	CHECKLISTS	NUMBER OF INDIVIDUALS
Alexandrine Parakeet	Psittacula eupatria	New South Wales	8	8	8
		Northern Territory	1	1	1
		Queensland	7	7	7
		Victoria	9	9	12
Blue-and-yellow Macaw	Ara ararauna	New South Wales	3	3	14
		Queensland	1	1	2
Blue-throated Macaw	Ara glaucogularis	New South Wales	1	1	1
Fischer's Lovebird	Agapornis fischeri	Victoria	1	1	1
Jandaya Parakeet	Aratinga jandaya	Queensland	1	1	1
Nanday Parakeet	Aratinga nenday	Victoria	1	1	1
Rose-ringed Parakeet	Psittacula krameri	New South Wales	12	12	19
		Northern Territory	1	1	1
		Queensland	23	23	24
		South Australia	1	1	1
		Victoria	36	36	39
		Western Australia	1	1	1
Rosy-faced Lovebird	Agapornis roseicollis	New South Wales	1	1	1
		Northern Territory	3	3	6
		Queensland	2	2	2
		Victoria	1	1	1
Slaty-headed Parakeet	Psittacula himalayana	Victoria	1	1	1
Yellow-collared Lovebird	Agapornis personatus	Victoria	1	1	1

Understanding the role of increased competition in evaluating invasive species' impacts is difficult due to a poor understanding of the mechanisms and drivers of competition between Australian birds more generally. The importance of competitive interactions may be unique in Australia due to the presence of abundant, hyperaggressive species such as the Noisy Miner (*Manorina melanocephala*). Noisy Miners aggressively control large territories, from which they exclude many species, with significant consequences for bird abundance and composition (Parsons et al. 2006; Maron et al. 2013). Any non-native species that will establish in urban areas must be able to tolerate high levels of aggressive interactions with Noisy Miners. While Noisy Miners have an especially large impact on smaller species (<70g; Mac Nally et al. 2012; Howes et al. 2014), it is not clear what traits allow some native species to tolerate Noisy Miner aggression. As Noisy Miners also do well in modified environments where invasive species are likely to establish, the high levels of baseline competition between native species need to be taken into account, so the impacts are not misattributed to invasive birds (Grarock et al. 2012). A better understanding of the traits that shape native species' competitive hierarchies (Miller et al. 2017) will improve our ability to predict the competitive

AUSTRALIA'S URBAN CAVITY NESTERS AND INTRODUCED PARROTS: PATTERNS, PROCESSES, AND IMPACTS

TABLE 18.5

Summary of parrot introductions in Australia based on the GAVIA database (Dyer et al. 2017b). The majority of species introductions are of native species outside their range or reported escaped pets (non-Australian species are marked with *). Region is the state within Australia: ACT = Australian Capital Territory, NSW = New South Wales, NT = Northern Territory, Qld = Queensland, SA = South Australia, Tas = Tasmania, Vic = Victoria, WA = Western Australia. Status as noted in GAVIA: Brd = breeding, DO = died out, Est = established, Unk = unknown, Uns = unsuccessful. Introduction method: Esc = escaped, Rel = released, Unk = unknown. The eBird presence (Yes or No) is based on whether the species was reported within the past 10 years in that state (eBird 2019). Species are listed in alphabetical order by common name. The only breeding record of a non-Australian species was of the Rose-ringed Parakeet in WA, but there are no recent studies confirming breeding activity.

COMMON NAME	SCIENTIFIC NAME	REGION	INTRO. DATE	STATUS	INTRO. METHOD	EBIRD
Australian Ringneck	Barnardius zonarius	Vic	Unk	Unk	Esc	Yes
		ACT	Unk	Unk	Esc	No
		NSW	Unk	Unk	Esc	No
		NT	Unk	Unk	Esc	No
		Tas	Unk	Unk	Esc	No
		Qld	Unk	Unk	Esc	No
Bourke's Parrot	Neopsephotus bourkii	Qld	Unk	Unk	Esc	No
Budgerigar	Melopsittacus undulatus	Tas	Unk	Unk	Unk	No
Cockatiel	Nymphicus hollandicus	Tas	Unk	Est	Unk	No
		NSW	Unk	Est	Unk	No
Crimson Rosella	Platycercus elegans	NSW	Unk	Uns	Unk	No
		WA	Unk	Unk	Unk	No
		Tas	Unk	Est	Esc	No
Eastern Bluebonnet	Northiella haematogaster	SA	1970s	Unk	Rel	Yes
Eastern Rosella	Platycercus eximius	SA	Unk	Est	Unk	Yes
		Tas	Unk	Uns	Rel	No
Galah	Eolophus roseicapilla	Tas	Unk	Est	Unk	Yes
Gang-gang Cockatoo	Callocephalon fimbriatum	SA	1940	Uns	Unk	No
		SA	1940	Unk	Rel	No
Green Rosella	Platycercus caledonicus	NSW	Unk	Unk	Esc	No
Little Corella	Cacatua sanguinea	WA	Unk	Est	Unk	Yes
		NSW	Unk	Est	Unk	Yes
		Vic	Unk	Est	Unk	Yes
		Tas	Unk	Unk	Unk	Yes
Long-billed Corella	Cacatua tenuirostris	SA	Unk	Est	Unk	Yes
		NSW	Unk	Est	Unk	Yes
		Qld	Unk	Unk	Unk	Yes
		Tas	Unk	Unk	Esc	Yes
		SA	1975	Unk	Rel	Yes
		WA	Unk	Est	Rel	Yes
		Vic	Unk	Unk	Unk	Yes

COMMON NAME	SCIENTIFIC NAME	REGION	INTRO. DATE	STATUS	INTRO. METHOD	EBIRD
Major Mitchell's Cockatoo	Lophochroa leadbeateri	NSW	Unk	Est	Unk	Yes
Monk Parakeet*	Myiopsitta monachus	NSW	Unk	Est	Esc	No
Musk Lorikeet	Glossopsitta concinna	WA	Unk	Est	Unk	No
		WA	Unk	Est	Unk	No
Northern Rosella	Platycercus venustus	NT	Unk	Unk	Esc	No
Pale-headed Rosella	Platycercus adscitus	NSW	Unk	Unk	Esc	No
		Vic	Unk	Unk	Esc	No
Rainbow Lorikeet	Trichoglossus moluccanus	Tas	Unk	Unk	Unk	Yes
		WA	1960s	Est	Rel	Yes
Red-winged Parrot	Aprosmictus erythropterus	NSW	Unk	Unk	Esc	Yes
		Vic	Unk	Unk	Esc	No
Rose-ringed Parakeet*	Psittacula krameri	WA	Unk	Brd1	Unk	Yes
Rosy-faced Lovebird*	Agapornis roseicollis	WA	Unk	Uns	Unk	No
Scaly-breasted Lorikeet	Trichoglossus chlorolepidotus	Vic	Unk	Est	Unk	Yes
		Vic	Unk	Est	Rel	Yes
Sulphur-crested Cockatoo	Cacatua galerita	WA	Unk	Est	Unk	Yes
		WA	<1935	Est	Unk	Yes
		Tas	Unk	Unk	Unk	Yes
		WA	Unk	Est	Unk	Yes
		Tas	Unk	DO	Rel	Yes
Superb Parrot	Polytelis swainsonii	NSW	Unk	Unk	Esc	No
		Vic	Unk	Unk	Esc	No

impacts of current and future potentially invasive species, especially in Australia, where only limited work has been done (Rogers et al. 2020).

Despite the lack of established non-native parrots in Australia, the large number of native parrots that have been moved around (Koch et al. 2008) should provide an interesting test of the likely impacts of adding more parrots to local communities. However, few studies have quantified how introductions of new species have changed interactions and competition between local species. One of the only Australian parrots assumed to be a problem outside its historic range is the Rainbow Lorikeet in Tasmania and Western Australia. In both locations, the bird is considered an agricultural pest, with most of the impacts associated with damage to crops, although it is also assumed to compete with native parrots of a similar size (Chapman 2005). While this assumption has not been tested as far as we could find, it is used as part of the justification for ongoing control efforts in the places where it has recently been introduced (Government of Western Australia 2018; Tasmanian Government 2018).

While future invasive species are likely to establish in cities first, they are unlikely to be confined to cities in Australia (McKinney and Kark 2017). Addressing key gaps in our understanding of the potential impact of non-Australian parrots will be an important challenge for the conservation of all native cavity-nesting species. Important first steps should include: to map the potential range of emerging invasive species across the continent; quantify preferences for breeding and foraging space; examine the factors shaping biotic interactions and especially competition with other species (Cooper et al. 2007); and identify which native species are likely to be most impacted.

CONSERVATION IMPLICATIONS

Australia's parrot assemblages continue to change due to ongoing habitat change and introductions of both Australian and non-Australian parrots. Significant challenges in understanding the impact of such changes include the poor ability to detect non-native species early in their establishment and to predict where impacts are likely to be the greatest. While the species most likely to establish are better known, there remain serious gaps in our understanding of what happens to birds as they escape into the wild. The lack of information on what is currently limiting the establishment of emerging invasive species hampers the ability of managers to act to prioritize control efforts. Beyond establishment, assessing the potential impact of non-native parrots has not received much research attention, despite the many Australian species that have been moved outside their historic range, which could be used as case studies.

A view of the future of invasive species' impacts on Australia's native parrots may be found in Tasmania. The parrot community in Tasmania is likely the most altered cavity-breeding community documented in Australia. On the island, there are 27 obligate cavity-nesting bird species, of which three are endangered and seven are introduced non-native species (Gibbons and Lindenmayer 2002; Koch et al. 2008). There are 12 parrot species on the island, including four introduced species: Little Corella, Long-billed Corella, Galah, and Rainbow Lorikeet (Koch et al. 2008). The ecological impact of such introductions remains poorly understood for most species; however, the influence of competition on native species' access to nesting hollows and breeding success has been noted for a few species (Holdsworth 2006; Edworthy 2016).

All of mainland Australia's established invasive cavity-nesting species also occur in Tasmania, but only the Common Starling has been studied in the field. The research suggests that starlings compete with native species for nest sites (Higgins 1999; Holdsworth 2006). Other invasive species such as the Rainbow Lorikeet have become increasingly abundant in modified habitats in Tasmania, but the listed impacts (at least based on research in Western Australia) focus mostly on its potential impact to agriculture (Chapman 2005). The lack of research on the ecological impacts of such species introductions is likely to significantly limit attempts to prioritize invasive species control efforts. Importantly, ongoing conservation efforts for Tasmania's threatened parrots have focused on improving breeding opportunities (Holdsworth 2006; Stojanovic et al. 2012; Webb et al. 2012), but the impact of increased competition by introduced species on Tasmania's other endemic and common species is unknown.

Across the continent, significant changes have occurred in the parrot community due to extensive habitat change and the number of Australian species that have moved around. Further changes are likely, given the rates of non-Australian species introductions. These introductions include many species, such as the Monk Parakeet and the Rose-ringed Parakeet, that are invasive in other parts of the world. With little understanding of the barriers to their establishment in Australia, it seems a matter of time before these species establish on the continent. The lack of data on how the cavity-breeding communities have changed over time and the role that competition plays in structuring current communities limits our ability to predict the impacts of the ongoing introductions of non-Australian parrots. Such efforts are essential for assessing the biosecurity risk to non-Australian species and can be a focus of future studies on both native and invasive parrots across the vast and diverse Australian continent and its islands.

REFERENCES

Anderies, J. M., Katti, M., and Shochat, E. 2007. Living in the city: Resource availability, predation, and bird population dynamics in urban areas. *Journal of Theoretical Biology* 247:36–49.

Atlas of Living Australia. 2018. Accessed Nov. 2018. http://www.ala.org.au.

Australian Bureau of Statistics. 2016. 2016 Census. Accessed June 2019. https://www.abs.gov.au/websitedbs/censushome.nsf/home/2016.

Bacher, S., Carlton, J. T., Blackburn, T. M., Pys, P., Wilson, J.R.U., Duncan, R. P., and Richardson, D. M. 2011. A proposed unified framework for biological invasions. *Trends in Ecology & Evolution* 26:333–339.

Baker, J., Harvey, K. J., and French, K. 2013. Threats from introduced birds to native birds. *Emu* 114:1–12.

Bennett, P. M., and Harvey, P. H. 1985. Relative brain size and ecology in birds. *Journal of Zoology* 207:151–169.

Birdlife Australia. 2015. *The State of Australia's Birds*. Melbourne: BirdLife Australia.

Birdlife Australia. 2018. 2018 Aussie backyard bird count results. https://aussiebirdcount.org.au/2018-results/.

Blackburn, T. M., Delean, S., Pyšek, P., and Cassey, P. 2015a. On the island biogeography of aliens: A global analysis of the richness of plant and bird species on oceanic islands. *Global Ecology and Biogeography* 25:859–868.

Blackburn, T. M., and Duncan, R. P. 2001a. Determinants of establishment success in introduced birds. *Nature* 414:195–197.

Blackburn, T. M., and Duncan, R. P. 2001b. Establishment patterns of exotic birds are constrained by non-random patterns in introduction. *Journal of Biogeography* 28:927–939.

Blackburn, T. M., Dyer, E., Su, S., and Cassey, P. 2015b. Long after the event, or four things we (should) know about bird invasions. *Journal of Ornithology* 156:15–25.

Brazill-Boast, J., Pryke, S. R., and Griffith, S. C. 2010. Nest-site utilization and niche overlap in two sympatric, cavity-nesting finches. *Emu* 110:170–177.

Buckmaster, L., and Simon-Davies, J. 2018. Australia's future population. Parliament of Australia. Accessed June 2018. https://www.aph.gov.au/About_Parliament/Parliamentary_Departments/Parliamentary_Library/pubs/BriefingBook43p/futurepopulation.

Cameron, M. 2006. Nesting habitat of the glossy black-cockatoo in central New South Wales. *Biological Conservation* 127:402–410.

Case, T. J. 1996. Global patterns in the establishment and distribution of exotic birds. *Biological Conservation* 78:69–96.

Cassey, P., Blackburn, T. M., Duncan, R. P., and Lockwood, J. L. 2005. Lessons from the establishment of exotic species: A meta-analytical case study using birds. *Journal of Animal Ecology* 74:250–258.

Cassey, P., Blackburn, T. M., Sol, D., Duncan, R. P., and Lockwood, J. L. 2004. Global patterns of introduction effort and establishment success in birds. *Proceedings of the Royal Society of London B: Biological Sciences* 271:S405–S408.

Chapman, T. 2005. *The Status and Impact of the Rainbow Lorikeet (Trichoglossus haematodus moluccanus) in South-west Western Australia*. Perth, Western Australia: Dept. of Agriculture and Food.

Chiron, F., Shirley, S., and Kark, S. 2009. Human-related processes drive the richness of exotic birds in Europe. *Proceedings of the Royal Society B: Biological Sciences* 276: 47–53.

Clavero, M., and García-Bethou, E. 2005. Invasive species are a leading cause of animal extinctions. *Trends in Ecology & Evolution* 20:110.

Clergeau, P., and Vergnes, A. 2011. Bird feeders may sustain feral rose-ringed parakeets *Psittacula krameri* in temperate Europe. *Wildlife Biology* 17:248–252.

Cooper, C. B., Hochachka, W. M., and Dhondt, A. A. 2007. Contrasting natural experiments confirm competition between house finches and house sparrows. *Ecology* 88:864–870.

Daniels, G. D., and Kirkpatrick, J. B. 2006. Does variation in garden characteristics influence the conservation of birds in suburbia? *Biological Conservation* 133:326–335.

Davis, A. 2013. Habitat and resource utilisation by an urban parrot community. PhD diss., Univ. of Sydney, Australia.

Davis, A., Major, R. E., and Taylor, C. E. 2013. Housing shortages in urban regions: Aggressive interactions at tree hollows in forest remnants. *PLoS One* 8:e59332.

Davis, A., Major, R. E., and Taylor, C. E. 2014. Distribution of tree-hollows and hollow preferences by parrots in an urban landscape. *Emu* 114:295–303.

Davis, A., Major, R. E., and Taylor, C. E. 2015. The association between nectar availability and nectarivore density in urban and natural environments. *Urban Ecosystems* 18:503–515.

Davis, A., Taylor, C. E., and Major, R. E. 2011. Do fire and rainfall drive spatial and temporal population shifts in parrots? A case study using urban parrot populations. *Landscape and Urban Planning* 100:295–301.

Davis, A., Taylor, C. E., and Major, R. E. 2012. Seasonal abundance and habitat use of Australian parrots in an urbanized landscape. *Landscape and Urban Planning* 106:191–198.

Dearborn, D. C., and Kark, S. 2009. Motivations for conserving urban biodiversity. *Conservation Biology* 24:1–9.3

Diamond, J. M., and Ross, M. S. 2020. Overlap in reproductive phenology increases the likelihood of cavity nest usurpation by invasive species in a tropical city. *Condor* 122.

Dooley, S., and Vine, S. 2013. Parrots in peril. *Australian Birdlife*, 19 Sept. http://www.birdlife.org.au/australian-birdlife/detail/parrots-in-peril.

Duncan, R. P., Blackburn, T. M., Rossinelli, S., and Bacher, S. 2014. Quantifying invasion risk: The relationship between establishment probability and founding population size. *Methods in Ecology and Evolution* 5:1255–1263.

Duncan, R. P., Bomford, M., Forsyth, D. M., and Conibear, L. 2001. High predictability in introduction outcomes and the geographical range size of introduced Australian birds: A role for climate. *Journal of Animal Ecology* 70:621–632.

Dyer, E. E., Cassey, P., Redding, D. W., Collen, B., Franks, V., Gaston, K., Jones, K. E., Kark, S., Orme, C.D.L., and Blackburn, T. M. 2017a. The global distribution and drivers of introduced bird species richness. *PLoS Biology* 15:e2000942.

Dyer, E. E., Redding, D. W., and Blackburn, T. M. 2017b. The Global Avian Invasions Atlas: A database of alien bird distributions worldwide. *Scientific Data* 4:170041.

eBird. 2018. Accessed Nov. 2018. https://www.ebird.org.

eBird. 2019. Accessed Dec. 2019. https://www.ebird.org.

Edworthy, A. B. 2016. Competition and aggression for nest cavities between striated pardalotes and endangered forty-spotted pardalotes. *Condor* 118:1–11.

Evans, T., Kumschick, S., Dyer, E., and Blackburn, T. 2014. Comparing determinants of alien bird impacts across two continents: Implications for risk assessment and management. *Ecology and Evolution* 4:2957–2967.

Ford, H. A., Walters, J. R., Cooper, C. B., Debus, S.J.S., and Doerr, V. A. J. 2009. Extinction debt or habitat change? Ongoing losses of woodland birds in north-eastern New South Wales, Australia. *Biological Conservation* 142:3182–3190.

Gaertner, M., Wilson, J.R.U., Cadotte, M. W., MacIvor, J. S., Zenni, R. D., and Richardson, D. M. 2017. Non-native species in urban environments: Patterns, processes, impacts and challenges. *Biological Invasions* 19:3461–3469.

Garnett, S. T., Duursma, D. E., Ehmke, G., Guay, P. J., Stewart, A., Szabo, J. K., Weston, M. A., Bennett, S., Crowley, G. M., Drynan, D., et al. 2015. Biological, ecological, conservation and legal information for all species and subspecies of Australian bird. *Scientific Data* 2:1–6.

Gibbons, P., and Lindenmayer, D. B. 2002. *Tree Hollows and Wildlife Conservation in Australia*. Collingwood, Victoria: CSIRO Publishing.

Gibbons, P., Lindenmayer, D. B., Barry, S. C., and Tanton, M. T. 2000. Hollow formation in eucalypts from temperate forests in southeastern Australia. *Pacific Conservation Biology* 6:218–228.

Goldingay, R. L., Rueegger, N. N., Grimson, M. J., and Taylor, B. D. 2015. Specific designs can improve habitat restoration for cavity-dependent arboreal mammals. *Restoration Ecology* 23:482–490.

Government of Western Australia. 2018. Rainbow lorikeet: Management. Dept. of Primary Industries and Regional Development: Agriculture and Food. https://www.agric.wa.gov.au/birds/rainbow-lorikeet-management.

Grarock, K., Lindenmayer, D. B., Wood, J. T., and Tidemann, C. R. 2013. Does human-induced habitat modification influence the impact of introduced species? A case study on cavity-nesting by the introduced common myna (*Acridotheres tristis*) and two Australian native parrots. *Environmental Management* 52:958–970.

Grarock, K., Tidemann, C. R., Wood, J., and Lindenmayer, D. B. 2012. Is it benign or is it a pariah? Empirical evidence for the impact of the common myna (*Acridotheres tristis*) on Australian birds. *PLoS One* 7:e40622.

Grarock, K., Tidemann, C. R., Wood, J. T., and Lindenmayer, D. B. 2014. Are invasive species drivers of native species decline or passengers of habitat modification? A case study of the impact of the common myna (*Acridotheres tristis*) on Australian bird species. *Austral Ecology* 39:106–114.

Griffin, A. S., Sol, D., and Bartomeus, I. 2012. The paradox of invasion in birds: Competitive superiority or ecological opportunism? *Oecologia* 169:553–564.

Haythorpe, K. M. 2013. Competitive behaviour in Common Mynas: An investigation into potential impacts on native fauna and solutions for management. PhD diss., Charles Sturt Univ., Australia.

Heinsohn, R., Murphy, S., and Legge, S. 2003. Overlap and competition for nest holes among eclectus parrots, palm cockatoos and sulphur-crested cockatoos. *Australian Journal of Zoology* 51:81–94.

Hernández-Brito, D., Carrete, M., Popa-Lisseanu, A. G., Ibáñez, C., Tella, J. L. 2014. Crowding in the city: Losing and winning competitors of an invasive bird. *PLoS One* 9:e100593.

Higgins, P. J. 1999. *Handbook of Australian, New Zealand and Antarctic Birds*. Ed. P. J. Higgins. Melbourne, Australia: Oxford Univ. Press.

Holdsworth, M. 2006. Reproductive success and demography of the orange-bellied parrot *Neophema chrysogaster*. MS thesis, Univ. of Tasmania.

Howes, A., Mac Nally, R., Loyn, R., Kath, J., Bowen, M., Mcalpine, C., and Maron, M. 2014. Foraging guild perturbations and ecological homogenization driven by a despotic native bird species. *Ibis* 156:341–354.

Ikin, K., Beaty, R. M., Lindenmayer, D. B., Knight, E., Fischer, J., and Manning, A. D. 2013. Pocket parks in a compact city: How do birds respond to increasing residential density? *Landscape Ecology* 28:45–56.

Ingold, D. J. 1998. The influence of starlings on flicker reproduction when both naturally excavated cavities and artificial nest boxes are available. *Wilson Bulletin* 110:218–225.

Joyce, M., Barnes, M. D., Possingham, H. P., and van Rensburg, B. J. 2018. Understanding of avian assemblage change within anthropogenic environments using citizen science data. *Landscape and Urban Planning* 179:81–89.

Kark, S., and Sol, D. 2005. Establishment success across convergent Mediterranean ecosystems: An analysis of bird introductions. *Conservation Biology* 19:1519–1527.

Kath, J., Maron, M., and Dunn, P. K. 2009. Interspecific competition and small bird diversity in an urbanizing landscape. *Landscape and Urban Planning* 92:72–79.

REFERENCES

Keast, A. 1995. Habitat loss and species loss: The birds of Sydney 50 years ago and now. *Australian Zoologist* 30:3–25.

Koch, A. J., Chuter, A., Barmuta, L. A., Turner, P., and Munks, S. A. 2018. Long-term survival of trees retained for hollow-using fauna in partially harvested forest in Tasmania, Australia. *Forest Ecology and Management* 422:263–272.

Koch, A. J., Munks, S. A., and Woehler, E. J. 2008. Hollow-using vertebrate fauna of Tasmania: Distribution, hollow requirements and conservation status. *Australian Journal of Zoology* 56:323–349.

Koch, A. J., and Woehler, E. J. 2007. Results of a survey to gather information on the use of tree hollows by birds in Tasmania. *Tasmanian Naturalist* 129:37–46.

Lefebvre, L., Reader, M., Sol, D., and Sol, D. 2013. Innovating innovation rate and its relationship with brains, ecology and general intelligence. *Brain, Behavior and Evolution* 1:1–3.

Le Roux, D. S., Ikin, K., Lindenmayer, D. B., Bistricer, G., Manning, A. D., and Gibbons, P. 2016a. Effects of entrance size, tree size and landscape context on nest box occupancy: Considerations for management and biodiversity offsets. *Forest Ecology and Management* 366:135–142.

Le Roux, D. S., Ikin, K., Lindenmayer, D. B., Bistricer, G., Manning, A. D., and Gibbons, P. 2016b. Enriching small trees with artificial nest boxes cannot mimic the value of large trees for hollow-nesting birds. *Restoration Ecology* 24:252–258.

Lindenmayer, D. B., Crane, M., Bekessy, S., Blanchard, W., Evans, M. C., Maron, M., and Gibbons, P. 2017. The anatomy of a failed offset. *Biological Conservation* 210:286–292.

Lindenmayer, D. B., Cunningham, R. B., and Pope, M. L. 2000. Cavity sizes and types in Australian eucalypts from wet and dry forest types—a simple of rule of thumb for estimating size and number of cavities. *Forest Ecology and Management* 137:139–150.

Lindenmayer, D. B., Laurance, W. F., and Franklin, J. F. 2012. Global decline in large old trees. *Science* 338:1305–1306.

Linz, G. M., Homan, H. J., Gaulker, S. M., Penry, L. B., and Bleier, W. J. 2007. European starlings: A review of an invasive species with far-reaching impacts. In *Managing Vertebrate Invasive Species: Proceedings of an International Symposium*, ed. G. W. Witmer, W. C. Pitt, and K. A. Fagerstone, 378–386. Fort Collins, CO: USDA/APHIS Wildlife Services, National Wildlife Research Center.

MacDonald, M. A, and Kirkpatrick, J. B. 2003. Explaining bird species composition and richness in eucalypt-dominated remnants in subhumid Tasmania. *Journal of Biogeography* 30:1415–1426.

Mac Nally, R., Bowen, M., Howes, A., McApline, C. A., and Maron, M. 2012. Despotic, high-impact species and the subcontinental scale control of avian assemblage structure. *Ecology* 93:668–678.

Mac Nally, R., Kutt, A. S., Eyre, T. J., Perry, J. J., Vanderduys, E. P., Mathieson, M., Ferguson, D. J., and Thomson, J. R. 2014. The hegemony of the "despots": The control of avifaunas over vast continental areas. *Diversity and Distributions* 20:1071–1083.

Manning, A. D., Gibbons, P., Fischer, J., Oliver, D. L., and Lindenmayer, D. B. 2013. Hollow futures? Tree decline, lag effects and hollow-dependent species. *Animal Conservation* 16:395–403.

Maron, M., Grey, M. J., Catterall, C. P., Major, R. E., Oliver, D. L., Clarke, M. F., Loyn, R. H., Mac Nally, R., Davidson, I., and Thomson, J. R. 2013. Avifaunal disarray due to a single despotic species. *Diversity and Distributions* 19:1468–1479.

Martin, W. K. 1996. The current and potential distribution of the common myna *Acridotheres tristis* in Australia. *Emu* 96:166–173.

Mawson, P. R., and Long, J. L. 1995. Changes in the status and distribution of four species of parrot in the south of Western Australia during 1970–90. *Pacific Conservation Biology* 2:191–199.

McKinney, M., and Kark, S. 2017. Factors shaping avian alien species richness in Australia vs Europe. *Diversity and Distributions* 23:1334–1342.

McKinney, M. L., and Lockwood, J. L. 1999. Biotic homogenization: A few winners replacing many losers in the next mass extinction. *Trends in Ecology & Evolution* 14:450–453.

Menchetti, M., and Mori, E. 2014. Worldwide impact of alien parrots (Aves Psittaciformes) on native biodiversity and environment: A review. *Ethology Ecology & Evolution* 26:172–194.

Miller, E. T., Bonter, D. N., Eldermire, C., Freeman, B. G., Greig, E. I., Harmon, L. J., Lisle, C., and Hochachka, W. M. 2017. Fighting over food unites the birds of North America in a continental dominance hierarchy. *Behavioral Ecology* 28:1454–1463.

Miller, J. R., and Hobbs, R. J. 2013. Conservation where people live. *Conservation Biology* 16:330–337.

Orchan, Y., Chiron, F., Schwartz, A., and Kark, S. 2013. The complex interaction network among multiple invasive bird species in a cavity-nesting community. *Biological Invasions* 15:429–445.

Parsons, H., French, K., and Major, R. E. 2003. The influence of remnant bushland on the composition of suburban bird assemblages in Australia. *Landscape and Urban Planning* 66:43–56.

Parsons, H., Major, R. E., and French, K. 2006. Species interactions and habitat associations of birds inhabiting urban areas of Sydney, Australia. *Austral Ecology* 31:217–227.

Peck, H. L. 2013. Investigating ecological impacts of the non-native population of rose-ringed parakeets (*Psittacula krameri*) in the UK. PhD diss., Imperial College, London.

Pell, A. S., and Tidemann, C. R. 1997a. The ecology of the common myna in urban nature reserves in the Australian capital territory. *Emu* 97:141–149.

Pell, A. S., and Tidemann, C. R. 1997b. The impact of two exotic hollow-nesting birds on two native parrots in savannah and woodland in eastern Australia. *Biological Conservation* 79:145–153.

Rogers, A. M. 2019. The role of habitat variability and interactions around nesting cavities in shaping urban bird communities. PhD diss., Univ. of Queensland, Australia.

Rogers, A. M., Griffins, A. S., van Rensburg, B. J., Kark, S. 2020. Noisy neighbours and myna problems: Interaction webs and aggression around tree hollows in urban habitats. *Journal of Applied Ecology* 57: 1891–1901.

Shukuroglou, P., and McCarthy, M. A. 2006. Modelling the occurrence of rainbow lorikeets (*Trichoglossus haematodus*) in Melbourne. *Austral Ecology* 31:240–253.

Sol, D., Griffin, A. S., Bartomeus, I., and Boyce, H. 2011. Exploring or avoiding novel food resources? The novelty conflict in an invasive bird. *PLoS One* 6:e19535.

Sol, D., Timmermans, S., and Lefebvre, L. 2002. Behavioural flexibility and invasion success in birds. *Animal Behaviour* 63:495–502.

Stojanovic, D., Koch, A. J., Webb, M., Cunningham, R., Roshier, D., and Heinsohn, R. 2014a. Validation of a landscape-scale planning tool for cavity-dependent wildlife. *Austral Ecology* 39:579–586.

Stojanovic, D., Webb, M. H., Alderman, R., Porfirio, L. L., and Heinsohn, R. 2014b. Discovery of a novel predator reveals extreme but highly variable mortality for an endangered migratory bird. *Diversity and Distributions* 20:1200–1207.

Stojanovic, D., Webb, M., Roshier, D., Saunders, D., and Heinsohn, R. 2012. Ground-based survey methods both overestimate and underestimate the abundance of suitable tree-cavities for the endangered swift parrot. *Emu* 112:350–356.

Stojanovic, D., Webb nee Voogdt, J., Webb, M., Cook, H., and Heinsohn, R. 2016. Loss of habitat for a secondary cavity nesting bird after wildfire. *Forest Ecology and Management* 360:235–241.

Tasmanian Government. 2018. Rainbow lorikeet. Dept. of Primary Industries, Parks, Water and Environment. Invasive Species. https://dpipwe.tas.gov.au/invasive-species/invasive-animals/invasive-birds/rainbow-lorikeet.

Tidemann, S. C., Boyden, J., Elvish, R., Elvish, J., and O'Gorman, B. 1992. Comparison of the breeding sites and habitat of two hole-nesting estrildid finches, one endangered, in northern Australia. *Journal of Tropical Ecology* 8:373–388.

Tulloch, A.I.T., Barnes, M. D., Ringma, J., Fuller, R. A., and Watson, J.E.M. 2016. Understanding the importance of small patches of habitat for conservation. *Journal of Applied Ecology* 53:418–429.

Vall-llosera, M., and Cassey, P. 2017a. Leaky doors: Private captivity as a prominent source of bird introductions in Australia. *PLoS One* 12:1–18.

Vall-llosera, M., and Cassey, P. 2017b. Physical attractiveness, constraints to the trade and handling requirements drive the variation in species availability in the Australian cagebird trade. *Ecological Economics* 131:407–413.

Vall-llosera, M., Woolnough, A. P., Anderson, D., and Cassey, P. 2017. Improved surveillance for early detection of a potential invasive species: The alien rose-ringed parakeet *Psittacula krameri* in Australia. *Biological Invasions* 19:1273–1284.

van Rensburg, B. J., Peacock, D. S., and Robertson, M. P. 2009. Biotic homogenization and alien bird species along an urban gradient in South Africa. *Landscape and Urban Planning* 92:233–241.

Webb, M. H., Holdsworth, M. C., and Webb, J. 2012. Nesting requirements of the endangered swift parrot (*Lathamus discolor*). *Emu* 112:181–188.

White, J. G., Antos, M. J., Fitzsimons, J. A., and Palmer, G. C. 2005. Non-uniform bird assemblages in urban environments: The influence of streetscape vegetation. *Landscape and Urban Planning* 71:123–135.

Woodall, P. F. 2002. The Birds Queensland garden bird survey, 1999–2000. *Sunbird* 32:37–51.

19

THE FUTURE FOR NATURALIZED PARROTS

Stephen Pruett-Jones

This edited volume was begun as an effort to bring together researchers active in the study of naturalized parrots and summarize our current knowledge about the distribution and ecology of this increasingly diverse, common, and important group of birds. An additional goal was to provide different perspectives about ongoing controversies involving naturalized parrots, including whether they should be considered invasive or not. In this last chapter, I hope to highlight some of the major points raised in the 18 preceding chapters as they relate to the future of naturalized parrots.

As a starting point, humans and parrots have had a long history of interactions. The last chapter/epilogue of the book *Parrots of the Wild* (Toft and Wright 2015) is entitled "Parrots as the Most Human of Birds." Toft and Wright argue that parrots exhibit a diverse adaptive syndrome that, while generally rare in the animal kingdom, is common to humans, nonhuman primates, cetaceans (whales and dolphins), elephants, and to a lesser degree corvids (crows and ravens). This "syndrome" encompasses the following characteristics: parrots are long-lived, reproduce slowly, usually engage in lifelong monogamous relationships, care for their dependent young, engage in play throughout their lives, communicate with sounds that they sometimes invent, and depend on locally available but unpredictable resources. In other words, parrots are not "normal" birds.

These characteristics of parrots, plus their beautiful plumages and coloration, can explain the fact that parrots are one of the most popular groups of cage birds in the world. Indeed, parrots have been kept as pets for at least 5,000 years and possibly longer. Although the international trade of parrots has waxed and waned over time, the long-term trend has been upward and increasingly so in the 19th and 20th centuries (Global Avian Invasions Atlas, GAVIA, Dyer et al. 2017). Our knowledge about the international trade in parrots increased dramatically after 1975, when the Convention on International Trade in Endangered Species (CITES 2018) began to keep records of legal exports and imports between signatory countries to the convention. Our recent examination of CITES data between 1975 and 2018 (C. Calzada Preston and S. Pruett-Jones, unpubl. data; see also Cassey et al. 2004) showed that of the approximately 380 species of extant parrots in the world, 330 (>80%) species or subspecies have been traded internationally to more than 100 countries. This recorded trade has involved more than 16 million individual birds since 1975. Surely, many tens of millions more birds were involved in trade before accurate records were kept, and even today the illegal trade involves countless more.

Of all of the species of parrots transported around the world, it is hard to know exactly how many individuals or species have escaped captivity or been released into the wild. For one area, the contiguous United States (US), relevant data are available. First, based on CITES data, approximately 290 species of parrots were transported to the US from 1981 to 2018 (Calzada Preston and Pruett-Jones, unpubl. data). Records from citizen science data (eBird, Christmas Bird Counts [CBC], etc.) document that during the 15-year period 2002–2016, more than 56 species of parrots were seen in the wild in the US, comprising more than 19,000 observations (Uehling et al. 2019). Thus, as a rough estimate, approximately one-fifth of all parrot species transported to the contiguous US were eventually observed in the wild. This is almost certainly an underestimate of the actual number. Many escaped birds never get recorded on eBird or CBC, and there must have been many sightings outside of the 2002–2016 period considered by Uehling et al. (2019).

Our interest here is focused on those introduced (that is, escaped or released) parrots that have been able to establish successful (breeding) populations in their new environments. Worldwide, based on the GAVIA database (Dyer et al. 2017) and data from the Global Biodiversity Information Facility (GBIF 2019) no fewer than 75 species (23% of the 330 transported taxa, 20% of 380 species total) are now breeding or established in at least one country. These figures are almost double the estimate of 10% as suggested by data in Cassey et al. (2004) and Abellán et al. (2017), although these differences are likely due to the data set being examined and the criteria for identification of a species as established. Even more striking, perhaps, is the fact that parrots have established naturalized populations in a minimum of 89 countries, more than 45% of 193 recognized countries, and as documented by Royle and Donner (chap. 2 this vol.), these countries are spread around the globe.

Many of the data presented in this volume come from citizen science databases (now preferably termed *community science databases*; National Audubon Society 2018). As important and critical as these sources of data are, one of the consistent themes here is that detailed, standardized surveys are needed to fully understand the diversity, distribution, and ecology of naturalized parrots. Data in Uehling et al. (chap. 12 this vol.) highlight the different perspectives one gets solely from publicly available databases and more comprehensive surveys. For the US, the GAVIA database lists up to 15 established species of parrots. In contrast, Uehling et al. (2019) document at least 25 species of naturalized parrots on the US mainland. Furthermore, there are an additional four unique species naturalized on Puerto Rico (Falcón and Tremblay 2018; Uehling et al., chap. 12 this vol.). Combined, the US now supports 29 species of naturalized parrots and an additional 11+ species breeding in the US (mostly Puerto Rico) that may soon be naturalized. It is possible that within the next decade, the US may be home to 40 species of naturalized parrots, more than 10% of all parrot species in the world.

The pet trade and the transport of species around the globe have dominated human and parrot interactions. Nevertheless, also important has been the persecution of parrots related to agricultural practices. It is likely that ever since humans developed agriculture within the natural range of parrots, there has been conflict. Bucher (chap. 8 this vol.) addresses this issue from the South American and historical perspective, and Mori and Menchetti (chap. 6 this vol.) as well as Brightsmith and Kiacz (chap. 9 this vol.) address it from a modern perspective.

The authors of these chapters do not necessarily agree on whether naturalized parrots should be considered invasive or the extent of their impacts. Mori and Menchetti (chap. 6 this vol.) take a precautionary and conservative approach highlighting the diverse interactions between introduced parrots and native species, as well as the various ecological effects that have been attributed to introduced parrots. Indeed, in the literature there are dozens of such recorded interactions; naturalized parrots can compete with other species for cavity nest sites, and they can compete with other species for food resources. A similar precautionary approach to naturalized parrots is taken by the research collective ParrotNet (2018). From its website:

> ParrotNet is a European network of scientists, practitioners and policy-makers dedicated to research on invasive parrots, their impacts and the challenges they present. Our network is working to better understand why parakeets are highly successful invaders, how we can predict their agricultural, economic, societal and ecological impacts across Europe, and the means to mitigate them, and to assist policy-makers with managing the challenges that Invasive Alien Species pose.

Focusing on the various interactions between naturalized parrots and native species, or on the parrots' economic impacts, it is easy to argue that populations of naturalized parrots should be actively controlled. Nevertheless, in comparison with other introduced species, and on the balance between their positive and negative effects, an argument can equally be made that naturalized parrots should be left alone or even promoted in some cases. Brightsmith and Kiacz (chap. 9 this vol.) present the most complete review to date on the possible negative effects of naturalized parrots. Their review highlights that: (1) Most populations of naturalized parrots have no or little ecological or economic impact; (2) The only well-documented example of direct impact of naturalized parrots has been the effect of Rose-ringed Parakeets (*Psittacula krameri*) on Greater Noctule Bats (*Nyctalus lasiopterus*) in Spain (Hernández-Brito et al. 2014; also Mori and Menchetti, chap. 6 this vol.); and (3) Monk Parakeets (*Myiopsitta monachus*) and Rose-ringed Parakeets are increasing in numbers in specific areas around the globe and can cause economic damage. In particular, Monk Parakeets can cause economic damage through the effects of constructing their nests on electrical structures. Despite these few examples of ecological or economic damage, for Brightsmith and Kiacz (chap. 9 this vol.), there is no instance where introduced or naturalized parrots rise to the level of a priority invasive species as established by the Convention on Biological Diversity (CBD 2010).

Regardless of whether or not naturalized parrots should be considered priority species, there are cases where populations are currently being controlled. In Spain, populations of Monk Parakeets have grown exponentially in recent years, population trends that have been exacerbated by continued legal importations (Souviron-Priego et al. 2018). In late 2019, authorities in Spain announced that they were going to cull as many as 12,000 Monk Parakeets from the city of Madrid (Woodyatt 2019), at a total cost of ~$3.25 million (~€3 million) or ~$270 (€228) for each parakeet (Today 2020). This culling program has already begun, but with the ongoing COVID-19 pandemic, it is presently put on hold. It is also the case that there are active petitions aimed at convincing the government to abandon this program. As documented by Senar et al. (chap. 7 this vol.) and Pruett-Jones et al. (2007), the only effective and efficient means of reducing population sizes in Monk Parakeets is by direct removal and euthanasia of adult birds. Knocking down nests or removal of eggs or nestlings can reduce the rate of population growth in Monk Parakeets, but these methods are generally ineffective at reducing population numbers (Pruett-Jones et al. 2007).

Naturalized Monk Parakeets are also controlled in the US because of their damage to electrical power generation and transmission infrastructure. In every large city in the continental US where Monk Parakeets are common, the birds' nests are known to interfere with electrical generation or transmission and in some cases cause frequent power outages (Avery et al. 2002, 2006; Newman et al. 2008; Brightsmith and Kiacz, chap. 9 this vol.). As emphasized by Calzada Preston et al. (chap. 11 this vol.), research is needed to understand the interaction between nest-site preferences (trees vs. man-made structures) and habitat preferences (natural vs. human-modified) in Monk Parakeets, and how this interaction varies in different geographical areas. For example, we don't know whether nest-site preferences are inherited or socially transmitted. Are nestlings raised in a nest on an electrical structure more likely to choose an electrical structure when they reproduce, or is nest-site selection random? Long-terms studies of banded individuals will be necessary to answer this question, but this is a critical aspect that needs to be addressed. An alternative view on mitigating damage by Monk Parakeets is to modify electrical structures to dissuade the parakeets from choosing that site for nesting. Burgio et al. (2014) suggest that intervention during the nest-building stages of Monk Parakeets can mitigate damage to electrical structures, but this idea has not been adequately tested.

Naturalized parrots represent novel additions to avifauna in areas in which parrots have never occurred naturally, e.g., northern Europe. In other areas, naturalized parrots may act as a species replacement for extinct native species. One possible example of this is Monk Parakeets and Carolina Parakeets (*Conuropsis carolinensis*) in the continental US. The Carolina Parakeet was the only endemic parrot species in the continental US, its distribution entirely confined to the US (Burgio et al. 2017). In fact, Carolina Parakeets had the most northerly native distribution of any species of parrot in the world. Despite abundant population size and a widespread distribution, by the end of the 19th century Carolina Parakeets were rare; the last official record of the species in the wild was in 1910, and the last captive individual died in 1918. The species was officially declared extinct in 1939. Although multiple causes likely contributed to the extinction of the Carolina Parakeet, a recent genomic analysis by Gelabert et al. (2020) suggests that it was a combination of habitat loss and hunting by humans. These authors reconstructed the species' genome and found no genetic evidence of a declining population, such as a loss of heterozygosity. In fact, the authors found no genetic signal of an extinction event at all. This suggests a very rapid decline in population numbers preceded the eventual extinction of Carolina Parakeets. The likely explanation of this was widespread and rapid habitat destruction of the bird's forest habitat, and then persecution through shooting. It is also possible that disease played a role in the eventual extinction (Snyder and Keith 2002).

Garber (1993) addressed the question of whether naturalized Monk Parakeets are the ecological equivalent to the Carolina Parakeet. This is an interesting question given that, although Monk Parakeets are not nearly as abundant in the US as Carolina Parakeets once were, they do occur over much of the same geographic area Carolina Parakeets once occupied. It is, however, hard to argue that Monk Parakeets are ecological equivalents of Carolina Parakeets. The latter species occupied old-growth continuous forests and fed on native plant and tree seeds. Carolina Parakeets did eventually occupy human-modified habitats, but it appears this occurred only once the continuous forest was destroyed. In contrast, Monk Parakeets are never found in undisturbed, continuous forests; they are a species of open and savanna habitats. Furthermore, their diet is extremely diverse (South and Pruett-Jones 2000) and likely not as restricted as that of Carolina Parakeets.

The argument that Monk Parakeets are equivalent to Carolina Parakeets is also sometimes made in an effort to persuade public opinion toward a more tolerant view of Monk Parakeets as an introduced species. As Crowley (chap. 3 this vol.) notes, the view toward naturalized parrots is often polarized, with most of the general public, which is happy to have new and colorful species in the environment, on one end of the extreme vs. land managers, policy makers, and many biologists, who are concerned with introduced and invasive species, on the other. This polarization is at its height in the US, where there is a patchwork of laws and regulations across the country. In some states, e.g., California, Monk Parakeets are strictly prohibited, and ownership, breeding, transport, and sale are illegal. In contrast, in Florida, birds can be owned, bought and sold, bred in captivity, and offspring from breeding efforts can be released into the wild.

This tension between public impression of a wildlife situation and administrative policies is not unique to naturalized parrots, but this very visible group of birds highlights this difference as few other groups do. Furthermore, it is likely that this tension will increase as naturalized parrots become even more widespread than they are today. Despite import bans on certain species of parrots by the US, the European Union, and other countries, transport of parrots around the world continues and in some areas is actually increasing. With the behavioral flexibility exhibited by parrots, it is likely that the number of species that establish naturalized populations around the globe will also continue to increase, as will the number of areas with established parrots. Along with some seabirds, parrots as a group are among the longest-lived of any birds. Thus, when parrots establish themselves in an urban habitat, they could live there as long or even longer than the human residents of the city. The naturalized parrots of the world will certainly be present for years to come, and in the future will become even more common and abundant.

Given that the future of the world is likely one with many naturalized species of parrots, what is needed is excellent research on populations, the biology of individual species, and environmental impacts. Several chapters in this volume highlight examples of the type of research that is needed on a much wider basis. This research can illuminate the origins of introduced populations, patterns of genetic diversity in current populations, patterns of dispersal, adaptation to new environments, social behavior and mating systems, and the genetic changes associated with the process of naturalization and invasion (Russello et al. 2008; Gonçalves da Silva et al. 2010; Raisin et al. 2012; Martínez et al. 2013). Genetic studies of naturalized parrots will become ever more important in the future as introduced populations become reservoirs for species endangered in their natural habitats, and as hybridization between naturalized species increases. In contrast to most introduced and widespread vertebrate species, and except in a few situations, naturalized parrots on the whole appear to offer more positive benefits than negative impacts (Brightsmith and Kiacz, chap. 9 this vol.). The research opportunities that naturalized parrots offer are beginning to be better understood and appreciated, and it will be exciting to see this research continue.

REFERENCES

Abellán, P., Tella, J. L., Carrete, M., Cardador, L., and Anadón, J. D. 2017. Climate matching drives spread rate but not establishment success in recent unintentional bird introductions. *Proceedings of the National Academy of Sciences (USA)* 114:9385–9390.

Avery, M. L., Greiner, E. C., Lindsay, J. R., Newman, J. R., and Pruett-Jones, S. 2002. Monk parakeet management at electric utility facilities in south Florida. In *Proceedings of the 20th Vertebrate Pest Conference*, ed. R. M. Timm and R. H. Schmidt, 140–145. Davis: Univ. of California.

Avery, M. L., Lindsay, J. R., Newman, J. R., Pruett-Jones, S., and Tillman, E. A. 2006. Reducing monk parakeet impacts to electric utility facilities in south Florida. In *Advances in Vertebrate Pest Management*, ed. C. J. Feare and D. P. Cowan, 4:125–136. Furth, Germany: Filander Verlag.

Burgio, K. R., Carlson, C. J., Tingley, M. W. 2017. Lazarus ecology: Recovering the distribution and migratory patterns of the extinct Carolina parakeet. *Ecology and Evolution* 7:5467–5475.

Burgio, K. R., Rubega, M. A., and Sustaita, D. 2014. Nest-building behavior of monk parakeets and insights into potential mechanisms for reducing damage to utility poles. *PeerJ* 2014:2:e601.

Cassey, P., Blackburn, T. M., Russell, G. J., Jones, K. E., and Lockwood, J. L. 2004. Influences on the transport and establishment of exotic bird species: An analysis of the parrots (Psittaciformes) of the world. *Global Change Biology* 10:417–426.

CBD. 2010. COP 10 Decision X/2. Strategic plan for biodiversity 2011–2020. Convention on Biological Diversity. https://www.cbd.int/decision/cop/?id=12268.

CITES. 2018. CITES Trade Database. Accessed Dec. 2018. https://trade.cites.org.

Dyer, E. E., Redding, D. W., and Blackburn, T. M. 2017. The Global Avian Invasions Atlas: A database of alien bird distributions worldwide. *Scientific Data* 4:170041.

Falcón, W., and R. L. Tremblay. 2018. From the cage to the wild: Introductions of Psittaciformes to Puerto Rico. *PeerJ* 6:e5669.

Garber, S. D. 1993. Is the monk parakeet the ecological equivalent of North America's extinct Carolina parakeet? *Focus* 43:26–35.

GBIF. 2019. Global Biodiversity Information Facility. Accessed Dec. 2019. https://doi.org/10.15468/dl.j1fzvm.

Gelabert, P., Sandoval-Velasco, M., Serres, A., de Manuel, M., Renom, P., Margaryan, A., Stiller, J., de-Dios, T., Fang, Q., Feng, S., et al. 2020. Evolutionary history, genomic adaptation to toxic diet, and extinction of the Carolina parakeet. *Current Biology* 30:P108–114.E5.

Gonçalves da Silva, A., Eberhard, J. R., Wright, T. F., Avery, M. L., and Russello, M. A. 2010. Genetic evidence for high propagule pressure and long-distance dispersal in monk parakeet (*Myiopsitta monachus*) invasive populations. *Molecular Ecology* 19:3336–3350.

Hernández-Brito, D., Carrete, M., Popa-Lisseanu, A. G., Ibáñez, C., and Tella, J. L. 2014. Crowding in the city: Losing and winning competitors of an invasive bird. *PLoS One* 9:e100593.

Martínez, J. J., de Aranzamendi, M. C., Masello, J. F., and Bucher, E. H. 2013. Genetic evidence of extra-pair paternity and intraspecific brood parasitism in the monk parakeet. *Frontiers in Zoology* 10:68.

National Audubon Society. 2018. Why we're changing from "citizen science" to "community science." Audubon Center at Debs Park. https://debspark.audubon.org/news/why-were-changing-citizen-science-community-science.

Newman, J. R., Newman, C. M., Lindsay, J. R., Merchant, B., Avery, M. L., and Pruett-Jones, S. 2008. Monk parakeets: An expanding problem on power lines and other electrical utility structures. In *8th International Symposium on Environmental Concerns in Rights-of-Way Management*, ed. J. W. Goodrich-Mahoney, L. P. Abrahamson, J. L. Ballard, and S. M. Tikalsky, 343–354. New York: Elsevier Science.

ParrotNet. 2018. European Network on Invasive Parakeets. Univ. of Kent. Accessed June 2020. https://www.kent.ac.uk/parrotnet/.

Pruett-Jones, S., Newman, J. R., Newman, C. M., Avery, M. L., and Lindsay, J. R. 2007. Population viability analysis of monk parakeets in the United States and examination of alternative management strategies. *Human-Wildlife Conflicts* 1:35–44.

Raisin, C., Frantz, A. C., Kundu, S., Greenwood, A. G., Jones, C. G., Zuel, N., and Groombridge, J. J. 2012. Genetic consequences of intensive conservation management for the Mauritius parakeet. *Conservation Genetics* 13:707–715.

Russello, M. A., Avery, M. L., and Wright, T. F. 2008. Genetic evidence links invasive monk parakeet populations in the United States to the international pet trade. *BMC Evolutionary Biology* 8:217.

Snyder, N. F., and Keith, R. 2002. Carolina parakeet (*Conuropsis carolinensis*). In *The Birds of North America*, ed. A. Poole. Ithaca, NY: Cornell Lab of Ornithology.

South, J., and Pruett-Jones, S. 2000. Patterns of flock size, diet, and vigilance of naturalized monk parakeets in Hyde Park, Chicago. *Condor* 102:848–854.

Souviron-Priego, L., Muñoz, A. R., Oliverso, J., Vargas, J. M., and Fa, J. E. 2018. The legal international wildlife trade favors invasive species establishment: The monk and ring-necked parakeets in Spain. *Ardeola* 65:233–246.

Today. 2020. Plenty to squawk about: Spanish capital plans for huge cull of parrots. *Today*, 13 Feb. https://www.todayonline.com/world/plenty-squawk-about-spanish-capital-plans-huge-cull-parrots-0.

Toft, C. A., and Wright, T. F. 2015. *Parrots of the Wild: A Natural History of the World's Most Captivating Birds*. Oakland, CA: Univ. of California Press.

Uehling, J. J., Tallant, J., and S. Pruett-Jones, S. 2019. Status of naturalized parrots in the United States. *Journal of Ornithology* 160:907–921.

Woodyatt, A. 2019. Madrid plans "ethical cull" of city's parakeets. CNN Travel, 8 Oct. https://www.cnn.com/2019/10/08/europe/madrid-parrot-cull-intl-scli/index.html.

INDEX

Abbreviations:
f = Figure
t = Table
a = Appendix
p = Photograph

Accipiter gentilis. See Northern Goshawk
Accipiter nisus. See Eurasian Sparrowhawk
Accipitriformes, 243f15.3
acclimatization societies, 227, 278
Acridotheres burmannicus. See Vinous-breasted Starling
Acridotheres tristis. See Common Myna
Aegithalos caudatus. See Long-tailed Tit
African Grey Parrot. See Grey Parrot
African Palm Swift, 90, 96t6.2
African Queen, The 164, 250
African Sacred Ibis, 87, 92
African Spoonbill, 87
Agapornis, 265t17.2, 267, 268t17.3, 269f17.4, 272
Agapornis canus. See Grey-headed Lovebird
Agapornis fischeri. See Fischer's Lovebird
Agapornis lilianae. See Lilian's Lovebird
Agapornis roseicollis. See Rosy-faced Lovebird
Agapornis personatus. See Yellow-collared Lovebird
Agapornis pullarius. See Red-headed Lovebird
Agapornis taranta. See Black-winged Lovebird
'Alalā, 73
Alexander the Great, 13, 227, 241
Alexandrine Parakeet, 25t2.2, 29, 31f2.3, 33, 34a2.1, 91–92, 139, 168p, 230–231, 233–236, 244, 249, 252, 266f17.3, 283–285
alien invasive species. See invasive species
alien species. See invasive species
Alisterus scapularis. See Australian King Parrot
Allee effect, 245, 254
Alopochen aegyptiacus. See Egyptian Goose
Amazona, 23, 75, 138, 265–266, 268–269
Amazona aestiva. See Turquoise-fronted Amazon
Amazona albifrons. See White-fronted Amazon
Amazona amazonica. See Orange-winged Amazon
Amazona auropalliata. See Yellow-naped Amazon
Amazona autumnalis. See Red-lored Amazon
Amazona brasilensis. See Red-tailed Amazon
Amazona farinosa. See Southern Mealy Amazon

Amazona finschi. See Lilac-crowned Amazon
Amazona guildingii. See St. Vincent Amazon
Amazona leucocephala. See Cuban Amazon
Amazona ochrocephala. See Yellow-crowned Amazon
Amazona oratrix. See Yellow-headed Amazon
Amazona ventralis. See Hispaniolan Amazon
Amazona viridigenalis. See Red-crowned Amazon
Amazona vittata. See Puerto Rican Amazon
American Crow, 90, 96t6.2
American Kestrel, 187
American Robin, 90, 96t6.2
Anas platyrhynchos. See Mallard
Anodorhynchus hyacinthinus. See Hyacinth Macaw
Anodorhynchus leari. See Lear's Macaw
Anoplophora glabripennis. See Asian Long-horned Beetle
Anous minutus melanogenys. See Hawaiian Black Noddy
Anseriformes, 243f15.3
Antipodes Parakeet, 34a2.1
Anthochaera carunculata. See Red Wattlebird
Anthracoceros albirostris. See Oriental Pied Hornbill
Anumbius annumbi. See Firewood-gathers
Apis mellifera. See European Honeybee
applied bird conservation, 227
Aprosmictus erythropterus. See Red-winged Parrot
Apus apus. See Common Swift
Aquila pennata. See Booted Eagle
Ara, 267–268
Ara ararauna. See Blue-and-yellow Macaw
Ara chloropterus. See Red-and-green Macaw
Ara glaucogularis. See Blue-throated Macaw
Ara macao. A. m. cyanopterus. See Scarlet Macaw
Ara militaris. See Military Macaw
Ara rubrogenys. See Red-fronted Macaw
Ara severus. See Chestnut-fronted Macaw
Aratinga, 78, 267–269, 272
Aratinga auricapillus. See Golden-capped Parakeet
Aratinga jandaya. See Jandaya Parakeet
Aratinga nenday. See Nanday Parakeet
Aratinga solstitialis. See Sun Parakeet

Aratinga weddellii. See Dusky-headed Parakeet
Ardea cinerea. See Grey Heron
Asio otus. See Long-eared Owl
Athene noctua. See Little Owl
Australian King Parrot, 232, 281t18.1, 283t18.2
Australian Ringneck, 34a2.1, 90, 109, 111, 117, 281t18.1, 283t18.2, 286t18.5
avian chlamydiosis, 45, 75, 92, 138, 236
avian influenza, 46, 92
avian malaria, 225
avian poxvirus, 224

Bald Eagle, 92, 138
Barnardius, 268t17.3
Barnardius zonarius. See Australian Ringneck
Baudin's Black Cockatoo, 61, 283t18.2
biological invasions. See invasion biology
biotic homogenization, 169
biotic resistance theory, 29
Black Kite, 90, 95t6.1
Black Lory, 34a2.1
Black Noddy, 96t6.2
Black Rat, 90, 93, 96t6.1, 97t6.2
Black Robin, 73
Black Stilt, 74
Black Stork, 91, 95t6.1
Black Woodpecker, 236
Black-capped Lory, 284t18.3
Black-capped Parakeet, 266
Black-headed Parrot, 34a2.1
Black-winged Lovebird, 232
Blue-winged Parakeet
Blossom-headed Parakeet, 34a2.1
Blue Jay, 90, 96t6.2
Blue Lorikeet, 34a2.1
Blue-and-yellow Macaw, 34a2.1, 59–60, 80, 197t12.1, 212t13.1, 232, 253p, 271, 284t18.3, 285t18.4
Blue-crowned Hanging Parrot, 34a2.1
Blue-crowned Parakeet, 25t2.2, 34a2.1, 44p, 90, 197t12.1, 200f12.5, 203, 211–212, 215–218, 225, 230, 244, 247, 249, 253
Blue-eyed Cockatoo, 34a2.1
Blue-headed Parrot, 34a2.1
Blue-naped Parrot, 25t2.2, 34a2.1
Blue-streaked Lory, 34a2.1
Blue-throated Macaw, 285t18.4
Blue-winged Macaw, 229, 266
Blue-winged Parrakeet, 284t18.3
Blue-winged Parrot, 283t18.2
Blue-winged Parrotlet, 34a2.1
Booted Eagle, 90–91, 95t6.1
Bourke's Parrot, 34a2.1, 281t18.1, 283t18.2, 286t18.5
Branta sandvicensis. See Nene
bromethalin toxicosis, 78
Brotogeris chiriri. See Yellow-crowned Parakeet
Brotogeris jugularis. See Orange-chinned Parakeet

INDEX

Brotogeris pyrrhoptera. See Grey-cheeked Parakeet
Brotogeris tirica. See Plain Parakeet
Brotogeris versicolurus. See White-winged Parakeet
Brown Cacholote, 183
Brown Tree Snake, 145t9.3
Brown-headed Parrot, 264
Brown-necked Parrot, 266
Brown-throated Parakeet, 25t2.2, 31f2.3, 34a2.1, 266
Budgerigar, 15, 24–25, 29, 31f2.3, 34a2.1, 78–79, 93, 96t6.2, 196–197, 200–201, 204, 212t13.1, 227, 231, 241, 244, 249, 252–254, 266, 281t18.1, 283t18.2, 286t18.5
Burrowing Parrot, 34a2.1, 59–60, 124, 126–129, 211–212, 216, 224, 229, 247, 266

Cacatua, 267, 268t17.3
Cacatua alba. See White Cockatoo
Cacatua galerita. See Sulphur-crested Cockatoo
Cacatua goffiniana. See Tanimbar Corella
Cacatua moluccensis. See Salmon-crested Cockatoo
Cacatua pastinator. See Western Corella
Cacatua sulphurea. See Yellow-crested Cockatoo
Cacatua sanguinea. See Little Corella
Cacatua tenuirostris. See Long-billed Corella
Cacatuidae, 278
California Condor, 73
Callocephalon fimbriatum. See Gang-gang Cockatoo
Calyptorhychus banksii. See Red-tailed Black Cockatoo
Calptorhynchus baudinii. See Baudin's Black Cockatoo
Calyptorhynchus funereus. See Yellow-tailed Black Cockatoo
Calyptorhynchus lathami. See Glossy Black Cockatoo
Calyptorhynchus latirostris. See Carnaby's Black Cockatoo
Cape Parrot, 138, 272
Carnaby's Black Cockatoo, 61, 89–90, 95t6.1, 283t18.2
Carolina Parakeet, 227, 232, 237, 295–296
Carrion Crow, 93
Caucasian Squirrel, 91, 96t6.1
Carduelis carduelis. See Goldfinch
CBD. See Convention on Biological Diversity
Chalinolobus gouldii. See Gould's Wattled Bat
Chattering Lory, 34a2.1
Chlamydophila. C. psittaci. See avian chlamydiosis
charisma: ecological, aesthetic, and corporeal, 47–48
Chatham Parakeet, 139
Chestnut-fronted Macaw, 25t2.2, 34a2.1, 197t12.1

Ciconia Ciconia. See White Stork
Ciconia nigra. See Black Stork
Ciconiiformes, 243f15.3
CITES. See Convention on International Trade in Endangered Species of Wild Fauna and Flora
citizen science/community science, 25, 51 193, 241, 280, 282, 293–294
Cockatiel, 34a2.1, 212t13.1, 228, 244, 265–266, 281t18.1, 283t18.2, 286t18.5
cockatoos, Cacatuidae, 269f17.4, 278
Coconut Lorikeet, 25t2.2, 34a2.1, 89–90, 96t6.2
Coloeus monedula. See Western Jackdaw
Columba livia, C. l. domestica. See Rock Dove
Columba oenas. See Stock Dove
Columba palumbus. See Common Wood Pigeon
Columbiformes, 243f15.3
Columbus, Christopher, 241
Common Blackbird, 95t6.1, 96t6.2, 188
Common Flameback, 91, 96t6.2
Common Kestrel, 96t6.2
Common Myna, 77–79, 87, 89, 92, 95–96, 139, 270–271, 279–280, 284
Common Noctule Bat, 96t6.1
Common Starling, 55, 89–90, 95–96, 167, 235, 270, 279–280
Common Swift, 88, 135–136
Common Wood Pigeon, 90, 188
concorde fallacy, 130
conures, 269f17.4
Conuropsis carolinensis. See Carolina Parakeet
Convention on Biological Diversity, 19, 133–134, 148, 198, 295
Convention on International Trade in Endangered Species of Wild Fauna and Flora, 16, 22, 63, 126, 174, 198, 241, 250, 261, 293
Coracopsis barklyi. See Seychelles Black Parrot
Coracopsis nigra. See Lesser Vasa Parrot
Corvus brachyrhynchos. See American crow
Corvus cornix. See Hooded Crow
Corvus corone. See Carrion Crow
Corvus hawaiiensis. See 'Alalā
Corvus monedula. See Wesern Jackdaw
Corvus splendens. See House Crow
Corvus splendens. See Indian House Crow
Crimson Rosella, 25t2.2, 34a2.1, 93, 96t6.2, 138, 281t18.1, 283t18.2, 286t18.5
Crimson Shining Parrot, 25t2.2, 34a2.1
Crimson-bellied Parakeet, 284t18.3
cryptic diversity, 58–60
Cuban Amazon, 34a2.1, 59, 62, 241
culling. See population control
Cyanistes caeruleus. See Eurasian Blue Tit
Cyanocitta cristata. See Blue Jay

Cyanoliseus patagonus. See Burrowing Parrot
Cyanoramphus auriceps. See Yellow-crowned Parakeet
Cyanoramphus cooki. See Norfolk Parakeet
Cyanoramphus forbesi. See Chatham Parakeet
Cyanoramphus novaezelandiae. See Red-crowned Parakeet
Cyclopsitta diophthalma. See Double-eyed Fig Parrot
Cypsiurus parvus. See African Palm Swift

Dacelo novaeguineae. See Laughing Kookaburras
Dendrocopos major. See Great Spotted Woodpecker
Dendrocopos syriacus. See Syrian Woodpecker
DiazaCon, 114–115
Diopsittaca nobilis. See Red-shouldered Macaw
Double-eyed Fig Parrot, 281t18.1
Drepanis coccinea. See Iiwi
Duke of Bedford, 227–228, 231–232
Dusky Lory, 34a2.1
Dusky-headed Parakeet, 25t2.2, 34a2.1, 266
Dryocopus martius. See Black Woodpecker

Eastern Bluebonnet, 34a2.1, 283t18.2, 286t18.5
Eastern Gray Squirrel, 90, 93, 96t6.1, 145t9.3
Eastern Ground Parrot, 59
Eastern Rosella, 25t2.2, 34a2.1, 93, 148, 281t18.1, 283t18.2, 286t18.5
Echo Parakeet, 61, 78, 89, 95t6.1, 135–138, 141, 148, 167
Eclectus, 268t17.3
Eclectus Parrot, 25t2.2, 34a2.1, 212t13.1, 266f17.3, 283t18.2
Eclectus roratus. See Electus Parrot
Edible Dormouse, 89, 145t9.3
Egyptian Goose, 87
Elegant Parrot, 283t18.2
Eliomys quercinus. See German Doormouse
END. See exotic Newcastle disease
enemy release hypothesis, 31, 166
Enicognathus leptorhynchus. See Slender-billed Parakeet
Eolophus, 268t17.3
Eolophus roseicapilla. See Galah
Eos bornea. See Red Lory
escaped parrots/pets. See invasion biology: introductions
EU ban. See Wild Bird Declaration
EU Wild Bird Declaration. See Wild Bird Declaration
Eupsittula canicularis. See Orange-fronted Parakeet
Eupsittula nana. See Olive-throated Parakeet
Eupsittula pertinax. See Brown-throated Parakeet
Eurasian Blue Tit, 90, 95t6.1

INDEX

Eurasian Collared Dove, 90
Eurasian Hobby, 93, 166
Eurasian Hoopoe, 88, 95t6.1, 136, 167
Eurasian Magpie, 109, 183
Eurasian Nuthatch, 88, 95t6.1, 135–136, 141, 167, 235, 237, 270
Eurasian Scops Owl, 88, 95t6.1, 135–136
Eurasian Sparrowhawk, 93, 166
Eurasian Tree Sparrow, 95t6.1
Eurasian Wryneck, 91, 95t6.1
European Green Woodpecker, 91, 95t6.1, 254
European wild bird trade ban. See Wild Bird Declaration
Eurystomus orientalis. See Oriental Dollarbirds
exotic Newcastle disease, 46, 75, 92, 137
extinction of experience, 79

Falco naumanni. See Lesser Kestrel
Falco peregrinus. See Peregrine Falcon
Falco sparverius. See American Kestrel
Falco subbuteo. See Eurasian Hobby
Falco tinnunculus. See Common Kestrel
Falconiformes, 243f15.3
feral pigeons. See Rock Dove
Festive Amazon, 34a2.1
Finsch's Parakeet, 34a2.1, 197t12.1
Firewood-gatherers, 182
Fischer's Lovebird, 16, 25t2.2, 34a2.1, 90–91, 96t6.2, 199, 200, 232, 244, 284–285
Forpus, 268t17.3
Forpus passerinus. See Green-rumped Parrotlet
founder population size. See biological invasion: propagule size

Galah, 25t2.2, 35a2.1, 90, 212t13.1, 265–266, 279, 281t18.1, 283t18.2, 286t18.5, 288
Galliformes, 243f15.3
Gang-gang Cockatoo, 35a2.1, 283t18.2, 286t18.5
GBIF. See Global Biodiversity Information Facility
GAVIA. See Global Avian Invasions Atlas
genetic bottlenecks, 14, 57, 62, 63, 65, 72t5.1, 77
genetic diversity and variation, 14, 33, 57–60, 62, 65, 77, 164–165,169
Geoffroy's Bat, 89, 96t6.1
German Doormouse, 89
Glis glis. See Edible Dormouse
Global Avian Invasions Atlas, 23–24, 26, 30f2.1, 33, 174, 178, 283, 293
Global Biodiversity Information Facility, 160f10.1, 174, 294
Glossopsitta, 268t17.3
Glossopsitta concinna. See Musk Lorikeet
Glossy Black Cockatoo, 281t18.1
Golden-capped Parakeet, 284t18.3
Golden-collared Macaw, 35a2.1
Golden-shouldered Parrot, 283t18.2
Gould's Wattled Bat, 89, 97t6.2

Great Spotted Woodpecker, 165, 235, 236, 254
Great Tit, 90, 95t6.1, 167
Great-billed Parrot, 35a2.1
Greater Noctule Bat, 88, 89p, 95t6.1, 135–137, 141, 148, 295
Greater Vasa Parrot, 35a2.1
Green Parakeet, 25t2.2, 35a2.1, 159, 197, 200f12.5, 204, 236
Green Rosella, 35a2.1, 286t18.5
Green-cheeked Parakeet, 35a2.1, 283, 284t18.3
Green-rumped Parrotlet, 25t2.2, 35a2.1, 266
Grey Heron, 95t6.1
Grey Parrot, 16, 22, 35a2.1, 76, 199–201, 212t13.1, 232, 241, 247, 260, 266–267, 284t18.3
Grey-cheeked Parakeet, 27, 35a2.1
Grey-crested Cacholote, 183
Grey-headed Lovebird, 25t2.2, 27, 231, 266
Gruiformes, 243f15.3
Gymnogyps californianus. See California Condor
guilty until proven innocent assumption, 45

Haliaeetus leucocephalus. See Bald Eagle
Harmonia axyridis. See Harlequin Ladybeetle
Hawaiian Black Noddy, 224
Hendrix hypothesis. See Hendrix, Jimi
Hendrix, Jimi, 164, 250
Himantopus novaezelandiae. See Black Stilt
Hispaniolan Amazon, 25t2.2, 35a2.1, 207
Hispaniolan Parakeet, 35a2.1, 207
Hooded Crow, 90–91, 97t6.2, 188
Hooded Parrot, 283t18.2
House Crow, 118
House Mouse, 145t9.3
House Sparrow, 55, 57, 77, 90–91, 95t6.1, 97t6.2, 186–188, 270, 279–280
Hyacinth Macaw, 35a2.1, 59, 266
Hypsugo savii. See Savi's Pipistrelle

Iiwi, 224
importer/exporter countries. See parrot (pet) trade
inbreeding depression, 62, 164
Indian House Crow, 87
international parrot trade. See parrot trade
introduction effort. See biological invasions: propagule size
invasion biology, 14, 19–20, 41, 54, 77–79, 87, 92, 140, 245
 epigenetics and, 67
 evolutionary mechanisms, 55, 164
 pathways, 63, 66, 163, 240
 propagule pressure/size and, 14, 29, 32–33, 42, 57, 77, 79, 129, 173, 199–201, 207, 245, 260, 272, 278, 284
 success or failure (of parrots) and importance of: abundant resources, 166; behavior and physiology, 15, 19, 28, 32, 165, 202; body size, 27, 43; brain size, 27; community assembly, 31; competition for nesting sites, 278; diet, 28, 174, 202; environmental similarity/matching, 16, 79, 162f10.4, 174, 201, 246; genetic diversity, 164; interspecific competition, 31; lack of predators, 166; latitude of native range, 29; local species diversity, 278; life history, 27, 43, 174; migratory habits, 27–28; occupying vacant niches, 270; phenological flexibility, 165; origin source (wild vs. captive), 15, 19, 55, 245; thermal tolerance, 30, 202
 transport, introduction, establishment, and spread as processes of, 7, 14, 16, 28, 43–44, 54–55, 57, 63, 65–66, 75, 78–79, 92, 159, 161, 164–166, 173, 178, 188, 194, 196–199, 201–203, 213, 215, 217–218, 221, 227, 231–233, 240, 243–246, 249–250, 254–255, 262, 264–272, 278, 281t18.1, 282, 286–7t18.5, 293
invasion genetics, 54, 62
invasive bridgehead effect, 57, 60, 65
invasive species, 18–19, 26–27, 32–33, 45, 54–55, 57, 62, 65–67, 87, 117–118, 133–135, 141, 145, 148–149, 159, 162–169, 173, 176, 202, 207, 223, 240, 244–245, 271, 273, 277–280, 284–285, 287–288. Also see naturalized parrots
Italian Sparrow, 92, 95t6.1, 97t6.2, 135–136, 140, 187

Java Sparrow, 166p
Jabiru, 183
Jabiru mycteria. See Jabiru
Jandaya Parakeet, 35a2.1, 212t13.1, 266, 284–285
Jynx torquilla. See Eurasian Wryneck

Kakapo, 24, 35a2.1, 62, 74
Kuhl's Lorikeet, 25t2.2, 35a2.1, 76

Lanius collaris. See Southern Fiscal
Larus michahellis. See Yellow-legged Gull
Lathamus discolor. See Swift Parrot
Laughing Dove, 88, 95t6.1
Laughing Kookaburra, 90, 97t6.2
Layard's Parakeet, 35a2.1
leaky doors, 43
Lear's Macaw, 246
Lesser Kestrel, 88, 95t6.1, 135t9.1, 136
Lesser Noctule Bat, 89, 96t6.1
Lesser Vasa Parrot, 35a2.1, 95t6.1
Lilac-crowned Amazon, 25t2.2, 35a2.1, 75, 135t9.1, 138, 197t12.1, 200f12.5, 204–205
Lilac-tailed Parrotlet, 35a2.1
Lilian's Lovebird, 25t2.2, 35a2.1
Little Corella, 25t2.2, 35a2.1, 96t6.2, 279, 281–283, 286t18.5, 288

INDEX

Little Lorikeet, 281t18.1
Little Owl, 89, 95t6.1
Little Wattlebird, 97t6.2
Lonchura oryzivora. See Java Sparrow
Long-billed Corella, 35a2.1, 96t6.2, 279, 281t18.1, 283t18.2, 286t18.5, 288
Long-eared Owl, 93, 95t6.1
Lophochroa leadbeateri. See Major Mitchell's Cockatoo
Lorius lory. See Black-capped Lory

macaws, 269f17.4
Madrid culling program, 295
Major Mitchell's Cockatoo, 35a2.1, 283t18.2, 286t18.5
Mallard, 270
Manorina melanocephala. See Noisy Miner
Maroon Shining Parrot, 25t2.2, 35a2.1
Maroon-bellied Parakeet, 35a2.1, 232
Maroon-fronted Parrot, 35a2.1
Melopsittacus undulatus. See Budgerigar
Meyer's Parrot, 35a2.1, 212t13.1, 264
Military Macaw, 35a2.1, 212t13.1
Milvus migrans. See Black Kite
Mitred Parakeet, 25t2.2, 35a2.1, 91, 93, 96t6.2, 197t12.1, 200f12.5, 203, 206, 211–212, 217–219, 223–224, 230
Monk Parakeet, 16, 18, 23p, 24–25, 28–32, 35a2.1, 42–51, 54–55, 62–66, 72, 77, 79–82, 87, 90, 92–96, 102–118, 123–131, 135–138, 140–142, 144–148, 159, 169, 173–188, 193–197, 199–200, 202–207, 212t13.1, 227–228, 233, 237, 244–245, 249, 252–253, 256, 265–266, 283–284, 286t18.5, 288, 295–296
Mourning Dove, 97t6.2
Mulga Parrot, 281t18.1, 283t18.2
Musk Lorikeet, 35a2.1, 281t18.1, 283t18.2, 286t18.5
Myiopsitta, 268t17.3, 269f17.4
Myiopsitta monachus; M. m. monachus; M. m. calita; M. m. cotorra; M. m. luchsi. See Monk Parakeet
Myotis emarginatus. See Geoffroy's Bat

Nanday Parakeet, 25t2.2, 31f2.3, 35a2.1, 197t12.1, 200f12.5, 203, 212t13.1, 229, 247, 266, 284t18.3, 285t18.4
nature deficit disorder, 79, 81
naturalized parrots:
  benefits of: development of research techniques, 72, 78; disease screening, 75–76; economic, 81, 144, 205; ecotourism and social, 48, 79–81; indirect effects, 270–271; niche expansion, 92; nest modification, 140; population and genetic reservoirs, 72–73; research opportunities in ecology, evolution, and behavior, 77–79, 140, 257; social, 147; source of food for predators, 140; sources for translocations and rewilding, 32, 73, 76, 140; surrogate parents in captive breeding, 78
cavity-nesting bird communities and, 278–280
charismatic attraction of, 41, 47–49, 80–81, 102, 169, 260, 270
control of. See population control
definition of, 22, 211, 213
distribution and numbers of species of, 23, 29, 30f2.1–2, 31f2.3, 55, 63, 87, 160–161, 176–177, 193–195, 203–207, 212–213, 220f13.1, 227–234, 245f15.4, 283t18.2, 285t18.4, 293–294
diseases and parasites of: See avian chlamydiosis; avian influenza; avian malaria; avian poxvirus; bromethalin toxicosis; exotic Newcastle disease; psittacosis; salmonella; West Nile virus
ecological niche expansion, 236–237
economics of. See naturalized parrots: benefits of
human reaction/responses to, 41–51, 79–80, 146–147, 169, 203, 205
hybrids between, 75, 91–92, 138–139, 214–215, 218, 228–229
impact costs of, 141–145, 205, 256
impacts of: airline strikes, 142t9.2, 144, 256; agricultural damage, 42, 46, 87, 123–126, 141–143, 167–168, 182, 219–220, 223–224, 256, 270, 287; behavioral interference, 88–90, 139–140; competition with native species, 88, 90–91, 135–136, 167, 223–224, 235–236, 270, 288; damage to buildings, 142t9.2, 144; disease transmission, 92, 135t9.1, 137–138, 167, 260, 272; ecosystem damage, 223–224, 256, 260; electrical infrastructure and transmission, 46, 87, 112–113, 144, 146, 205, 295; facilitation of introduced/invasive species, 92–93, 135t9.1, 223; harassment, 91; herbivory, 135t9.1, 139–140, 224; hybridization, 75, 87, 91–92, 135t9.1, 138–139; killing other species, 89–90; nest displacement, 90, 235–236; noise pollution, 45–47, 168, 256; predation by, 89–90; social conflicts, 49–50, 146; threats to human health or well-being, 223, 260
native species beyond natural limits as, 277, 280–282
population estimates and trajectories, 13, 22–24, 87, 160–161, 197t12.1, 200f12.5, 203–204, 213–223, 227–234, 249–255, 265t17.2.
predators of, 93, 166
seed dispersal and, 139–140
transport, imports and exports of. See parrot trade
Nannopsittaca panychlora. See Tepui Parrotlet
Nene, 73
Neophema, 268t17.3
Neophema chrysogaster. See Orange-bellied Parrot
Neophema chrysostoma. See Blue-winged Parrot
Neophema elegans. See Elegant Parrot
Neophema pulchella. See Turquoise Parrot
Neophema splendida. See Scarlet-chested Parrot
Neopsephotus bourkii. See Bourke's Parrot
New Holland Honeyeater, 97t6.2
Newcastle disease. See exotic Newcastle disease
Niam-niam Parrot, 25t2.2, 35a2.1
Noisy Miner, 285
Norfolk Parakeet, 89, 95t6.1
Northern Goshawk, 93, 227, 236
Northern Rosella, 35a2.1, 281t18.1, 286t18.5
Northiella haematogaster. See Eastern Bluebonnet
Nyctalus lasiopterus. See Greater Noctule Bat
Nyctalus leisleri. See Lesser Noctule Bat
Nyctalus noctula. See Common Noctule Bat
Nymphicus, 269f17.4
Nymphicus hollandicus. See Cockatiel

Ochre-marked Parakeet, 284t18.3
Olive-throated Parakeet, 25t2.2, 35a2.1
Orange-bellied Parrot, 76, 278
Orange-chinned Parakeet, 36a2.1, 212t13.1
Orange-fronted Parakeet, 25t2.2, 36a2.1, 207, 212t13.1
Orange-winged Amazon, 25t2.2, 36a2.1, 159, 197t12.1, 212t13.1, 229
Oriental Dollarbird, 91, 97t6.2
Oriental Pied Hornbill, 90, 97t6.2
origin stories, 43
Ornate Lorikeet, 36a2.1
Osprey, 92
Otus scops. See Eurasian Scops Owl
Oxyura jamaicensis. See Ruddy Duck
Oxyura leucocephala. See White-headed Duck

Pale-headed Rosella, 36a2.1, 211–212, 281t18.1, 283t18.2, 286t18.5
Palm Cockatoo, 36a2.1, 212t13.1
Pandion haliaetus. See Osprey
Passeriformes, 243f15.3
parrot fever. See psittacosis
parrot party bus, 81
parrot (pet) trade, 13–17, 20, 22, 27–29, 41–43, 63, 87, 102, 126–127, 164, 175–177, 198f12.4, 200, 240–243, 250, 267, 268–272, 293–294
ParrotNet, 94, 294
parrots as pets, 7, 13, 42, 46, 79, 87, 102, 159, 169, 173, 194, 199, 201–202, 223, 227, 240–241, 247, 260, 293

INDEX

parrots as the most human of birds, 293
Parus major. See Great Tit
Parvipsitta pusilla. See Little Lorikeet
Passer domesticus. See House Sparrow
Passer italiae. See Italian Sparrow
Passer montanus. See Eurasian Tree Sparrow
PBFD. See psittacine beak and feather disease
Peach-fronted Parakeet, 36a2.1
Pearly Parakeet, 284t18.3
Pelicaniformes, 243f15.3
Peregrine Falcon, 91, 93, 95t6.1, 166, 227
Peruvian Incas, 123
pet trade. See parrot trade
Petroica traversi. See Black Robin
Pezoporus wallicus. See Eastern ground Parrot
Phaethon rubricauda. See Red-tailed Tropicbird
phylogeography, 58–62
Pica pica. See Eurasian Magpie
Picus viridis. See European Green Woodpecker
Pied Currawong, 97t6.2
Pink-necked Green Pigeon, 91, 95t6.1, 97t6.2
Pionus, 266f17.3, 268t17.3
Pionus maximiliani. See Scaly-headed Parrot
Plain Parakeet, 27, 36a2.1, 232
Platalea alba. See African Spoonbill
Platycercus, 268t17.3
Platycercus adscitus. See Pale-headed Rosella
Platycercus elegans. See Crimson Rosella
Platycercus eximius. See Eastern Rosella
Platycercus icterotis. See Western Rosella
Platycercus venustus. See Northern Rosella
Plum-headed Parakeet, 36a2.1, 212t13.1, 264, 266, 284t18.3
Poicephalus, 267, 268t17.3, 269f17.4
Poicephalus crassus. See Niam-niam Parrot
Poicephalus cryptoxanthus. See Brown-headed Parrot
Poicephalus fuscicollis. See Brown-necked Parrot
Poicephalus gulielmi. See Red-fronted Parrot
Poicephalus meyeri. See Meyer's Parrot
Poicephalus robustus. See Cape Parrot
Poicephalus rueppellii. See Rüppell's Parrot
Poicephalus rufiventris. See Red-bellied Parrot
Poicephalus senegalus. See Senegal Parrot
Polytelis alexandrae. See Princess Parrot
Polytelis anthopeplus. See Regent Parrot
Polytelis swainsonii. See Superb Parrot

population control
  alternatives to, 108; barriers, 130; compensation, 131; crop substitution, 130; habitat change, 130; lethal control as last measure, 131; repellants, 131
  costs of, 109, 110t7.3, 111, 113–114, 144, 256, 295
  effectiveness of, 108f7.3, 110t7.3, 111f7.4, 113–114, 115t67.5, 118, 125
  integrated strategies for, 129–130
  methods of: biological control, 112; cannon netting, 107, 109–110; contraception, via chemosterilants 114–115, 118; drop-in traps, 109; effigies, 108; funnel traps, 109; international pet trade, 125–127; lasers, 108; limiting/reducing access to food, 115; mist nets, 109, 218; nest poisoning, 125; oiling eggs, 114; poking eggs, 110–111, 114–115; remote trigger traps, 109; removing or destroying nests, 110–111, 113–114, 118; scare-eye balloons, 108; shooting, 107, 110–111, 118, 218; toxic substances or insecticides, 112, 124; trapping, 107, 110–111; visual scare devices, 108
  of native species, 124–127
  optimization model of, 116–118
population genetic structure, 59–61
precautionary principle in conservation science, 20, 45, 47, 50, 133
Primolius maracana. See Blue-winged Macaw
Princess Parrot, 283t18.2
priority invasive species, 133, 134f9.1, 135, 148–149
Probosciger aterrimus. See Palm Cockatoo
propagule pressure/size. See invasion biology
Prosopeia splendens. See Crimson Shining Parrot
Prosopeia tabuensis. See Maroon Shining Parrot
Psephotellus chrysopterygius. See Golden-shouldered Parrot
Psephotellus varius. See Mulga Parrot
Psephotus, 268t17.3
Psephotus dissimilis. See Hooded Parrot
Psephotus haematonotus. See Red-rumped Parrot
Pseudoseisura lophotes. See Brown Cacholote
Pseudoseisura unirufa. See Grey-crested Cacholote
Psittacara, 139
Psittacara chloropterus. See Hispaniolan Parakeet
Psittacara erythrogenys. See Red-masked Parakeet
Psittacara finschi. See Finsch's Parakeet
Psittacara holochlorus. See Green Parakeet
Psittacara leucophthalmus. See White-eyed Parakeet

Psittacara mitratus. P. m. mitratus. See Mitred Parakeet
Psittacara wagleri. See Scarlet-fronted Parakeet
Psittaciformes, 13, 22, 24, 87, 102, 159, 173, 198, 240, 243f15.3, 271
psittacine beak and feather disease, 75, 78, 92, 137–138, 167, 272
Psittacoidea, 278
psittacosis, 45–46, 131, 137
Psittacula, 267, 268–269
Psittacula alexandri. See Red-masked Parakeet
Psittacula columboides. See Blue-winged Parakeet
Psittacula cyanocephala. See Plum-headed Parakeet
Psittacula eques. See Echo Parakeet
Psittacula eupatria. See Alexandrine Parakeet
Psittacula himalayana. See Slaty-headed Parakeet
Psittacula krameri; P. k. borealis; P. k. manillensis; P. k. krameri; P. k. parvirostris. See Rose-ringed Parakeet
Psittacus, 268–269, 272
Psittacus erithacus. See Grey Parrot
Psittacus timneh. See Timneh Parrot
Psitteuteles versicolor. See Varied Lorikeet
Puerto Rican Amazon, 73–74, 78, 138
Purple Martin, 97t6.2
Purpureicephalus spurius. See Red-capped Parrot
Pycnonotus goiavier. See Yellow-vented Bulbul
Pyrrhura, 267–268
Pyrrhura cruentata. See Ochre-marked Parakeet
Pyrrhura frontalis. See Maroon-bellied Parakeet
Pyrrhura lepida. See Pearly Parakeet
Pyrrhura molinae. See Green-cheeked Parakeet
Pyrrhura perlata. See Crimson-bellied Parakeet
Pyrrhura rupicola. See Black-capped Parakeet

Rainbow Lorikeet, 139, 146, 279–281, 283t18.2, 286t18.5, 288
Rana catesbeiana. See American Bullfrog
Rattus rattus. See Black Rat
Red Lory, 25t2.2, 36a2.1
Red Squirrel, 90, 93, 96–97
Red Wattlebird, 90, 97t6.2
Red-and-green Macaw, 27, 36a2.1, 232, 284t18.3
Red-bellied Parrot, 266
Red-breasted Parakeet, 25t2.2, 31f2.3, 36a2.1, 90, 96t6.2, 146p, 284t18.3
Red-capped Parrot, 283t18.2
Red-crowned Amazon, 24–25, 36a2.1, 71, 73, 75–76, 135t9.1, 138, 144, 197, 200–201, 203–204, 206, 211–212, 215–217

INDEX

Red-crowned Parakeet, 25t2.2, 36a2.1, 138, 212t13.1, 284t18.3
Red-fronted Macaw, 36a2.1, 266
Red-fronted Parrot, 266
Red-headed Lovebird, 36a2.1, 246
Red-lored Amazon, 25t2.2, 36a2.1, 135t9.1, 138, 197t12.1, 200f12.5, 203, 212t13.1
Red-masked Parakeet, 24–25, 36a2.1, 48–49, 80, 139p, 197t12.1, 200f12.5, 203, 206–207, 211–212, 217–219, 224, 230
Red-rumped Parrot, 36a2.1, 231, 249, 266, 281t18.1, 283t18.2
Red-shouldered Macaw, 36a2.1, 284t18.3
Red-spectacled Amazon, 36a2.1
Red-tailed Amazon, 61
Red-tailed Black Cockatoo, 60, 281t18.1, 283t18.2
Red-tailed Tropicbird, 224
Red-winged Parrot, 36a2.1, 281t18.1, 283t18.2, 286t18.5
Regent Parrot, 283t18.2
Rhynchopsitta pachyrhyncha. See Thick-billed Parrot
Rock Dove, 90–92, 95t6.1, 97t6.2, 115, 118, 135t9.1, 138, 187–188, 270
Rose-ringed Parakeet, 13p, 16, 23–25, 27–29, 31f2.3, 32, 36a2.1, 43–47, 50–51, 55–58, 60, 66, 72, 78–79, 81–82, 87–94, 109, 135t9.1, 136–145, 148, 159–169, 173, 188, 197t12.1, 200f12.5, 203, 206, 211–212, 219–225, 227–228, 230–237, 241, 244–245, 246, 249–252, 254–257, 260–264, 266, 269–270, 272, 283–286, 288, 295
Rosy-faced Lovebird, 16, 25t2.2, 36a2.1, 159, 196, 197t12.1, 200, 206, 211–212, 221–222 , 225, 231, 249, 253, 261p, 264–266, 284t18.3, 285t18.4, 286t18.5
Ruddy Duck, 87
Rüppell's Parrot, 36a2.1, 266

Salmonella, 236
Salmon-crested Cockatoo, 36a2.1, 91p, 207, 213–215
Savi's Pipistrelle, 89, 96t6.1
Scaly-breasted Lorikeet, 36a2.1, 281t18.1, 283t18.2, 286t18.5
Scaly-headed Parrot, 36a2.1, 244
Scarlet Macaw, 36a2.1, 61, 207, 213t13.1, 229–230, 253p, 284t18.3
Scarlet-chested Parrot, 283t18.2
Scarlet-fronted Parakeet, 26p, 36a2.1, 197t12.1, 213t13.1
scientific denialism, 134
Sciurus anomalus. See Caucasian Squirrel
Sciurus carolinensis. See Eastern Grey Squirrel
Sciurus vulgaris. See Red Squirrel
Senegal Parrot, 16, 25t2.2, 36a2.1, 90, 213t13.1, 228, 244, 246–247, 266, 284t18.3

Seychelles Black Parrot, 89, 93, 167
Sciurus vulgaris. See Red Squirrel
Sitta europaea. See Eurasian Nuthatch
Slaty-headed Parakeet, 285t18.4
Slender-billed Parakeet, 213t13.1
Spiziapteryx cirumcincta. See Spot-winged Falconet
SoCal parrot, 204
Southern Fiscal, 88, 95t6.1
Southern Mealy Amazon, 36a2.1, 58–59
Spilopelia senegalensis. See Laughing Dove
Spot-winged Falconet, 186–187
Spotless Starling, 90, 96t6.1, 187
St. Vincent Amazon, 36a2.1, 62
Stock Dove, 91, 96t6.1, 140, 188, 236–237
Streptopelia decaocto. See Eurasian Collared Dove
Strigops habroptilus. See Kakapo
Strix aluco. See Tawny Owl
Sturnus unicolor. See Spotless Starling
Sturnus vulgaris. See Common Starling
Sulphur-crested Cockatoo, 25t2.2, 31f2.3, 36a2.1, 93, 96t6.2, 138, 140, 213t13.1, 228, 279p, 280, 281t18.1, 283t18.2, 286t18.5
Sun Parakeet, 266, 283–284
Superb Parrot, 36a2.1, 96t6.2, 283t18.2, 286t18.5
synanthropy/synanthropic species, 28, 32, 43, 87, 159
Syrian Woodpecker, 91, 96t6.1, 271
Swift Parrot, 36a2.1, 61, 278, 281t18.1

Tanimbar Corella, 25t2.2, 36a2.1, 90–91, 96t6.2, 213–214
Tanygnathus lucionensis. See Blue-naped Parrot
Tawny Owl, 93, 166
Tejano Parrot Project, 204
Tepui Parrotlet, 27, 36a2.1
the 10s rule, 240
Thectocercus acuticaudata. See Blue-crowned Parakeet
Thick-billed Parrot, 32, 37a2.1, 76
Threskiornis aethiopicus. See African Sacred Ibis
Timneh Parrot, 22
Treron vernans. See Pink-necked Green Pigeons
Trichoglossus, 268t17.3
Trichoglossus chlorolepidotus. See Scaly-breasted Lorikeet
Trichoglossus haematodus. See Coconut Lorikeet
Trichoglassus moluccanus. See Rainbow Lorikeet
Tui Parakeet, 37a2.1
Turdus merula. See Common Blackbird
Turdus migratorius. See American Robin
Turquoise Parrot, 283t18.2
Turquoise-fronted Amazon, 25t2.2, 37a2.1, 58–59, 90, 92, 96t6.2, 124, 126–129, 138, 197t12.1, 203, 213t13.1, 229, 266, 284t18.3

Ultramarine Lorikeet, 37a2.1
US ban. See Wild Bird Conservation Act
US Wild Bird Conservation Act. See Wild Bird Conservation Act
Upupa epops. See Eurasian Hoopoe

Varied Lorikeet, 281t18.1
Vernal Hanging Parrot, 37a2.1
Vini kuhlii. See Kuhl's Lorikeet
Vinous-breasted Starling, 271

WBCA. See Wild Bird Conservation Act
West Nile virus, 179, 206
Western Corella, 283t18.2
Western Jackdaw, 90, 93, 96–97, 187
Western Rosella, 283t18.2
White Cockatoo, 25t2.2, 37a2.1, 91p, 207, 213–215, 232
White Stork, 92, 245
White-crowned Parrot, 37a2.1
White-eyed Parakeet, 37a2.1, 197t12.1
White-fronted Amazon, 27, 37a2.1, 197t12.1, 203, 213t13.1
White-headed Duck, 87
White-winged Parakeet, 24–26, 31f2.3, 37a2.1, 197t12.1, 203, 207
Wild Bird Conservation Act, 17, 18, 42, 164, 175, 198
Wild Bird Declaration, 17–18, 20, 164, 175, 227, 246
Wild Parrots of Telegraph Hill, The 48, 49p

Yellow-chevroned Parakeet, 25t2.2, 37a2.1, 197t12.1, 200f12.5
Yellow-collared Lovebird, 16, 25t2.2, 37a2.1, 90, 96t6.2, 197t12.1, 213t13.1, 232, 244, 246p, 264–265, 284–285
Yellow-crested Cockatoo, 15p, 23, 25t2.2, 31–32, 37a2.1, 71, 73, 76, 213t13.1, 228
Yellow-crowned Amazon, 25t2.2, 37a2.1, 58, 60, 138, 197t12.1, 203
Yellow-crowned Parakeet, 138
Yellow-eared Parrot, 37a2.1
Yellow-headed Amazon, 25t2.2, 27, 37a2.1, 58, 60, 76, 92, 197t12.1, 200f12.5, 203–204, 213t13.1, 229, 234, 284t18.3
Yellow-legged Gull, 91, 96t6.1
Yellow-naped Amazon, 37a2.1, 58, 60–61
Yellow-shouldered Amazon, 37a2.1
Yellow-tailed Black Cockatoo, 281t18.1, 283t18.2
Yellow-vented Bulbul, 90, 97t6.2
Yucatan Amazon, 37a2.1